GENETICS OF COMPLEX DISEASE

GENETICS OF COMPLEX DISEASE

Peter Donaldson // Ann Daly
Luca Ermini // Debra Bevitt

LONDON AND NEW YORK

Vice President: Denise Schanck
Senior Editor: Elizabeth Owen
Assitant Editor: David Borrowdale
Production Assistant: Deepa Divakaran
Illustrator: Oxford Designers & Publishers
Layout: Techset Composition Ltd
Cover Designer: Andrew McGee
Copyeditor: Ray Loughlin
Proofreader: Susan Wood

ISBN 9780815344919

Library of Congress Cataloging-in-Publication Data

Donaldson, Peter, 1959-, author.
Genetics of complex disease/Peter Donaldson, Ann Daly, Luca Ermini, Debra Bevitt.
p. ; cm.
Includes bibliographical references.
ISBN 978-0-8153-4491-9 (pbk.)
I. Daly, Ann K., author. II. Ermini, Luca, 1978-, author. III. Bevitt, Debra, 1966-, author. IV. Title.
[DNLM: 1. Disease–genetics. 2. Genetic Diseases, Inborn–genetics. 3. Genetic Predisposition to Disease. QZ 50]
RB155
616'.042–dc23

2015022195

Published by Garland Science, Taylor & Francis Group, LLC, an informa business,
711 Third Avenue, New York, NY, 10017, USA, and 3 Park Square, Milton Park, Abingdon, OX14 4RN, UK.

Printed in Great Britain by Ashford Colour Press Ltd

15 14 13 12 11 10 9 8 7 6 5 4 3 2 1

Visit our web site at http://www.garlandscience.com

Preface

There is a scientific revolution happening in biomedical genetics. The new genetics does not just apply to the well-known and well-described Mendelian diseases with clear patterns of inheritance, nor is it limited to major chromosomal abnormalities. What makes the revolution so exciting is that it includes all human diseases and all aspects of human disease. Diseases that have been largely, but not entirely, ignored in the past are the main focus of this revolution. The potential arising from this work is astounding. It is already having an impact and the impact will only grow over time. There are many books on genetics, but few concentrate on complex diseases—those that do not fit the simple patterns of Mendelian disease and cannot be described as chromosomal abnormalities.

Over the past 15–20 years interest in these genetically complex diseases has taken full flight. Though earlier studies had identified some important genetic links and associations, many of the early studies had failed to be replicated and studies in this area of genetics had developed a poor reputation. There were some good studies and many bad studies. The difference between good and bad studies is quite well known. However, developments in the last 20 years have restored interest and confidence in studies of complex disease.

A number of important developments were the keys to opening up this area for high-quality research. The two most important developments have been the Human Genome Mapping Project and the development of supercomputers along with the necessary systems capable of handling the data that very-large-scale studies produce. These two developments go cap-in-hand, one is not possible without the other. In 2015, we have the human genome sequence, the SNP Map and the HapMap. Of course array platforms for genotyping and application of this knowledge as well as more sophisticated statistical analysis have also filled an essential gap. Indeed, the genetics of today is as much about statistics as it is about biology and there are Professors of Statistical Genetics in our academic institutions who dedicate their research to extracting important facts from the mountains of data that current studies can generate.

This book addresses the subject of genetics of complex disease and is designed in two parts. The first part (Chapters 1–5) provides a basic background to genetically complex diseases, and why and how we study them. The second part (Chapters 6–12) focuses on specific sub-branches of genetics of complex disease and specific examples to highlight the application of genetic data in complex disease and the extent to which this data is fulfilling the promises of the Human Genome Project.

Chapter 1 covers the necessary background to genetic variation in the human population, i.e. our evolutionary past and how genetic variation arises. Chapter 2 goes on to define complex diseases and compare them with Mendelian and chromosomal diseases. Chapter 3 looks at how we investigate complex diseases, including the different plans and strategies available to us. Do we chose a single gene or region to study, or do we throw the net wider and investigate the whole genome in a genome-wide association or linkage study? Chapter 4 considers why we are interested in complex diseases, focusing on the major promises of the Human Genome Project in relation to complex disease. These suggested that genetic testing will be used in disease diagnosis, patient treatment and management, and in understanding disease pathology. Chapter 5 looks at how data from the studies described below is handled in a range of different statistical tests.

Sufficient information is given in each of Chapters 1–5 to enable students to understand the major points and, where appropriate, examples are used to illustrate the key concepts (e.g. in Chapter 2, where Crohn's disease and Hirschsprung's disease are discussed as two different models of genetically complex diseases). Chapter 4 uses quite a few disease examples to illustrate how the genetic information may be used to meet the promises of the Human Genome Project.

After Chapter 5, the book goes on to look at three specific areas: immunogenetics (Chapter 6), infectious disease (Chapter 7), and pharmacogenetics (Chapter 8).

Chapter 6 on immunogenetics deals with how common variation in genes that regulate the immune response can increase or reduce susceptibility to common diseases. The chapter concentrates on the major histocompatibility complex on chromosome 6p21.3. The chapter includes a considerable number of recently studied examples and discusses the different interpretations that can be applied to the data. In each case, the extent to which these examples do or do not fulfill the promises of the genome project is considered. There are positive examples of how genetics can be used as an aid to diagnosis (e.g. in ankylosing spondylitis), and also how associations and linkage with certain risk alleles may be helping us to understand disease pathogenesis (e.g. in autoimmune liver disease).

Chapter 7 on infectious disease looks at the past and the present considering how genetic variations may influence the likelihood of infection per se and the outcome following exposure to infectious agents. The discussion provides interesting links with mankind's early history. The chapter concentrates on a few selected examples to illustrate the concept and demonstrate how the studies discussed are helping to fulfill the promises of the Human Genome Project. Once again, there are clear examples of genetic investigations impacting on our understanding of disease pathology.

Chapter 8 on pharmacogenetics discusses past and present developments in a fast-expanding field that is at present providing some of the most promising results in complex disease genetics. Studies have shown that responses to commonly used pharmacological agents can be determined by common genetic variation. The impact of this variation ranges from failure to respond to a drug to life-threatening toxic reactions. The potential to use genetics to tailor therapy and also to develop new therapeutic agents is a real possibility in this sub-branch of complex disease genetics, and one that major pharmaceutical companies and academic institutions are aware of.

Chapters 9 and 10 focus on specific disease groups: cancer (Chapter 9) and diabetes (Chapter 10). These two chapters stand alone because one group of diseases (cancer) has a very significant impact in terms of morbidity and mortality in the developed and developing world and the other group (diabetes) is for the most part a perfect example of a complex disease. The potential medical impact of genetic studies in these diseases is vast. More rapid diagnosis, better patient care, personal life planning, and personal treatment planning are all possible. As we gain a greater knowledge of the genetics of these diseases we will start to have a better grasp on the underlying pathology of each disease, which will open up doors for diagnosis, treatment, and management. In some cases, this will mean simple things like changes to a person's diet; in others, selecting the appropriate chemotherapeutic agent to use for a patient. To some extent some of these aims have already been achieved, but as this book indicates, there is still much to be done.

In Chapter 9 (cancer), a selected number of examples are discussed. These include breast cancer, prostate cancer, and lung cancer. The selection is based on the most common cancers, which are also, to some extent, those about which we know the most. Links to useful websites are given for further information and updates. Diabetes (Chapter 10) is discussed in its various forms, especially type 1 and type 2 diabetes and is specifically used to illustrate the difference in the genetic portfolios in type 1 and type 2 diabetes. Here, the question is why are two diseases that have so much in common so different in terms of their genetic profiles?

The last two chapters deal with societal and ethical issues in the new genetic era and the future of genetics in complex disease. This is a fast-moving area of science. The facts being produced today will be marketed as diagnostic or prognostic indicators almost as quickly as they are identified. Genetic testing for risk alleles will soon be normal practice, but this will have ethical and social consequences. The potential for misuse of genetics is discussed in Chapter 11, highlighting the importance of understanding what a genetic test in complex disease is really telling you. You will need to know what a genetic profile is telling you before getting tested. There is considerable commercial interest in genetic profiling and this has ethical and societal impact. Other points discussed include who owns your genome and who can access your genetic data?

Chapter 12 closes the book by looking at the techniques and technologies that have been used and those that will be used in the future. The chapter reminds us that technologies used in the past will also be used in the future, but it also highlights some fascinating new possibilities. Most important will be direct sequencing either at the level of the exome (i.e. protein-coding genes only) or the whole genome.

The structure of the book is designed to provide a basic platform on which students can build their knowledge base. Each of the chapters (including the basic chapters) uses examples of disease to illustrate key specific points and provides a reasonable level of basic current data on each example used. In particular, the book focuses on the promises of the Human Genome Project that suggested genetics will be used to improve disease diagnosis, to develop individual treatment and management plans for patients, and to inform the debate on disease pathogenesis. At each stage and after each example, the text reflects on the extent to which these promises have been or will be met, looking at both the present and, if possible, the future. Links to the web are also provided for access to updates and further information throughout the book. There is an extensive Glossary at the end of the book.

These are very exciting times for genetics, especially in complex disease. They are also fast-moving times. The book is written as a starting point (a first block) and for the most part it is written in an historical style to ensure it remains in date whatever develops in the future.

This book provides a good starting point for anyone studying the genetics of so-called complex diseases. It is written for the undergraduate student and early postgraduate student alike. It is written for the medical and non-medically minded individual. This era is one of the most exciting eras in modern genetics, perhaps as exciting as when the structure of DNA was first revealed to the scientific community.

We would like to thank the staff at Garland Science, Liz Owen, David Borrowdale and Deepa Divakaran, for their support and encouragement in producing this book.

Peter Donaldson, Ann Daly, Luca Ermini, and Debra Bevitt

Acknowledgments

As senior author I would like to give specific thanks to: Robert Taylor (Newcastle University) who provided advice on the mitochondrial genome, John Mansfield (Newcastle University) who provided necessary background on inflammatory bowel disease, Roger Williams (King's College, London) and Oliver James (Newcastle University) both of whom provided a supporting environment within which to learn and develop a background in liver disease and genetics as well as the necessary skills to produce this book, Derek Doherty (Trinity College, Dublin) who worked with me on the molecular genetics of the MHC in liver disease at King's College Hospital, London and the many members of different research teams who have contributed to my research between 1982 and 2015. In addition I would like to give special thanks to the hundreds of students who, through their positive interaction and feedback, have encouraged the writing of this book. Finally, I would like to give very special thanks to Carolyn Donaldson who encouraged and supported production of this book from start to finish, especially during difficult times.

Peter Donaldson

The authors and publisher would like to thank external advisers and reviewers for their suggestions and advice in preparing the text and figures.

Geoffrey Bosson (Newcastle University, UK); Margit Burmeister (University of Michigan, USA); Angela Cox (University of Sheffield, UK); Rachelle Donn (University of Manchester, UK); Yalda Jamshidi (St George's, University of London, UK); Martin Kennedy (University of Otago, New Zealand); Andrew Knight (Newcastle University, UK); Hao Mei (Tulane University, USA); John Pearson (University of Otago, New Zealand); Logan Walker (University of Otago, New Zealand); Kai Wang (University of Iowa, USA); Yun Zhang (Oxford Brookes University, UK).

Contents

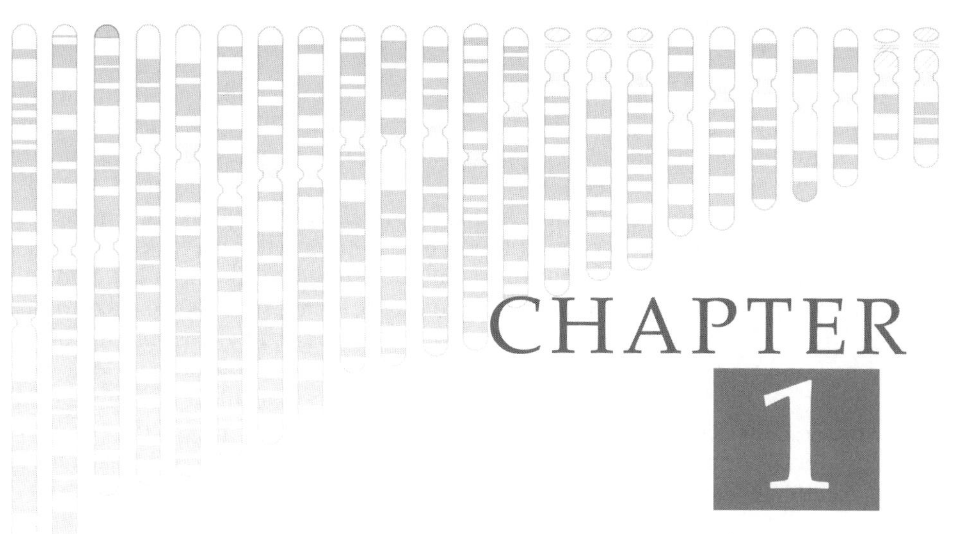

CHAPTER 1

Genetic Diversity

Human evolution is driven by a number of different factors, including migration and settlement in different environments, genetic **mutation**, **natural selection**, and **genetic drift**. The product of these different forces is genetic diversity within a population, and understanding this genetic diversity and the reasons for it are essential when considering the genetic basis of common human diseases.

Though the origin of modern humans is relatively recent, humans have managed to colonize almost all possible environments and in doing so have been exposed to considerable **selective pressure**. Consequently, there is extensive variation in the human genome and in the phenotypic traits (e.g. skin color) expressed. In this chapter, we will review the basic background information on mutation, natural selection, and evolution, and the way this helps us to understand the importance of genetic variation in the human genome. We will pinpoint the reasons why genetic variation arises in a population and introduce phenomena such as **epigenetics**. We will also consider the **mitochondrial genome**.

The genetic variation described here creates a basis for genetic risk in the majority of human diseases. Understanding this genetic diversity and how it has arisen is a necessary precursor to understanding the genetics of complex disease. Genetic variations between individuals determine individual susceptibility or protection from a variety of common diseases. This is the basic subject of this book, the idea that common genetic variation gives rise to different levels of susceptibility to common disease. The evolutionary forces that created this genetic variation have enabled populations to thrive, throughout human history, because some population members are likely to be less susceptible to a given illness than others and are thereby more likely to survive even the most catastrophic event.

1.1 GENETIC TERMINOLOGY

As with many scientific disciplines, genetics employs a large number of specific terms and this terminology is given in the Glossary at the back of the book. The term **genome** refers to the complete set of genetic information found in a cell and includes 22 pairs of the autosomal chromosomes plus either XX (females) or XY (males) (**Figure 1.1**) and a small amount of DNA found in the **mitochondria (mtDNA)**. Human **chromosomes** are the organized packages of **DNA** found in the nucleus of a cell. DNA is comprised of linear double-stranded molecules that form a helix. The strands of the helix are made up of alternative sugars (deoxyribose) and phosphate groups. Each sugar is attached to one of four bases A, C, G, and T, and the whole molecule is stabilized by cross-linking of the bases A with T and C with G. The DNA structure we are most familiar with looks like a twisted ladder, though when packaged DNA is wound around histone proteins into compressed units. The sequence A–T/C–G provides a code for the production of RNA and RNA production may lead to protein production. The human genome is made up of more than 20,000 **genes**; each gene being a single unit of inheritance that is transmitted from parent to offspring. The location of a gene on a chromosome is referred to as a **locus** (plural: **loci**) and genetic variation at a locus is referred to as **allelic variation**, where the different forms are known as **alleles**. On average, human genes encode approximately 28 kilobases (kb) of DNA with a series of small **exons** (protein-coding sequences) separated by long **introns** (non-coding sequences). Primary transcripts can be differentially spliced into alternative proteins, adding yet another level of complexity to the story of genetics in complex disease. In his book *The Language of Life: DNA and the Revolution in Personalised Medicine* (2010), Francis Collins refers to a single gene in the brain that is capable of making 38,000 different proteins. This is a remarkable and unusual figure. The total impact of intronic genetic variation on common disease is only just beginning to be investigated, but this figure is most likely to be an exception rather than the rule.

The use of the terms genes and alleles varies, though they do have precise definitions

The terms gene and allele are often used as though they are the same, but it is important to note that this is incorrect and that the correct term to use when considering genetic variation is allele. A gene is, as stated above, the basic unit of inheritance. The scientific literature is peppered with examples of incorrect use of this terminology. Writers often refer to the "cystic fibrosis gene" and the "**hemochromatosis** gene" as though only patients with these diseases possess the gene, when actually all members of the population possess these genes. In these two examples, which are both Mendelian autosomal recessive disorders, the difference between affected patients and healthy members of the population is that patients possess two copies of the disease-causing alleles. Unaffected population members may have a single copy of the disease-causing allele or may not carry this allele at all. Instead, they will have one or two copies of the non-disease-causing allele. Thus, it is the possession of the requisite alleles that causes the disease and not the possession of the gene. Finally, the term allele is sometimes used to include any genetic variation within a region,

Figure 1.1: Karyotypes of human chromosomes. The figure illustrates the entire autosome showing banding patterns for each chromosome in size order. Chromosomal banding was (and is) traditionally used to identify chromosomes and chromosomal sites for clinical diagnosis. To obtain these patterns it is necessary to first denature the DNA with enzymes, and then dye the sample to produce light and dark bands. Karyotypes are assigned based on the chromosome length, banding pattern, and position of the centromere. Chromosome 1 is the longest, and chromosome 22 is the shortest among the autosomal chromosomes. (From Strachan T & Read A [2011] Human Molecular Genetics, 4th ed. Garland Science.)

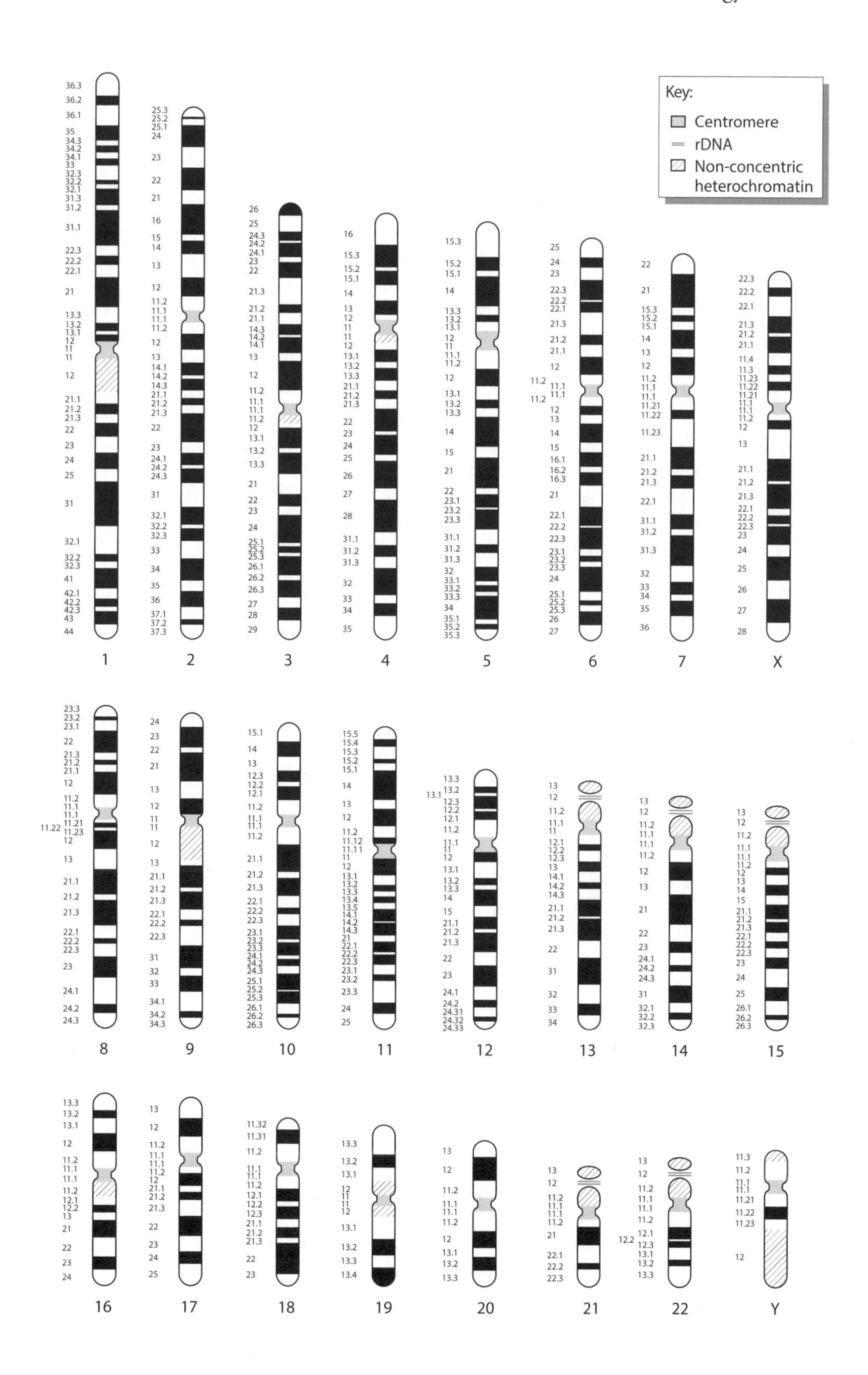
Key:
Centromere
rDNA
Non-concentric heterochromatin
1
2
3
4
5
6
7
X
8
9
10
11
12
13
14
15
16
17
18
19
20
21
22
Y

whether or not it is part of the exome or intronic sequence. This would not be acceptable to all readers of this book, but those with a focused interest in this area may consider this correct. The use of terminology changes over time.

Most individuals have two copies of a given gene – one inherited from the mother and one from the father. As a result, they are **diploid**. The **genotype** is the set of alleles that an individual possesses. An individual may have two identical alleles, in which case they are **homozygous**, or two different copies, in which case they are **heterozygous** (**Figure 1.2**). When we consider the expression of a genetic variant we use the term **phenotype**. A phenotype can also be referred to as a **trait** or characteristic and may be either physical, physiological, biochemical, or behavioral. Thus, the condition of having blue eyes or dark hair is a phenotype, but so is having sickle cell anemia. Phenotypes are most often referred to as traits or characteristics when they do not relate to an illness or disease.

In 2001, the first draft of the map of the human genome was published. Even though this was not the complete sequence, it marked the beginning of a new era in genetics frequently referred to as the post-genome era. The great advantage of working in the post-genome era is that we have access to the genome map, and the majority of human genetic variation is known and available through websites such Human Genome Resources (http://www.ncbi.nlm.nih.gov/projects/genome/guide/human/index.shtml) and the **SNP Map** (http://www.ncbi.nlm.nih.gov/SNP/).

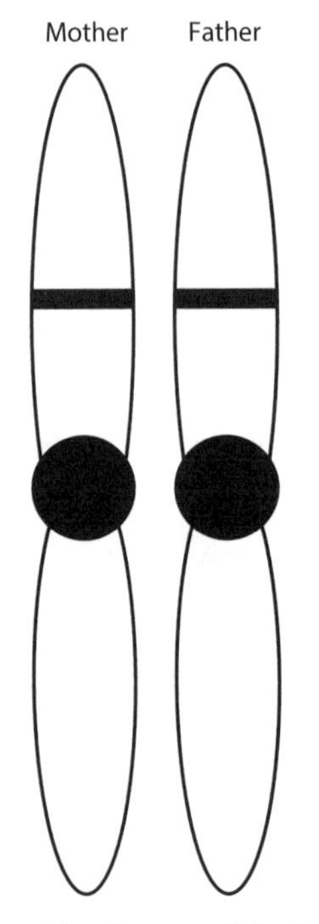

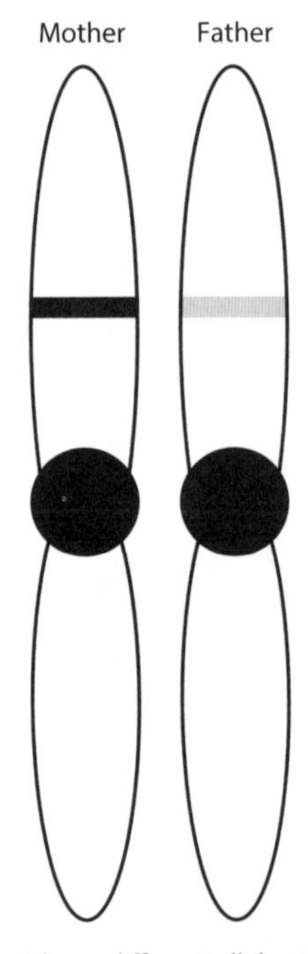

Figure 1.2: Homozygous versus heterozygous. The figure illustrates a single pair of chromosomes in two individuals (A and B). In contrast to the picture in Figure 1.1, the band represents a single gene. Individual A inherits the same allele from both parents (i.e. both black) and is therefore homozygous for this genotype, while individual B inherits different alleles (gray and black) for this gene and is therefore heterozygous.

1.2 GENETIC VARIATION

Genetic variation is by convention discussed in terms of allele frequencies. A frequency is simply a proportion or a percentage usually expressed as a decimal fraction. For example, if 20 out of 100 of the alleles at a particular locus in a population are of the *A* type, we would say that the frequency of the *A* allele in the population is 20% or 0.2.

The term **population** in human genetic studies refers to the group of individuals occupying a defined area such as a country, county, city, or town. Occasionally, a population will be defined by other characteristics, including age, ethnicity, and even in some cases by a particular disease. The complete set of genetic information contained within a population is called the **gene pool**. The gene pool includes all alleles present in the population. Some genes do not encode variation (i.e. they are monomorphic). A monomorphic gene only exists in a single form and therefore has a single allele at a frequency of 100% or 1. However, the majority of our genes are **polymorphic**, existing in two or more (poly) forms in a population. In a population a gene may encode a limited or small number of different alleles or it may encode a much larger number of alleles. One example of the latter is the *HLA-B* gene, which encodes over 3000 **polymorphisms** and mutations. The two terms, mutation and polymorphism, are defined and used differently by different groups. Classical geneticists, particularly those associated with the use of genetics in a clinical setting, who are involved in diagnosis and screening for Mendelian traits, use the term mutation to refer to genetic variations that have a causative effect [i.e. a **disease-causing mutation (DCM)**] and use the term polymorphisms to describe other variations found in the population. Many evolutionary geneticists also prefer to use this definition in this way. However, other geneticists prefer to use the definition provided by Cavalli-Sforza and Bodmer (1971), whereby genetic polymorphism was defined as "the occurrence in the same population of two or more alleles at one locus, each with an **appreciable frequency**." Most authors who apply this definition agree that polymorphic loci are those for which the frequency of the least common allele is greater than 1%. This works well when there are only two alleles. It can become very complicated when considering some of our most complex genes, such as the cystic fibrosis gene ***CFTR*** with over 1910 variations, some of which are common (e.g. Δ508), but the majority of which are rare. In this situation it is difficult to decide which terminology applies; Δ508 is a polymorphism, while most of the other *CFTR* alleles are found at frequencies of less than 1% and are therefore mutations. The dilema is should we call all the *CFTR* variants, including the Δ508 mutations, polymorphisms, or should we apply a mixture of terms as implied in the definition above? There are similar problems with the naming in the **major histocompatibility complex (MHC)** (see Chapter 6).

The problem with the use of this terminology is not simply a matter of choice. Nearly all genetic variation arises through mutation (deletions, **duplication**, insertions, and unrepaired DNA damage). Therefore, most polymorphisms are simply common mutations and, as a consequence, it is not possible to insist on the strict application of this terminology. Though the debate on the correct use of these terms continues, they are used interchangeably in the literature on complex disease. Both terms will be applied throughout this book: polymorphism when describing common variations associated with specific diseases or traits, and mutations when discussing rare variations and evolutionary principals.

A common change in a single base pair (point mutation) is called a **single nucleotide polymorphism (SNP)**. The site at which a SNP is encoded is marked by the "rs" number (ref-SNP cluster ID number) – a unique ID number based on its position on a chromosome. SNPs are the smallest and most common type of genetic change in humans, and account for an estimated 90% of all variation in the genome. There are currently thought to be more than 38 million SNPs in the genome. Consequently, SNPs are most frequently used as markers to identify genetic variation in human disease. The high frequency of SNPs in the genome enables high-density profiling to be undertaken, which increases the likelihood of accurate identification of disease-promoting alleles. When SNPs were first used for screening for disease alleles on a genome-wide basis, only common SNPs (i.e. those where the least frequent allele was present in 5% or more of the population) were used. The reason for this was that rare SNPs were considered to be less statistically informative. Therefore, it was considered that larger numbers would have to be included in the study to test rare alleles in order to have adequate statistical power. The problem with excluding rare alleles is that potentially important associations with rare SNPs may have been missed. However, as sample collections have become larger the potential to identify statistically significant associations with less common SNPs has grown and the lowest applied limit for SNP frequency has been adjusted downward. For example, instead of a lowest frequency of 5%, a 1% limit can now be applied. The potential for using even lower frequency SNPs will increase as study cohort sizes increase.

Another form of genetic variation that is quite common in the population is **copy number variations (CNVs)**. These occur when there are multiple numbers or copies of a specific gene on a chromosome. CNVs are structural variations that can occur through deletions, duplication, insertions, and translocations. They may represent large or small areas of the chromosome. Good examples can be seen in Chapter 8 on pharmacogenetics.

Genetic variation can be measured by several methods

Though SNPs are the preferred markers for measuring genetic variation, other markers have been used in the past, including **microsatellites**. These are **variable number tandem repeat (VNTR)** sequences in the genome. VNTRs can be "short" (involving two to five nucleotide repeats) or "long" (involving more substantial repeat sequences). VNTRs are still used in studies today and are especially useful where the candidate gene is known or a specific region is being scanned. Earlier studies used **restriction enzymes** to identify different VNTRs and SNPs. To determine VNTR genotypes, one or more restriction enzymes that cut the DNA sequence above and below the region encoding the VNTR sequences can be used and DNA fragments of different sizes can be obtained. After digestion with the appropriate enzyme(s), the DNA sample can be run by electrophoresis on either an agarose gel or a polyacrylamide gel to reveal the size(s) of the fragments and thus the number of sequence repeats in each individual sample. Genotypes can be assigned based on the pattern obtained on the gel. This method is known as **restriction fragment length polymorphism (RFLP) analysis**. RFLP analysis was also used to detect SNPs where the differences in the DNA sequence can be detected by use of restriction enzymes that cut the DNA at a particular sequence encoded by one allele, but not the other. Multiple enzymes were often used when genotyping SNPs in order to obtain readable accurate results. Different enzymes are used to detect different polymorphisms. Later studies substituted RFLP genotyping for more reliable **polymerase chain reaction (PCR)** genotyping using primers specific for the gene sequence of interest. This method uses a polymerase enzyme purified from the hot-springs "thermophilic" bacteria *Thermus aquaticus* to amplify multiple copies of the gene sequence. These amplified sequences are then run out on a gel using the same process as that used with RFLP fragments and genotypes can be assigned from the specific banding patterns obtained for each sample (**Figure 1.3**).

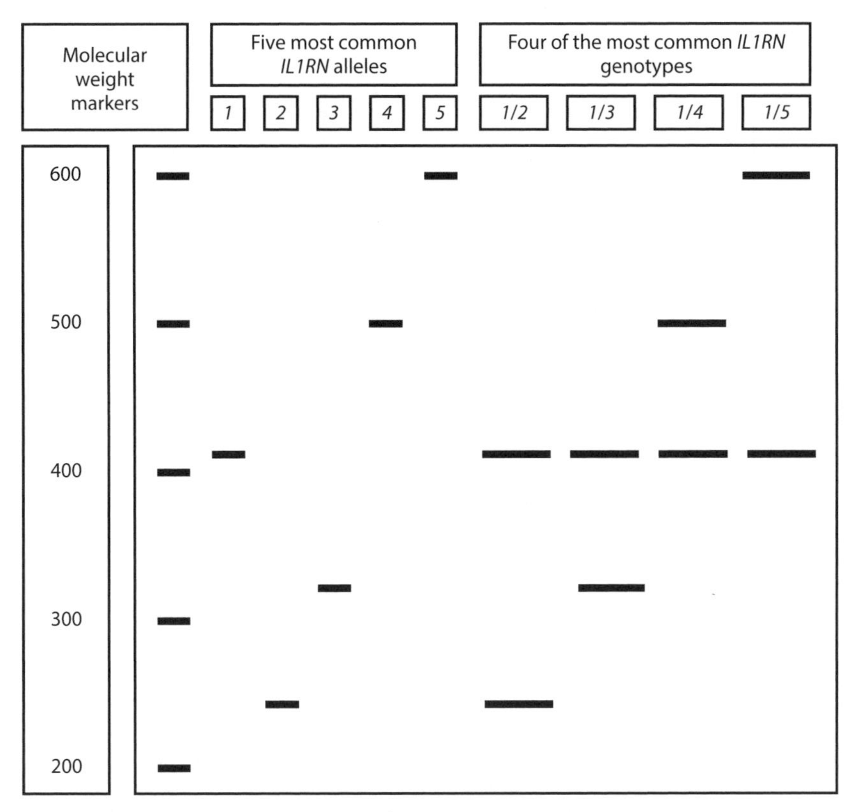

Figure 1.3: VNTRs used to genotype the interleukin (IL)-1 receptor antagonist gene (*IL1RN*). Genotyping the IL-1 receptor antagonist 86 bp VNTR sequence using PCR and agarose electrophoresis. The figure shows the five most common alleles (*1*–*5*) and the four most common genotypes (*1/2*, *1/3*, *1/4*, and *1/5*). The molecular weight markers indicate the approximate band sizes for each allele on the agarose gel as follows: allele *1*, 410 bp; allele *2*, 240 bp; allele *3*, 325 bp; allele *4*, 500 bp; allele *5*, 600 bp. Note the figure does not show the precise position in the gel and the ladder is illustrative only. The genotypes for each sample can be assigned using the band sizes obtained.

Alleles on the same chromosome are physically linked and inherited as haplotypes

As genes are inherited on chromosomes and each chromosome carries a large number of genes, genetic variations on a specific chromosome are inherited en masse as **haplotypes**. Haplotypes do not change from one generation to the next because mutation rates are low, but will change due to recombination during crossover. The potential for change is based on the distance between the genes. One of the very interesting observations to emerge from analysis of haplotypes is that for any small region of a chromosome, most people in a population will carry one of approximately six different haplotypes that can be traced back through history to a shared ancestry in the distant past. However, because **recombination** events exchange pieces of DNA between chromosomes during meiosis, person A may share the same haplotype with person B for a region at one end of a chromosome, yet have a different haplotype compared with person B at a position 1 million base pairs further down the same chromosome. Person B, however, may share the same haplotype in the second region with person C. By studying these haplotypes, it is possible to look back at genetic events that may have happened thousands of years ago.

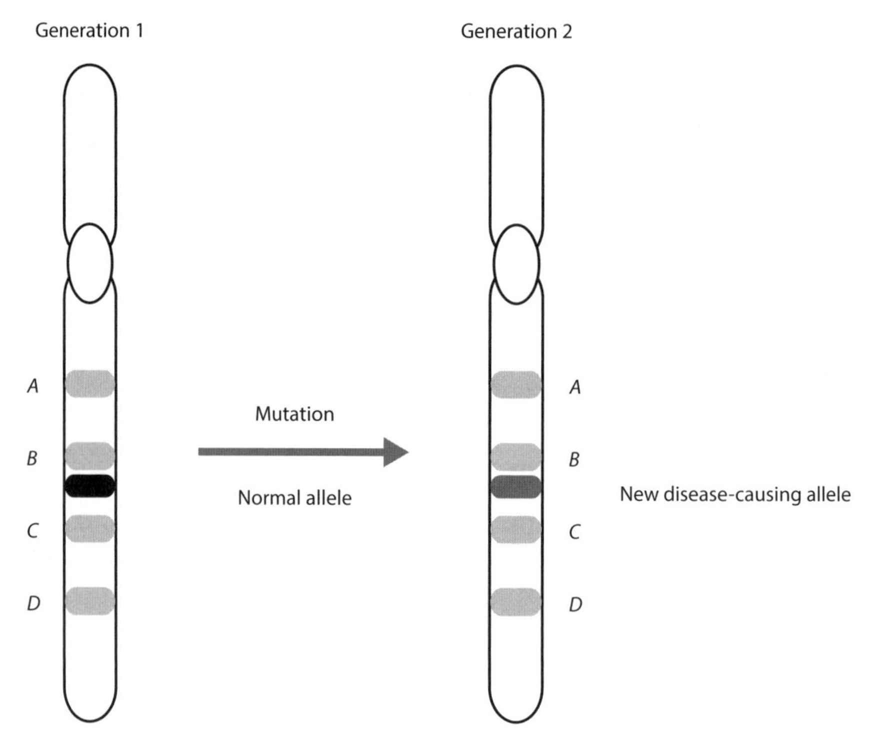

Figure 1.4: The development of a new mutation or allele in the *A–B–C–D* haplotype. The figure illustrates the same chromosome in two individuals in two different generations (generations 1 and 2). On the two chromosomes there are two patterns for the haplotype *A–B–C–D*. One includes a black normal band (representing the normal allele) and one includes a dark gray band (representing a "new" mutated allele). In this illustration the mutation may have arisen through recombination in meiosis. The bands illustrated are not the same as those seen in Figure 1.1, which are the bands based on chromosome staining for karyotyping.

African populations tend to have a greater variety of haplotypes in any given region than other populations. This is expected for a population that is older than all others, and therefore has had more time to diversify and develop more haplotype variations (**Figure 1.4**). In younger populations, such as those in Europe and Asia, fewer haplotypes would be expected because these populations have descended from smaller **founder populations** in which a small subset of the total available haplotypes were present and there has also been less time for new combinations to develop.

Linkage disequilibrium promotes conservation of haplotypes in populations

The term **linkage** refers to the physical association (link) between two alleles that are on the same chromosome. **Linkage disequilibrium** is a population genetics phenomenon whereby alleles on the same chromosome are transmitted together over generations within a population and such pairs or groups of alleles are found together more frequently than expected by chance. In other words, there is **non-random segregation** of the alleles. This is due to the physical proximity of the alleles in question and the low rate of segregation at meiosis. The phenomenon of linkage disequilibrium is common when disease-causing alleles arise in a founder and the alleles are closely linked to other markers along a chromosome. **Crossover**, however, may break up this disequilibrium. When the loci are further apart, linkage disequilibrium breaks down quickly;

Figure 1.5: Extreme linkage disequilibrium. The MHC illustrates extreme linkage disequilibrium whereby alleles at closely linked gene loci are inherited together more often than expected by chance. One example of this is the HLA 8.1 ancestral haplotype (shown above), which is associated with an increased risk of many different autoimmune diseases, but may also convey some survival advantage. The individual alleles are all common in the normal northern European population, but occur together at frequencies far greater than expected by chance. Thus, 60% or more of *HLA-B8*-positives have *HLA-A1*, and 90% or more of *HLA-B8*-positives have *HLA-DRB1*03* and *DQB1*02*. *HLA-B8* is the least common of all of these alleles at around 16%, and if the assortment were random then these pairings would be in equilibrium and the likelihood of finding *HLA-B8* and *HLA-DRB1*03* together would be the sum of their individual frequencies. In this case, the values are approximately 20% for *HLA-DRB1*03* and 16% for *HLA-B8*. This would mean that instead of seeing approximately 14% of the population with the combination *HLA-DRB1*03–HLA-B8*, we would see approximately 3%.

when the loci are close together, crossover is less common and linkage disequilibrium is more likely to persist. Linkage disequilibrium can be used to provide useful information about the distance between genes. Where there is extreme linkage disequilibrium, haplotypes may be conserved and in many cases these conserved haplotypes are common in the population. The human MHC (**Figure 1.5**) illustrates these concepts well (see also Chapter 6). Linkage disequilibrium is a major tool in understanding modern **genome-wide linkage/association studies** (**GWLS/GWAS**).

1.3 GENETICS AND EVOLUTION

Evolution in population genetics refers to changes in the gene pool resulting in the progressive adaptation of populations to their environment. Four main processes account for most of the changes in allele frequency in populations: mutation, migration, natural selection, and random genetic drift. Together these form the basis of cumulative change in the genetic characteristics of populations, leading to the descent with modification that characterizes the process of evolution.

If the population is large enough, allele frequencies remain stable and do not change significantly as a result of random reproduction, and therefore other processes must be responsible for changes in allele frequency. Genetic variation within populations can be increased by migrations and mutations that introduce new alleles into the population. Variation within populations can also be increased by some types of natural selection, such as **over-dominance**, in which both alleles are favored. These evolutionary forces that act to maintain or increase genetic variation are shown in the upper-left quadrant of **Table 1.1**. The lower-left quadrant of Table 1.1 shows evolutionary forces that decrease genetic variation within populations. These forces include genetic drift, which decreases variation through the **fixation** of alleles, and some forms of natural selection, such as **directional selection**, which selectively favors one allele over the other.

Evolutionary forces also affect the **genetic divergence** between populations and are shown on the right quadrants of Table 1.1. Genetic divergence between populations is increased by mutation, genetic drift, and natural selection. Different mutations can arise within each population and therefore mutations almost always increase divergence between populations. **Positive natural selection** can increase or decrease divergence between populations depending on the favored alleles. If different alleles are favored, populations will diverge; however, if natural selection favors the same allele in different populations, genetic divergence between populations will decrease. Migration reduces divergence between populations

Table 1.1 Mutation, migration, genetic drift, and natural selection have different effects on genetic variation within populations and on genetic divergence between populations.

	Within populations	Between populations
Increase genetic variation	Migration Mutation Natural selection	Genetic drift Natural selection Mutation
Decrease genetic variation	Genetic drift Natural (directional) selection	Migration Natural selection

because blending the total gene pool makes populations similar in terms of their genetic composition. Note that migration and genetic drift act in opposite directions: migration increases genetic variation within populations and reduces divergence between populations, whereas genetic drift reduces genetic variation within populations and increases divergence among populations. Mutations mostly increase the genetic variability both within and between populations, though they can occasionally restore the **wild-type**. Natural selection, by contrast, can either increase or reduce genetic variation within a population and increase or reduce genetic divergence between populations.

Finally, before considering each of these processes in turn, it is important to make it clear that populations are simultaneously affected by many evolutionary forces acting at the same time and that evolution results from the complex interplay of these processes.

Mutation is the major cause of genetic variation

Almost all genetic variants arise through some form of mutation. New combinations of these mutations may then arise through recombination in **meiosis**. Meiosis is a process through which cells are able to divide and produce haploid gametes. It is sometimes called a reductive division because there are two stages of cell division, but only one round of DNA replication. Thus, four haploid gametes are created for each diploid spermatocyte (i.e. sperm cell). In oocytes (i.e. egg cells) the situation is different. Division is asymmetric, unlike that for spermatocytes, and the cell division results in a large secondary oocyte and a smaller **polar body** that is discarded. Evolution through natural selection depends on these processes because there has to be genetic variation in the population before evolution can take place. There can be no selection without genetic variation in a population.

A mutation is a heritable change in the DNA sequence. This means that the structure of DNA has been changed permanently and this alteration can be passed from mother to daughter cells during cell division. If a mutation occurs in reproductive cells, it may also be passed from parent to offspring. This kind of mutation is responsible for changing allele frequencies in a population and is an essential process in evolution, as mutations provide the variation that enables humans to change and adapt to their environment when selective pressure is applied. Some mutations may be selectively neutral, which means they do not affect the ability of the organism to survive and reproduce. Only a very few mutations are favorable for the organism and contribute to evolution.

Mutation rates are typically low. The mutation rate (μ) is the frequency of such change and it is expressed as the number of mutations per locus per gamete per generation. Estimating the mutation rate is difficult because mutations are rare. In humans, most information on mutation rates comes from studies of rare **Mendelian autosomal dominant** diseases where it is much easier to estimate mutation rates than it is for **Mendelian autosomal recessive** diseases or for **non-Mendelian complex diseases**. Estimates of mutation rates for a variety of human genes lie between 10^{-6} and 10^{-5} mutations per locus per gamete per generation. However, the estimated mutation rate is higher for some **Mendelian diseases**. For example, in type 1 neurofibromatosis and Duchenne muscular dystrophy the estimated mutation rate is as high as 10^{-4}. This is 10–100 times greater than the general mutation rates.

The **OMIM (Online Mendelian Inheritance in Man)** database (http://www.ncbi.nlm.nih.gov/omim) lists human genes and it is an excellent source for information on specific genetic diseases. For many diseases, a larger number of genetic mutations have been identified than those listed on OMIM, but this is a good starting point to catalog genetic variations that are linked to or associated with a specific disease and it also has a very good bibliography for each disease.

Introducing the Hardy–Weinberg Principle

The **Hardy–Weinberg Principle (HWP)** or **Hardy–Weinberg Equilibrium (HWE)** test is one of the central pillars of statistical analysis in population genetics (**Table 1.2**). The term equilibrium in population genetics refers to something (an allele or gene) that is in a state of balance. Equilibrium arises when alleles remain unchanged over time. The HWE test assesses how allele frequencies have changed from generation to generation. The HWP states that in a large breeding population, provided none of the evolutionary forces described below are operating, allele frequencies will remain the same from generation to generation. In practice, the HWE test can be used to understand the change in allele frequencies over time and indicate whether evolution has taken place. HWE is also used in studies of complex disease to determine whether there is bias in the study sample and in the qualitative assessment of studies. The HWP is a complex principle and the basic concept and its application are discussed in more detail in Section 1.4.

Genetic variation caused by mutation alters allele frequencies in populations

The rate at which a genetic variation increases or decreases is determined by the mutation rate. Consider the example of a single locus with two alleles *A1* and *A2* with frequencies p and q, respectively, in a population of 10 diploid individuals. In this example, the pool of alleles for this gene within the population will consist of 20 allele copies. If there are 15 copies of *A1* and five copies of *A2* in the population, then the frequency of each allele is $p = 0.75$ and $q = 0.25$. If we suppose that a mutation changes one *A1* allele into an *A2* allele, after one mutation there will be 14 copies of *A1* and six copies of *A2*, and the frequency of *A2* will increase from 0.25 to 0.30; a mutation has therefore changed the allele frequency for the population. If copies of *A1* continue to mutate to *A2*, the frequency of *A2* will continue to increase, while the frequency of *A1* will decrease.

Table 1.2 The HWE ($p^2 + 2pq + q^2 = 1$) dictates that the sum of allele genotypes is always 100% and this formula can be used to determine the expected frequency of the different genotypes in a population.

Maternal gamete	Paternal gamete	
	A (p)	a (q)
A (p)	AA (p^2)	Aa (pq)
a (q)	Aa (pq)	aa (q^2)

Thus changes in the frequency of the *A2* allele (Δq) depend on:

- μ: the mutation rate *A1* to *A2*
- p: the frequency of the *A1* allele in the population

When p is large, many copies of *A1* are available to mutate to *A2* and the amount of change will be relatively large. However, as more mutations occur and p decreases, fewer copies of *A1* will be available to mutate to *A2*. The change in *A2* frequency as a result of mutation equals the mutation rate multiplied by the allele frequency:

$$\Delta q = \mu p$$

So far, we have considered only the effects of forward mutations (*A1* → *A2*); however, reverse mutations (*A2* → *A1*) can also occur. Reverse mutations will occur at a rate ν, which will probably be different from the forward mutation rate μ. When a reverse mutation occurs, the frequency of the *A2* allele decreases while the frequency of *A1* increases. The overall change in allele frequency (*A1* and *A2*) is a balance between the two opposing forces of forward and reverse mutations:

$$\Delta q = \mu p - \nu q$$

These allele frequencies are determined only by the forward and reverse mutation rates, and they will increase or decrease until the HWE is established. When the equilibrium is established, the HWP indicates that genotype frequencies will remain the same.

The mutation rates of most human genes are low and changes in allele frequencies due to mutation in one generation are very small. Therefore, it may take a long time to reach the HWE. For example, consider a locus where the forward and reverse mutation rates for alleles are $\mu = 1 \times 10^{-5}$ and $\nu = 0.5 \times 10^{-5}$ per generation, respectively, and the allelic frequencies are $p = 0.85$ and $q = 0.15$. The change in allele frequency per generation due to mutation is:

$$\Delta q = \mu p - \nu q = (1 \times 10^{-5})(0.85) - (0.5 \times 10^{-5})(0.15) = 7.75 \times 10^{-5} = 0.0000775$$

This shows that the change due to mutation in a single generation is extremely small and because the frequency of p drops as a result of each mutation, the frequency of change will become even smaller over time, as shown in **Figure 1.6**.

Migration and dispersal cause gene flow

Another process that introduces change in the allele frequencies is the **gene flow**. Gene flow is the result of migration where many individuals of one population move en masse from one geographic location to another. Though migration is the main cause of gene flow, it can also result from population dispersal, i.e. the spreading of individuals away from others. Migration has a similar impact to mutation as new alleles are introduced into a local gene pool by the migrants. In this case, however, the new alleles are new only to the population into which the migrants move and they are not the result of new mutations.

In the absence of migration, the allele frequencies in each local population can change independently through genetic divergence. As a consequence there will be differing frequencies of common alleles among local populations and some local populations will possess certain rare alleles not found in others. This effect of the accumulation of genetic differences among subpopulations can be reduced if subpopulations undergo migration. Human population migration leads to mixing of the gene pool, preventing populations from becoming too different from one another. A relatively small amount of migration among subpopulations, in the order of just a few migrant individuals in each local

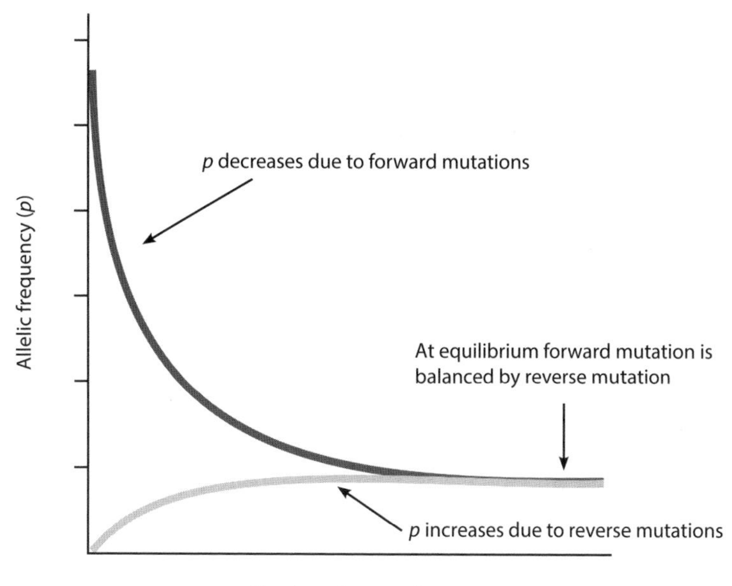

Figure 1.6: Changes due to recurrent mutations slows as the frequency *p* of an allele drops. The figure shows the influence of mutations on the frequency (p) of a single allele. The mutation rate in a single generation is exceedingly small and because the frequency of allele p drops with each mutation, the rate of change will become even slower over time. Reverse mutations will increase the frequency of allele p. Eventually the actions of the opposing forces, i.e. forward and reverse mutations, will establish equilibrium in the frequencies of the alleles p and q.

population in each generation, can be sufficient to prevent the accumulation of high levels of genetic differentiation between populations. Migration adds genetic variation to populations and increases genetic differences within the recipient population. However, genetic diversification can also occur in spite of migration when other evolutionary forces such as natural selection are sufficiently strong.

Allele frequencies can change randomly via genetic drift

Sewell Wright (1931) introduced the concept of **random genetic drift** into the study of population genetics. Genetic drift refers to changes in allele frequencies in a population due to random fluctuations. These are the frequencies of alleles found in gametes that unite to form **zygotes**. Zygotes are single diploid cells formed by the combination of a single haploid sperm and a single haploid egg, and the alleles found in these gametes vary from generation to generation simply by chance. The zygote referred to is the fertilized egg cell and is the cell from which all other cells in the body are derived. Over time genetic drift usually results in either the loss of an allele or preservation of the allele in the population and fixation at 100%. The rate at which genetic drift occurs depends on the population size and on the initial allele frequencies.

To illustrate the concept of genetic drift we can consider the following hypothetical simulation of changes in allele frequencies for a single gene in five populations of 20 individuals each ($N = 20$) over many generations (**Figure 1.7**). Suppose there are only two alleles, *A* and *B*, and the allele frequencies are identical in all five populations, each with a frequency of 0.5. In the five small populations, the allele frequencies will fluctuate from generation to generation. Eventually, in each of the five populations one of the alleles will be eliminated

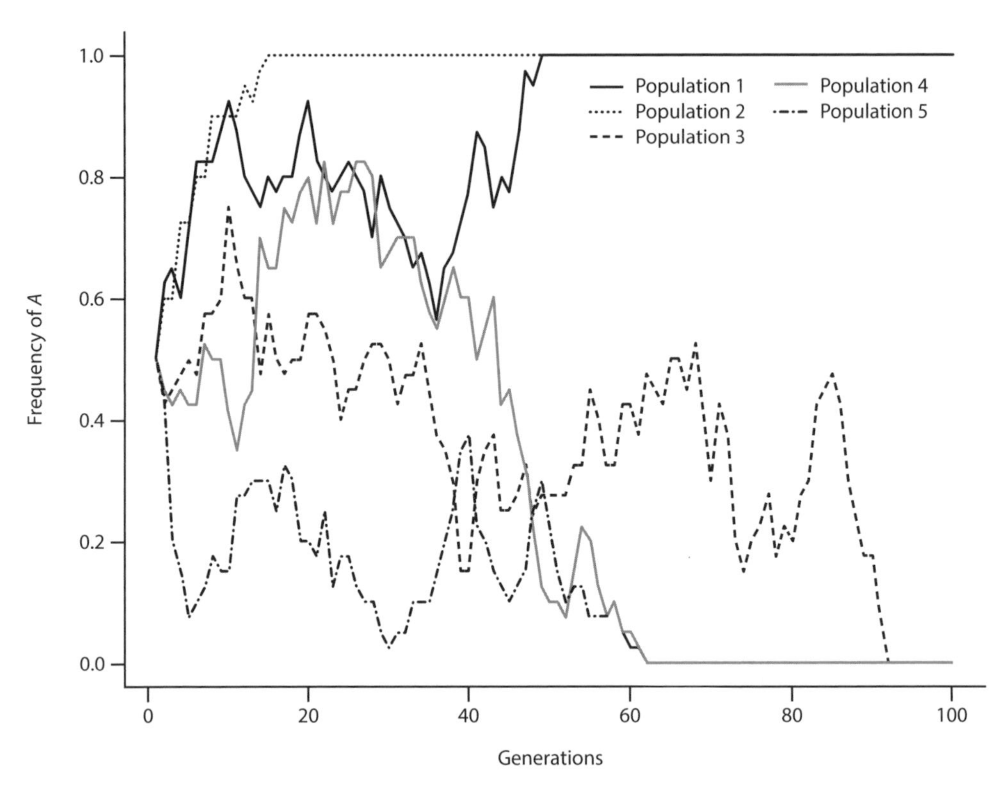

Figure 1.7: Hypothetical model of genetic drift in five different populations. The figure illustrates the potential influence of genetic drift in five populations. The model considers the different outcomes over 100 generations. In all cases the starting allele frequencies are 0.5 for *A* and 0.5 for *a* and each population is assigned 20 individuals ($N = 20$). In all cases the frequency of allele *A* only is considered. The simulations obtained over 100 generations indicate a variety of outcomes with peaks and troughs moving towards frequencies of 1 or 0 for *A* in every case.

and the other will be fixed at 100%. As in this case the gene is now monomorphic (i.e. there is only one allele), all individuals are homozygous for the predominant allele and there can be no further fluctuation in that population. Genetic drift can lead to homozygosity even in large populations, but this will take many more generations to occur.

Figure 1.7 also illustrates another effect of genetic drift. In the example all five populations begin with the same allele frequencies (50% or 0.5 for both alleles), but because genetic drift is random, the frequencies in different populations do not change in the same way and so populations gradually acquire genetic differences. Consequently genetic drift will increase the genetic variation between different populations and there will be genetic divergence over time. In contrast, the opposite effect may also be seen whereby there is reduced genetic variation within populations. Through random change, an allele may eventually reach a frequency of either 100% or 0, at which point all individuals in the population are homozygous for one allele. When an allele has reached a frequency of 1, we say that it has reached fixation. The other allele is lost (reaching a frequency of 0) and can be restored only by migration from another population or by mutation. Fixation leads to a loss of genetic variation within a population. Given enough time, all small populations will become fixed for one allele or the other. Which allele becomes fixed is random in the absence of other forms of selection pressure, though it may be determined by the initial frequency of the allele. If the initial frequency of two alleles is 0.5, both alleles have an equal probability of fixation; however, if one allele is initially more common, it is more likely to become fixed.

Genetic drift can lead to the fixation of deleterious, neutral, or beneficial alleles, but the effect is greatly influenced by the population size. Allele loss and fixation due to genetic drift occur more rapidly in small populations. Therefore, in nature, both population size and geography can influence genetic drift, and consequently the genetic composition of a population. Some human populations have settled on small islands or in geographically isolated areas and the allele frequencies within these small isolated populations are more susceptible to genetic drift. A population may be reduced in size for a number of generations because of epidemic disease, famine, or other natural or even man-made disasters. As genetic drift is a random process, small isolated populations tend to be more genetically dissimilar to other populations. Geography and population size can influence the effect of genetic drift by creating either a bottleneck or a founder effect.

Bottleneck effect

Changes in population size may influence genetic drift via the **bottleneck effect**. Natural and man-made disasters such as famines or war may reduce the size of the founder population. Depending on the size of the effect and the original population, this can change the degree of genetic variability within the population. Such events may randomly eliminate most of the members of the population with or without regard to the genetic composition or through selection of a group, for example favoring those with specific alleles following infectious epidemics. This can create a bottleneck effect within a population whereby the level of genetic variation is extremely limited (**Figure 1.8**).

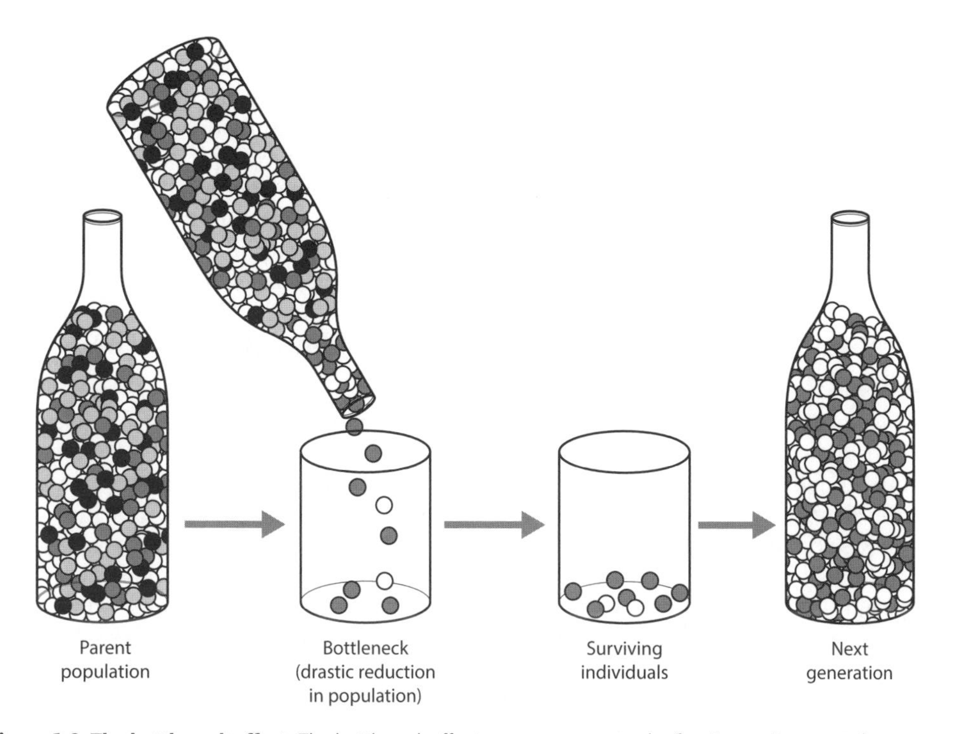

Figure 1.8: The bottleneck effect. The bottleneck effect can occur as a result of major environmental events such as famine or plague whereby the founder or parent population is drastically reduced. This may affect the degree of genetic variability within a population. Natural selection may also operate under these circumstances, favoring those with specific alleles, especially when the disaster involves infectious disease.

The thrifty gene hypothesis

The **thrifty gene hypothesis** was proposed by J. V. Neel in 1962 to explain the growing incidence of diabetes in the Western world. Neel suggested that a thrifty genotype that was more capable of modifying insulin release and glucose storage may have a survival advantage. Though this worked well for our ancestors who had to survive periods of famine, possession of the thrifty genotype in a modern Western society with a plentiful food supply may be a disadvantage as it may cause elevated insulin levels and excessive energy stores. This is seen in clinical cases of type 2 diabetes (T2D).

This hypothesis has been supported by a number of research groups. Work on late-Paleolithic human ancestors indicates alternating periods of abundance and famine and recently two genes or gene sequences have been said to be associated with thrifty characteristics. These are the insulin (*INS*) VNTRs and the apolipoprotein E (*APOE*) gene. More variation has been recorded in the *INS* VNTR genes in African versus non-African populations (27 versus three variants, respectively). *APOE* has been linked with Alzheimer's disease and cardiovascular disease. *APOE2,* which is common in Mediterranean populations, is less common in African and Native American populations. Interestingly, studies have shown that women who possess *APOE4* tend to have more children than those with *APOE2.*

Despite this emerging support for the thrifty gene hypothesis it is still controversial. One reason for this may be that the effects seen are not genetic, but are determined by other factors. Some authors have gone as far as to suggest this may be a thrifty phenotype as opposed to a thrifty genotype effect. In this latter hypothesis the authors suggest that the environment is responsible for the phenotypic variation seen and not the genes, with nutrition in newborn and infant children being particularly important.

Founder effect

Geography and population size may also influence genetic drift via the founder effect. The founder effect involves migration, where a small group of individuals separate from a larger population and establish a colony in a new location. For example, a few individuals may migrate from a large continental population and become the founders of an island population. The founding population is likely to have less genetic variation than the original population from which it was derived and consequently the allele frequencies in the founding population may differ markedly from those of their original population.

Natural selection acting on different levels of fitness affects the gene pool

The final process that brings about changes in allele frequencies is natural selection. Selection is the differential reproduction of genotypes. Selection represents the action of environmental factors on a particular phenotype and genotype through selective pressure. Natural selection is the consequence of differences in the biological fitness of individual phenotypes. Biological fitness is a measure of fertility and reproductive success of a genotype compared with other genotypes in a population. Genotypes with a greater level of biological fitness contribute more to the gene pool of succeeding generations. Differential fitness among genotypes leads to changes in the frequencies of the genotypes over time, which in turn leads to changes in the frequencies of the alleles that make up the gene pool. The effect of natural selection on the gene pool of a population depends on the fitness values of the genotypes in the population. Thus, selection may operate at any time from conception to the end of the reproductive period. There are three major forms of natural selection: purifying or negative selection, positive or adaptive Darwinian selection, and balancing selection (**Figure 1.9**)

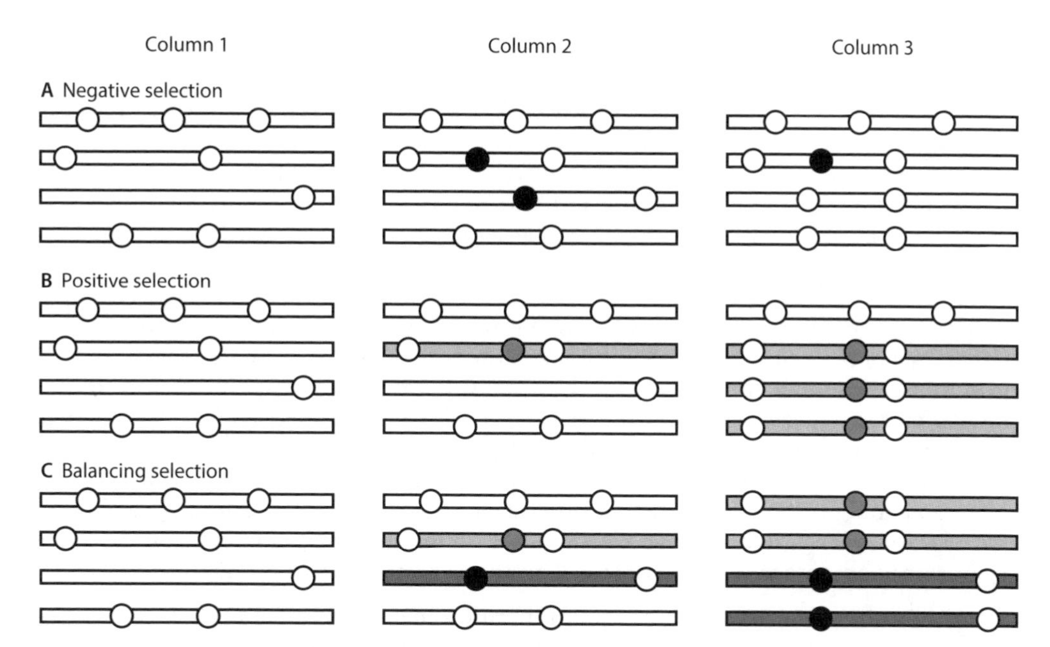

Figure 1.9: Three different models of natural selection. Rows A, B and C illustrate three patterns of natural selection (selection signatures). The columns represent three generations: the first column shows the starting group of four individuals looking at the same chromosome in each, the second column shows the first generation with mutations, and the third column shows the final outcome for the chromosomes in the three different patterns of natural selection. Each circle represents a polymorphism, within a haplotype. White circles represent mutations under neutrality, black circles indicate deleterious mutations, and gray circles indicate advantageous mutations. Pattern A illustrates genetic polymorphisms under negative selection. Deleterious mutations arise (black dot) and they can be removed immediately (if severely deleterious, e.g. line 3 in column 3) or kept at low frequencies (if weakly deleterious, e.g. line 2 in column 3). Linked neutral polymorphism will also disappear (or be kept at low frequencies, e.g. line 3 in column 3). Pattern B illustrates genetic polymorphism under positive selection. When a new advantageous mutation arises (shaded circle in line 2, column 2), the allele increases in frequency (in the population) along with linked neutral polymorphisms (lines 3 and 4 in column 3, which now resemble line 2, column 2). Pattern C illustrates balancing selection. Two new alleles are shown (shaded and black circles) and, if they confer advantage in the heterozygous state, they will increase to intermediate frequencies. Linked neutral polymorphisms will also increase to intermediate frequencies. (From Ermini L, Wilson IJ, Goodship TH & Sheerin NS [2012] *Immunobiology* 217:265–271. With permission from Elsevier.)

Purifying selection

Purifying natural selection (also called negative selection) reduces the frequency of detrimental alleles in a population. New mutants often have detrimental effects on biological fitness and purifying selection reduces the number of new mutations in the gene pool. In humans, 38–75% of all new non-synonymous mutations are thought to be affected by moderate to strong negative selection. Deleterious mutations are generally found at low frequencies because of the adverse effect they may have on biological fitness. Negative selection is responsible for the removal (or maintenance at low frequencies) of mutations associated with severe Mendelian disorders. Mendelian disease genes come under widespread purifying selection, especially when the disease mutations are dominant.

Positive Darwinian selection

Some mutant alleles introduced to a population by gene flow may be advantageous. In this case a directional genetic change may allow a population to adapt to its environment and new, better adapted alleles may replace old, less well adapted alleles. Such selection of alleles that are advantageous is called adaptive Darwinian selection or **positive Darwinian selection**. Under the action of positive selection advantageous alleles rapidly achieve high frequencies within the population. This occurs at a rate much faster than that of a neutrally selected allele. As a consequence of this rapid increase few recombination events will take place and any neutral variation linked to selected variants will also increase in frequency within the population. This process often results in a transitory increase in the strength of linkage disequilibrium between alleles on the same haplotype.

Balancing selection

A third form of natural selection is **balancing selection**, whereby **heterozygotes** show a higher level of biological fitness than **homozygotes**. This leads to the maintenance of two or multiple alleles in a population at a given locus. Polymorphisms are maintained in the population for a longer period of time than expected. Balancing selection is often referred to as **heterozygote advantage**, especially in cases where a mutant allele known to cause a disease in homozygotes is found at a high frequency in heterozygous healthy members of the population. Genome scans suggest that balancing selection is less extensive than positive selection. However, balancing selection does occur. The two examples below illustrate heterozygous advantage in two **autosomal recessive Mendelian** diseases.

Cystic fibrosis is one of the most common autosomal recessive diseases in Northern European populations, affecting approximately 1:2500 new born children. The causative gene in cystic fibrosis is the cystic fibrosis transmembrane conductive regulator gene (*CFTR*) and there are currently 1910 mutations on the *CFTR* mutation database (http://www.genet.sickkids.on.ca/StatisticsPage.html) (**Figure 1.10**). Carriers of the *CFTR* mutations (heterozygotes) appear to have, or have had in the past, some reproductive advantage over wild-type normal homozygotes. There has been debate over what this advantage might be. The *CFTR* gene encodes a membrane chloride channel protein that is required by some bacteria such as those belonging to the genus *Salmonella*

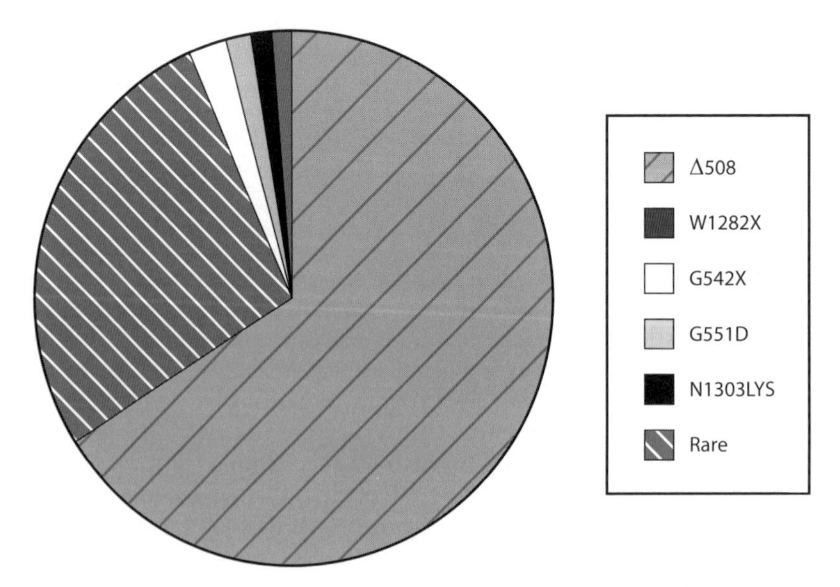

Figure 1.10: The five most common *CFTR* gene mutations. The five mutations listed account for over 70% of the overall mutations and Δ508 is the most common of all, accounting for approximately two-thirds of all cases. All of the other mutations, of which there are at least 1905, are found at frequencies of less than 1% and together these account for approximately 30% of all *CFTR* mutations.

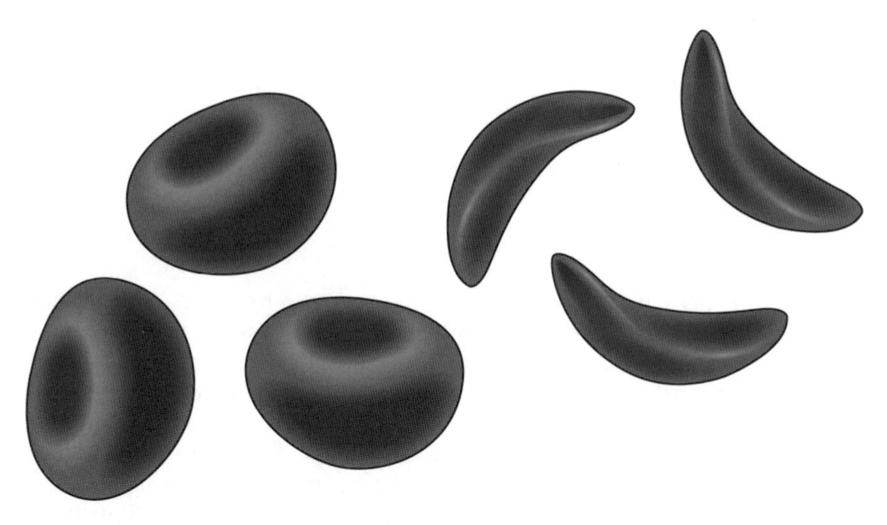

Figure 1.11: Red blood cells in sickle cell disease. Sickle cells are shaped like a harvesting sickle and, unlike the normal doughnut-shaped red blood cells, these cells can be hard with sharp edges that can damage the wall of small blood vessels as they passage through the body. They will often clog the flow of blood and break up as they pass through the small blood vessels.

(e.g. *Salmonella typhi*) to enter into epithelial cells. One explanation is that carriers of a mutant *CFTR* allele may be more resistant to infection by such bacteria than those with two copies of the wild-type gene.

Sickle cell anemia is another example. This is a genetic autosomal recessive blood disorder that is characterized by red blood cells that occasionally assume an abnormal, rigid, sickle shape (**Figure 1.11**). The β-globin allele variant, called *HbS*, is responsible for the sickling of red blood cells seen in the disease. Despite the high mortality associated with homozygosity the sickling allele *HbS* is found at high frequencies in Africa (up to 30%). One possible explanation for the abundance of the *HbS* allele in Africa is that heterozygosity confers some resistance to malaria.

1.4 CALCULATING GENETIC DIVERSITY: DETERMINING POPULATION VARIABILITY

Genotype and allele frequencies illustrate genetic diversity

The genetic diversity of a population can be described using genotype or allele frequencies. A large number of samples from a population are usually collected, and the genotype and allele frequencies are calculated. The genotype and allele frequencies of the sample population are then used to estimate the diversity of the population. To calculate a genotype frequency, the number of individuals having the same genotype is divided by the total number of individuals in the sample (N). For a locus with three genotypes, AA, Aa, and aa, the frequency (f) of each genotype is:

$$f(AA) = \frac{\text{number of } AA \text{ individuals}}{N}$$

$$f(Aa) = \frac{\text{number of } Aa \text{ individuals}}{N}$$

$$f(aa) = \frac{\text{number of } aa \text{ individuals}}{N}$$

The sum of all the genotype frequencies always equals 1 (or 100%).

Genotypes are not permanent. They are disrupted in the processes of segregation and recombination that take place when individual alleles are passed to the next generation through the gametes. Alleles, in contrast, are not broken down and the same allele may be passed from one generation to the next. For this reason the calculation of allele frequencies is often the preferred choice when determining the genetic variability of a population. In addition, there are always fewer alleles than genotypes, e.g. for the gene with two alleles *A* and *a* above, there are two alleles, but there are three genotypes. By using alleles, population diversity can be described in fewer terms than by using genotypes. Finally, by using allele frequencies in case control population studies rather than genotype frequencies, no assumptions about the impact of homozygosity or of heterozygote advantage are being made. This is especially important in the context of complex disease where in the absence of a clear pattern of inheritance it would not be appropriate to make any such assumption, at least in the initial stages of analysis.

Allele frequency refers to the numbers of alleles present in a population

The number of copies of an allele at a locus is divided by the total number of all alleles in the sample:

$$\text{Frequency of an allele} = \frac{\text{number of copies of the allele}}{\text{number of copies of all alleles at the locus}}$$

If we consider a gene with only two alleles *A* and *a* and we suppose the frequencies are *p* for allele *A* and *q* for allele *a*; then *p* and *q* can be calculated as:

$$p = f(A) = \frac{2n_{AA} + n_{Aa}}{2N}$$

$$q = f(a) = \frac{2n_{aa} + n_{Aa}}{2N}$$

In this equation n_{AA}, n_{Aa}, and n_{aa} represent the numbers of *AA*, *Aa*, and *aa* individuals, and *N* represents the total number of individuals in the sample it is necessary to divide by *2N* because being diploid means each individual has two alleles for each gene (one from the maternal locus and one from the paternal locus).

The sum of the allele frequencies is always 1 (100%) ($p + q = 1$); therefore where there are only two alleles, *q* can be determined by simple subtraction after *p* has been calculated:

$$q = 1 - p$$

These calculations apply only where there are two alleles. In cases where there are several different alleles at a locus the calculation used is based on the same principle, but is more complicated. Statistical software will usually be used to perform complex calculations, but it is important to understand the underlying principles in any analysis.

Heterozygosity provides a quantitative estimation of genetic variation

Knowing the frequency of heterozygotes (i.e. those carrying both wild-type and mutant alleles for the same gene) can be a very useful tool for the quantitative estimation of genetic variation in a population. Where mutations are common, heterozygotes are common and homozygotes can be quite rare. Heterozygosity can provide information on the structure and even the history of a population. High levels of heterozygosity reflect high levels of genetic variability, while low levels of heterozygosity indicate low levels of genetic variability. Very low levels of heterozygosity can indicate the effects of small population sizes created by population bottlenecks. Often the observed levels of heterozygosity are compared with what is expected under HWE (see below). If the observed heterozygosity deviates from HWE or is lower than expected, this discrepancy may be attributed to non-random mating. This can occur in small isolated populations when individuals select a closely related mate more often than would be expected by chance in a larger population.

Non-random mating does not change the allele frequencies, but leads to an increase in homozygous offspring over time because the parents are more likely to be genetically similar. Consequently, there will be a decrease in heterozygosity in such populations. This places individuals and the population at a greater risk from Mendelian recessive diseases. The impact of accumulating deleterious homozygous traits is called **inbreeding depression**. This term refers to the loss of **population vigor** due to reduced genetic variability or reduced biological fitness in a given population as a result of the breeding between related individuals. This phenomenon is often the result of a population bottleneck. If heterozygosity is higher than expected, an isolated breakout may have taken place through contact with individuals from another population, which can introduce a temporary excess of heterozygotes.

Expected heterozygosity can be measured using the simple formula:

$$H_E = 1 - \sum_{i=1}^{n} p_i^2$$

In this equation n is the number of alleles and p_i is the frequency of the ith allele at a locus.

The value of this measure ranges from 0 for no heterozygosity to nearly 1 (i.e. 100%) for a system with a large number of equally frequent alleles.

The HWP is a complex but essential concept in population genetics

The HWP, which was introduced earlier in this chapter, is one of the most important statistical principles in population genetics, and because it is an abstract and quantitative principle it is one of the hardest concepts to understand. Therefore, it needs to be discussed in some detail. We may wonder why a recessive trait is not gradually eliminated over the course of time or how the O blood type can be the most common blood type if it is a recessive trait. These questions reflect the assumption that the dominant allele in a population will always be found at the highest frequency and the recessive allele will always be less common. The HWP addresses these questions and enables us to consider the frequency of alleles over the time.

The HWP depends on certain assumptions, of which the most important are:

- Mating in a population is random – there are no subpopulations that differ in allele frequency.
- Allele frequencies are the same in males and females.
- All the genotypes are equal in viability and fertility – selection does not operate.
- Mutation does not occur.
- Migration into the population is absent – gene flow does not occur.
- Genetic drift does not occur.
- The population is sufficiently large that the frequencies of alleles do not change from generation to generation.

The HWP states that after one generation of random mating, in a large breeding population, where the restrictions listed above all apply, single-locus genotype frequencies can be presented as a binomial function (where there are only two alleles) or multinomial function (where there are multiple alleles). Under the above conditions and over time, allele frequencies will reach equilibrium and remain constant from generation to generation.

Calculating expected genotype frequencies using the HWP

The HWE can be calculated using the simple mathematical formula:

$$p^2 + 2pq + q^2 = 1$$

In this equation p and q represent the frequencies of alleles (**Figure 1.12**). It is important to note that the sum of the allele frequencies ($p + q$) *is* always equal to 1. To illustrate HWE, we can consider a single gene with two alleles A and a in a large population with frequencies p and q, respectively (**Box 1.1**). First, we must assume that male and female gametes interact randomly, and all of the major assumptions above remain true (i.e. there is no mutation, selection, random genetic drift, or gene flow). We can then use a simple calculation based on the **Punnett Square** illustrated in Table 1.2. The Punnett square is perhaps the most common of all mathematical representations used in the study of the genetics of complex disease. The data entered in the table can be used to generate odds ratios (ORs) and significance values via the χ^2 test. It is important to note that the exact numbers and not the percentages must be used in the calculations to generate accurate outcomes.

Different populations may have different allele frequencies

Note that heterozygotes are more common when allele frequencies are intermediate; however, when one allele is more common than the other, homozygosity for that allele is increased and heterozygosity reduced. In this illustration heterozygotes have a maximum frequency of 50%, which is achieved when $p = q = 0.5$. When either locus is monomorphic ($p = 1$ or $q = 1$), there are no heterozygotes.

HWE allows us to describe a population only considering the frequencies of n alleles at a particular locus. For the less statistically inclined geneticists, the HWE principle is mostly applied to ensure validation of data from population studies. Departure from the expected distribution of genotypes generally indicates problems in sample recruitment or some other form of population bias. Conversely, there is more confidence in the results when there is equilibrium.

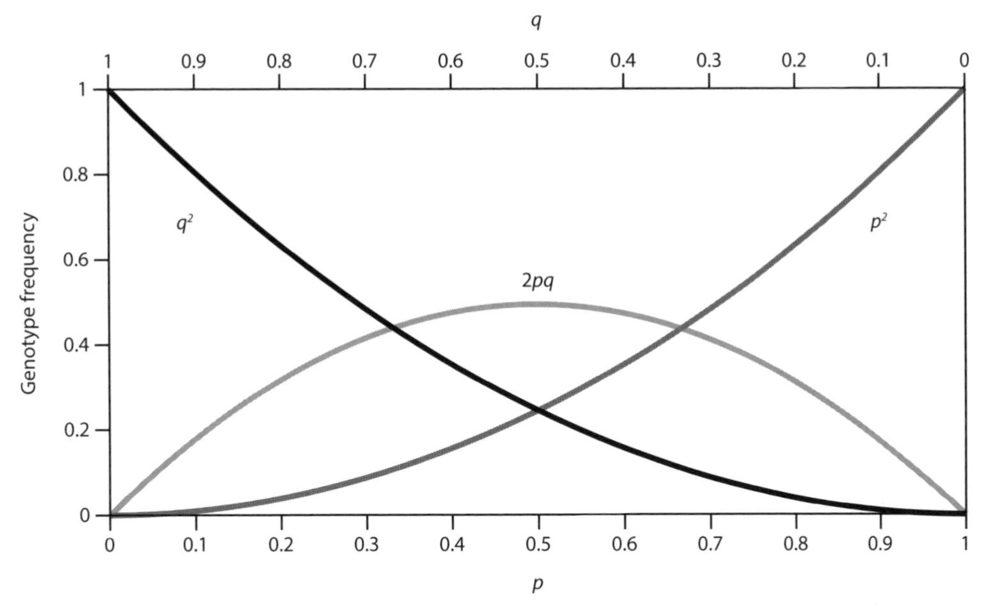

Figure 1.12: A plot of the HWE-based genotype frequencies (p^2, $2pq$, and q^2) as a mathematical function of allele frequencies (p and q). The plot illustrates the influence of allele frequencies on genotype frequencies and shows what we can expect from the HWE test. The plot shows how the two alleles p and q determine genotype frequencies and these change as the allele frequencies change. For example the closer q is to 1, the lower the value of p and the higher the value for the q^2 genotype (homozygous q). When p and q are both set at 0.5 (50%), then the frequency of pq heterozygotes is high.

BOX 1.1 CALCULATING EXPECTED GENOTYPE FREQUENCIES USING THE HWP

The HWE can be calculated using the simple mathematical formula:

$$p^2 + 2pq + q^2 = 1$$

In this equation p and q represent the frequencies of alleles. It is important to note that the sum of the allele frequencies ($p + q$) is always equal to 1.

To illustrate HWE, we can consider a single gene with two alleles *A* and *a* in a large population with frequencies p and q, respectively. First, we must assume that male and female gametes interact randomly and all of the following assumptions are true:

- Mating in a population is random – there are no subpopulations that differ in allele frequency.
- Allele frequencies are the same in males and females.
- All the genotypes are equal in viability and fertility – selection does not operate.
- Mutation does not occur.
- Migration into the population is absent – gene flow does not occur.
- Genetic drift does not occur.
- The population is sufficiently large that the frequencies of alleles do not change from generation to generation.

BOX 1.1 CALCULATING EXPECTED GENOTYPE FREQUENCIES USING THE HWP (*Continued*)

Then we can use a simple calculation based on the Punnett Square illustrated in Table 1.2. The top side of the square is divided into proportions p and q representing the frequencies of the male alleles A and a, respectively. The left side represents the same proportions, but for the female alleles. If we assume there is random union of gametes we can apply the product rule of probabilities. Imagine a pool with male and female gametes, p with A alleles and q with a alleles, and where zygote formation occurs by random union. The upper-left square represents the frequency of the homozygous genotype AA. The expected frequency is simply the product of the separate allele frequencies.

$$\text{Frequency of } AA = p \times p = p^2 \text{ (Homozygous for } A\text{)}$$

The frequency of homozygous genotype aa is shown in the lower-right square:

$$\text{Frequency of } aa = q \times q = q^2 \text{ (Homozygous for } a\text{)}$$

The other two squares illustrate the third possibility, i.e. Aa heterozygotes. The total proportion of Aa heterozygotes can be calculated:

$$\text{Frequency of upper-right square } Aa = pq \text{(Heterozygous)} \qquad (1)$$

$$\text{Frequency of lower-left square } Aa = pq \text{(Heterozygous)} \qquad (2)$$

Total frequency of $Aa = Aa\ (1) + Aa\ (2) = 2pq$ (Heterozygous Aa)

The three different genotypes AA, Aa, and aa are formed in proportions p^2, $2pq$, and q^2, respectively. The sum of allelic frequencies is:

$$(p + q)^2 = p^2 + 2pq + q^2$$

This illustrates the HWE. It is important to note that Hardy–Weinberg proportions are binomial in this case (i.e. there are only two alleles). Given any set of genotype frequencies (AA, Aa, and aa), the HWE predicts that after one generation of random mating, provided the assumptions above are all met, the genotypic frequencies will be in the proportions p^2, $2pq$, and q^2. For example, given the initial genotype frequencies of $AA = 0.4$, $Aa = 0.4$, and $aa = 0.2$, where $p = 0.6$ (frequency of allele A) and $q = 0.4$ (frequency of allele a), after one generation the genotype frequencies become:

$$p^2, 2pq, q^2 = (0.6)^2, 2(0.6)(0.4), (0.4)^2 = 0.36, 0.48, 0.16.$$

The genotype frequencies will stay in these proportions generation after generation provided mating is random and the assumptions above are all met. Deviation from any of the above conditions can led to an increase or decrease in allele frequencies from one generation to another and this will impact on the genotype distribution. Finally, it is important to note that this example deviates from the statement earlier in this chapter that states it takes a long time to reach HWE. This is because mutation rates in human genes are low; a full explanation of this is given earlier in this chapter.

Probabilities

Probability represents the chance of a given event (or outcome) to occur. It is a measure of the uncertainty and can be a number between 0 and 1. There are different schools of thought regarding the concept of probability. The use of probability is illustrated in **Box 1.2**. In the case of mutually exclusive events (e.g. tails or heads when the same coin is flipped at the same time), the combined probabilities of the outcomes (i.e. heads or tails) can be calculated by summing the individual probabilities of each event. This is known as the **sum rule**. When two (or more) independent outcomes can occur simultaneously (i.e. they are not mutually exclusive), the joint probability of the outcomes is expressed by the **product rule**. The sum rule and product rule are also illustrated in Box 1.2.

BOX 1.2 PROBABILITY, SUM RULE, AND PRODUCT RULE

Probability

Probability represents the chance of a given event occurring. It is a measure of the uncertainty and can be a number between 0 and 1. If probability is equal to 0 the event cannot take place, in contrast if it is 1 the event must occur. Although there are different schools of thought regarding the concept of probability, we prefer to define the probability as belief in future events. For example, when we flip a coin, if we state that the probability of observing heads (A) is 0.5 and the probability of observing tails (B) is 0.5, we believe heads and tails have equal chance in the next flip. The probability of heads [$P(A)$] can be calculated using:

$$P(A) = A/N$$

where N is the total number of outcomes (i.e. the number of times the coin is flipped). In statistical terms, the total number of times the coin is flipped is equal to 1 and thus the probability of heads [$P(A)$] is 0.5. Instead of flipping a coin, we may want to throw a dice and estimate the probability of throwing a 5. In this case, the number of throws is 1. If there is only one 5 on a dice and we are using a six-sided dice, the probability of throwing a 5 is 0.167.

Sum rule

In the case of mutually exclusive events (e.g. tails or heads when the same coin is flipped at the same time), the combined probabilities of the events can be calculated by summing the individual probabilities of each event. This is known as the addition law of probability, or the sum rule:

$$P(A \text{ or } B) = P(A) + P(B)$$

where A and B are two mutually exclusive events, and P represents the probability.

If we flip the same coin twice, the combined expected probability is:

$P(\text{heads or tails}) = P(\text{head}) + P(\text{tails}) = 0.5 + 0.5 = 1$

Product rule

When two (or more) independent events can occur simultaneously, meaning that they are not mutually exclusive, the joint probability of the events is expressed by the product

BOX 1.2 PROBABILITY, SUM RULE, AND PRODUCT RULE *(Continued)*

rule. The product rule states that the joint probability of two independent events occurring is the product of the individual probabilities:

$$P(A \text{ and } B) = P(A) \times P(B)$$

where A and B are two events, and P represents the probability.

If we flip two different coins, the expected probability of a head from coin 1 is 0.5 and the probability of a head from the coin 2 is 0.5, the joint probability of two heads is:

$$P(\text{two heads}) = P(\text{head coin 1}) \times P(\text{head coin 2}) = 0.5 \times 0.5 = 0.25$$

1.5 POPULATION SIZE AND STRUCTURE

The term population has already been defined at the beginning of this chapter. Here, we consider a population through two important parameters: population size and population stratification. Populations are continually being modified by increase (births and immigrations) and decrease (deaths and emigrations). Some geneticists therefore consider a population to be best defined as the area within which individuals are likely to find a mate. Sometimes, however, human populations may be geographically widespread and be subdivided into local groups, called subpopulations.

Breeding population size is important in evolution

When considering population size in evolutionary terms, the relevant information concerns the number of breeding individuals, which may be quite different from the total number of individuals in the population. In some cases the breeding population may be a small proportion of the total. In developed countries, the proportion of older people is rapidly increasing due to improvements in healthcare. As a consequence of an expanding aging population, overall population fertility inevitably declines and the proportion of the population that can be counted as members of the breeding population decreases. In addition, even if the size of a breeding population can be estimated with reasonable accuracy, the breeding population number may not be indicative of the actual breeding population size. For example, factors such as sex ratio of breeding individuals, social status, and disease may influence their genetic contribution to the next generation. As a result, the concept of effective population size, an ideal population of size N in which all adults have an equal expectation of becoming parents, tends to be used. The effective population size (usually indicated as N_e) is the number of individuals in a population who contribute offspring to the next generation, and it can determine the amount of genetic variation, genetic drift, and linkage disequilibrium in populations.

Genetic variation is not always uniform in a population

A population may have substructural differences in the genetic variation among its constituent parts. Population substructure or stratification may be due to several different evolutionary reasons. For example, a population may have localized subpopulations in which there is genetic drift. Exchange of genotypes may not have equal probabilities throughout a population or selection may have different effects in different parts of the population.

Migrations from one population into another may also be responsible for stratification. Thus, a population is considered stratified if:

- Genetic drift occurs in some, but not all, subpopulations.
- Migration does not happen uniformly throughout the population.
- Mating is not genetically random throughout the population.

All of the evolutionary factors that we have already discussed can contribute to the structure of a population. This structure affects the extent of genetic variation and the pattern of distribution.

Wahlund's principle

A population may appear to be homogeneous, but this may be a deception. This can lead to false-positive associations, as we will see in Chapter 5. Subpopulations and population stratification may not be obvious in studies of populations and population structure, and as a result the study samples may include heterogeneous subsamples or clusters from the study population. When data from these subpopulations are grouped together and differences in allele frequencies among them are inferred, a deficiency of heterozygotes and an excess of homozygotes will be found, even if Hardy–Weinberg proportions exist within each subsample. This effect is known as **Wahlund's principle** or the Wahlund effect and is one of the major problems in genetic association studies.

1.6 THE MITOCHONDRIAL GENOME

In humans and most other eukaryotes the mitochondria are large intercellular organelles approximately 0.5–1 μm in diameter. Their main function is to generate energy in the form of adenosine triphosphate (ATP). As a result, the mitochondria control and contribute to a range of cellular functions, including cell signaling, differentiation, and even cell death. The mitochondria are composed of inner and outer membranes, and are packed with multiple copies of their own DNA genome (**mtDNA**).

The mitochondrial genome has been extensively studied in relation to human evolution. Generally, the mitochondrial genome is easier to analyze than the autosome by virtue of the much shorter sequence (16.6 kb of circular DNA), greater abundance of mtDNA, and the fact that there are only 37 genes encoded in the mtDNA genome. In addition, because mtDNA is maternally inherited and does not recombine at meiosis, the mtDNA genes are not reshuffled every generation through recombination (a process that tends to obscure genetic relationships). Therefore, genetic relationships are more easily identified in mtDNA. Usually, all the copies in an individual's mitochondrial genome are identical (i.e. individuals are **homoplasmic**), but occasional **heteroplasmy** is observed where there is a mixture of two or more mitochondrial genotypes. In most individuals, though there is no evidence of heteroplasmy, there is strong evidence that the mtDNA is undergoing constant mutation. However, because these mutations are present at a low level they are often not detected. When it comes to considering populations, however, there is often considerable variation with high levels of polymorphism within and between populations.

The general low level of heteroplasmy in the population suggests that, at some point in the germ line, the effective number of copies of the mitochondrial genome must have been very small otherwise a greater level of diversity would be seen. More than 10,000 complete mitochondrial genomes have now been sequenced (http://www.mitomap.org). These include mtDNA from a number of different populations and also from some ancient individuals of whom the first three were sequenced in 2008, including one Neanderthal and two *Homo sapiens* (the Neolithic Tyrolean Iceman and a Paleo Eskimo).

The substitution rate for the entire mitochondrial genome has been estimated as 1.665×10^{-8} ($\pm 1.479 \times 10^{-9}$) substitutions per nucleotide per year or one mutation every 3624 years per nucleotide. Though the overall substitution rate is low, substitutions at some positions occur more frequently than at others. These are referred to as **hot-spot positions** and are mostly located in the control regions (positions 16,362, 16,311, 16,189, 16,129, 16,093, 195, 152, 150, and 146). The mutation rate has also been estimated for both the control and coding regions, and this higher rate of substitution makes mtDNA particularly valuable in studying relationships in recently diverged lineages.

When applied to modern populations, mtDNA studies clearly support the theory that modern humans originated in Africa and spread from that continent approximately 56,000–73,000 years ago. This fits with the observation of greater genetic diversity in African populations and can be observed through the construction of a **phylogenetic tree** based on mtDNA variations from populations from all parts of the world. By applying the **molecular clock** to the tree, it has been possible to demonstrate that the ancestral mtDNA, i.e. that one from which all modern mitochondrial genomes are descended, existed between 140,000 and 290,000 years ago. Phylogenetic reconstruction shows that this mitochondrial genome was located in Africa and the person who possessed it must have been African. Thus, we can conclude that the most recent common ancestor for modern humans is African and because of the matrilineal inheritance of mtDNA, she has been called the mitochondrial Eve (**Figure 1.13**). This theory also relies on the observation that less variation in mtDNA occurs among humans than would be expected. Perhaps this reflects the importance of mitochondrial gene function, which may drive conservation of the mitochondrial genome or alternatively genetic variation may be reduced by genetic drift, and thus a small population size 50,000 years ago could have the same effect as the bottleneck from a single common ancestor 200,000 years ago.

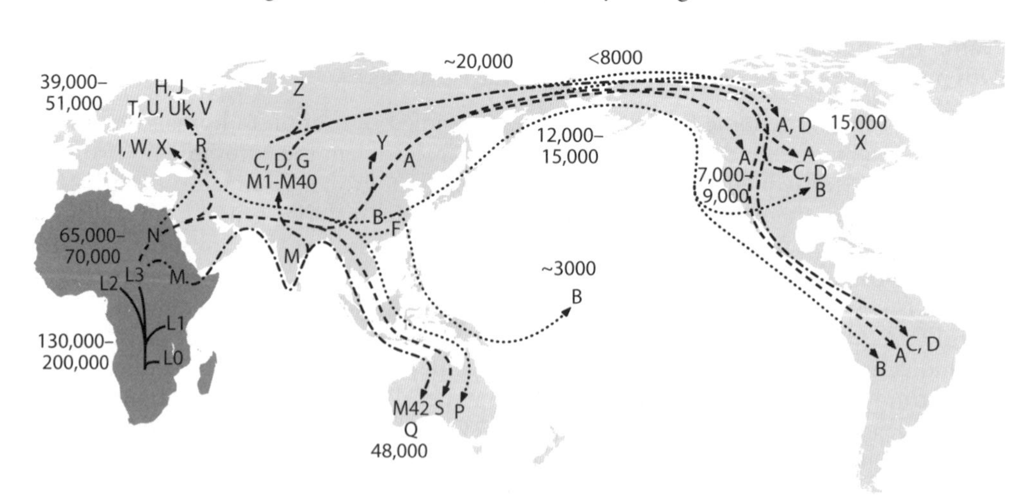

Figure 1.13: Mitochondrial genotypes and human migration. The figure illustrates human migration out of Africa and subsequent migrations based on mtDNA genotypes. The "out of Africa" principle of human evolution and the idea of the mitochondrial Eve are important in understanding human diversity. Based on this figure, expansion began around 120,000–150,000 years ago in Africa and 56,000–73,000 years ago out of Africa, but estimates vary and quotes depend on the hominid species being discussed. (From the MITOMAP database; http://www.mitomap.org.)

1.7 GENE EXPRESSION AND PHENOTYPE

Genetic variation is manifested in the phenotype

The term phenotype is the result of the coordinated expression of approximately 20,000 genes that make up the human genome and the interaction with the environment. A phenotype can be continuous, e.g. eye or hair color, or discontinuous, e.g. phenotypes that relate to health or physiology where there is complex interplay between alleles of different genes and between the genotype and the environment. The balance of gene expression among many different genes usually produces an individual with a normal (healthy) phenotype.

Phenotypes are influenced by the environment

A specific set of environmental circumstances can strongly influence the phenotype. In order to better illustrate this concept we can consider a classical example from population genetics. The Himalayan allele in rabbits produces dark fur on the nose, ears, and feet. The dark pigment develops only when a rabbit is bred at a temperature of 25°C or lower; if a Himalayan rabbit is bred at 30°C, no dark patches develop. Pigmentation is dependent on the environment, which modifies the effect of possessing the Himalayan allele. In other words, the rabbit may possess the allele, but possession of the allele is not sufficient for the phenotype to be expressed. The phenotype is only manifested when the environmental conditions are correct. The enzyme necessary for the production of dark pigment is inactivated at higher temperatures. This pigment is found only in the extremities of the body. The animal's core body temperature is normally above 25°C, i.e. a temperature where the enzyme is not functional.

Environmental factors also play an important role in the phenotypic expression of a number of genetic diseases. Phenylketonuria (PKU) is caused by a gene (also called *PKU*) encoding a dysfunctional phenylalanine hydroxylase enzyme (PAH) that does not convert phenylalanine in tyrosine, resulting in the accumulation of abnormal levels of phenylalanine that can lead to brain damage in children. A simple environmental change in those homozygous for the *PKU* allele, such as a low-phenylalanine diet, reduces the risk of brain injury.

These examples illustrate the point that genes and their products do not act in isolation; rather, they frequently interact with other factors, including environmental factors.

In order to compare the impact of the environment versus the genome on the phenotype, we need to construct a model that will allow us to breakdown the phenotype into genetic and environmental components. We can accomplish this for a **quantitative trait** using the equation:

$$P_{ij} = G_i + E_j$$

In this equation P_{ij} is the phenotype, G_i is the genetic contribution, and E_j is the environmental component. E_j may be either positive or negative depending upon the effect of the environment. Individuals with a particular genotype may do well in a specific environment depending on the level of interaction with the environment. If such a specific interaction occurs, then the basic model can be expanded to include a term for genotype–environment interaction, resulting in:

$$P_{ij} = G_i + E_j + GE_{ij}$$

In this equation GE_{ij} measures the interaction between genotype i and environment j. The model could be further expanded by splitting the genetic component into additive or dominant.

1.8 EPIGENETICS

Epigenetics means above or in addition ("epi-") to genetics and it may be typically defined as the study of heritable changes in gene expression that are not due to changes in the DNA sequence. Epigenetics can involve chemical modifications of DNA or proteins that are closely associated with DNA (e.g. histones) and a prominent role for RNA is also emerging. The structure of the chromosome in the eukaryotic cell is highly ordered and it undergoes a process of compaction. To achieve these compact structures DNA is combined with various proteins, and then coiled and **super-coiled** to form **chromatin**. The basic unit of chromatin [the nucleosome core particle (NCP)] is composed of a 147-bp DNA chain in a 1.7 left-handed super-helical turn around an eight sectioned octamer composed of two copies each of four different histones (**Figure 1.14**). The fundamental unit of this packaging is called the **nucleosome**. This involves a complex interaction with histone proteins. This coiling to form chromatin and the interactions with **histone** proteins are key elements in epigenetics.

DNA is also modified by biochemical processes such as methylation and two alleles with the same sequence may have different states of methylation that confer a different phenotype. Though these changes do not alter the DNA sequence, they may have major effects on the expression of the gene. Some of these changes are heritable, though they do not affect the DNA structure. Methylation is an important factor for post-translational modification of histones and subsequent formation of nucleosomes for packing DNA in the nucleus. It is thought that methylation establishes epigenetic

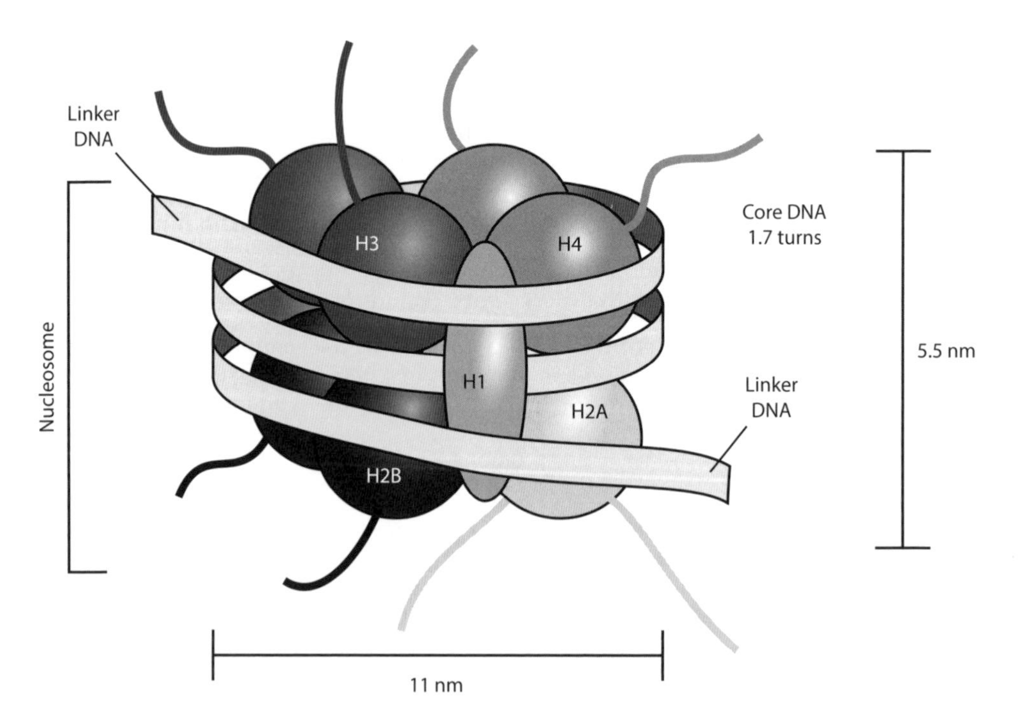

Figure 1.14: The NCP: the basic unit of chromatin. The figure shows the structure of the histone octamer with the N-terminal tails and DNA wrapped around the histone structure. This core protein comprises four different histones: H2A, H2B, H3, and H4. The NCP organizes 147 bp of DNA in a 1.7 left-handed helical coil. Small DNA sections called linker DNA help to stabilize the structure by association with a linker histone H1. (From Armstrong L [2014] Epigenetics. Garland Science.)

inheritance as long as the maintenance methylase acts to restore the methylated state after each cycle of replication. Thus, a methylated state can be perpetuated through an indefinite series of somatic meiosis. Methylation is an important factor in histone formation for packing DNA in the nucleus. A useful site for epigenetics can be found at http://www.ncbi.nlm.nih.gov/epigenomics/.

Not all geneticists interested in epigenetics are focused on biochemical processes—some are concerned with the wider meaning of the word epigenetics as "epi-," i.e. outside genetics, in more general terms. This group concerns themselves with the environment, diet, lifestyle, and the interaction with our genes. There is significant evidence to link these phenotypic characteristics with genetic variation, as noted in Section 1.7.

1.9 GENOMIC IMPRINTING

Genomic imprinting is also considered to be an epigenetic phenomenon, though to some extent it is more complex than any of the above. In some families with autosomal dominant disease the trait is only expressed if inherited from a parent of one particular sex. This occurs despite the fact that the disease-causing mutation can be present in both male and female parents. Examples of genetic imprinting are Beckwith–Wiedemann syndrome, Prader–Willi syndrome, Angelman syndrome, and Silver syndrome (discussed in Section 2.3). Imprinted genes harbor regions known as **imprint control elements** that act over long distances. The imprint control elements have been found to be restricted to small areas known as imprinting centers. In the case of Prader–Willi syndrome and Angelman syndrome, the imprinting centers have been identified at the 5′ end of the *SNURF–SNRPN* gene. Both syndromes are associated with the loss of the same small region on the long arm of chromosome 15 (see Chapter 2).

It is tempting to ask why genomic imprinting occurs. One possibility is that throughout evolution there are different and conflicting evolutionary pressures acting on maternal and paternal alleles for genes (such as *IGF2*, which affects fetal growth). From an evolutionary standpoint, paternal alleles that maximize the survival of offspring are favored and because low birth weight is strongly associated with infant mortality and adult health, infant size is a major factor. Thus, it is to the advantage of the male parent to pass on alleles that promote maximum fetal growth of their offspring. In contrast, a maternal allele with more limited fetal growth is favored from a maternal standpoint. The fetus takes nutrients from the mother and unlimited fetal growth may limit her ability to reproduce in the future. In addition, giving birth to very large babies is difficult and risky for the mother.

CONCLUSIONS

This chapter illustrates some of the basic science and statistical concepts that are a prerequisite for understanding genetic variations in human populations. Human evolution has created a diverse and complex gene pool, and a number of different interacting evolutionary forces have played a key role in the generation of this diversity. These include mutation, migration, genetic drift, and the bottleneck and founder effects, to name but a few. Considering genetic diversity is important in the study of complex diseases. The level of genetic diversity in a population varies depending on the population size and structure. Genetic diversity is not confined to the autosome, but is also found in the mtDNA.

Other factors that are important in considering complex disease are gene expression and phenotype. Epigenetics is important both in the narrow sense, where we consider such things as methylation of genes, and in a broad sense, where we consider the impact of the environment (e.g. nutrition) on disease.

Statistics plays an increasingly important and complex role in modern genetics. Though there are many excellent statistical programs to use for analysis of data, all those with an interest in this field are advised to have a basic understanding of the statistical concepts that apply to studies of complex disease.

Diversity in the gene pool confers both advantages and disadvantages to individuals within a population. Diversity though modification of common genes can lead to a variety of diseases, such as those described by the classic genetic models: chromosomal, mitochondrial, and Mendelian traits. However, these are rare. Diversity also leads to increased risk of more common diseases (genetically complex diseases), whereby the inheritance of an allele confers an increased or reduced risk. This is important in evolution because a diverse gene pool increases the likelihood that some of the population will survive even the most severe disease.

The research that is being applied to complex disease will be increasingly applied in medical practice, but there are also societal and ethical issues arising from our increased knowledge of the genome and the increasing role it is beginning to play in modern medical practice.

FURTHER READING

Books

Armstrong L (2014) Epigenetics. Garland Science. This book is a useful new guide to epigenetics.

Cavalli-Sforza LL & Bodmer WF (1971) The Genetics of Human Populations. W. H. Freeman.

Collins F (2010) The Language of Life: DNA and the Revolution in Personalised Medicine. Harper-Collins.

Crawford MH (2007) Foundations of Anthropological Genetics. In Anthropological Genetics: Theory, Methods and Applications (Crawford MH ed.), pp 1–16. Cambridge University Press.

Darwin C (1859) The Origin of Species: By Means of Natural Selection or the Preservation of Favoured Races in the Struggle for Life. J Murray. This is perhaps the most important book in the history of genetics.

Hedrick PW (2011) Genetics of Populations, 4th ed. Jones & Bartlett.

Jones S (2000) The Language of the Genes, 2nd ed. Harper Collins.

Lewis R (2011) Human Genetics – The Basics. Routledge. An alternative very short textbook on genetics that covers the basics quite well. This is a good source for students wishing to extend their knowledge to include more on classical genetics in a short time.

Strachan T & Read AP (2011) Human Molecular Genetics, 4th ed. Garland Science. This is an excellent basic genetics textbook that goes into more depth on many of issues in this chapter of this book. Chapters 2 and 3 and also chapters 15 and 16 of HMG are particularly useful.

Articles

Barreiro LB, Laval G, Quach H et al. (2008) Natural selection has driven population differentiation in modern humans. *Nat Genet* 40:340–345.

Cann RL, Stoneking M & Wilson AC (1987) Mitochondrial DNA and human evolution. *Nature* 325:31–36.

Ermini L, Olivieri C, Rizzi E et al. (2008) Complete mitochondrial genome sequence of the Tyrolean Iceman. *Curr Biol* 18:1687–1693.

Ermini L, Wilson IJ, Goodship TH & Sheerin NS (2012) Complement polymorphisms: geographical distribution and relevance to disease. *Immunobiology* 217:265–271.

Ewing B & Green P (2000) Analysis of expressed sequence tags indicates 35,000 human genes. *Nat Genet* 25:232–234.

Gilbert MT, Kivisild T, Grønnow B et al. (2008) Paleo-Eskimo mtDNA genome reveals matrilineal discontinuity in Greenland. *Science* 320:1787–1789.

Green RE, Malaspinas AS, Krause J et al. (2008) A complete Neanderthal mitochondrial genome sequence determined by high-throughput sequencing. *Cell* 134:416–426. This is a fascinating report of mtDNA sequencing in an ancient mtDNA sample.

Hales CN & Barker DJP (1992) Type 2 (non-insulin dependent) diabetes mellitus: the thrifty phenotype hypothesis. *Diabetologica* 35:595–601. This is an original paper that contradicts Neel's thrifty gene hypothesis; see Neel (1962).

Hardy GH (1908) Mendelian proportions in a mixed population. *Science* 28:49–50. This is the original report of the Hardy–Weinberg Principal.

Mayo O (2008) A century of Hardy–Weinberg equilibrium. *Twin Res Hum Genet* 11:249–256. This is a review of this difficult concept, and is of particular use for students and researchers needing to understand the principal in more detail.

Mueller JC (2004) Linkage disequilibrium for different scales and applications. *Brief Bioinform* 5:355–364.

Nachman MW & Crowell SL (2000) Estimate of the mutation rate per nucleotide in humans. *Genetics* 156:297–304.

Neel JV (1962) Diabetes mellitus: a "thrifty" genotype rendered detrimental by "progress"? *Am J Hum Genet* 14:353–362. This is an original paper with an interesting original hypothesis. The hypothesis is contested by Hales and Barker (1992).

Reich DE, Cargill M, Bolk S et al. (2001) Linkage disequilibrium in the human genome. *Nature* 411:199–204.

Soares P, Ermini L, Thomson N et al. (2009) Correcting for purifying selection: an improved human mitochondrial molecular clock. *Am J Hum Genet* 84:740–759.

The International HapMap Consortium (2007) A second generation human haplotype map of over 3.1 million SNPs. *Nature* 449:851–861.

Tishkoff SA, Reed FA, Friedlaender FR et al. (2009) The genetic structure and history of Africans and African Americans. *Science* 324:1035–1044.

Wahlund S (1928) Zusammensetzung von Population und Korrelationserscheinung vom Standpunkt der Vererbungslehre aus betrachtet. *Hereditas* 11:65–106. This is an original report of the Wahlund principal and reminds the reader that not all major papers are published in English.

Weber JL & Broman KW (2001) Genotyping for human whole-genome scans: past, present, and future. *Adv Genet* 42:77–96

Wright S (1931) Evolution in Mendelian populations. *Genetics* 16:97–159.

Online sources

http://www.ncbi.nlm.nih.gov/projects/genome/guide/human/index.shtml
This provides a very useful guide for those studying the human genome.

http://www.ncbi.nlm.nih.gov/SNP
These two sites are essential for updates on SNPs.

http://www.ncbi.nlm.nih.gov/mapview
This is an excellent resource for those wishing to stay up-to-date on the genome.

http://www.ncbi.nlm.nih.gov/projects/SNP/snp_summary.cgi
This site now contains more information about more than 38 million of validated reference SNPs clusters in the genome.

http://www.ncbi.nlm.nih.gov/epigenomics
This site is another excellent source for information from NIH.

http://www.ncbi.nlm.nih.gov/omim
This is an essential site for updates and information on genetic disease of all types.

http://www.genet.sickkids.on.ca/StatisticsPage.html
This is an excellent place to search for updates on cystic fibrosis and the *CFTR* gene.

http://www.mitomap.org
This is a great resource for those wishing to investigate the mitochondrial genome further.

http://www.ncbi.nlm.nih.gov/omim/
On-line inheritance in man (OMIM) is an essential site for updates and information on genetic disease of all types.

CHAPTER

2

Defining Complex Disease

Until recently, healthcare professionals considered medical genetics to be the province of specialists seeking to understand rare cases of **Mendelian disorders**, birth defects, and **chromosomal abnormalities**. **Complex diseases** were more or less excluded from consideration as affected (informative) families were comparatively rare and genetic linkage was difficult to prove even if families were available for testing. However, following the publication of the **Human Genome Map (HGM)** and the **haplotype map** (**HapMap**) as well as developments in genetic technology, the focus for research in medical genetics has changed. Collins and McKusick writing in 2001 even went so far as to suggest that "except for some cases of trauma, it is fair to say that virtually every human illness has a hereditary component."

However, the term hereditary (or **heritability**) does not necessarily imply genetic. For example, heritable traits may be related to diet and alcohol consumption which could impact on families and populations, causing changes in the frequency of a disease or trait that may be falsely attributed to genetics. Heritability considered in this way is complex. It is necessary to distinguish between heritability in the narrow sense and heritability in the broad sense. It is also important to define the components, such as additive dominant alleles, and **epistasis**. Epistasis involves gene–gene interaction. It occurs when the expression of a gene requires the involvement of one or other modifier genes. Epistasis is also important in epigenetics—a term that refers to things outside of genetics and is discussed in Chapter 1 and also later in this book.

However, for the purpose of this book we are concerned with the genetics of disease and in that sense heritability can be defined as the proportion of variation in a phenotype in a population that can be accounted for by our genes. Some authors have suggested we overestimate the impact of our genes on disease. In the current millennium it is hard to disagree with the concept that our genes play a major role in disease and disease susceptibility. Even if not all of assigned heritability is genetic, it is likely that a great deal of this heritability will be genetic and, since most diseases are not Mendelian or chromosomal, a very significant proportion will fit into the classification of genetically complex disease. In fact, chromosomal abnormalities account for less than 1% and Mendelian disease for no more than 4% of human disease. Does this mean that 95% of human diseases are genetically complex diseases? Geneticists do not agree on how many diseases may have a genetic component. However, all of the evidence suggests it may be a significant number of different diseases, and these include some infectious, immune and metabolic diseases as well as some cancers and some forms of toxicity.

In this chapter, we will consider the definition of genetically complex disease. We will compare complex diseases to diseases that arise either as a result of major chromosomal abnormalities, single inherited genetic mutations (Mendelian diseases), or genetic abnormalities in the mitochondrial genome. In addition, we will consider in detail three different models for genetically complex disease using key examples to illustrate how each model is different.

2.1 DEFINITION OF A GENETICALLY COMPLEX DISEASE

When constructing a definition it is often useful to consider what is not included. Complex diseases are not simple. They do not lend themselves to simple analysis and do not conform to the expected patterns of inheritance that define Mendelian diseases, i.e. autosomal recessive, autosomal dominant, and **sex-linked**. Complex diseases do not arise from major chromosomal abnormalities. The terms characteristic or trait often get used interchangeably with the term disease and though this is not strictly correct, this has become quite common. One of the reasons for this is defining disease is itself difficult. Whatever terminology is used, complex traits or diseases are perhaps best defined in the words of Haines and Pericak-Vance (1998) as those where "alterations in more than one gene that alone or in concert either increase or decrease the risk of developing a trait." We would however modify this to say allele rather than gene, and insert disease in place of trait frequently (but not in all cases) as we are mostly interested in diseases and the characteristic clinical variations seen in different diseases. Thus, we will use the term disease as much as possible in the book and not the term trait.

The problem with this definition is that it not only applies to complex disease, but can also be applied to some Mendelian disorders such as ataxia and retinitis pigmentosa that may be caused by more than one gene. In these cases, the genetic mutation acts in a simple Mendelian fashion (i.e. alone) and they are not considered complex diseases. It is also the case that some traits and diseases are known to have both pure Mendelian families and pure sporadic complex cases.

To fully understand complex disease it is important to deconstruct this definition

Based on the definition of Haines and Pericak-Vance, three things are immediately apparent when comparing genetically complex diseases with classical Mendelian and chromosomal diseases. Complex diseases show complex patterns of inheritance and we are dealing with alleles as determinants of risk rather than as absolute causes of disease. Possession of a particular **risk allele** may affect disease outcome or phenotype, e.g. signs, symptoms, or response to treatment. Genetic variations and the terminology that is applied to them is discussed at length in Chapter 1 and also in the Glossary at the end of the book, which lists some of the appropriate terminology that applies to this area of genetic research.

Complex diseases show complex patterns of inheritance

In complex disease there may be a single risk allele or several risk alleles. In their definition, Haines and Pericak-Vance say "more than one", we would say one or more. They also say that these "genes", whereas we would use the term alleles, may act "either alone or in concert" to "increase or decrease the risk of developing a trait." We would replace the word trait with disease in most circumstances, but the definition remains solid and the implications of this definition are the same. This implies that there may be one or more than one risk allele and the patterns of inheritance seen in complex disease are likely to be complex (**Figure 2.1**). This suggests that there is heterogeneity in complex disease whereby different individuals with the same disease may inherit different alleles, i.e. a **genetic portfolio**, associated with an increased or reduced risk of the disease. By inheriting different risk alleles there is genetic heterogeneity for a given disease. At this stage it is reasonable to say we understand little about the interactions that occur within many of the portfolios that have been investigated, though some understanding is beginning to emerge. It is also true to say that the interaction of the different risk alleles is likely to be complex and is currently poorly understood for a number of reasons dealt with at the end of this book.

Studies so far have shown that there is considerable genetic sharing of risk alleles between diseases whereby one or a group of alleles may predispose to more than one disease. Some consider this to be pleiotrophy, but others consider this to be clinical heterogeneity. Strictly speaking, pleiotrophy refers to mutations that have multiple effects; in these cases, the final clinical outcome may be determined by the complement of risk alleles inherited and by the environmental risk. Technically, pleiotrophy and clinical heterogeneity are distinct entities, but pleiotrophic alleles may be involved in clinically heterogeneous disease. Overall, this is why complex diseases do not comply with simple Mendelian patterns of inheritance. However, within complex disease there are recognizable patterns. These are best defined as **monogenic**, **oligogenic**, and **polygenic**, depending on the number of risk alleles. Despite this, there is no simple mathematical test to determine which group each disease belongs to. The terms simply translate as single (mono), several (oligo), and many (poly). Monogenic disease is relatively easy to define in Mendelian disease, but rare in complex disease, whereas oligogenic and polygenic, though not as easily defined, are more common. The idea that monogenetic disease can be genetically complex is very difficult for some. However, where there is sporadic disease and a single identified common risk allele, there is no alternative explanation at present. Future studies may prove this to be naive, but we must allow for this possibility at present with the hope that a better explanation awaits for these rare monogenic cases.

Complex diseases involve genetic variation that increases or decreases risk

Perhaps the most important detail in this definition is the use of the word "risk." In complex disease, inherited genetic variation gives rise to an increased or reduced risk of a disease. This is the crucial concept of this book and much of modern medical genetics. Risk alleles are those that have been found to be present at a statistically significant increased or reduced frequency in a subpopulation with a disease or trait compared with the healthy or normal population, either through linkage or genetic **association analysis**. To date, most studies have concentrated on identifying single nucleotide polymorphisms (SNPs) as markers for risk alleles in complex disease. The choice of SNPs is based on their

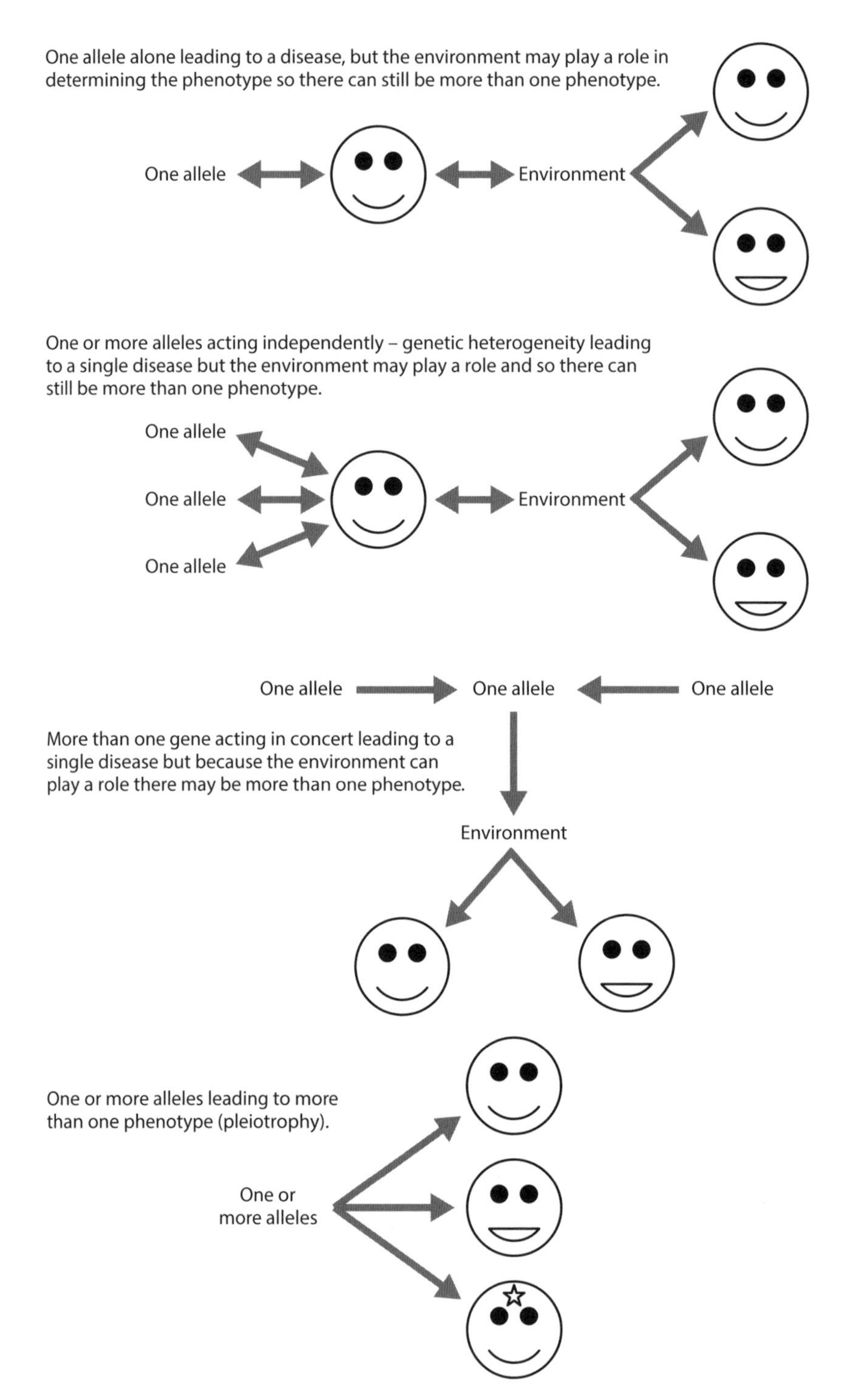

Figure 2.1: One or more alleles acting alone or in concert illustrates the key concept in the genetics of complex disease. The first three patterns illustrate one allele alone, more than one allele acting independently, and several alleles acting in concert. In each case the final outcome is complicated because the phenotype is determined by both the genotype and the environment in most cases, and so there can be more than one phenotype as illustrated by the different facial expressions. The final example illustrates several different

frequency in the population. Most studies chose SNPs with a frequency of no less than 5% for the rarest alleles. However, as study groups are getting larger, the thresholds for the lowest frequency SNPs are being set at lower levels (i.e. 1% or less). This is because greater sample numbers are now available for testing which increases the statistical power of the study enabling less common SNPs to be included. Assessing less common SNPs may pick up hitherto missed genetic associations.

Other forms of genetic variation have been investigated including insertions, deletions, and copy number variations (CNVs). Only a few complex diseases arising from insertions and deletions have been identified so far. One major example where CNVs have been reported is in autism.

It should be noted that statistical significance is based on a threshold model whereby the statistical value obtained is expected to equal or exceed the threshold set prior to any investigation. This is discussed at greater length in Chapter 5.

The inheritance of a specific risk allele or group of alleles does not necessarily cause the disease. Inheriting the risk allele or alleles simply increases the likelihood (risk) of disease. We may say that inheritance of the risk allele is neither necessary nor sufficient for the disease to occur. Note also that the definition used by Haines and Pericak-Vance includes the word "decrease." This indicates the possibility that inheritance of some alleles may protect from disease. Considering **genetic protection** is as important as considering susceptibility, especially when we consider infectious disease (see Chapter 7). This protective effect is seen quite frequently with different **human leukocyte antigen (HLA)** alleles and haplotypes. Many autoimmune diseases have genetic associations with groups of HLA alleles or haplotypes, some of which increase disease susceptibility, while others reduce susceptibility. Comparison of the susceptibility versus the protective alleles has enabled investigation of the molecular mechanisms that underpin these HLA associations (see Chapter 6). This latter development is one of the keys in moving from simple genetic investigation to understanding basic disease pathology.

It is also essential to dispel any idea of good and bad alleles. This idea oversimplifies the concept of complex disease. Many of the known risk alleles that have been identified as potent inducers of disease for some clinical syndromes have also been identified as protective for other diseases. This is illustrated by the example of the impact of the C-chemokine receptor 5 Δ32 deletion (*CCR5*-Δ32), which protects from human immunodeficiency virus (HIV) infection, but predisposes to infection with West Nile virus (see Chapter 7).

Genetic variation also affects the outcome of disease

Finally, genetic variation does not only determine susceptibility and resistance to disease per se, but also determines various outcomes following disease onset. Therefore, it is appropriate to use the term "trait" in place of the word "disease" in this definition. By restricting the definition to disease we narrow the field, and exclude the possibilities that common genetic variations influence disease severity and outcome both before and following treatment. However, it is quite clear that **morbidity**, mortality, and response to certain commonly used pharmacological agents (including adverse drug reactions) are all

Figure 2.1: (*Continued*) phenotypes resulting from the same risk portfolio. Phenotypes are identified by different facial expressions: smile, open mouth, and star decoration on temple (in the last case only). Variable phenotype and pleiotrophy are similar. Note that some authors use alleles and genes as interchangeable terms. However, when talking about variation of a gene at a locus, the correct term is allele. Phenotype is a complex issue because although each of these models indicates a single disease, the clinical phenotype is the product of both genetic and environmental factors. Therefore, the phenotype can vary even with a single allele.

influenced by common genetic variation. The use of the word trait as opposed to disease is important throughout the study of complex disease, but it is particularly important when considering the role of genetic variation in infectious diseases (see Chapter 7) and also in **pharmacogenetics** (see Chapter 8).

The take-home message from all of the points above is that the simple definition of a genetically complex disease as *a disease or trait where "alterations in more than one gene that acting alone or in concert either increase or reduce the risk"* needs to be deconstructed and considered word for word. It is essential to remember we have, in the past, been dealing with alleles that for the most part are common. However, more recently we have started to investigate risk alleles with low frequencies. Developments in technology, statistical analysis, and study design have all played important roles in focusing our attention on these less common alleles. Finally, before we move on to concentrate on complex disease, let us first consider genetic diseases that are not classified as genetically complex diseases: chromosomal, Mendelian, and **mitochondrial diseases**.

2.2 CHROMOSOMAL DISEASES

Chromosomal and Mendelian diseases include some of the cruelest of all human diseases, with early onset, high levels of morbidity, and high levels of mortality. The ease with which the heritable nature of these diseases can be identified has made them perfect (if simple) models upon which to develop genetic theory and practice, some of which is now being applied to the study of complex diseases. Chromosomal abnormalities are fortunately quite rare. They are visible (often very large) alterations of the chromosome causing abnormalities in specific chromosomal regions. Early cytogenetic studies were restricted to detecting changes of greater than 4 Mb of DNA, but more recent developments such as **fluorescence *in situ* hybridization (FISH)** have enabled much smaller changes to be identified.

The most common causative mechanisms for chromosomal abnormalities are misrepair of broken chromosomes or **incorrect segregation** during **mitosis** and meiosis. Chromosomal abnormalities can be classified as either **constitutional chromosomal abnormalities** (occurring in all cells) or **somatic chromosomal abnormalities** (occurring in a selected group of cells or tissues only) and are further classified as numerical or **structural abnormalities**.

Changes in chromosome number cause serious genetic diseases

Numerical abnormalities are referred to using different forms of the word "ploid," which means number. There are several forms of **ploidy** or changes in the chromosome number. Most humans are diploid with two copies of the entire genome and our gametes are **haploid**. Therefore, most individuals inherit one copy of the genome from each parent, and thus the normal genome is made up of 22 pairs of chromosomes (**autosomes**) and either XX (females) or XY (males). However, some individuals inherit either additional copies or fewer copies of the whole genome (**polyploidy**), or they inherit additional or fewer copies of a specific chromosome or chromosomes (**aneuploidy**).

Polyploidy

Polyploidy involves the transmission from parent to offspring of multiple copies of the genome. It can be subclassified depending on the number of copies involved. These subgroups include **tetraploidy**, **triploidy** (**Figure 2.2**), **monosomy**, and **mosaicism**.

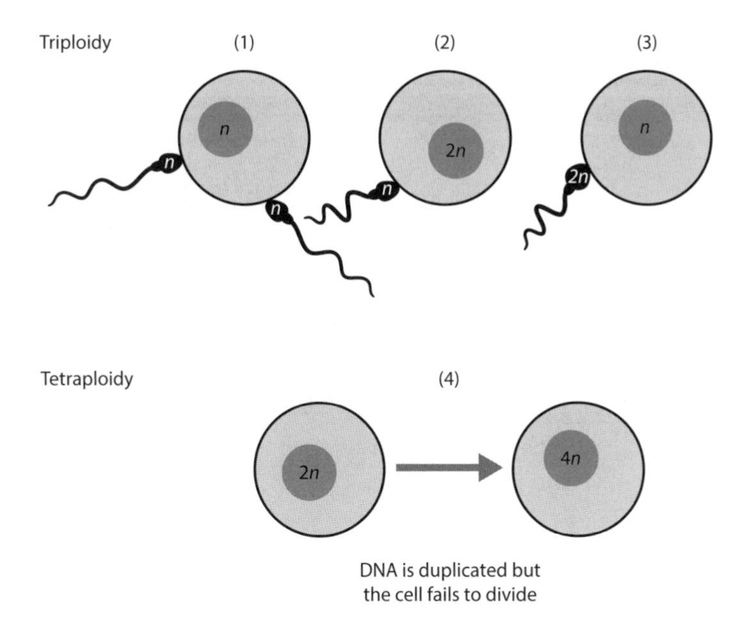

Figure 2.2: Numerical chromosomal abnormalities. The top of this figure illustrates three different models of triploidy: (1) dispermy, (2) diploid egg, and (3) diploid sperm. At the bottom of this figure, (4) illustrates tetraploidy where DNA is duplicated, but there is no cell division. (Adapted from Strachan T & Read A [2011] Human Molecular Genetics, 4th ed. Garland Science.)

Tetraploidy occurs when offspring inherit four copies of the whole genome. Tetraploidy is rare and is most commonly caused by the failure of the cell to divide after the first DNA duplication. Tetraploidy is not compatible with life.

Triploidy can occur through several mechanisms, the most common of which is two sperm fertilizing a single egg (**dispermy**) and, as with tetraploidy, triploidy is not compatible with life.

Aneuploidy

Individuals with aneuploidy have one or more chromosomes present either with an extra copy or with a missing copy. Aneuploidy involving the autosomes is usually lethal, but it can be compatible with life. Key examples include: trisomy 13 associated with Patau syndrome; trisomy 18 associated with Edward's syndrome and trisomy 21 associated with Down syndrome. Aneuploidy associated with additional copies of the sex chromosomes (e.g. XXX, XXY, XYY) is mostly associated with a normal lifespan and relatively few minor clinical problems. By contrast, aneuploidy associated with missing sex chromosomes is frequently lethal, though monosomy X (associated with Turner's syndrome) is an exception. Ninety-nine percent of all cases involving missing sex chromosomes abort spontaneously (**Table 2.1**).

Changes in chromosome structure can cause serious illness

Structural abnormalities include deletions, **inversions**, duplications, insertions, formation of **ring chromosomes**, and **translocations**. Inversions are subclassified as **paracentric** (not involving the centromere) and **pericentric** (involving the centromere) (**Figure 2.3**). All of these abnormalities may result from the misrepair of chromosome breaks or from failures in recombination. Chromosomal breaks may occur naturally as a result of recombination

Table 2.1 Terminology applied to numerical and structural abnormalities.

Numeric abnormalities		
Polyploidy	Tetraploidy	Four copies of whole genome
	Triploidy	Three copies of whole genome
	Monosomy	One copy of whole genome
	Mosaicism	Variable number of copies in different cells
Aneuploidy	Trisomy	Three copies of one or more chromosomes
	Monosomy	Only one copy of one or more chromosomes
Structural abnormalities		
Deletions	Unbalanced abnormalities	
Inversions	Paracentric or pericentric balanced abnormalities	
Duplications	Unbalanced abnormalities	
Insertions	Unbalanced abnormalities	
Ring formation		
Translocations	Balanced abnormalities	

and can occur at different phases of the cell cycle. There are natural mechanisms to prevent cells with unrepaired chromosomes from entering mitosis. Structural abnormalities occur when breaks are repaired incorrectly. For example, the broken ends of the chromosome are sometimes joined incorrectly and this can lead to chromosomes without a **centromere** (**acentric**) or with two centromeres (**dicentric**). Such abnormal chromosomes will not **segregate** stably in mitosis and are eliminated. However, chromosomes with a single centromere carrying structural abnormalities can be propagated through mitosis.

Balanced abnormalities

When structural abnormalities are associated with no overall gain or loss of chromosomal material they are said to be balanced. Provided the **balanced structural abnormalities** do not result in disruption of the function of a gene through interrupted expression, control, or activation of the gene, they are unlikely to have an adverse effect on the phenotype. Balanced abnormalities are those involving inversions and translocations.

Unbalanced abnormalities

When structural abnormalities are associated with gains or losses of chromosomal material they are said to be unbalanced. In contrast to balanced abnormalities, **unbalanced chromosomal abnormalities** are more likely to give rise to problems in meiosis, and they result from deletions, insertions, and duplications. As a consequence, these abnormalities are likely to have adverse effects on the phenotype.

As stated above, chromosomal abnormalities are rare. It is important for the reader to note that, unlike complex disease, all carriers of these abnormalities are affected and symptoms most often occur from birth. However, these genetic illnesses are not the subject of this book.

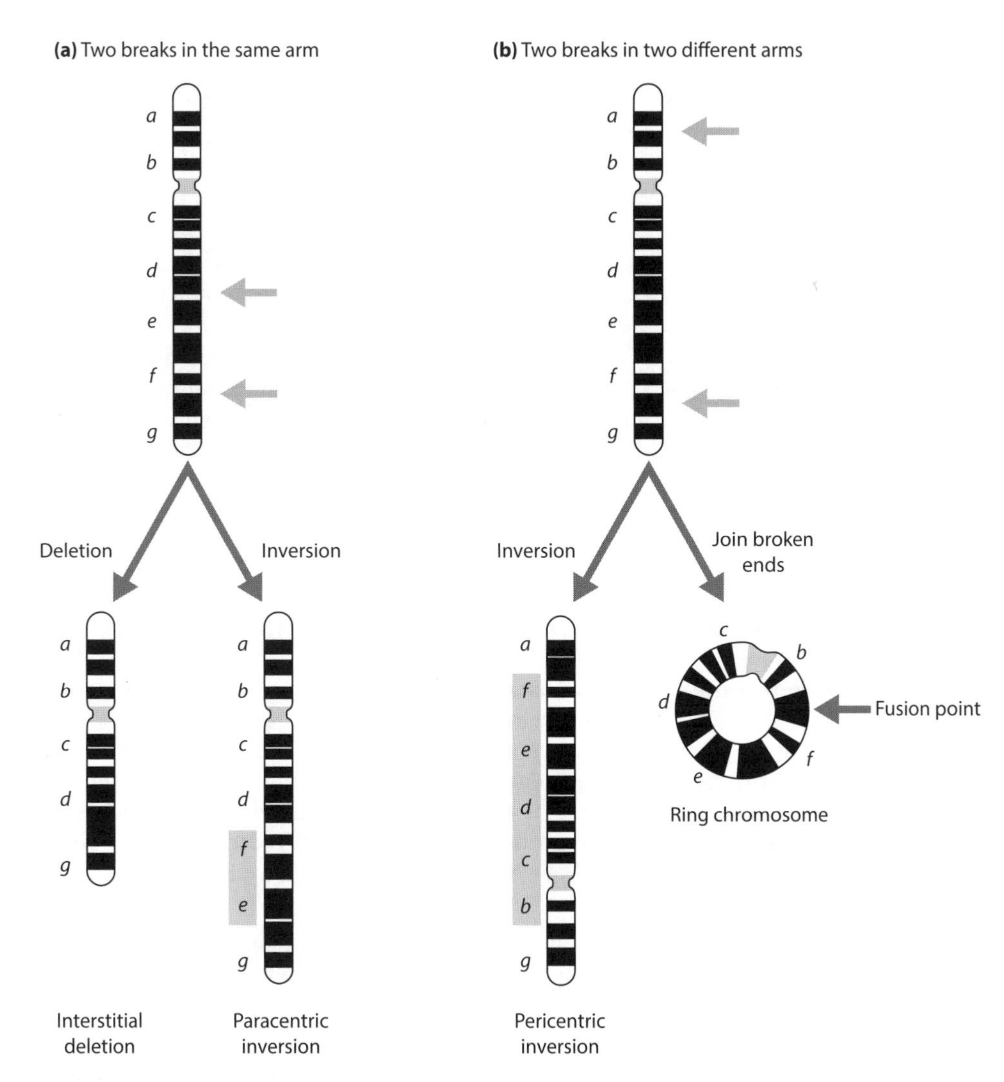

Figure 2.3: Outcomes after incorrect repair of two breaks on a chromosome. (a) Incorrect repair of two breaks in the same arm of the chromosome can lead to either a parametric inversion (such as *e* and *f* inversion) or interstitial deletion (i.e. deletion of *e* and *f*) of the fragments. In the former case, the inversion is paracentric as it does not involve the centromere. (b) Incorrect repair occurs following breaks in two different arms of the same chromosome. Here, the fragments coded from *b*–*f* are inverted, and in this case the inversion involves the centromere and is referred to as a pericentric inversion. As the fragment involves the centromere, the fragment could also rejoin to form a stable ring chromosome. In this latter model, the fragments *b*–*f* are included in the ring structure, but fragments *a* and *g* are missing. The banding patterns represent the karyotype represented in Figure 1.1. (From Strachan T, Goodship J & Chinnery P [2014] Genetics and Genomics in Medicine. Garland Science.)

2.3 MENDELIAN DISEASES

Mendelian diseases involve a single gene and show simple patterns of inheritance

The rediscovery of the papers of Gregor Mendel (**Box 2.1**) in 1900 and the different patterns of genetic inheritance that the experiments illustrated gave rise to what are

BOX 2.1 A SHORT HISTORY OF GREGOR MENDEL

Gregor Mendel's research, or more specifically the rediscovery of his papers in 1900, is rightly credited as the time at which modern genetics was born. However, people were aware of the passage of traits from generation to generation long before the discovery of the papers reporting Mendel's experiments with pea plants. Indeed, the bible gives the example of Jacob's sheep. Gregor Mendel was born Johann Mendel in a village in Moravia (later Czechoslovakia) in 1822; he entered an Augustinian Monastery in 1843. He was a gifted teacher, though he failed his teaching certificate. Mendel's experiments with peas enabled him to identify different patterns of inheritance for specific selected traits. The terminology applied to his work post-dates his discoveries because despite the ground-breaking significance of his work initial interest in his papers rapidly faded, and disappointed and burdened with administration he retired into monastic life and died in 1884. Nevertheless, he is and was the founding father of modern genetics, and we refer to his name every time we look at a disease with a known pattern of inheritance.

called Mendelian patterns of inheritance, and today these patterns are used in classifying genetic diseases and traits. Over 4609 traits are currently identified on the OMIM (Online Mendelian Inheritance in Man) database for which the molecular basis of the risk gene is known (http://www.ncbi.nlm.nih.gov/omim). Over 12,000 genes are also listed on OMIM. Not all of these traits and genes are for Mendelian diseases, many are for complex diseases, but those which are Mendelian disorders all involve a single gene (i.e. they are monogenic) and all conform to one of five readily identifiable patterns: autosomal dominant, autosomal recessive, **X-linked dominant**, **X-linked recessive**, and **Y-linked** (illustrated by the pedigrees in **Figure 2.4**). Sex-linked disorders may also be referred to as gonosomal as opposed to autosomal.

In addition, occasionally geneticists will use terms such as **semi-dominant** to describe a family where heterozygotes have a phenotype. In many cases it may not be clear if the heterozygote has a less severe phenotype than the homozygote, as would be expected in a truly semi-dominant disease. However, for the most part, Mendelian diseases conform to the expected patterns. Mendelian diseases can be described as those where a particular genotype at one locus is both necessary and sufficient for the character to be expressed.

Mendelian genotypes have variable phenotypes

It is important to understand that biological systems are complex and interactive. Therefore, though these Mendelian diseases are considered simple in as much as they involve a single dysfunctional genetic character at a single gene locus, the biological character affected by this disability is likely to be programmed by a large number of interacting genes and also by environmental factors. This complexity may explain the different levels of phenotypic variation in individuals and families with the same Mendelian genotype. Expression of the trait even in Mendelian disease is not always absolute. There are variable levels of **penetrance** whereby possession of the mutant gene does not give rise to the disease.

An extreme example of this **incomplete penetrance** is found in the autosomal recessive disease hemochromatosis. This disease is characterized (as the name suggests) by iron overload in the liver and other tissues. The gene responsible for the disease has been identified and is referred to as *HFE*. The causative gene is located on chromosome 6p21.3 at the

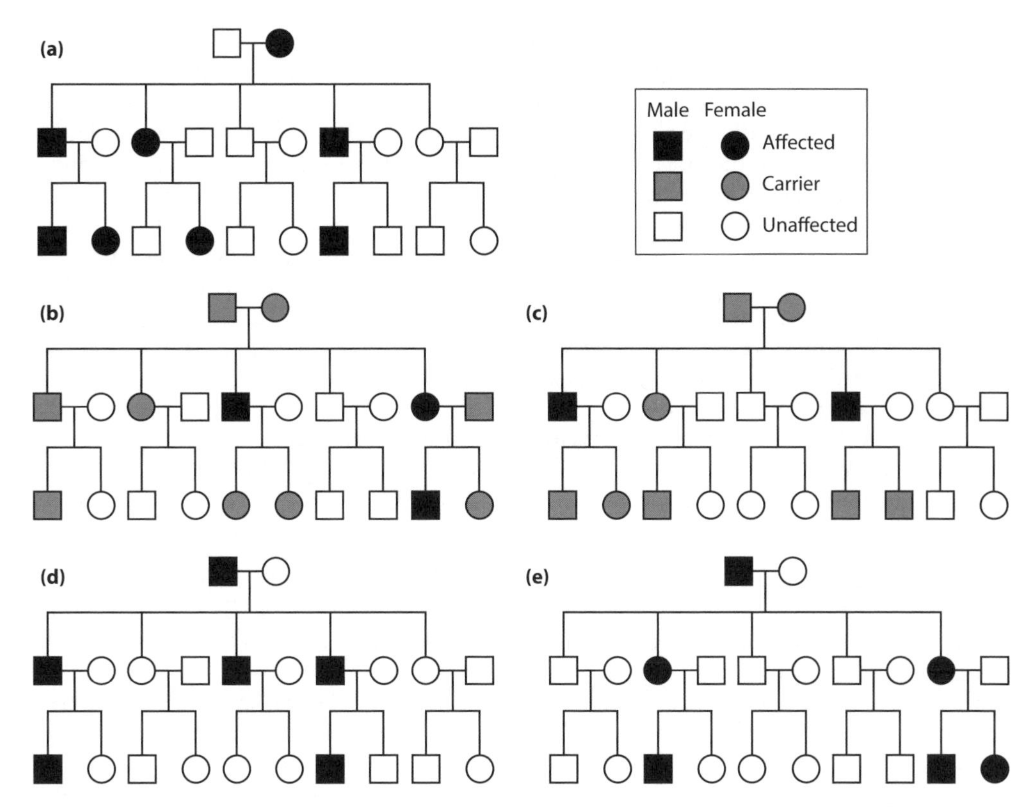

Figure 2.4: Genetic pedigrees in Mendelian diseases. The five different patterns displayed represent three generations all modeled on the same family structure. (a) Autosomal dominant: both males and females are affected (marked in black) and the trait does not skip generations. (b) Autosomal recessive: both males and females are affected, but here not all generations are affected (e.g. the grandparents are unaffected carriers), carriers are marked in gray. (c) X-linked recessive: as with autosomal recessive not present in all generations; mostly males are affected as they carry only one copy of X, but females can be affected. (d) Y-linked: males only are affected and as they inherit a single Y chromosome only, all Y-linked Mendelian disease assumes a dominant pattern. (e) X-linked dominant: passed from fathers to daughters in the first generation (males receive the Y chromosome from the affected father). When these daughters have offspring, the trait is passed to both sons and daughters, but equally sons as daughters may inherit the wild-type gene from their mothers and not develop the trait.

extreme **telomeric** end of the major histocompatibility complex (MHC) and was for a short while referred to as *HLA-H*. A number of studies indicate that the penetrance in those with two copies of the *HFE* mutation can be extremely low, with the majority of homozygous carriers being asymptomatic possibly unaffected. The pattern of homozygosity for the disease causing mutation with extremely low levels of penetrance (disease expression), is unusual. Of course in a late onset disease such as this one, we cannot be certain that some of the homozygotes will not develop the disease at a later date.

Penetrance is an important difference between Mendelian and complex diseases

In Mendelian traits, we expect the trait to be expressed when one or both copies of the disease gene (depending on a recessive or dominant pattern) are dysfunctional. The example of hemochromatosis above is an extreme exception. For the most part, penetrance in Mendelian disease is high and in many, but not all, cases penetrance is 100%. This is in stark contrast to complex disease where penetrance is mostly low. In many cases of complex disease the majority of those who inherit the risk allele do not express the disease phenotype and some members of the population develop the disease in the absence of the risk allele. In complex disease the presence of an allele confers risk, but the risk allele is neither necessary nor sufficient for the disease to occur. Examples of this are to be found throughout this book. One key example is ankylosing spondylitis—a degenerative disease of the lower spine that occurs in young adults and is associated with the *HLA-B*27* family of HLA alleles. In some populations it has been found that up to 94% of patients with this illness have *HLA-B*27*. The **odds ratio (OR)** (a crude measure of the **population risk**; see **Box 2.2**) for ankylosing spondylitis in those with *HLA-B*27*, is estimated to be as high as 171 times greater compared with those without *HLA-B*27*. From a personal point of view this increased risk seems to be high until we consider that 6% of ankylosing spondylitis patients do not have *HLA-B*27* and that on average only one out of every 25 *HLA-B*27*-positive individuals will develop the disease (**Figure 2.5**). Interestingly, recent genome-wide association studies (GWAS) have identified a range of other risk alleles for ankylosing spondylitis outside the MHC (see Chapter 3).

The example above is an illustration of incomplete penetrance at work in complex diseases. Penetrance is the key to understanding the difference between complex and Mendelian disease. Whereas in Mendelian disease the presence of the allele (whether a mutation, polymorphism, or deletion) is considered to be necessary and sufficient to cause the disease, this is not true for complex disease. This difference is at the heart of understanding the genetics of complex disease and the potential applications of this evolving branch of biomedical science.

BOX 2.2 A SHORT DISCUSSION ON HOW TO CALCULATE THE ODDS RATIO IN A CASE CONTROL ASSOCIATION STUDY

The odds ratio (OR) is a frequently used simple statistical measure of the likelihood of risk based on the ratio of an allele in healthy controls compared with affected cases. It is usually calculated from the figures entered in the Punnett Square (Table 1.2) as $A \times D/B \times C$. To understand this test, consider a population in which we test a single gene with two alleles (α and β), in 100 cases and in 100 healthy controls. First remember that in a diploid species we have two copies of each chromosome, therefore in a group of 100 individuals there are 200 alleles for this single gene. We perform a genetic test on this population and the results of our testing show that the frequency of the α allele is higher in patients than in healthy controls, e.g. there are 150 α alleles in the patients and only 100 α alleles in the healthy controls. Because we already know the total number of individuals in each group is 100, we can assume that the frequencies of the β alleles for the two groups are 50 for patients and 100 for healthy controls. The OR is calculated as above with $A \times D/B \times C$, where the A box value is the number of α alleles in patients and the D box value is the number of β alleles in controls, while the B box value is the number of α alleles in healthy controls and the C box value is the number of β alleles in patients. Numerically, $A \times D/B \times C = 150 \times 100/100 \times 50$; this can be simplified by deleting the common values on either side of the equation, i.e. 100, to give 150/50 and also removing the 0 on either side to give a simple calculation of 15/5 from which the OR value of 3 is derived.

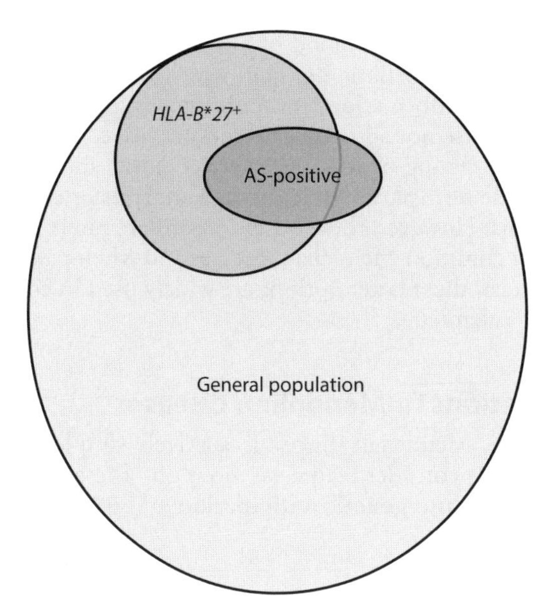

Figure 2.5: *HLA-B*27* and ankylosing spondylitis. The figure illustrates incomplete penetrance for *HLA-B*27* in ankylosing spondylitis (AS). Most *B*27*-positives are ankylosing spondylitis-negative even though most ankylosing spondylitis-positives have *B*27*.

Some diseases have both Mendelian and complex characteristics

One complication not considered above can be seen when Mendelian disease becomes mixed with complex disease. Comparing early- and late-onset Alzheimer's disease (EOAD and LOAD) illustrates this problem. EOAD mostly conforms to a Mendelian pattern of inheritance and is marked by clear familial inheritance within the bounds of the proscribed Mendelian patterns. LOAD is mostly marked by the sporadic occurrence of the disease in (mostly) hitherto unaffected families. However, there are individual family cases that map to a gray area in between, with some cases being Mendelian and some cases being complex. In this situation it is important to use all of the available information to identify the genotype and phenotype of the individual cases in the family concerned.

Modifier genes may also confuse the picture

Another problem can arise when a Mendelian disease with a clear pattern of inheritance has low penetrance for the disease-causing mutation. This may be due to a high level of expression of a modifier gene. Changes in penetrance can lead to the absence of the expressed phenotype even in a Mendelian disease. This can mean that cases look more like complex disease cases than Mendelian disease cases and may also explain some of the cases of monogenic complex disease.

Mendelian traits can be studied in families

Studying Mendelian traits is relatively easy compared with the study of complex diseases. Mendelian diseases occur in families and they tend to present at an early age with a few notable exceptions, including hemochromatosis (which presents later in life) and also EOAD (which though called early onset, is not a genuine example of early-onset disease as it does not occur until the affected individual is well into adult life). In Mendelian disease, early onset increases the likelihood of availability of family members for genetic testing and also influences the likelihood of family members engaging in genetic research. Families are a rich source of information and traditional (**parametric**) linkage-based analysis can be applied. Of course not all families are informative. There may be too few family members available for testing or too few affected family members. Other options in Mendelian disease include **non-parametric linkage analysis** or **case control** population-based association analysis. However, these other options are rarely used in Mendelian disease, except in the more common Mendelian diseases and when parametric linkage testing is not possible. In contrast, these latter options are widely used in complex disease studies, as discussed later in this chapter.

There are complications to Mendelian diseases

Though, as we have seen, Mendelian disease is relatively simple, there are a number of additional complications to consider before we move on. These include locus, allelic, and **clinical heterogeneity**, and also **genetic anticipation** and **genetic imprinting**.

Locus heterogeneity

Locus heterogeneity refers to the occurrence of the same trait arising from genetic variation at different loci. It is not uncommon for the same Mendelian disease to result from mutations in different genes in different families.

Allelic heterogeneity

Allelic heterogeneity refers to the occurrence of the same trait/phenotype from different mutations at the same locus. In these cases, a single gene is involved, but there are multiple potential disease-causing mutations. An example of allelic heterogeneity is the autosomal recessive disease cystic fibrosis, for which there are more than 1910 disease-causing mutations listed on the *CFTR* mutation database (http://www.genet.sickkids.on.ca/StatisticsPage.html). Therefore, each patient may have a different *CFTR* genotype and still have the same phenotype. Note, however, as with many of these Mendelian diseases where there are multiple disease-causing mutations at a single locus, in cystic fibrosis, there is one very common mutation, the Δ508 mutation, that accounts for the majority of cases and a host of less common mutations that account for the remainder (**Figure 2.6**).

Clinical heterogeneity

Most human diseases are very variable. Even in Mendelian disease patients with the same genetic mutation display variation in clinical symptoms. This is referred to as clinical heterogeneity. It should not be assumed that this is entirely due to genetics. The environment is likely to play a significant role in the determination of clinical outcomes.

Genetic anticipation

Genetic anticipation describes the tendency of some Mendelian dominant diseases to become more severe in successive generations. Examples of this phenomenon include fragile X syndrome, myotonic dystrophy, and Huntington's disease. All three of these diseases arise from the inheritance of specific unstable trinucleotide repeats. In each case, severity or age of onset is associated with the number of repeat sequences. The number of repeats has been shown to increase in successive generations. Genetic anticipation in other diseases remains controversial.

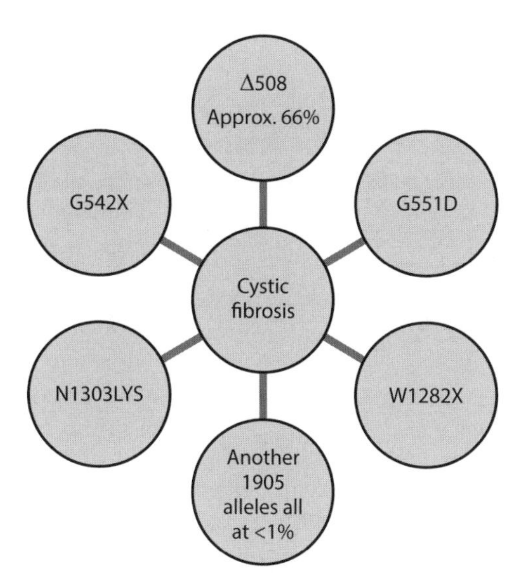

Figure 2.6: Mutations in the *CFTR* gene that cause cystic fibrosis. There are over 1910 listed mutations in the *CFTR* gene. Of these, one accounts for approximately two-thirds of all mutations, while most others (with a few exceptions) are found at frequencies of less than 1%.

Genomic imprinting

Genomic imprinting is considered to be an epigenetic phenomenon and is more complex than any of the above. In some families with autosomal dominant disease the trait is only expressed if inherited from a parent of one particular sex. This occurs despite the fact that the disease-causing mutation can be present in both male and female parents. An example of genetic imprinting is Beckwith–Wiedemann syndrome, which is only expressed in children who inherit the mutated gene from their mother. There are other examples of human disease where imprinting has been identified as a key factor, including Prader–Willi syndrome, Angelman syndrome, and Silver syndrome (**Table 2.2**). Research has identified approximately 100 different imprinted loci on 11 mouse chromosomes (http://www.mousebook.org/mousebook-catalogs/imprinting-resource). Even so, imprinting is the least well understood of the many complications that apply to Mendelian disease.

An example of genomic imprinting is given by *IFG2* gene encoding insulin-like growth factor 2. Offspring inherit one *IGF2* allele from their mother and one from their father, but the paternal allele of *IGF2* is expressed, while the maternal one is completely silent. If both alleles begin to be expressed in a cell, that cell may develop into a cancer.

Children with Prader–Willi syndrome have small hands and feet, short stature, poor sexual development, and mental retardation, while children with Angelman syndrome exhibit frequent laughter, uncontrolled muscle movement, a large mouth, and unusual seizures. The imprint control centers have been found to restrict small areas known as imprinting centers. In the case of the disorders above, the imprinting centers have been identified at the 5′ end of the *SNURF–SNRPN* gene. Both syndromes are associated with the loss of the same small region on the long arm of chromosome 15.

The most common maternal form of Angelman syndrome is a 4 kb mutation on chromosome 15q11–q13 that contains the *UBE3A* gene encoding for E6AP ubiquitin ligase. The mutation prevents methylation of the *UBE3A* gene and also silences the *SNRPN*

Table 2.2 Some examples of disease thought to involve genomic imprinting and some of the genes involved.

Chromosomal location of gene	Candidate gene or genes if known	Disease/syndrome/phenotype
11q33	*PGL1*	Paragangliomar 1: hearing loss
7p11.2 11p15.5	*H19*	Silver–Russell syndrome
5q35	*NSD1*/*ARA263*/*STO*	Beckwith–Wiedemann syndrome (also: Soto syndrome, Weavers syndrome, acute myeloid leukemia)
11p15.5	*H19*: cyclin-dependent kinase inhibitor IC ($p57^{Kip2}$)	Beckwith–Wiedemann syndrome
11p15.5	*KCNQ1* overlap transcript 1	Beckwith–Wiedemann syndrome
11p15.5	*H19*	Wilms' tumor 2
13q14.1–q14.2	*RB1*	Retinoblastoma
Microdeletion of 15q11–q13	*SNURF–SNRPN*	Prader–Willi syndrome: hypotonic, male hypogonadism, mental retardation, and severe obesity
Microdeletion 15q11–q13	*UBE3A*	Angelman syndrome: small size, severe retardation, lack speech, and characteristic behavioral pattern
20q31		Pseudohypothyroidism type 1A
20q31		Pseudohypothyroidism type 1B

gene. If the deletion of this region is inherited from the mother, the offspring will develop Angelman syndrome because the paternal gene is silenced and the maternal copy is almost exclusively expressed.

Prader–Willi syndrome involves a similar mechanism, but in these cases the maternal genes are silenced and not expressed, and the paternal genes are mutated. There may be a number of different genes involved in cases of Prader–Willi syndrome; however, the same pathway is affected, and the key to understanding the mechanism of the disease rests with *SNRPN* and *UBE3A*.

It is tempting to ask why genomic imprinting occurs. This type of pressure may have significance in evolutionary terms, and there may be conflicting pressures acting on maternal and paternal alleles for genes, such as those that affect fetal growth (see Section 1.9).

2.4 VARIATION IN THE MITOCHONDRIAL GENOME IS ASSOCIATED WITH DISEASE

Not all of the human genome is inherited as autosomal DNA from the nucleus either in our autosomes (chromosomes 1–22) or in our gonosomes (X and Y). A small proportion of our DNA is inherited from the mitochondria, the so-called mitochondrial genome (mtDNA). The human mitochondrial genome is extremely small compared with the **nuclear genome** (i.e. the autosome). The mitochondrial genome is a circular molecule of 16,569 bases in length (**Figure 2.7**). Mitochondrial genetics is very different to Mendelian genetics. The mitochondrial genome is passed from generation to generation down the maternal line only and each cell may contain several thousand copies.

The mitochondrial genome encodes a total of 37 genes: 13 encoding for **proteins** that function within mitochondria, two **ribosomal RNAs** (rRNA, 16S and 23S) and 22 **transfer RNA** (tRNA) genes necessary for the synthesis of the 13 encoded polypeptides. There are two nucleotide strands of the mtDNA: a heavy strand (H) rich in guanine and a light strand (L) rich in cytosine. The H strand provides a template coding for the two rRNAs, 14 out of the 22 tRNAs, and 12 out of the 13 proteins. The L strand is the template for the remaining eight tRNAs and one protein. The mitochondrial genome is extremely compact with approximately 93% of the DNA sequence representing coding sequences. Consequently, all 37 mitochondrial genes lack introns and are tightly packed within the coding region. The remaining 7% of the mitochondrial genome encodes the displacement D loop or control region where the replication of both the H and L strands begins. This region also encodes the promoters for transcription of both H and L strands. The primary role of mitochondria is to provide cells with the bulk of their adenosine triphosphate (ATP), which is used as an energy source to drive cellular reactions. The proteins produced by the mitochondrial genes work together to create this energy by turning complex molecules such as sugars into simple substances such as carbon dioxide and water.

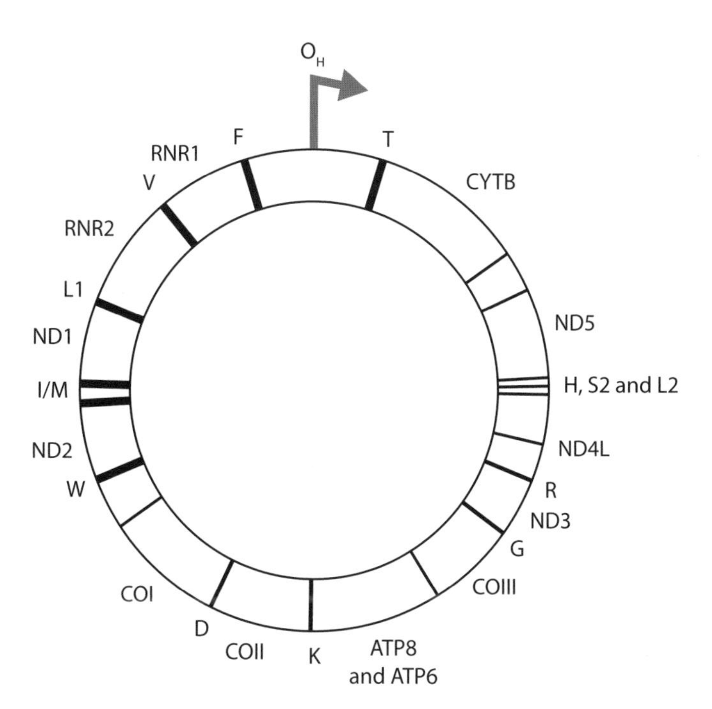

Figure 2.7: An abridged map of the human mitochondrial genome. The genome is 16.6 kb and encodes 13 essential polypeptides of the oxidative phosphorylation (OXPHOS) system (ND1-ND6, NDL4, the cytochrome *c* oxidases COI-COIII, cytochrome *b*, and the subunits of ATP synthase ATPase 6 and 8) and the necessary RNA components (two rRNAs and 22 tRNAs) for their translation.

Variation in the mtDNA has been widely associated and linked with many different diseases

The mitochondrial genome has been associated with many different diseases in humans (**Table 2.3**). The majority of early studies concentrated on rare diseases caused by inherited mutations at specific sites, such as Kearns–Sayre syndrome where there are large-scale deletions in the mtDNA, Pearson's syndrome, **Leber hereditary optic neuropathy**, mitochondrial myopathy, and many others, most of which occur in infants. More recently, studies have begun to identify genetic associations with common diseases and illnesses, including migraines, epilepsy, dementia, cardiomyopathy, renal tubular defects, adrenal failure, and liver failure, which occur both in infants and adults (**Figure 2.8**). Diagnosing and managing mtDNA disease is a major clinical challenge in the second decade of the new millennium. One strategy that is being applied is to manipulate the mtDNA mutation rates through exercise. A lack of exercise leads to a reduction in mitochondrial enzyme activity. In contrast, endurance training improves enzyme activity. It has been suggested that increased exercise may help to reduce the proportion of sporadic mutations, and may be used to combat some of the clinical symptoms and illness caused by mtDNA mutations. Identification of the dysfunctional mtDNA genes and sequences in mitochondrial syndromes enables the development of new therapies based on targeting specific gene

Table 2.3 Some of the diseases that are encoded by genetic mutations and variations in the human mitochondrial genome.

Disorder	Genotype	Phenotype	Inheritance
Kearns–Sayre syndrome	Large-scale deletion	Ophthalmoplegia and cardiomyopathy	Sporadic
Chronic progressive external ophthalmoplegia (CPEO)	Large-scale deletion	Ophthalmoplegia	Sporadic
Pearson's syndrome	Large-scale deletion	Lactic acidosis and pancytopenia	Sporadic
Mitochondrial myopathy	3243A > G *TRNL1* 3271R > C *ND1* and *ND5*	Stroke like	Maternal
Myoclonic epilepsy associated with ragged red fibers (MERRF)	8344A > G *TRNK* 8356T > C *TRNK*	Myoclonic epilepsy	Maternal
Neuropathy, ataxia and retinitis pigmentosa (NARP)	8993T > C *ATP6*	Neuropathy	Maternal
Mitochondrial complex V (ATP deficiency) (MILS)	8993T > C *ATP6*	Brain stem	Maternal
Maternally inherited diabetes and deafness (MIDD)	3243A > G *TRNL1*	Diabetes and deafness	Maternal
Leber hereditary optic neuropathy	3460G > *A ND1* 1178G > *A ND4* 1448T > *C ND6*	Ophthalmic neuropathy	Maternal

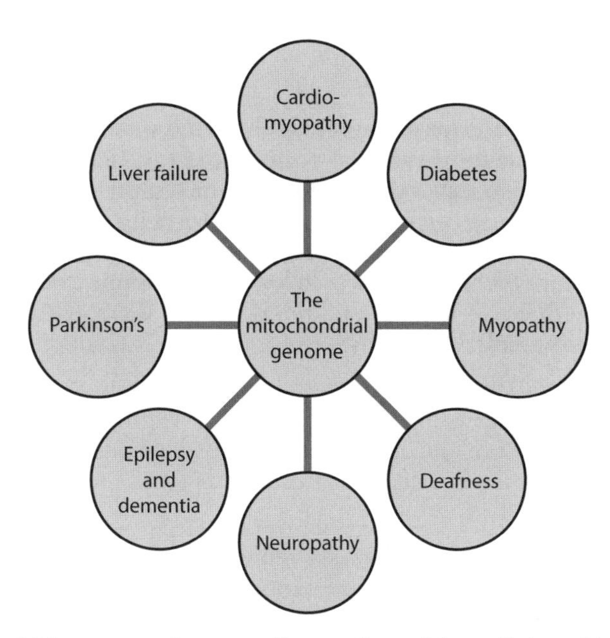

Figure 2.8: Mitochondrial genome and common diseases. Some of the medical conditions that have been claimed to be related to variation in the mitochondrial genome.

sequences using procedures like **allotropic expression**, whereby it is possible to express the wild-type using transfected hybrid cell lines.

2.5 *DE NOVO* MUTATIONS AND HUMAN DISEASE

One group of human conditions rarely discussed is that caused by *de novo* mutations. These diseases and traits are only just being understood. *De novo* mutations are new mutations, so there are no other affected family members and therefore they are not Mendelian traits. They are also caused by simple mutations, so these are not chromosomal disorders involving large sections of the genome. The rate of mutation in the genome is low (as discussed in Chapter 1) and so these mutations are rare. However, these mutations when they are disease-causing mutations, that is having the mutation is both necessary and sufficient for the disease to occur, cannot themselves be considered with genetically complex diseases because carrying the mutation will cause the disease. In this case we are not talking about mutations that simply increase disease risk. However, *de novo* mutations that are not disease causing, but may be associated with disease risk, can be considered with complex disease (as occurs in some cancers). The former group are a category unto themselves.

One example of a trait caused by a *de novo* mutation is Hutchinson–Gilford progeria syndrome (HGPS). HGPS is incredibly rare, occurring in around one per 4 million live births in the USA. There is no family history with this trait because the mutation occurs in the gametes, most often in the sperm. The mutation is not found in the parent's autosome. The symptoms of HGPS are rapid aging; life expectancy can be very short at around 12–13 years. Patients show extremely early signs of aging, hair loss and aging of the skin, and severe osteoporosis. We see the traits that we expect to be expressed in an average quite healthy octogenarian in an HGPS patient often before the age of 10 years. This is a cruel condition for parents and affected children alike.

Genetic testing has shown that the major mutation for this syndrome is a mutation encoding an exchange of cytosine for thymine at position 1824 (C1824T) in the lamin A gene (*LMNA*), which gives rise to a shortening of the lamin A protein. The gene is located on chromosome 1q22. There is a degree of clinical heterogeneity in HGPS and this may be in part due to genetic variation. Other, less frequent mutations have also been identified in *LMNA* and these have been associated with longer life in some cases. *LMNA* mutations are also associated with other genetic conditions, such as Charcot–Marie–Tooth disease and Emery–Dreifuss muscular dystrophy.

Lamin A is an important structural protein and one of a family of intermediate filament proteins. Intermediate filaments provide stability and strength to all our cells. Lamin A is a scaffolding component of the nuclear envelope that surrounds the nucleus in cells, creating a mesh-like layer of intermediate filaments attached to the inner membrane of the nuclear envelope of the cell.

In HGPS, the C1824T mutation encodes a short version of the lamin A protein called progerin. This protein cannot be processed correctly within the cell. This can disrupt the nuclear envelope, and over time damage the structure and function of the nucleus, causing cells to die prematurely. However, there is hope in HGPS and clinical trials have been started, though it is early days.

2.6 THREE DIFFERENT TYPES OF COMPLEX DISEASE

Complex diseases as defined above do not lend themselves to simple analysis or display predictable patterns of inheritance (i.e. Mendelian patterns). However, we can distinguish different levels of complexity. The greatest problem is identifying genuine monogenic complex disease. These diseases are near-Mendelian, involving only a single risk allele, but do not behave in a simple pattern. Whether this is due to incomplete penetrance of the Mendelian disease-causing mutation is unclear. However, until these cases have been defined the concept of a monogenic complex disease remains open. In many of these diseases there are also Mendelian families with the same disease, which adds a further layer of complication. Even with these caveats, true monogenic complex diseases are rare; most genetically complex diseases are associated with multiple risk alleles and are more correctly referred to as either oligogenic or polygenic. The difference between these last two categories is very gray. Oligogenic implies that there are several risk alleles and polygenic implies that there are many risk alleles. In practice, in oligogenic complex disease, risk alleles are easier to identify and there are likely to be more individual Mendelian families and non-Mendelian families for analysis.

Studying complex disease is different from studying Mendelian disease

In contrast to Mendelian disease, which relies mostly on linkage analysis of informative families, in complex disease even if informative families are available they have often dispersed (due to late onset) and even when DNA samples can be obtained the only options are to use non-parametric linkage analysis or case control population-based association analysis. The former is a linkage-based method that does not require the

Table 2.4 Some of the common major differences between genetically complex diseases compared with Mendelian disease.

Mendelian traits	Complex traits
Mostly early onset	Mostly late onset
High penetrance	Low penetrance
Mostly familial	Some familial, but mostly sporadic
Known pattern of inheritance	Unknown pattern of inheritance
Suitable for parametric linkage analysis	Not suitable for parametric analysis Use non-parametric linkage analysis Use mostly association analysis
Low/no environmental input (exceptions include phenylketonuria, Wilson's disease, and hemochromatosis)	Moderate/high environmental input

use of a genetic (Mendelian) model. Therefore, there are no preset parameters as there are in more conventional parametric linkage analysis. Non-parametric linkage analysis can be based on families with multiple cases of disease or on affected sibling pairs. Case control population-based association analysis is a quite different method that simply compares the frequency of genetic variations between groups of patients or cases and healthy (matched) individuals from the same population. Until recently, most genetic studies in complex disease were based on association analysis and this was thought to be a poor mechanism for the identification of disease genes. However, for various reasons, opinions have changed about the value of association studies in complex disease (see Chapter 3). A number of other differences between studies of Mendelian versus complex disease are listed in **Table 2.4**.

Monogenic complex diseases involve a single risk allele

The idea of a monogenic complex disease is difficult to explain, as discussed above. However, the use of the term monogenic does not by itself suggest a single allele at a locus is both necessary and sufficient for the disease to be expressed. If this were the case, then this would be a Mendelian disease. However, to argue that all monogenic disease must be Mendelian is incorrect because such ideas ignore incomplete penetrance. In monogenic complex disease there is always incomplete penetrance in families, therefore the disease-promoting allele is not sufficient for the disease to occur. In addition, because there are cases that do not carry the disease-promoting allele, we can say that the disease-promoting allele is not necessary for the disease to occur. A further complication is that in many disease populations where there are monogenic cases the risk allele for each case may be on a different gene. In a disease population, different individuals may have different genetic **risk portfolios**, so that in a population of patients with a specific disease there could be some familial Mendelian cases, some complex monogenic cases, and some oligogenic or polygenic cases. Altogether, monogenic complex disease is a gray area between Mendelian and complex disease, and though individual cases do occur, the majority of complex disease fits into an oligogenic or polygenic pattern.

We must also consider the possibility that in some cases they are in fact Mendelian. In these cases we may be fooled by two things. First, expression may not be seen in all individuals with the risk allele, because expression is modified by another gene causing incomplete penetrance. Second, the presence of the disease in some without the risk allele may simply indicate a different disease-causing mutation is active in those individuals. Taken altogether this could explain why the disease-causing mutation is neither sufficient nor necessary for the disease to occur and may explain a large proportion of monogenic complex disease.

Oligogenic complex diseases involve several alleles

Risk alleles are easier to identify in oligogenic disease than in polygenic disease because, by their nature, in oligogenic diseases there are less risk alleles involved (as the name implies) and there tends to be a higher heritable component compared with polygenic disease. As a consequence families are more frequent, onset is more likely to be early, and the risk alleles tend to have large effects as measured by OR, **relative risk**, or **LOD score** (these different statistical terms are explained in Chapter 5). Hirschsprung's disease (HSCR) is a good example of an oligogenic disease (see Section 2.8) with at least 13 identified risk alleles: two major risk alleles, three minor risk alleles, and a group that may be modifier alleles, some of which may as their implied role suggests only act in the presence of the major risk allele (or alleles). A further five chromosomal regions have been identified in HSCR bringing the total to 18, but no specific genes have yet been identified as potential risk alleles.

Polygenic complex diseases involve many risk alleles

One of the difficulties of defining polygenic disease is the word "poly." The word itself means many, but how many is the question? There is no specific number for this term. However, it is quite clear that some diseases are linked or associated with a large number of different risk alleles. This does not mean that every case will inherit every risk allele or that there may not be occasional near-Mendelian cases in the disease population. In this regard it is important to consider patients in the context of populations and population subgroups. The term poly applies to the patients as a group. Overall, there are many risk alleles associated with the disease and in this case the term poly would apply. Generally, the contrast with oligogenic disease becomes clear when we consider the size of the heritable component and frequency of families in the two models. The example of Crohn's disease (see Section 2.9) is a good example of a polygenic disease because though there are **familial** cases, most cases are sporadic and the number of identified risk alleles is now considered to be greater than 163 and rising. Compare the latter figure of 163 potential risk alleles in Crohn's disease to the figure for HSCR, where there appears to be no more than approximately 18 different risk-associated alleles, including five chromosomal regions in which risk alleles have yet to be identified. If we then consider that out of this group of 18 risk alleles for HSCR, two mutations have very major strong effects and the remainder have weak effects, with most appearing to operate as modifiers in the presence of the major risk allele, then differences between oligogenic and polygenic disease are quite clearly demonstrated.

However, it would be incorrect to assume that all cases of a specific disease have the same genetic pattern. A disease population is best defined as a group of patients with the same or similar signs and symptoms. Though disease groups have the same clinical picture, the disease may arise through a different pathway and the genetic contribution to pathogenesis may be different in different cases. Nowhere is this better illustrated than in the three selected examples described in Sections 2.7–2.9. Alzheimer's disease occurs in two forms: EOAD and LOAD. Most EOAD cases are Mendelian families (but not all), while most LOAD cases are sporadic, genetically complex cases. A single risk allele is likely to be the total genetic contribution in some of the latter (i.e. they are monogenic). In HSCR, not all cases are complex—a significant proportion are due to chromosomal abnormalities or are familial with clear Mendelian inheritance and only 70% of cases fit the pattern of oligogenic complex diseases. In Crohn's disease, approximately 10% of cases may have a familial pattern of inheritance and some of these are likely to be Mendelian cases, but the majority of cases are sporadic and genetic susceptibility is determined by a large number of different risk alleles, indicating a polygenic model.

2.7 ALZHEIMER'S DISEASE MAY BE A MONOGENIC COMPLEX DISEASE

LOAD is an example of a complex disease where there may be a significant number of monogenic cases. Alzheimer's disease is a very **heterogeneous** disease and it is also very common, particularly in the developed world. This disease is characterized by progressive dementia in the elderly or aging population caused by the formation of **intracellular neurofibrillary tangles (NFT)** and extracellular **amyloid plaques** that accumulate in the vulnerable regions of the brain, causing damage to the brain as illustrated in **Figure 2.9**. It has been observed that reductions in microtubule-dependent transport occur in Alzheimer's disease and this may stimulate proteolytic processing of the β-amyloid precursor protein (APP), resulting in the development of senile plaques that are seen in Alzheimer's disease. As discussed above, Alzheimer's disease can be either Mendelian (EOAD, about 5% of cases) or complex (LOAD, about 95% of cases). The genes involved in EOAD are *APP*,

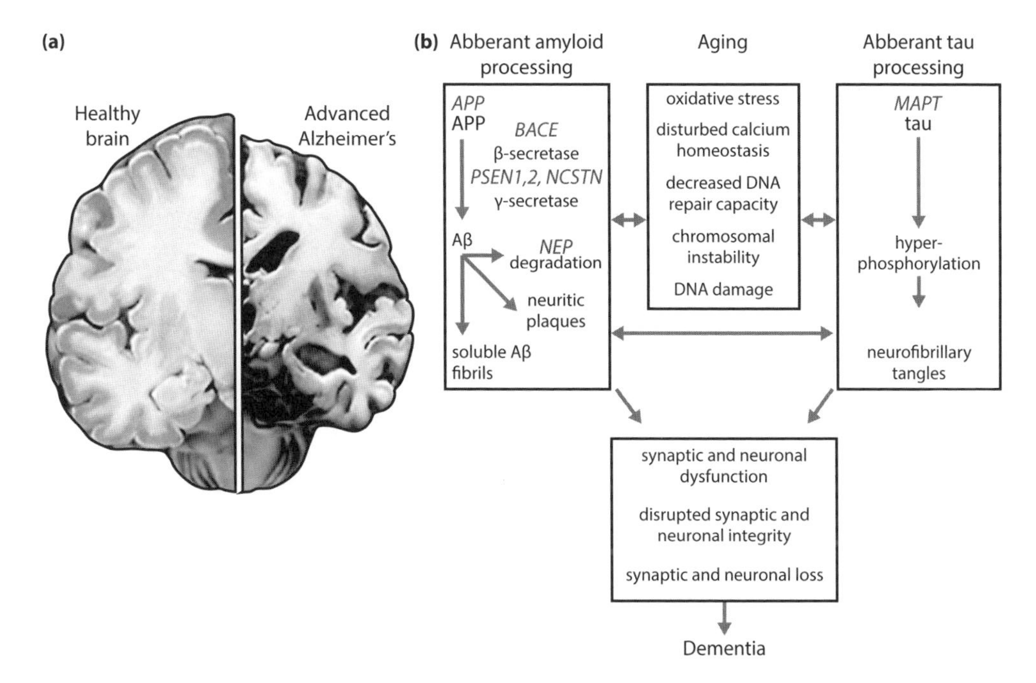

Figure 2.9: The mechanisms of Alzheimer's disease. (a) A healthy brain compared with a brain with advanced Alzheimer's disease. (b) The processes and some of the genes involved in the formation fibrils, plaques, DNA damage, and neurofibrillary tangles. (From Armstrong L [2014] Epigenetics. Garland Science.)

PSEN1, and *PSEN2*. In LOAD, it is the apolipoprotein E4 (*APOE4*) allele that confers the major risk. There is clear evidence for a causative link with *APOE4* and LOAD. Prince et al. (2004) found that the presence of the *APOE4* allele was strongly associated with reduced levels of the β-amyloid-42 protein in cerebral spinal fluid. However, there are multiple other risk alleles that have been identified for LOAD; OMIM currently lists at least 16 possible genetic sites and many more are listed in the literature that have yet to be added to this list. There are individual cases of LOAD where the *APOE4* allele is the only identified risk allele. *APOE4* may account for as much as 50% of the genetic risk in LOAD. However, new research may at anytime identify hitherto unknown risk alleles. In addition, we cannot ignore the potential effects of the environment. For example, Evans et al. (2004) reported an association between total serum cholesterol and disease progression in patients with LOAD who did not have the *APOE4* allele.

2.8 HSCR - AN OLIGOGENIC COMPLEX DISEASE

HSCR was described by Harald Hirschsprung in 1886 and it is defined by the congenital absence of ganglia in some or all of the large intestine or colon. HSCR occurs because there is a failure of development of the central nervous system or neural crest. The consequences of this are a lack of peristalsis in the gut, which leads to a toxic megacolon in the newborn. This can be fatal when left untreated. Most cases of HSCR are sporadic and these fit the definition of oligogenic complex disease.

Sporadic HSCR illustrates the oligogenic model for complex disease

Amiel and Lyonnet (2001) reported 12% of cases are due to chromosomal abnormalities, 18% are part of a syndrome (mostly Mendelian), and 70% are isolated non-syndromic cases, often familial but not Mendelian (**Figure 2.10**). HSCR is associated with a **sibling relative risk** of 187. The major risk genes and alleles for HSCR have been

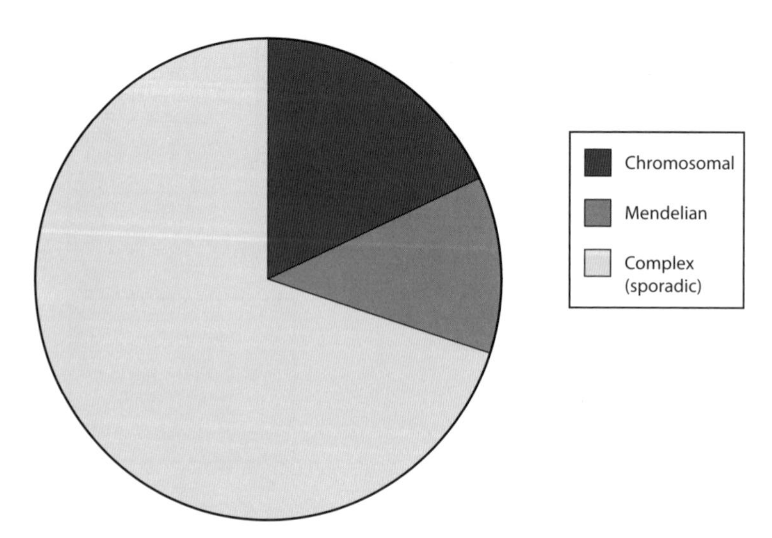

Figure 2.10: HSCR is either a chromosomal, Mendelian, or a complex disease. The majority of cases of HSCR are genetically complex (70%); however, a substantial number are familial cases (inherited in a Mendelian pattern, 12%) and some occur from chromosomal abnormalities (18%).

Table 2.5 Genetic associations with non-syndromic HSCR: Examples from OMIM and other sources December 2014.

Name	Location	Function
HSCR1	10q11–q21	*RET*
HSCR2	13q22.3	*EDNRB*
HSCR3	5p13.2	*GDNF*
HSCR4	20q13.32	*EDN3*
HSCR5	9q31	Unknown
HSCR6	3p21	Unknown
HSCR7	19q12	Unknown
HSCR8	16q23	Unknown
HSCR9	4q31	Unknown
Other minor associated genes	22q13 4p12 2q22	*SOX10* *PHOX2B* *ZFHX1B*
Other minor modifier genes	16p13.3 8p22–p11 Xq28 15q22.3–q23 2q31 4q27	*SOX8* *NRG1* *L1CAM* *BBS4* *BBS5* *BBS7*

identified. These include a deletion in the *RET* transforming sequence (oncogene *RET*) that encodes a receptor tyrosine kinase on chromosome 10q in the area 10q11–q21 (HSCR1) which is found in 50% of familial and 15–35% of sporadic cases, the gene encoding endothelin receptor type B (*EDNRB*) on chromosome 13q (HSCR2), as well as 11 other gene loci. Three of these 11 other genes appear to have a minor function and six of the 11 appear to be modifier genes. In addition, there are five identified chromosomal locations for which the candidate genes have yet to be identified (summarized in **Table 2.5**).

The *RET* oncogene deletion is the major genetic risk factor for the isolated non-syndromic form of the disease, but does not account for all cases. In addition, there is a male bias for cases with this deletion (65% males). Linkage analysis indicates that several other genes interact with the *RET* deletion to increase the risk of the disease as noted above.

The *EDNRB* gene was identified as a risk factor for HSCR through a series of linkage studies in a Mennonite kindred that identified deletions and mutations in chromosome 13q. Independent work in mouse models suggested a relationship between *EDNRB*, which is located within this region. As with the *RET* mutation, there is incomplete penetrance and also penetrance may be sex-specific.

Both *RET* and *EDNRB* encode receptors that are active in the process of neural crest stem cell regulation. It is therefore easy to understand how mutations in these genes can give rise to HSCR. As there are two pathways in the process, it is also possible to envisage why a mutation in one pathway does not always lead to the clinical syndrome being expressed.

Other susceptibility alleles for HSCR have been proposed and research suggests that mutations in the genes encoding the ligands for *RET* and *EDNRB* may be involved as well as a number of genes encoding for other components of the biological pathways that are involved in neural crest stem cell development (**Figure 2.11**). Thus far, studies have identified mutations in *GDNF* and *EDN3*, as well as *ECE1* and *NTN*. An association with *SOX10* (which may interact with *ZFHX1B* and *SOX8*) has been reported, though this remains controversial and may relate to other similar developmental disorders. Another gene implicated in HSCR is the *PHOX2B* gene, which interacts with the *RET* gene.

Some of the associated risk alleles above and a number of other more recently identified potential risk alleles (*L1CAM*, which interacts with *ENDRB*, and *NRG1, BBS4, BBS5,* and *BBS6*, which all interact with *RET*) may act as modifiers in the presence of the major risk alleles, which may explain why they are only seen to increase disease risk in those carrying the major risk allele.

Inheritance of multiple risk alleles may explain clinical heterogeneity and in a complex model may also explain incomplete penetrance. HSCR genetics illustrates the importance of considering the whole genome and studying complete biological systems. Recent developments have enabled both of these objectives to be possible, though much more work remains to be done before the work on HSCR genetics is complete.

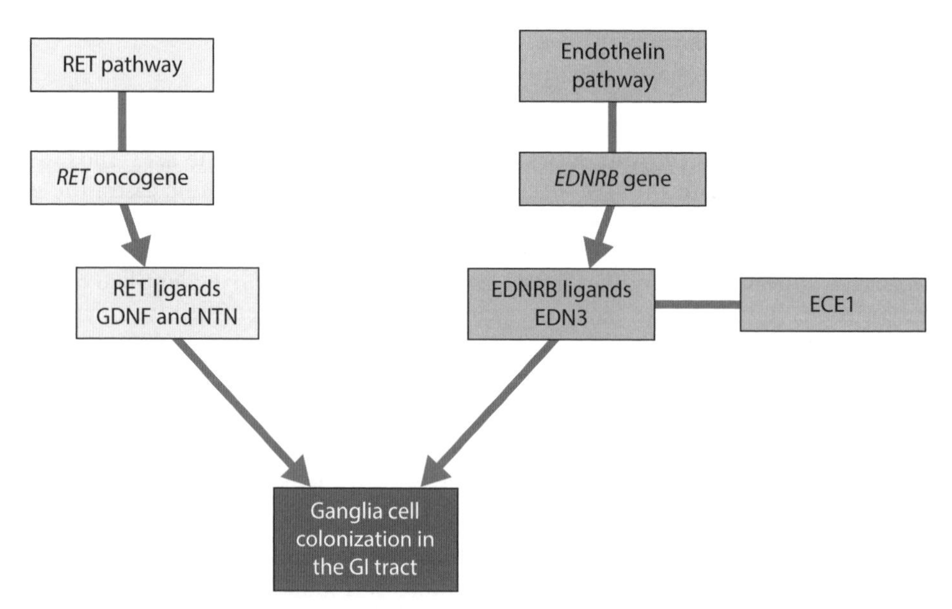

Figure 2.11: Major pathways in neural crest stem cell development with links to HSCR. This figure illustrates the major pathways involved in neural crest stem cell development implicated by linkage and association with susceptibility to HSCR. The major contributing genes are *RET* and *EDNBR*. Polymorphisms of other genes whose products are important in these *RET* and endothelin pathways have been identified as potential risk factors for HSCR. GI, gastrointestinal. (Adapted from Strachan T & Read A [2004] Human Molecular Genetics, 3rd ed. Garland Science.)

2.9 CROHN'S DISEASE IS MOSTLY A POLYGENIC COMPLEX DISEASE

Crohn's disease serves as a good model for polygenic complex disease. The disease was first described by Dalziel, a Glaswegian surgeon, in 1913, but the definitive paper was published by Crohn, Ginzburg, and Oppenheimer in 1932, and the name Crohn's disease was applied. Crohn's disease is one of a group of inflammatory bowel diseases (IBDs), though inflammation in Crohn's disease is not restricted to the bowel, but can be found throughout the gastrointestinal system, from the oral cavity to the anus. The other major form of IBD, known as ulcerative colitis, was described in 1888 by Hale-White. Ulcerative colitis, unlike Crohn's disease, is restricted to the lower gut. Despite this, there is frequent clinical overlap between Crohn's disease and ulcerative colitis, and there is also a level of shared pathology and genetic risk.

Early studies of Crohn's disease suggested a number of locations for risk alleles

Approximately 10% of cases with Crohn's disease are familial, with the remaining 90% being sporadic. Estimates of the sibling relative risk vary from 17- to 35-fold and there is 50% concordance for the diseases in **monozygotic** (identical) twins compared with 10% concordance in **dizygotic** (non-identical) twins. These values are considerably lower than those for HSCR which boasts a sibling relative risk of 187, reflecting the greater level of heritability for HSCR compared with Crohn's disease. Early genetic studies based on linkage analysis of families identified several (tagged) regions of the genome with significant linkage to Crohn's disease, a selection of which are listed in **Table 2.6**. Among the identified regions, chromosome 16q (tagged as IBD1) was associated with the greatest risk of Crohn's disease. A brief history of early and recent genetic studies is given in **Figure 2.12**.

Twin studies have always been valued in genetics. They are simple studies that investigate the degree of co-inheritance of a disease or trait in monozygotic and dizygotic twins. The difference in the level of **concordance** between the two groups is considered to be a measure of the degree of heritability for the disease or trait. The drawback of using twin data is that the twins share the same environment unless adopted into different families, and the quality of the data obtained depends on the numbers studied. As identical twins are rare, numbers are likely to be low in less common diseases.

Table 2.6 Major susceptibility regions for IBD identified by early linkage analysis and GWAS.

Region	IBD-LABEL	Crohn's disease	Ulcerative colitis	Putative candidate gene
16q12	IBD1	Strong	None	*CARD15* (2001)
12q13.2–q14.1	IBD2	Moderate	Moderate	
6p21.3	IBD3	Moderate	Moderate	MHC/HLA
14q11–q12	IBD4	Moderate	Moderate	
5q31	IBD5	Strong	Moderate	Cytokine cluster
19p13	IBD6	Moderate	Moderate	2007 WTCCC1 GWAS
1p36	IBD7	Moderate	Moderate	

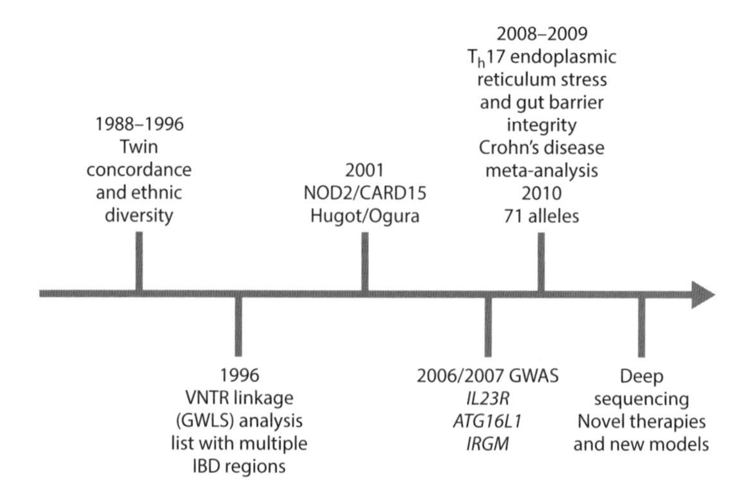

Figure 2.12: Time-line of genetic investigations in Crohn's disease. The figure shows the timing of major developments in our understanding of the genetics of Crohn's disease from 1988 to recent.

Genetic variations in the human equivalent of the plant *nod2* gene (*CARD15*) were the first identified and confirmed Crohn's disease risk alleles

Following on from the early studies referred to above, increasing levels of fine mapping were applied seeking to identify the specific susceptibility loci within each tagged region. The first major Crohn's disease risk gene was identified in 2001 as *CARD15* (caspase recruitment domain family member 15) on chromosome 16q12, which encodes the human homolog of the plant root nodule protein NOD2. These studies reported two common mutations and a deletion in the *CARD15* gene. Further studies identified more, less common mutations. As expected in complex disease, possession of the *CARD15* mutations confers an increased risk of Crohn's disease, but is neither necessary nor sufficient for the disease to occur (**Table 2.7**).

Table 2.7 *CARD15* mutations and variations encode major risk alleles in Crohn's disease - estimates from case control cohort studies.

Allele	Crohn's disease patients (*N* = 688) versus healthy controls (*N* = 250)	OR (95% confidence interval)
SNP8 (R702W)	11 versus 4.4	2.66 (1.59–4.46)
SNP12 (G908R)	4.1 versus 0.007	6.33 (2.15–18.59)
SNP13 (insertion C at 3020[a])	6.7 versus 2.8	2.59 (1.37–4.91)
Compound heterozygote	6.6 versus 0	24.85 (3.82–161.67)
Homozygotes	5.9 versus 0	22.19 (3.96–124.36)

[a]SNP13 is not a SNP, it is an insertion. Compound heterozygotes carry a combination of SNP8, SNP12, and SNP13. In 2002 Lesage et al. identified 27 rare variants not listed in this table.

NOD2 is a NACHT-LLR family member with two N-terminal caspase activation and recruitment domains (CARDs), a nucleotide-binding domain, and a series of leucine-rich repeats (LLRs) at the C-terminal. NOD2 expression is mostly restricted to the cytosol in **monocytes** and **Paneth cells**. Paneth cells, named after Joseph Paneth, are specialized cells found in the small intestine and appendix. Their precise function is unknown, though they are thought to have antibacterial properties, and may be important in immune regulation and responses to bacteria in the gut. NOD2 activates the nuclear factor NF-κB through interaction between the LRR domain and muramyl dipeptide (MDP) on **Gram-positive bacteria**. This interaction enables the CARD domains of NOD2 to interact with the protein kinase RICK (also RIP or TRAF). RICK induces K63-linked ubiquitinylation of the signaling molecule IKKY (NEMO), which in turn causes phosphorylation of the NF-κB inhibitors allowing NF-κB to initiate production of a series of pro-inflammatory and **regulatory cytokines**, including interleukin (IL)-1B, IL-8, IL-10, and tumor necrosis factor (TNF)-α (**Figure 2.13**). Under normal circumstances this process leads to host recognition and destruction of bacterial pathogens in the gut. However, when *CARD15* is present in the mutant form, the normal mechanism of signaling is interrupted and there is decreased NF-κB production. However, this is not the whole story. Crohn's disease has been associated with infection with *Mycobacterium tuberculosis*, *Pseudomonas*

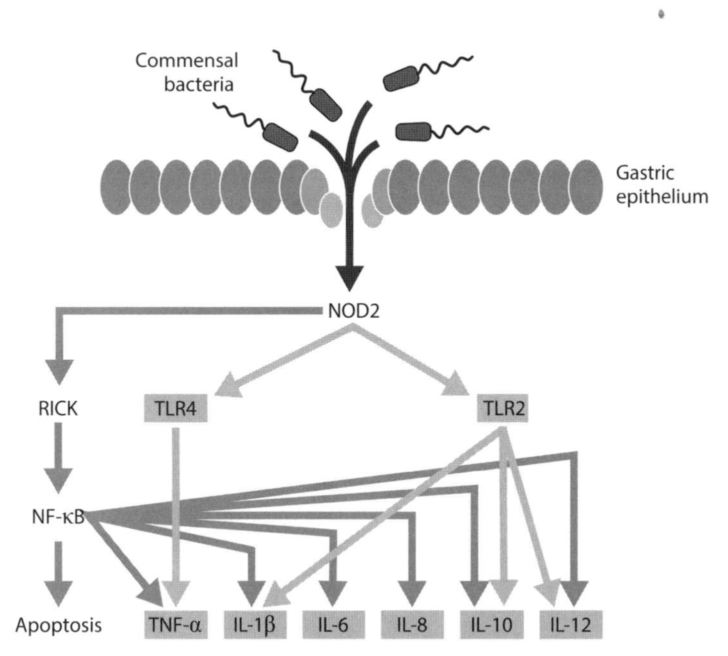

Figure 2.13: NOD2 immune interactions in maintenance of homeostasis for gut commensals. NOD2 polymorphism is the major susceptibility determinant for Crohn's disease. NOD acts through a series of pathways to influence the immune response to gut commensals. The TLR family are major components of the immune system involved in immune homeostasis of the gut. Missense mutations in TLR4 (D299G and T399I) are associated with susceptibility to Crohn's disease. TLR4 expression is increased in the intestinal epithelial cells in active disease, but found at low levels only in healthy intestine. The figure illustrates bacterial invasion (shown as tadpoles or small figures with tails) through the gut wall. Both TLR2 and TLR4 interact with NOD and each will cause up- or downregulation of pro-inflammatory and regulatory cytokines. Polymorphisms that create higher or lower activity of these immune compounds may have a significant influence on bacterial homeostasis in the gut. This figure represents only part of what is beginning to look like a very complex picture for Crohn's disease. *NOD2* also interacts with NF-κB, and NF-κ promotes apoptosis and stimulates the production of both pro-inflammatory and regulatory cytokines.

species and *Listeria* species. Crohn's disease patients have exaggerated immune responses in the gut and earlier genetic studies identified several other areas of significant linkage for Crohn's disease (Tables 2.6 and 2.7), suggesting other factors and pathways may also play a role in susceptibility, in particular genes that regulate the immune response to bacteria.

Genetic variations in other immune regulatory genes are important risk factors in Crohn's disease

One specific **immune regulatory pathway** involves the *ATG16L1* (autophagy-related 16-like 1) gene. Studies have suggested that a mutation in this gene is associated with increased risk of Crohn's disease (the region this gene is located in has been labeled IBD10). Activation of NOD2 through MDP interaction and interaction of the CARD domains with RICK has also been shown to induce **autophagy**. Other pathways involved in the recognition of bacterial lipopolysaccharide and peptidoglycan have been investigated in Crohn's disease. The **toll-like receptor (TLR) genes** encode a family of pattern recognition receptors that interact in a complex way to regulate immune responses, to self and non-self and genetic variation in these receptors may be crucial in determining susceptibility to a wide range of different diseases. TLR2 and TLR4 act as extracellular receptors, and are important in inducing **immune tolerance** to normal gut bacteria. TLR-2 interaction in *CARD15* mutants leads to decreased IL-1β, IL-10, and IL-12 production, whereas TLR-4 interaction in *CARD15* mutants leads to increased TNF-α release.

The Wellcome Trust Case Control Consortium (WTCCC1): Crohn's disease

The WTCCC1 study provided a gold standard model for case control studies in complex disease. Published in 2007, this study included Crohn's disease among the seven diseases investigated. The study was not the first genome-wide study in Crohn's disease, but it was the first GWAS. This study, as the name implies, was based on a study of the genetic variation across the whole genome using common SNPs (frequency of 5% or more) as tagged markers for susceptibility loci in a case control population-based study. The difference to previous studies is that GWAS are not restricted to candidate genes or areas of the chromosomes identified by previous analysis. This **hypothesis-non-restricted** (or **hypothesis-non-constrained** or **hypotheses-free**) approach aimed at identifying hitherto overlooked susceptibility alleles in large-scale studies with a high degree of resolution. Key differences are the size of these studies, usually 2000 or more cases with an equal (or greater) number of controls, and the number of markers used to identify associations. This means routinely using more than 100,000 and occasionally more than 1 million markers in the form of **tagged SNPs** for a study. The WTCCC1 study included analysis of more than 500,000 such tagged SNPs in 2000 cases from seven different common genetically complex diseases and 3000 shared controls. Analysis of Crohn's disease identified 10 independent genetic associations at probabilities $P < 5 \times 10^{-7}$, of which six were confirmations of previously reported associations:

- *CARD15* on chromosome 16q12 (previously IBD1).
- A cluster of potential candidates on chromosome 5q31 (previously IBD5, the precise identity of which is a matter of dispute).
- *IL23R* on chromosome 1p31 (previously IBD17).

- *ATG16L1* on chromosome 2q37 (previously IBD10, which had the strongest reported association for Crohn's disease in this study $P < 7.1 \times 10^{-14}$).
- An unknown locus in the chromosome 10q21 region around rs10761659 (previously IBD15, which is 14 kb telomeric of the *ZNF365* gene and 55 kb centromeric of the pseudogene antiquintin-like 4).
- An area around rs17234657 within a 1.2 Mb gene desert on chromosome 5q13.1 (previously IBD18).

In addition, four novel strong associations were identified:

- Chromosome 3p21 (previously IBD12). There are a number of candidate genes in the area, of which *MST1* (macrophage stimulating 1) is the most plausible. This gene encodes a protein that affects macrophage function and macrophages have a major role in phagocytosis – a key process in maintaining defense against bacterial invasion.
- The *IRGM* (immunity-related guanosine triphosphatase) gene on chromosome 5q33 (previously IBD19). *IRGM* encodes a GTP-binding protein that induces autophagy and reduced *IRGM* activity could lead to persistence of intracellular bacteria, which is consistent with the function of other genes identified as those carrying risk alleles for Crohn's disease.
- A cluster on chromosome 10q24.2 (previously IBD20). The likely candidate from this cluster is *NKX2-3* (NK2 transcription factor-related locus 3). Disruption of this gene homolog in mice leads to abnormal development of the intestine and secondary lymphoid organs. It has been suggested that mutations in this gene may disrupt gut migration of antigen-responsive lymphocytes and influence the inflammatory response in Crohn's disease.
- Close to the *PTPN2* gene on chromosome 18p11 (previously IBD21). PTPN2 (protein tyrosine phosphatase non-receptor type 2) is a key negative regulator of the immune response. Members of this family have been associated with increased risk of both rheumatoid arthritis and type 1 diabetes.

In addition to these 10 highly significant loci, there are a further eight chromosomal regions identified as having moderate associations: 1q24, 5q23, 6p21, 6p22, 6q23, 7q36, 10p15, and 19q13. Most of these associations are unsurprising, e.g. the association with the region of 6q23 which encodes the *TNFAIP3* gene (TNF-α-induced protein 3). The product of the *TNFAIP3* gene inhibits TNF-α-induced NF-κB-dependent gene expression through RIP and TRAF-2-mediated transactivation signaling in the same way that NOD2 works on NF-κB. This illustrates the common link between many of these genes and autophagy and **phagocytosis**, both of which are key processes in the defense of the gut from pathogenic bacteria and maintenance of homeostasis for normal gut bacteria. Even the association with the 6p21 region is not surprising. 6p21 is the home of the human MHC within which the HLA genes are to be found. HLA antigens are key players in both adaptive and innate immunity, and therefore an association between an immune inflammatory disease like Crohn's disease and genetic variation in the MHC is highly likely. Interestingly, the genetic association between Crohn's disease and HLA has been a matter of dispute, and is considerably weaker than that seen for many autoimmune and viral diseases as well as some adverse drug reactions. Some have suggested this association is due to the close proximity of the gene for TNF-α (*TNFA*), which maps within the MHC and not the HLA genes.

The current number of risk alleles for Crohn's disease may be as high as 163

Currently, OMIM lists a total of 29 IBD genes. Interestingly, the OMIM database lists the 10 highly significant associations reported by the WTCCC1 study (including superscript 1, 2 and only two of the eight moderately associated regions (IBD3 and IBD6). The remainder are not currently listed. However, this may be incorrect. A review in 2011 identified at least 14 major GWAS in Crohn's disease and recent studies suggested that there are at least 71 risk alleles for Crohn's disease, with a possible total over 163 if all of the chromosomal regions identified for IBD so far are considered.

Studying the genetics of complex disease is to some extent beginning to be similar to train spotting as more and more genes with small effects are added to the list. More important than the list of alleles and their locations is what do the associations tell us about the disease pathogenesis? To understand this better it may necessary to group some of the risk alleles according to function and pathways, seeking similarity and differences that may explain phenotypes. This is illustrated in **Figure 2.14**, where the different major risk alleles for Crohn's disease are grouped according to function. The different GWAS in Crohn's disease all identify candidate genes involved in the maintenance of the intestinal barrier and the activity of the different cells within the barrier. There are three major gene subgroups. First, *CARD15*, which is expressed in the gut wall in the Paneth cells and recognizes a variety of organisms, including commensal bacteria. The second group includes *ATG16L1* and *IRGM*, which are involved in autophagy, i.e. the degradation of unwanted cellular organelles and pathogens. The third group includes genes that regulate pathogenic T cell differentiation, such as members of the *IL12–IL23* pathway. Bacterial handling and MHC class II presentation appear to be regulated by *CARD15* through *ATG16L1*. This indicates how we can start to consider how different genes operate to form systems involved in disease pathogenesis.

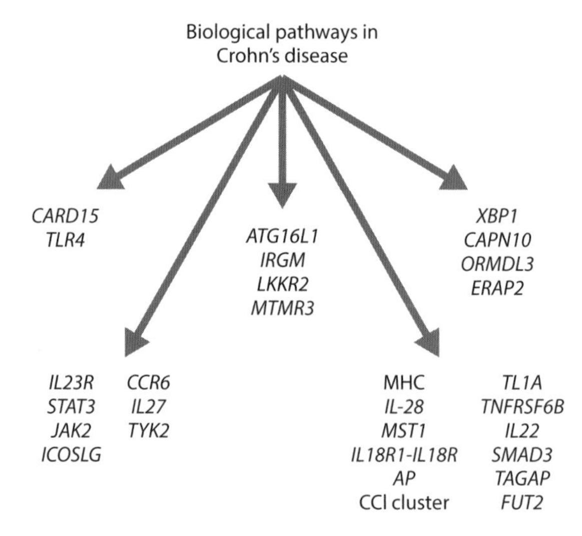

Figure 2.14: Genetically identified pathways associated with increased risk of Crohn's disease. The five arrows point to five different gene groups: innate immune signaling genes (*CARD15* or *NOD2* and *TLR4*), autophagy (*ATG16L1*, etc.), the endoplasmic reticulum biology (*XBP1*, etc.) and genes involved in T_h17 T regulatory cell balance (*IL23R*, etc.), as well as a group of genes involved in the secondary immune response (e.g. MHC, *IL12B*, etc.) The figure indicates the complexity of biological pathways and the likely interactive nature of susceptibility alleles.

The association of such a large number of susceptibility alleles, at so many loci, most with low relative risks, suggests that Crohn's disease is a polygenic complex disease rather than an oligogenic complex disease. The inherited component of the disease is low in Crohn's disease and it is possible that many cases involve the cumulative interaction of several mutant alleles. In contrast, in HSCR, the inherited component is greater, there are fewer risk alleles or mutations, and most cases involve only one major risk mutation. In some cases of HSCR, however, the effect of the major mutation can be modified by inheritance of other risk alleles.

2.10 APPLYING DISEASE MODELS TO POPULATIONS

A level of caution should be applied when considering these different models of genetically complex disease. In the analysis of populations we overlook the fact that cases are individuals and that each case is different from others. It is quite possible that many individual cases have only one risk allele. It is also possible that patients with the same clinical disorder can have different risk alleles on the same gene or on different genes. The variable severity of the disease phenotype seen in individual Crohn's disease and HSCR cases may be due to the particular portfolio of risk alleles inherited.

The size of the portfolio or genetic load on an individual is another consideration. Portfolios could be built up in different ways. Within the portfolio there may be a single common genetic variant that has a large effect in the majority of the cases or there may be a series of rare variants. The former is predicted by the **common disease common variant hypothesis (CDCVH)** and holds true for a disease like Alzheimer's disease and the *APOE4* allele. The latter is predicted by the **common disease rare variant hypothesis (CDRVH)**, which predicts that if a disease is common in the population, the genetic risk may not be common in the population and may be due to the effect of multiple interacting risk alleles. Both models apply in complex disease, even in the same disease where some cases are seen as near-Mendelian and others as sporadic. LOAD illustrates the possibility of both models. HSCR, which is associated with a very significant genetic component, illustrates the CDCVH model and most cases of Crohn's disease illustrate CDRVH. However, we need to be cautious applying models as each case may be different. Models are best applied to populations with no assumption that all cases will fit within the model. One other consideration is whether we are dealing with a small number of common variants or a large number of rare variants. Both can apply and indeed they do appear to.

Returning to the idea of heritability, one concept that is important to consider in complex disease is the idea of missing heritability. There is intense discussion of various hypotheses and models for complex disease. Some suggest that we could create a model of heritability whereby continuous inheritance of both genetic and environmental factors leads to a point whereby some individuals express the trait. This is not unlike the risk portfolio suggested earlier. Finally, rare alleles with small or moderate effects are likely to be missed in GWAS and it is also possible that heritability estimates are overinflated.

CONCLUSIONS

One of the key concepts in this chapter and throughout the book is that complex diseases are not simple and do not lend themselves to simple analysis. They are best defined using the modified definition of Haines and Pericak-Vance (1998) as those where one or more genetic variations (alleles) may increase or decrease the risk of a trait or disease. In this book we are discussing disease. In complex disease we exchange trait for disease from time to time. This exchange is not accepted by all, but using this term occasionally enables us to be flexible when considering overlapping diseases (syndromes) and when considering different subgroups within a disease population. It is therefore more useful to be flexible than proscriptive in the application of terminology. Chromosomal and Mendelian diseases are not genetically complex diseases; however, there are some gray areas where simple definitions cannot be so easily applied.

Dismantling the definition produces a clearer understanding of what this book is all about. The most important concept throughout the book is that the inheritance of an allele at a locus increases or reduces the risk of the disease. This means that the allele is neither necessary nor sufficient to cause the disease. Inheritance of an allele is simply associated with either an elevated or a reduced risk. In this definition, the possibility of protective as well as susceptibility alleles is considered. Not all risk alleles are associated with the occurrence or initial pathogenesis of a disease. Alleles may be associated with the risk of a specific disease subgroup within a population, e.g. severity defined by age of onset, disease progression (including morbidity and mortality), or response to therapy. This illustrates how not all genetic associations can be interpreted as tags (or markers) for disease causing genes.

Complex diseases are complex and we can consider three potential models: monogenic, involving a single gene, but not conforming to a Mendelian pattern, oligogenic involving several, but a limited group of genes, and polygenic involving a larger group. These models clearly overlap, but it is still useful to categorize disease like this if possible. The latter two models illustrate not only the basic concepts of the genetics of complex disease, but also indicate why we seek to identify risk alleles in complex diseases. Understanding the basic principles set out here is essential in order to understand current research and the potential applications of genetic investigations such as those described in this book in modern medical practice now and in the future.

FURTHER READING

Books

Collins F (2010) The Language of Life: DNA and the Revolution in Personalised Medicine. Harper-Collins.

Haines JL & Pericak-Vance MA (1998) Overview of mapping common and genetically complex human disease traits. In: Approaches to Gene Mapping in Complex Human Disease (Haines JL & Pericak-Vance MA eds), pp 1–16. Wiley. This book provides the best definition of complex disease and is a good starter text on major issues in complex disease.

Lewis R (2011) Human Genetics – The Basics. Routledge.

Parham P (2009) The Immune System, 3rd ed. Garland Science. This is an excellent textbook providing a good background on human immunology and some examples of immunogenetics.

Strachan T & Read AP (2011) Human Molecular Genetics, 4th ed. Garland Science. This is an excellent basic genetics textbook that goes into more depth on many of issues in this chapter, especially in chapters 2 and 3 and also in chapters 15 and 16.

Articles

Amiel J & Lyonnet S (2001) Hirschsprung disease, associated syndromes and genetics: a review. *J Med Genet* 38:729–739. This is an excellent review of HSCR and, together with Puffenberger et al. (1994), provides a basic starting point for students investigating this disease.

Beutler E, Felitti VJ, Koziol JA et al. (2002) Penetrance of 845G-A (C282Y) HFE hereditary haemachromatosis mutation in the USA. *Lancet* 359:211–218. This paper is a landmark paper on *HFE* and raises a number of important questions.

Brown MA, Pile KD, Kennedy LG et al. (1996) HLA class I associations of ankylosing spondylitis in the white population in the United Kingdom. *Ann Rheum Dis* 55:268–270. This paper is the gold standard paper for ankylosing spondylitis and *HLA-B*27*.

Cario E (2005). Bacterial interactions with cells of the intestinal mucosa: Toll-like receptors and NOD2. *Gut* 54:1182–1193. This paper provides some essential insight into the relationship between the genetics of Crohn's disease and the developing understanding of the disease pathogenesis that has followed from these genetic studies.

Collins FS & McKusick VA (2001) Implications of the human genome project for medical science. *JAMA* 285:540–544. This is a very important review that provides the reader with a guide to and an illustration of the potential for understanding complex disease in the immediate post-genome period.

Cooney R, Baker J, Brain O et al. (2010) NOD2 stimulation induces autophagy in dendritic cells influencing bacterial handling and antigen presentation. *Nat Med* 16:90–97. This paper relates genotype to phenotype – an essential next step in complex disease.

Cuthbert AP, Fisher SA, Muddassar MM et al. (2002) The contribution of NOD2 gene mutations to the risk and site of disease in inflammatory bowel disease. *Gastroenterology* 122:867–874. This is a good study for students to look at for critical analysis and to understand basic pre-GWAS association work in Crohn's disease.

Evans RM, Hui S, Perkins A et al. (2004) Cholesterol and APOE genotype interact to influence Alzheimer disease progression. *Neurology* 62:1869–1871.

Franke A, McGovern DP, Barrett JC et al. (2010) Genome-wide meta-analysis increases to 71 the number of confirmed Crohn's disease susceptibility loci. *Nat Genet* 42:1118–1125. This paper introduces meta-analysis and summarizes a series of GWAS of Crohn's disease.

Henderson P & Satsangi J (2011) Genes in inflammatory bowel disease: lessons from complex diseases. *Clin Med* 11:8–10.

Hugot J-P, Chamaillard M, Zouali H et al. (2001) Association of NOD2 leucine-rich repeat variants with susceptibility to Crohn's disease. *Nature* 411:599–603. This, together with the paper by Ogura et al. (2001), was the first to identify the IBD1 gene as *CARD15* on 16q12 – a major step forward in our understanding of the genetics of Crohn's disease.

Inohara N, Ogura Y, Fontalba A et al. (2003) Host recognition of muramyl dipeptide mediated through NOD2. *J Biol Chem* 278:5509–5512.

Jostins L, Ripke S, Weersma RK et al. (2012). Host–microbe interactions have shaped the genetic architecture of inflammatory bowel disease. *Nature* 491:119–124. This paper identifies 63 risk alleles for IBD.

Kullberg BJ, Ferwerda G, De Jong DJ et al. (2008) Crohn's disease patients homozygous for the 3020insC NOD2 mutation have a defective NOD2/TLR4 cross-tolerance to intestinal stimuli. *Immunology* 123:600–605.

Lala S, Ogura Y, Osborne C et al. (2003) Crohn's disease and the NOD2 gene: a role for paneth cells. *Gastroenterology* 125:47–57.

Lees CW, Barrett JC, Parkes M & Satsangi J (2011) New IBD genetics: common pathways with other diseases. *Gut* 60:1739–1753. This paper provides a current summary of the genetics of Crohn's disease and the wider connotations of this type of research.

Lesage S, Zouali H, Cezard J-P et al. (2002) CARD15/NOD2 mutational analysis and genotype-phenotype correlation in 612 patients with inflammatory bowel disease. *Am J Hum Genet* 70:845–857. This paper opened the door to further exploration of the *CARD15* gene in Crohn's disease and marked an important development in the study of Crohn's disease genetics.

Ogura Y, Bonen DK, Inohara N et al. (2001) A frameshift mutation in NOD2 associated with susceptibility to Crohn's disease. *Nature* 411:603–606. This, together with the paper by Hugot et al. (2001), was the first to identify the IBD1 gene as *CARD15* on 16q12—a major step forward in our understanding of the genetics of Crohn's disease.

Puffenberger EG, Hosada K, Washington SS et al. (1994) A missense mutation of the endothelin-B receptor gene in multigenic Hirschsprung's disease. *Cell* 79:1257–1266.

Prince JA, Zetterberg H, Andreasen N et al. (2004) APOE (epsilon)4 allele is associated with reduced cerebrospinal fluid levels of A (beta) 42. *Neurology* 62:2116–2118.

Schreiber S, Rosenstiel P, Albrecht M et al. (2005) Genetics of Crohn disease, an archetypal inflammatory barrier disease. *Nat Rev Genet* 6:376–388. This is an excellent review of the genetics of Crohn's disease. The information and discussion provides insight not only into Crohn's disease, but is also a useful model to understand other similar conditions.

Sennvik K, Fastbom J, Blomberg M et al. (2000) Levels of alpha- and beta-secretase cleaved amyloid precursor protein in the cerebrospinal fluid of Alzheimer's disease patients. *Neurosci Lett* 278:169–172.

Stokin GB, Lillo C, Falzone TL et al. (2005) Axonopathy and transport deficits early in the pathogenesis of Alzheimer's disease. *Science* 307:1282–1288.

Stoll M, Corneliussen B, Costello CM et al. (2004) Genetic variation in *DLG5* is associated with inflammatory bowel disease. *Nat Genet* 36:476–480.

Taylor RW & Turnbull DM (2005) Mitochondrial DNA mutations in human disease. *Nat Rev Genet* 6:389–402. This review is highly informative for the student or clinician trying to understand the nature of mitochondrial disease and the mitochondrial genome.

The Thousand Genomes Consortium (2012) An integrated map of genetic variation from 1,092 human genomes. *Nature* 491: 56–65.

The Wellcome Trust Case Control Consortium (2007) Genome-wide association study of 14,000 cases of seven common diseases and 3,000 shared controls. *Nature* 447: 661–678. This is another landmark paper highlighting the development and application of new technologies (GWAS) in complex disease research, and it provides a mass of useful information for those studying complex disease. Together with the other papers on Crohn's disease, this study provides some of the basic information for students studying Crohn's disease as well as a further six diseases not discussed in this chapter.

Online sources

http://hapmap.ncbi.nlm.nih.gov
This is the site of the results of the International HapMap project.

http://pngu.mgh.harvard.edu/~purcell/plink
PLINK is a freely available, open-source, whole-genome association analysis toolset designed for quality control and analysis of GWAS data.

http://www.genet.sickkids.on.ca/StatisticsPage.html

http://www.mousebook.org/mousebook-catalogs/imprinting-resource
This is an essential database for mouse imprint genes and current research on imprinting.

http://www.ncbi.nlm.nih.gov/omim
This is an essential site for updates and information on genetic disease of all types.

http://www.ncbi.nlm.nih.gov/projects/SNP/snp_summary.cgi
This site provides information about validated reference SNPs clusters in the genome. The 1000 genome project provided a validation map including 38 million SNPs in 2012 (reference above).

http://www.ncbi.nlm.nih.gov/SNP
This an essential site for updates on SNPs.

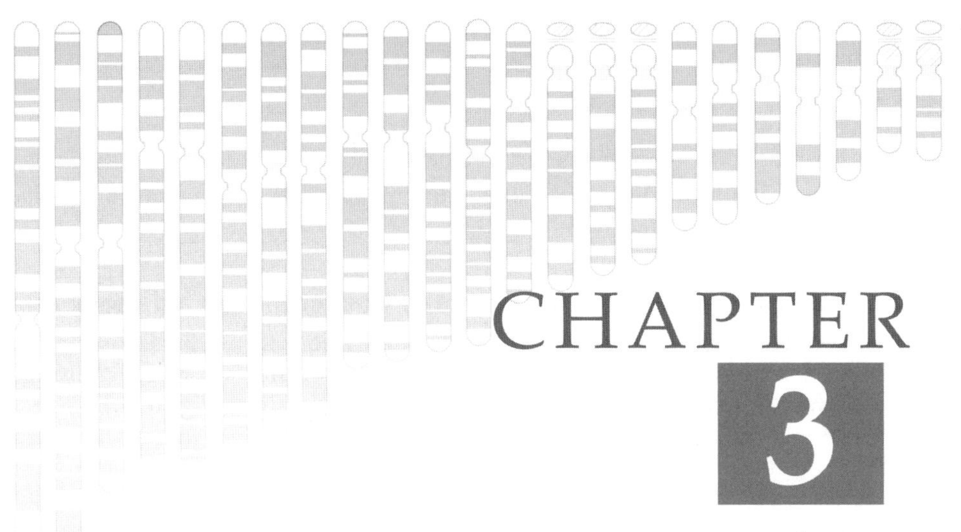

CHAPTER 3

How to Investigate Complex Disease Genetics

In this chapter, we will consider the questions of how to identify risk alleles in genetically complex disease, starting with the basic knowledge that informs study design and the different options that are available to us. We will look at study design from the past to the present day and consider some of the advances. Having already seen how complex diseases do not lend themselves to simple analysis and having been made aware in Chapter 2 that we are dealing with mutations and polymorphisms that increase or reduce the risk of a disease, it is possible to see how hard this task may be. In addition, it is also clear that there may be complex interactions between different risk alleles, and between risk alleles and the environment. Altogether this adds up to a difficult task and yet it is one that has attracted considerable attention in medical research. We shall see how study design has advanced to reach previously impossible heights.

3.1 PLANNING STAGE 1: GATHERING THE BASIC KNOWLEDGE

Before we investigate a disease we need to know what is possible. For example, rare diseases do not lend themselves to single-center studies, whereas common diseases may. It is difficult to perform genome-wide studies on small groups of samples and small sample sizes do not facilitate the identification of low-risk alleles. Good planning begins with a few simple questions:

- How common is the disease?
- What is the evidence for a genetic component to the disease?
- What is known about the disease pathology that may indicate associations or linkage with specific biological pathways?

Incidence and prevalence are measures of how common a disease is

It is essential before starting any study to be aware of the frequency of the disease within the population that we intend to study (**Box 3.1**). Disease frequency is normally measured as either **incidence** or **prevalence**. These two terms are very different (**Figure 3.1**).

Incidence

Incidence refers to the number of cases of a disease within a defined time period. Usually, incidence is measured as the number of new cases per year. However, for rare diseases, incidence may be measured over a longer period in order to obtain accurate figures. Incidence deals with new cases only and excludes all cases reported outside the time period.

BOX 3.1 INCIDENCE, PREVALENCE, AND FAMILY RISK RATIO (λ)

Incidence can be calculated from the simple equation:

$$\text{Incidence} = \frac{\text{number of new cases within a given time period}}{\text{population size}}$$

Thus, in a population of 1000 with 10 new cases per year the incidence = 1/100.

Prevalence can be calculated from the simple equation:

$$\text{Prevalence} = \frac{\text{number of cases in the population at time } x}{\text{population size}}$$

Thus, in a population of 1000 with 100 cases at time x, the prevalence = 1/10.

Family risk ratio (λ) can be calculated for different family members including siblings, children or by sex (females/males). The risk ratio for siblings (λ) can be calculated from the simple equation:

$$\lambda = \frac{\text{incidence or prevalence in the siblings of affected individuals}}{\text{incidence in the population}}$$

Thus, if the incidence in siblings is 1/50 and the incidence in the population is 1/1000, then $\lambda = 0.02/0.001 = 20$. Unless tested in very large populations, it is appropriate to consider more than one estimate of λ and quote a range. When calculating this value for the first time, we need to be aware of the problems associated with small numbers, which often lead to high estimates that may prove false on further scrutiny.

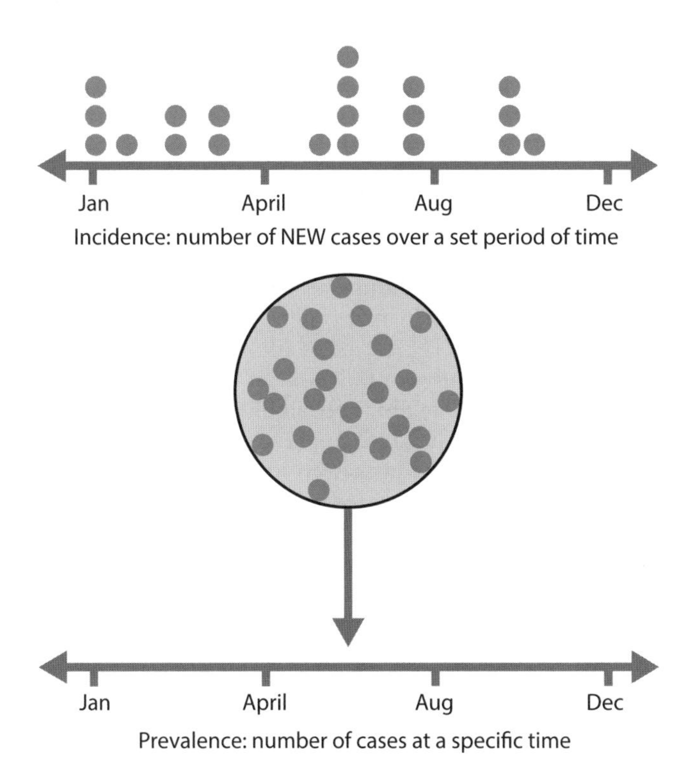

Figure 3.1: Incidence and prevalence. The gray dots represent individual cases. The horizontal arrows are time-lines. The figure illustrates the difference between incidence and prevalence. Incidence is a measure of the number of new cases in a set period of time (often 1 year). Prevalence refers to the total number of cases at a particular time. Incidence and prevalence can be very different or very similar depending on the severity of disease (prognosis). When the disease is severe, survival times can be low and prevalence will match incidence; where survival times or prognosis is good, then prevalence is often greater than incidence.

Prevalence

Prevalence refers to the number of cases at a given time point. On a specific date the number of reported cases may be 1000. This is different from incidence because it includes all cases irrespective of when they were reported.

Incidence and prevalence can be very different or very similar depending on the prognosis for the disease

In diseases with a poor prognosis, the survival time from diagnosis may be very short and therefore the two measurements may be very similar. However, where the prognosis is favorable, the survival time from diagnosis may be very long and the two measurements will reflect this. For example, if the estimated survival time from diagnosis for a disease is 20 years and the number of new cases per year (incidence) is 100, then at a given time point we would expect there to be up to 2000 cases in the study population. Thus prognosis is an important factor to be aware of when planning studies. There are many reasons for this. Recruitment may be difficult where prognosis is poor and, even where incidence is high, the poor prognosis means there is less time for sample collection and recruitment. There are also ethical issues regarding recruitment for all studies that may be exacerbated in these circumstances.

Incidence and prevalence of disease may vary in different populations

Population variation can be due to either genetic or environmental factors. For example, malaria is rarely seen in the UK. The reasons for the virtual absence of malaria in the UK are obvious and are clearly environmental. This example can be applied to many infectious diseases where the pathogen is found only at a specific location or where other environmental factors (e.g. clean water) play a role. Some geographic differences in incidence and prevalence are due to genetic variation between populations. For example, the disease autoimmune hepatitis (AIH) occurs in many different countries and is associated with specific human leukocyte antigen (HLA) alleles (discussed in Chapter 6). Interestingly, in the UK (and USA) there are early- and late-onset peaks for AIH within the population, whereas in Japan the disease is mainly seen in older members of the population. The difference between the UK/USA AIH and Japanese AIH populations is the presence of a specific HLA haplotype (**HLA 8.1**) that is seen in the former populations, but rarely if ever seen in the Japanese population. Such genetic differences are common between populations and these may be very important in the survival of the species (**Figure 3.2**). Other diseases such as schizophrenia are seen at similar frequencies across the globe, and though the frequencies are slightly higher in some developed compared with some developing countries, this difference is not as big as expected. Also contrary to popular belief, it has been shown by Bhugra (2005) that the illness is not more common in men than women, as some have reported.

What is the evidence for a genetic component to the disease?

It is an essential early step to gather information on the potential for a genetic component in the disease under study. There are several simple questions that can be asked that will inform the study design. The most important of these is to find out whether there are

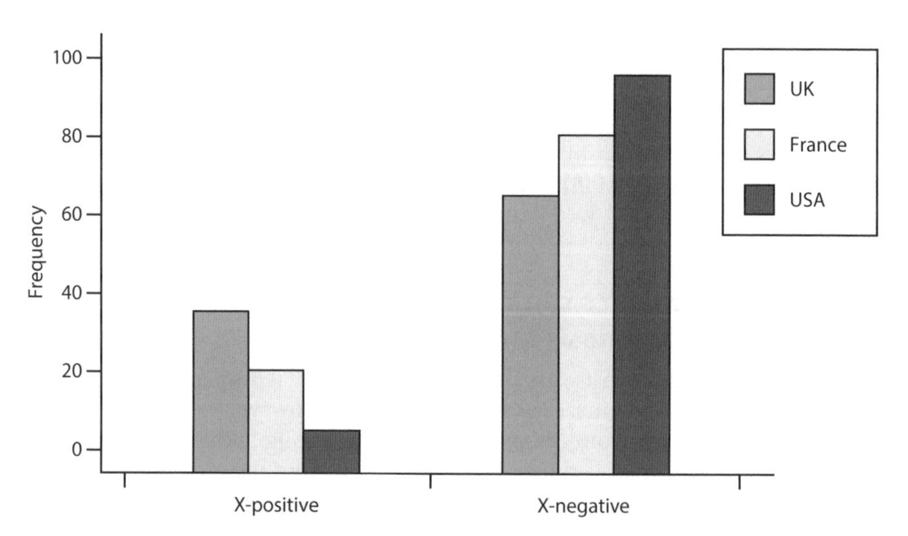

Figure 3.2: Incidence and prevalence in different populations. The figure illustrates three populations (e.g. UK, France, and USA) with different frequencies of X disease. These values could have been measured as either incidence (number of new cases in a set period) or prevalence (total number of cases at a time). The important point is that incidence and prevalence for a given trait may vary between populations. The further apart these populations are located geographically, the greater these differences are likely to be. Thus, the greatest difference in the frequency of X disease is between the USA and the UK, while France, which is closer to the UK, has a smaller difference.

any informative (i.e. multicase) families with the disease. At this stage it is also important to remember missing heritability. This reminds us that not all heritability is due to our genes—the proportion of our heritability that is not determined by our genes is referred to as missing heritability. The extent to which missing heritability occurs varies between diseases. Inflated values for some of the data collected in the tests referred to in this book (see Chapter 4 in particular) are the result of this missing heritability. For example, a sibling relative risk of 50 may not all be attributed to sharing of the same genetic variation—the shared environment can also play a significant role.

Family information

Any good geneticist will always look for informative families. To be informative, a family must have more than one affected member (**Figure 3.3**). If there are enough informative families, family studies can be performed. **Linkage analysis** based on multicase families has been and still is considered the gold standard for genetic studies; however, multicase families are rare in complex disease. In the absence of families, affected sibling pairs can be studied. However, even if there are families or large numbers of sibling pairs and linkage analysis is to be performed, the non-familial (sporadic) cases should not be overlooked. Linkage analysis in families and sibling pairs is mostly parametric linkage analysis. This means the parameters have to be known at the outset. One of these parameters is the pattern of inheritance. Therefore, this suits Mendelian disease, but not complex disease where the key parameter, i.e. pattern of inheritance, is by definition unknown. Non-parametric linkage analysis can be used instead for linkage studies in complex disease.

It is important to remember that there are a number of caveats with respect to families in genetic studies. Familial history is not always a sign of a genetic effect. Families share environments, which can have a strong effect on susceptibility to a variety of different diseases. The best example of this can be seen in an analysis of the likelihood of attending medical school in the USA. This study found attendance was an autosomal recessive trait, when in

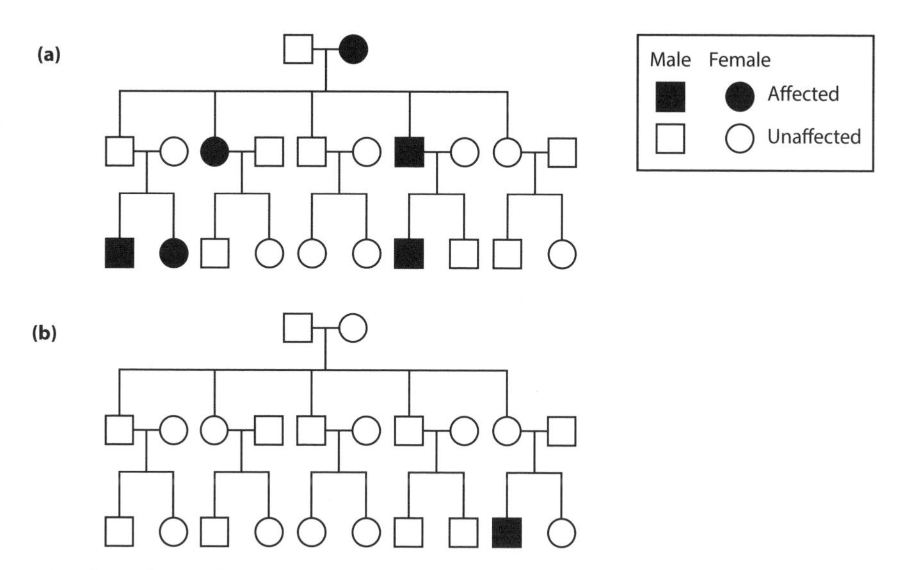

Figure 3.3: Informative families. This simple comparison between two families (a and b) illustrates the importance of there being multiple cases in a family if linkage analysis is to be applied. Here, family (a) is informative and family (b) is uninformative because for linkage there needs to be more than one affected family member in each generation. In practice, the picture may be more complicated and the decision of whether or not to use families may rest on the actual number of affected cases and the number of cases in each generation.

fact the real impacting factors were social status and parental income. In a sense, the trait has a significant heritable component, but that heritability is not genetic. The issue of the term heritability has been extensively discussed in Chapter 2.

The second caveat regarding family studies is that genetic risk in families may be attributable to different risk alleles compared with those involved in sporadic disease. However, caveats aside, families do represent the genetic high ground in such studies and for many diseases, especially near-**Mendelian** and oligogenic complex disease, identifying risk alleles in families has led to significant findings. These findings have in some, but not all cases, also led to findings in sporadic non-familial cases. Linkage analysis is no longer restricted to accessing single genes or chromosomal regions as in the past. In the present decade, linkage studies can include the whole genome. Indeed, many **genome-wide linkage studies (GWLS)** have been performed. However, this does require large family pools. It is also important to remember that not all families are informative (e.g. Figure 3.3b) and even where there are multiple cases in diseases with late onset or severe prognosis, recruiting family members for studies is often difficult.

Relative risk (λ)

Apart from using families in linkage analysis, we can perform a few basic tests to provide an indication of the size of the **heritable component** of a disease. One such test is based on relative risk, which can be calculated by comparing the known incidence or prevalence in families, with the incidence or prevalence in the population as a whole. The **risk (λ)** can be calculated for specific family members, for example λs is a measure of the risk for siblings of the affected (and this is the most commonly calculated risk). The risk may be different in different family members. Some diseases are known to be more common in the female members of the population, such as in many, though not all, autoimmune diseases. In these cases, calculating the risk for daughters of affected mothers may be more appropriate. The λ value gives a good, but crude measure of the relative risk within a population and in practice when looking at published data it is always better to quote ranges from several studies rather than from individual studies. Values quoted for Mendelian disease may be as high as several thousand for dominant conditions and several hundred for recessive diseases. Figures for complex diseases vary from zero to the low hundreds (see **Table 3.1**).

Table 3.1 Sibling relative risks for different diseases.

Disease	Type	Relative risk (λ)
Huntington's disease	Mendelian dominant	5000
Cystic fibrosis	Mendelian recessive	500
Hirschsprung's disease	Oligogenic complex	~185
Crohn's disease	Polygenic complex	13–36
Ulcerative colitis	Polygenic complex	7–17
Primary biliary cirrhosis	Polygenic complex	10.5 (daughters of affected mothers = 58)
Type 1 diabetes	Polygenic complex	~15
Multiple sclerosis	Polygenic complex	~50
Schizophrenia	Polygenic complex	~10

It is also important to remember that this is a measure of heritability and genetic variation may only make up part of this.

Twin studies

Twins represent a very useful study group in genetics. Twins may be either identical or non-identical. Identical twins are the product of a single fertilization and share the same genome, i.e. they are **monozygotic twins.** Non-identical twins occur when two fertilizations of different gametes occur at approximately the same time giving rise to two individuals who, though born at the same time and of the same parents, are unlikely to have a larger degree of shared genetic identity than any two other offspring of the same parents conceived on different occasions (**Figure 3.4**). Non-identical twins are referred to as **dizygotic twins.**

Twins can provide very useful information about the genetic component for a disease. Simple comparison of disease concordance rates for monozygotic versus dizygotic twins indicates the extent to which the genome plays a role in disease. Twin data has to be handled with caution because concordance data for large disease populations is more reliable

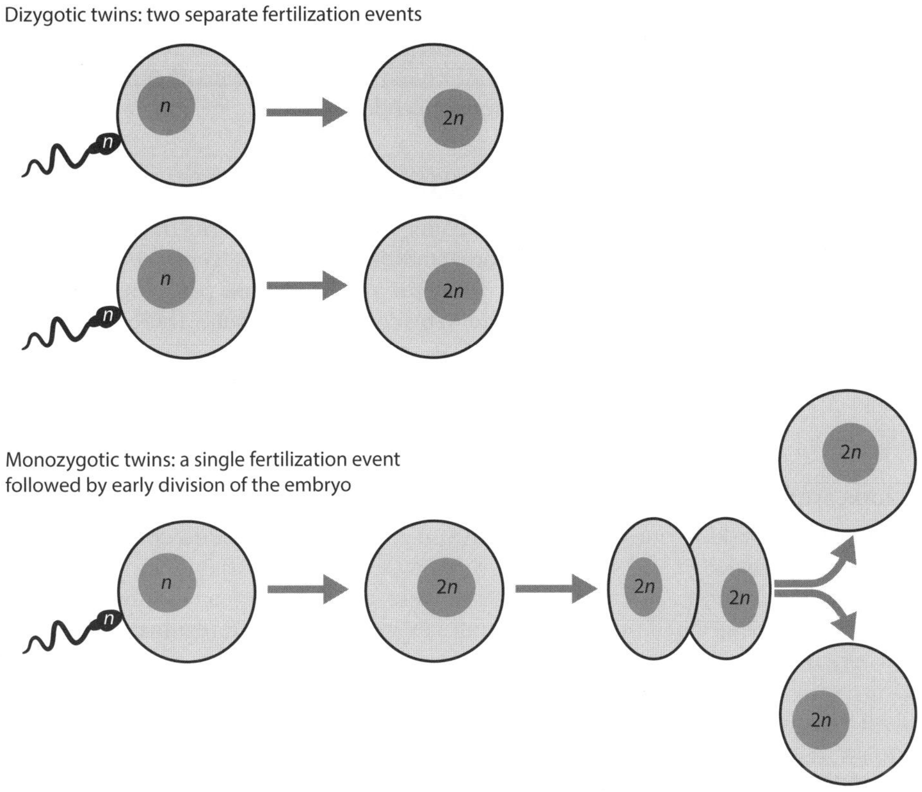

Dizygotic twins: two separate fertilization events

Monozygotic twins: a single fertilization event followed by early division of the embryo

Figure 3.4: Monozygotic and dizygotic twins. Twinning occurs in one of two ways. Fraternal (non-identical or dizygotic) twins occur when two eggs are independently fertilized at approximately the same time. Dizygotic twins develop in the womb at the same time, but have separate embryonic membranes. Dizygotic twins have the same level of genetic identity to each other as their siblings. Monozygotic twins arise from the same single fertilization event and are produced by the division of the embryo very early in gestation. The illustration shows this division after the two-cell stage, but it may occur anytime before, such as splitting at the morula or in the early blastocyst stage. Monozygotic twins are essentially genetically identical.

Table 3.2 Concordance data for monozygotic and dizygotic twins.

	Concordance (%)	
Disease	**Dizygotic twins**	**Monozygotic twins**
Type 1 diabetes	15	30–40
Crohn's disease	7	37
Type 2 diabetes	10	90
Rheumatoid arthritis	3.6	15.4
Multiple sclerosis	5.4	25.3
Bipolar disorder	6	43
Autism	0–21	~60
Schizophrenia	4	15–33

than for small patient populations. However, large numbers of twins are unlikely to be available in all but the most common of complex diseases because twins are themselves quite rare. Approximately one in every 200 pregnancies give rise to twins and most of these are dizygotic twins. Some examples of concordance data for different diseases are given in **Table 3.2** and elsewhere throughout this book.

What is known about the disease pathology?

It is often assumed that because diseases are diagnosed according to the presence of a series of signs and symptoms we have a clear understanding of all aspects of the **pathology** of these diseases. Nothing could be further from the truth. Disease pathology reflects the downstream events (**Figure 3.5**). Signs and symptoms arise because there is some form of malfunction. However, the human body can tolerate significant trauma before organs and structures fail. Signs and symptoms are in many cases endpoints of at least the early phases of disease. In practice, this means that there are gaps in our knowledge of disease pathology and much of what we do know is based on hypotheses generated from research.

Previous studies

When it comes to asking questions about disease pathology we need to consider the quality of current knowledge and the gaps in the jigsaw. In the 1990s and early years of the new millennium, studies of complex disease were mostly based on **hypothesis-driven research** investigating small numbers of candidate genes in favored regions. In the present decade, planners have cast aside the restrictions imposed by hypotheses and have investigated the whole genome in a **hypothesis-free** (or hypothesis-non-constrained) way. Some argue that there is still a hypothesis, but it is just very broad, i.e. there is a disease allele or alleles out there somewhere. However, most prefer the idea that these are hypothesis-free studies, or at least the initial stages are, though replication and fine mapping may not be.

This change in approach reflects the introduction of new technologies in genetics of complex disease. Though GWLS at low resolution in which microsatellites were used as markers were performed in 1994 and 1996, these were rare exceptions and modern **genome-wide association studies (GWAS)** and GWLS did not start appearing until after 2005. The change in favored choice of study reflects years of frustration regarding results from earlier traditional **candidate gene-based association studies** that were often contradictory. One

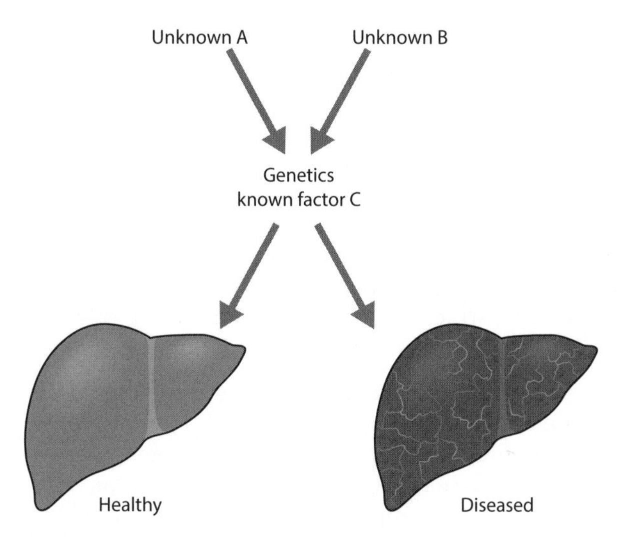

Figure 3.5: Disease pathology. The figure illustrates a healthy (light gray) and an unhealthy, diseased liver (dark gray with scarring). The figure shows that the many of the factors that cause early disease are often unknown to us, and reminds us that we tend to see patients (represented by the organs here) with specific signs and symptoms when they present at a clinic late in disease progression. This means the opportunity to gain knowledge of the early biological processes through which the disease arises are lost to the medical community, preventing earlier interaction. However, our inherited genes do not change and knowledge of the risk alleles, shown as factor C (the known factor), in complex disease is potentially the key to unraveling this conundrum.

group would report a significant association with a specific allele and another would fail to replicate the original findings. This happened because many studies were poorly designed using small numbers, poorly matched or inappropriate controls, and badly defined patient groups. In general, much of this has now been dealt with. Large-size, multicenter studies using well-defined patient and control groups have been able to identify significant risk alleles for many diseases and many of these reports have been replicated. However, studies still report weak associations that are not replicated by others. The reasons why this still happens vary. Some studies are still poorly designed, but in many cases the differences found reflect the different populations studied.

It is essential to understand the history of these studies in order to critically evaluate the literature prior to undertaking any study of the genetics of a complex disease. Questions about quality and reliability of previous studies are crucial if we intend to build on previous research. It is also important not to waste research resources by simply repeating previous studies unless there is a bona fide reason to do so. Information regarding previous studies and potential risk haplotypes is essential in study planning. There may be very little mileage in reconsidering serially confirmed associations and it may be more fruitful to look at previously ignored chromosomal regions or genes.

Disease pathology as a means to identifying candidate pathways

Though our knowledge of early disease pathology is often incomplete there is a great deal of information about disease pathology in the literature. Our task is to sift through the data and identify key biological pathways that, when stressed, may contribute to the disease expression. Biological pathways are the key to unpacking disease pathology and this process is itself the key to identifying potential candidate genes in complex disease. For example, pathways associated with bone formation and with inflammation may be critical in rheumatoid arthritis. Where there are well-established hypotheses, it may be appropriate to consider the constituent pathways that have led to the development of that hypothesis. For example, early hypotheses suggested that Crohn's disease may arise from either an abnormal immune inflammatory response to gut bacteria or as a food allergy. Recent evidence favors the former and indeed appropriate risk alleles have been identified that indicate the immune response to gut bacteria is a key process in Crohn's disease. Not all disease-related genes are involved in immunity. Genes impacting on lipid metabolism are known to be important in coronary artery disease and some of the same genes may also influence risk in Alzheimer's disease. In neurological diseases and some mental disorders (bipolar disorder and schizophrenia), genetic associations have been identified with genes such as *DISC1*, *GRM7*, and *GABRB1*, all of which are important in brain cell activity and neurotransmission (especially *GABRB1*). A more detailed consideration of these is given in Chapter 4.

When considering the pathology of a specific disease we should also consider what is known about the pathology of other similar diseases. Many diseases are grouped into syndromes or collections. Sharing of susceptibility alleles between diseases is not unusual, and therefore looking at related diseases with similar signs and symptoms may help to identify important pathways. A primary example of the latter is autoimmune disease where many of the major autoimmune diseases have associations with the HLA 8.1 haplotype or with members of the *HLA DRB1*04* family of alleles. Examples include type 1 diabetes mellitus, the autoimmune liver diseases, AIH and primary sclerosing cholangitis, rheumatoid arthritis, and systemic lupus erythematosus (SLE).

Before we get down to the hard business of study planning there are one or two other questions that it is important to ask

The remaining information required for planning a study is illustrated in the organizational chart **Figure 3.6** and includes:

- What is the average (mean) age of onset of the disease?
- Are there any other centers investigating the genetics of the disease?
- What is the likelihood of obtaining research funding to study the disease?

Age of onset

Age of onset is an important statistic in studying any disease. Working with children presents a number of moral and ethical issues (discussed in Chapter 11). However, with young families, recruiting parents and other family members for studies can be easier than working with older families where family members are more likely to be diversely spread and less likely to be in contact. In many cases, early-onset diseases are also more likely to have a Mendelian or near-Mendelian pattern of inheritance. However, there are always exceptions to every rule and several common forms of complex disease, such as autism, attention deficit hyperactivity disorder (ADHD), childhood obesity, and asthma, are all complex disorders that present in childhood.

Other centers

It is essential to consider the competition when planning studies. There is very little to be gained from competing with a rival who has access to larger patient numbers and better resources. It is more useful in such a scenario to collaborate. However, collaboration is not always possible or appropriate. Awareness of rivals can be very useful in planning. There are usually several ways to approach key questions and taking an alternative approach may be very fruitful. For example, rather than looking for genetic associations with the disease per se, it may be more appropriate to consider specific subgroups within the disease

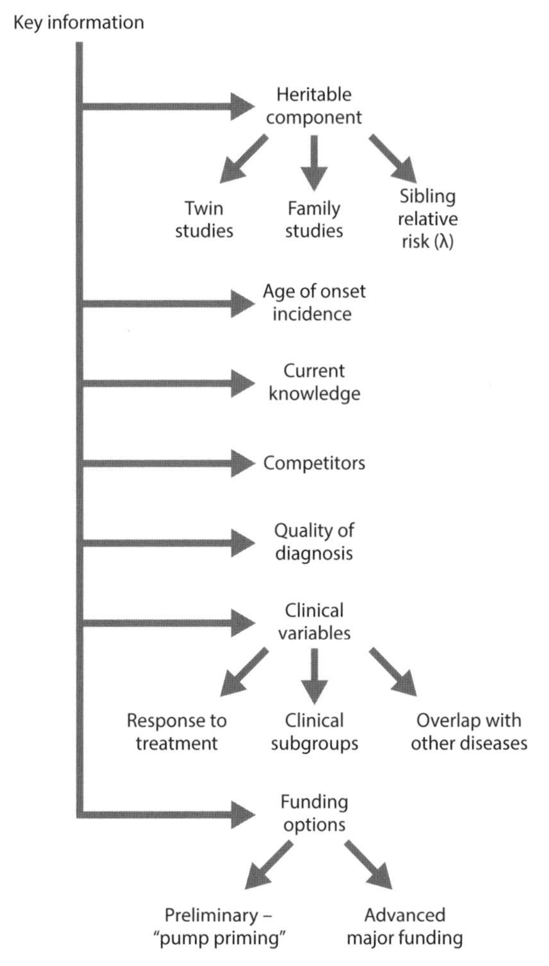

Figure 3.6: Planning a study: things to consider before proceeding. The organization chart illustrates the key points that need to be considered before commencing with the detailed planning of a study. Most are common sense, but it is important to consider each of these. This can prevent misuse of time and resources, and also lead to new approaches to a particular disease. For example, based on the information on subgroups, we may choose to investigate one particular group within a disease cohort rather than all cases.

population. Looking at disease phenotype is particularly important in infectious disease where transmission is not genetically determined, but post-infection outcome or response to therapy is.

Research funding

Research does not fund itself. Sources of funding are available, but are limited. Applications to major bodies require a considerable effort in terms of planning and writing, and are frequently unsuccessful. The process is competitive and is particularly difficult for those starting out in a career in medical science, whether they are clinical or non-clinical careers. However, the reward for perseverance in applications is worthwhile and the fruits of well-planned studies can be significant. We need to consider where the funding will come from when planning. Most countries have both central sources (often government funds) and also both large and small charities capable of funding different research projects for particular diseases. Smaller charities are particularly helpful in funding short, pump-priming projects as a first step in a research program.

3.2 PLANNING STAGE 2: CHOOSING A STRATEGY

Having considered all of the above, we are ready to put together a plan to study the genetics of a complex disease. The choices we make at this stage are dependent on the answers to the questions above. In terms of overall strategy, there are two basic methods—each with two variations.

Two basic strategies for identifying risk alleles in complex disease

These two strategies—linkage and **association** analysis—are discussed throughout this book. Linkage has traditionally been seen as the gold standard for genetic studies, but is dependent on the availability of multicase families. Even in complex diseases with significant numbers of such families, the majority of cases are non-familial (sporadic) cases. Therefore, if our primary strategy is to collect and study families through linkage analysis we would hope to consider the sporadic cases at a later time. In the absence of families, our primary strategy would be association analysis. In this instance non-familial cases would be analyzed. Samples from affected family members may be collected, but only one family member for each such family (usually the first diagnosed case, referred to as the **index case**) would be used in the final analysis to avoid introducing bias.

In terms of the history of genetic studies in complex disease there are two main periods: pre- and post-genome

Prior to the publication of the draft Human Genome Map (HGM) in 2001, studies were either based on linkage or association analysis just as at present; however, the precise location of the target genes was mostly unknown.

Linkage analysis in the pre-genome era

Linkage analysis identifies chromosomal regions that **co-segregate** with the disease. Linkage analysis establishes the likelihood of a physical link between an allele at a marker locus and a disease. The degree of linkage is expressed as the log (to base 10) of odds for linkage between the marker and the disease, and is referred to as the LOD score. The LOD score is discussed in detail in Chapter 5. Linkage has been traditionally used in **Mendelian monogenic** disease and has been very successful. Penetrance is a factor in linkage analysis. Linkage analysis is more likely to be successful if penetrance is high. Even in complex disease, where there are occasional familial cases, linkage can be useful provided penetrance is high. Linkage analysis does not identify a specific gene or group of genes or establish causal links; it simply establishes the statistical likelihood of a link with a marker or allele in a region of a chromosome. As the location of most human genes was unknown prior to 2001, researchers would have to best-guess gene locations based on whatever information they had available at the time. There are positional and positional-independent routes to identify disease genes. Positional studies use **positional cloning** to identify a disease gene based on its approximate chromosomal location. There are several steps in this process. First, a candidate region needs to be identified, **clones** from the region are produced, and all of the genes in the region identified and prioritized for mutation screening (**Figure 3.7**). This was a considerable undertaking prior to the publication of HGM, and positional

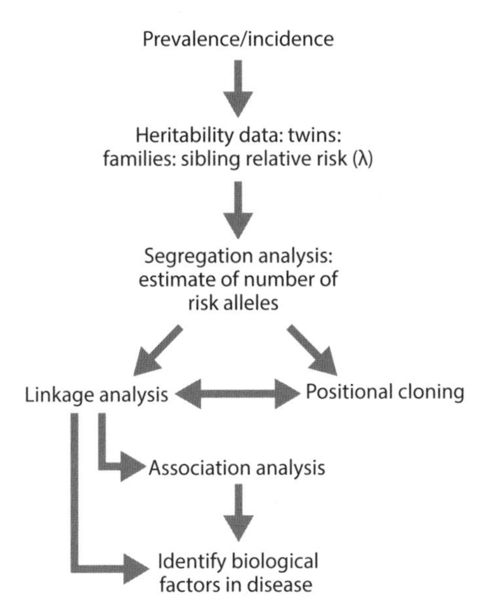

Figure 3.7: Logical sequence for disease gene identification prior to the HGM. This figure illustrates the normal sequence of events proposed for studies designed to identify disease genes prior to the publication of the HGM. However this sequence is more suitable for Mendelian disease than for complex diseases. Segregation analysis has rarely been performed in complex diseases. Indeed, many complex diseases have never even been subject to linkage analysis. The sequence that suggests association analysis should only be performed after linkage analysis is not practical in many complex diseases. This "logical" sequence suits the rare, highly penetrant near-Mendelian complex diseases, but does not work with diseases which are mostly sporadic or where transmission is horizontal, e.g. many infectious diseases. Only when there are sufficient family cases can linkage be performed. Nevertheless, this approach has been used in a number of diseases at least once (e.g. type 1 diabetes). This sequence is especially useful where there are a reasonable number of cases in the collection (i.e. the disease or trait is common) and when possible it remains a logical sequence.

cloning was mostly applied to Mendelian disease and rarely (if at all) used in complex disease. Today, in the post-genome period, the precise location of most human genes is known and the first steps in this method (cloning) have therefore become redundant. In addition, since the availability of the SNP Map (http://www.ncbi.nlm.nih.gov/SNP/) and the **HapMap** (http://hapmap.ncbi.nlm.nih.gov/) the majority of the mutations and polymorphisms are also known, so the whole process of testing candidate genes for relationships with disease susceptibility can be short circuited.

The problem with linkage analysis is that it requires informative families. Unfortunately, many complex disease cases are sporadic cases, without a family history, and linkage would not or could not be applied, and the number of familial cases available for linkage analysis would be insufficient to ensure adequate **statistical power** in any analysis undertaken. Even where there are significant numbers of family members, such as in cardiovascular disease, diabetes, and depression, cases do not cluster in a Mendelian fashion. This is partly because most common diseases are complex and phenotype is determined by the interaction of several factors, both genetic and environmental. Individual genetic variations often have relatively small effects on the disease risk and are more likely to be detected in large-scale case control association studies.

Association analysis in the pre-genome era

Association analysis is perhaps the most important method used to identify risk alleles in complex disease. Case control association analysis tests for correlation between the inheritance of one or more genetic markers or alleles and a disease. However, unlike linkage analysis, association analysis does not establish a physical link. This approach compares the two groups from the same population. These groups are usually (though not always) cases (patients) and healthy controls. The history of association analysis is one of mixed fortunes and to a conventional geneticist who deals with Mendelian diseases association analysis has had a very bad reputation. However, this bad reputation is not entirely deserved, as discussed below.

Some authors refer to the common disease common variant hypothesis (CDCVH) in considering association studies and even go so far as to suggest it is the foundation for association studies, especially GWAS. It may be useful to consider this idea before embarking on an association study. The hypothesis predicts that common disease-causing alleles will be found in all human populations with specific diseases.

Studies starting in the 1960s and 1970s began to identify both genetic linkage and genetic associations in complex disease. One of the major areas for association analysis was the human major histocompatibility complex (MHC) on chromosome 6p21.3, with the first association reported in 1967. The clinical need to HLA-match patients requiring either bone marrow or kidney transplants revealed an abundance of certain HLA phenotypes in patient groups with specific presenting diagnoses. Many of these genetic associations, though not all, have stood the test of time. We should note here that it is important to acknowledge that there have been significant developments in the methods used for HLA typing with a major shift from what was low (slow)-resolution phenotyping to very high (rapid)-resolution genotyping (see Chapter 6). It is also fair to say that the majority of these genetic associations were poorly understood at the time and to some extent geneticists are still cautious when considering the human MHC.

Expectations in the earlier association studies were quite different from those of today. A study reporting an association with an increased risk [measured as odds ratio (OR) or relative risk] of less than 3 would be unusual. Almost all studies expected large risks and studies were often small. Many of the statistical principles considered in Chapter 5 were not considered in those early studies. Consequently, association analysis developed the bad reputation referred to above, but some good things have arisen from this bad reputation. The high number of inconsistencies between studies created a platform from which statisticians were able to critically review the past, and launch new, better guidelines for research in this area.

There are a few exceptions to the statement above, for example large OR values above 2 or 3 were reported for *PTPN22* in type 2 diabetes and rheumatoid arthritis, and *CARD15* in Crohn's disease. Many of the early associations in pharmacogenetics also reported high OR values (see Chapter 8). Other than these, early studies outside the MHC rarely found strong associations with OR values of 3 or more. As a consequence, these studies were more susceptible to finding false positives that proved difficult to replicate. Of course with small studies many weak (but important) associations were also missed and for every false positive it is almost certain there were several false negatives.

It is important when considering data to remember that OR values are rough calculations at best and should always be considered within the range identified by the **confidence intervals (CIs)**. In addition, some of the early associations remain controversial even today despite multiple studies and meta-analysis.

Transmission disequilibrium testing and family-based association testing

The **transmission disequilibrium test (TDT)** or **family-based association test (FBAT)** are designed to detect the presence of a genetic association (albeit through linkage) between a genetic marker and a disease or disease subgroup. The TDT was originally developed for use in family-based studies. The test is based on the probability of equal transmission of an allele or genotype from parents to affected and unaffected offspring. If the marker allele in the group being tested is equally transmitted, then there is no disequilibrium (that is there is equilibrium); if they are not equally transmitted, then there is transmission disequilibrium. If sufficient families are tested, then the association can be statistically proven. Consider a single gene with alleles *A* and *B*. Using heterozygous parents in a TDT test, the genotype distribution among tested offspring should be 1:2:1, i.e. we would expect 25% to be homozygous genotype *AA*, 50% to be heterozygous genotype *AB*, and 25% to be homozygous genotype *BB*, if the transmission is in equilibrium. Otherwise there is transmission disequilibrium suggesting an association with one or other allele (*A* or *B*) with the disease.

Usually the sample consists of a set of families each with four individuals, i.e. the affected and unaffected offspring and the two parents, though in some studies trios can be used (affected cases and parents only) and occasionally families with more than one affected case can also be used. A variation on the original TDT is the so-called FBAT. Both TDT and FBAT are family-based association tests and can be used in most complex diseases. However, for many diseases they are rarely used because obtaining family material is difficult for the reasons discussed above (**Figure 3.8**).

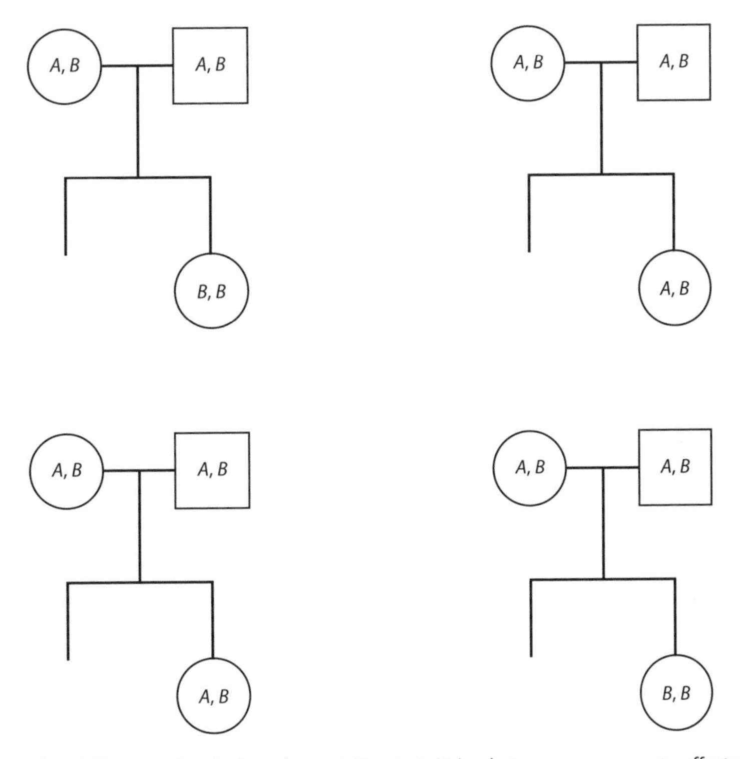

Figure 3.8: TDT. The TDT test is a family-based association test. Using heterozygous parents, affected and unaffected siblings are tested to see if the alleles are transmitted equally as expected. Unaffected siblings are not always required for TDT. This illustration shows four families with two potential alleles *A* and *B*. All of the parents are heterozygous *AB*. If the alleles are equally transmitted, then the frequency of the genotypes in the affected offspring should be 1:2:1 in the order *AA*, *AB*, and *BB*, respectively, or alternatively each allele should be found at a frequency of 50%. In this case, two of the offspring have the *BB* genotype and two are heterozygous *AB* positive (the frequency of the *B* allele is 75% compared with 25% for the *A* allele). This suggests disequilibrium with a preponderance of the *B* allele. In a larger cohort of cases it would be possible to analyze the data and be more precise about the statistical significance of this observation.

Choosing study strategies in the post-genome era

Linkage is and can be used when families are available. However, even when the primary strategy is to collect and study families, a good plan would include collection of sporadic cases for consideration at a later time. It must be stressed that linkage analysis is rare. In the absence of families our primary strategy would be association analysis. In this instance, non-familial cases would be collected with one affected from each family if families are present in the case group.

One justification for studying families when the majority of cases are sporadic is that families may have a stronger genetic predisposition than sporadic cases and hence identifying risk alleles may be easier in families. However, there is no guarantee that the same risk alleles will be found in familial cases and sporadic cases, but it is possible that the risk alleles for familial disease will identify pathways in which there is polymorphism in other genes that explains genetic susceptibility and resistance to sporadic disease. An example of such a difference is early- and late-onset Alzheimer's disease (EOAD and LOAD; see Chapter 4), where EOAD is associated with genetic variation in the *APP*, *PSEN1*, and *PSEN2* genes, and LOAD is associated with genetic variation in the *APOE* gene (among others).

In the past, linkage was favored because early studies of complex disease based on association analysis were given a bad press largely due to small numbers tested and the irreproducible results. This was addressed extensively in the late 1990s and at the start of the new millennium. This subject is addressed in published reviews by Cardon & Bell (2001) and by Colhoun et al. (2003). Association analysis is currently the strategy of choice in genetically complex disease. In particular, association studies are more appropriate for infectious diseases where transmission is horizontal and not vertical.

Each of these two strategies has a substrategy

Having decided which of the two main strategies to apply (i.e. linkage or association or both), it is important to gather together the information to determine our plan. Will we use a genome-wide approach or will we be selective? If we decide to be selective, what will we investigate? The options are:

- The whole genome.
- Genes within one or more chromosomal regions.
- Genes within one or more extended haplotypes.
- Genes identified within one or more specific biological pathways.
- A single candidate gene.

The choice of which option to take is dependent on our knowledge of previous studies, current studies and the disease pathology.

Genome-wide studies

Genome-wide studies can be either linkage (GWLS) or association (GWAS) depending on the availability of families. Both are mostly post-genome phenomena. The Human Genome Mapping Project (HGMP) followed by the SNP Map and HapMap has opened up the genome to extensive high-resolution study. It is now possible to study millions of genetic markers, mostly single nucleotide polymorphisms (SNPs) or tagged SNPs as they are called, but also copy number variants (CNVs), across the whole human genome (**Figure 3.9**). It is also possible to use databases with haplotype data to identify common linkage patterns between tagged SNPs and **impute** missing data from these databases, thereby extending the

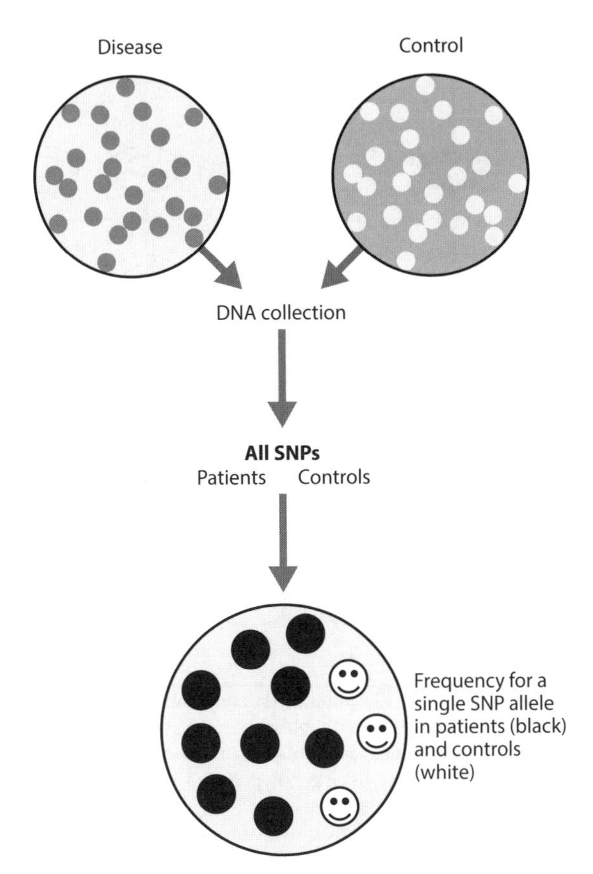

Figure 3.9: GWAS. The illustration shows two populations of diseased patients and healthy controls. The DNA samples for each individual (shown as small dots) in each group are collected. The frequency of a series of SNPs (around 500,000 to 1 million plus in some studies) across the whole genome is tested in each group on a commercial chip or platform. The frequency data for one SNP allele in each group is shown in the gray sphere at the bottom of the figure. Data for each SNP are subject to statistical testing to determine whether there are any statistically significant differences between them. The predominance of black dots in this illustration suggests an association with this allele.

analysis within the target haplotypes to include more data (see Chapters 5 and 12). Impute can be used provided there are sufficient fully genotyped samples with complete haplotypes in a database. One can impute the missing genotypes for a set of study samples based on the expected pattern of linkage. The method exploits linkage disequilibrium. Essentially, it involves replacing missing data with substituted values. This means that any bias generated by discarding data from samples that are not fully genotyped is avoided.

The level of resolution from a study can be significantly increased using impute software to identify SNPs through known linkage disequilibrium, so that two or even four times as many SNP genotypes can be assigned than were actually tested (**Figure 3.10**). Impute requires complex computer software such as **PLINK** and access to good-quality databases such as that generated by the 1000 Genomes Project. The greater the number of whole-genome sequences put into public access systems, the greater the quality of the imputed data will be. Of course the success of this method depends on the degree of linkage disequilibrium between tagged SNPs on the target haplotypes. The beauty of genome-wide studies, at least in the initial stages, is that they do not require a hypothesis. This allows us

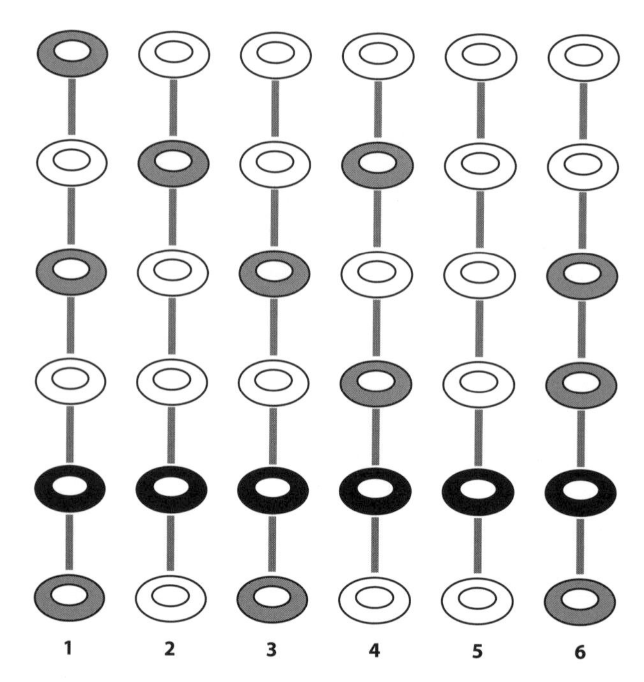

Figure 3.10: Imputation analysis. The figure illustrates six different haplotypes in six individuals numbered 1–6 for a combination of six biallelic genes illustrated as double rings. Each gene is polymorphic; five of the genes have been genotyped. There are two possibilities for all six genes (i.e. two alleles). The alleles are illustrated by gray shading or no shading in those that have been genotyped. The gene illustrated in black has not been genotyped. Imputation analysis can enable the genotype of the black gene on this haplotype to be estimated based on the known pattern of linkage disequilibrium for the surrounding genes for which this data has been obtained. Therefore, if we know that individuals with the haplotype which runs (top to bottom) gray, white, gray, white always carry the gray allele at the fifth position, then we can assign this allele to individual 1 at position 5. If we were to know that all carriers of white at position 4 and gray at position 6 also carry gray at position 5, we can also impute the gray allele for these individuals even though they have not been genotyped for this position. This imputed data can be included in the analysis. As linkage disequilibrium is so strong in certain areas of the human genome, imputation analysis from initial GWAS data can massively extend the studies. Thus, an initial GWAS study based on 300,000–400,000 SNPs can generate data for up to 1 million or more markers.

look more freely at the genome and identify hitherto unthinkable genetic relationships. This can lead to favored hypotheses being rejected or being accepted, or in extreme circumstances it can turn some favored hypotheses on their heads. In all cases it is likely to identify new risk alleles.

Selecting this approach in an association study would depend very heavily on previous studies and sample size (see below). In the absence of any data at all, provided sufficient candidates could be recruited, this strategy would be a good first choice. Even when numbers are small, GWAS can be performed, but it is essential to understand the limits of using these studies on small numbers. The reason GWAS are based on large numbers is that they are required to have sufficient statistical power (see Chapter 5) to detect weak genetic effects with a high degree of confidence. If we look at small numbers we may only be able to detect strong associations with a reasonable degree of confidence.

If we apply a non-hypothesis-constrained approach we cannot claim to be testing a specific hypothesis, we are simply looking for associations (or linkage) with tagged SNPs or other genetic variants that mark areas of statistical significance and therefore genetic interest. The great thing about this approach is that it is **hypothesis generating**. Thus, from the ashes of hypothesis-free GWAS a mass of data arise that can identify regions around genes that contribute to biological systems that may be of specific importance in understanding the pathology of the disease under study. This is very much a first phase and a first study would require replication to validate any findings.

Studying a selected chromosomal region

This is a second-phase process and would only be undertaken where prior studies had identified a region or regions of interest. GWAS is likely to generate interest in several chromosomal regions. The tagged SNPs may indicate association with a specific gene, but are more likely to identify associations across an area for further study. In this case, high-resolution analysis using selected tagged SNPs will identify peaks of association in or around genes with the region. However, just like initial GWAS, the findings of this type of study should be replicated in a second cohort and the results may not identify a specific gene, but may identify a group of genes or a haplotype. Currently, the application of GWAS to a single chromosome is not an option, as most commercial SNP chips are not set up to do this and the cost is too great to do a single study.

Studying extended haplotypes

Haplotypes in genetic terms are groups of alleles at different loci that are inherited as a unit. Extended haplotypes are particularly lengthy sections of DNA where multiple alleles are inherited in specific groups. One particular feature of haplotypes is that the alleles that compose them are found to be inherited together more often than expected by chance, i.e. they are in linkage disequilibrium. As a result of this, extended haplotypes can be both good and bad news. Linkage disequilibrium means that alleles of two or more genes are found together more often than they should be if there was independent (equal) segregation. These genetic packages may contain several important genes all related to the same pathway. For example, the human MHC contains many of the key genes involved in T cell immunity (among other functions). When polymorphism is widespread within a haplotype, as it is within the MHC, it can be very difficult to pinpoint the precise susceptibility allele, but it is not impossible if the data set is large enough. Combining haplotype data with GWAS data using **imputation** to assign HLA haplotypes has recently generated some very interesting results in the autoimmune liver disease primary biliary cirrhosis.

Investigating pathways

Proteins almost always work in groups or pathways. Systems biology is a new approach to biological processes that concentrates on the study of interaction of pathways. Those who study the genetics of complex disease can learn a lot from considering the ways in which pathways interact to form systems. If we take this approach to studying a complex disease, we would need to have a reasonable amount of prior data or a plausible hypothesis. This is likely to be a phase 2 or phase 3 investigation based on prior knowledge. Evidence from recent studies in Crohn's disease indicates the value of considering processes compared with considering individual genes in complex disease. The identification of *nod2* (*CARD15*) and the autophagy gene (*ATG16L1*) both indicate links to pathways associated with immune tolerance to Gram-positive commensal bacteria in the gut. Most studies of Crohn's disease now look for links to these and related pathways.

Investigating a single gene: candidate gene studies

In the past, most association studies considered variation at only one or two candidate gene loci. A few considered haplotypes with specific regions based on selected candidates. This approach is hypothesis driven and, as such, is limited to searching for associations at specified loci. Gene loci were selected where there was at least some potential functional relationship between the product of the gene under investigation and the disease. The problem with this candidate gene approach is that our knowledge of the biological relevance of most human genes is limited by our lack of understanding of human biology and disease pathology. If we take a narrow view and study only one or two genes, we will almost certainly miss a great deal. If we cast our nets wider, perhaps studying genes in whole pathways and systems, then there is more chance of finding significant associations.

Investigation of a single gene is less likely in the present era than it would have been 20 years ago. However, where prior studies indicate a strong relationship with a particular gene or region close to a gene, then it may be relevant to perform high-resolution genotyping of that gene to identify polymorphisms (alleles) associated with susceptibility to the disease under study. It may occasionally be prudent to investigate a region or haplotype one gene at a time. This can be useful if resources are limited and it can also be helpful to produce data for initial funding applications. Often researchers will use small funds to pump-prime preliminary studies to gain data for larger research funding applications.

Candidate gene studies have rarely been very successful in identifying associations with diseases. Most associations that were initially positive have failed to been confirmed. The reason for this is in the selection of the candidate, which depends on the validity of the biological hypothesis being tested, which in turn depends upon our knowledge of the disease pathogenesis. Our knowledge of the early pathogenesis for most diseases is relatively poor; patients present late in the disease process and identifying biological features of the early disease is impossible. However, the good news for geneticists is that while pathology changes, the genome does not. The future of medicine lies in early identification and prevention of disease. To achieve this goal it is necessary to identify risk alleles and use genetic studies to inform the debate on disease pathogenesis. In 2015, this almost always means throwing the net as wide as possible, i.e. genome-wide, at least in phase 1 of the study and coming back to regions, pathways, haplotypes, and candidate genes later in the study as our knowledge base progresses and we focus on the real disease-causing polymorphisms (**Table 3.3**).

Table 3.3 Current organization for genetic studies in complex disease including options for genome-wide linkage and genome-wide association analysis.

Linkage	Association
Collect families	Collect cases
Genome-wide studies (GWLS)	Genome-wide studies (GWAS)
Replication	Replication
Chromosomal regions High resolution	Chromosomal regions High resolution
Replication	Replication
Extended haplotype(s) High resolution	Extended haplotype(s) High resolution
Target gene	Target gene
Biological Function	Biological Function
Pathways and Systems	Pathways and Systems

3.3 GOOD AND BAD PRACTICE

It is important in planning to consider what constitutes good and bad practice. The published literature on genetics of complex disease is peppered with learned text on the problems we can encounter and it is also peppered with contradictory studies. Much of the critique is aimed at association studies. However, things have changed and association studies have become the method of choice. Why has there been such a change of heart?

The answer lies in the field of statistics, or statistical genetics, but what are the problems?

- Sample size
- Case selection
- Controls selection
- Sampling errors
- Statistical analysis
- SNP chip selection
- Publication bias

Accurately identifying true disease susceptibility alleles in GWAS (and other association studies) is dependent on sample size

This is a very basic idea and is common sense. Studying larger groups produces data for which there is a greater degree of **statistical confidence**. There is no doubt about the truth of this statement. The problem arises when we look at the practical issues of recruiting large numbers. Single centers may not have access to large numbers of patients. Indeed, if the incidence of the disease is low, then it may take many years to recruit reasonable numbers for a study. This does not excuse the use of small numbers, but it does to some extent explain why small numbers have been used.

Sample size is the biggest concern in all studies because the greater the sample size, the greater the statistical power of the study. For a standard GWAS, where between 500,000 and 1 million SNPs are tested, a sample of approximately 2000 cases and 2000 controls are considered reasonable, though more is considered to be even better. If fewer samples are used then confirmation of any associations in a second set is advised (replication, see below). The 2007 Wellcome Trust Case Control Consortium (WTCCC1) study provides a good example of how to design a GWAS. In planning any study it is important to consider the limits of the study. The numbers of cases and controls that are required in an association study to detect risk alleles, depends on the frequency of allele and the size of effect that the allele is expected or likely to have. This is considered in more detail in Chapter 5. In order to determine the appropriate sample size, we can perform **power calculations**.

Power calculations

In order to determine the appropriate size for a study, power calculations are recommended during the planning stages. This can be done quite simply using freely available computer software. The calculation in its most simple format determines the level of confidence that can be assigned to a study based on a predicted relative risk (or OR) and given sample size. Two variables are critical in this calculation: the size of the patient population being studied and the expected impact of the allele (or alleles) in the test. Where the expected

OR (a measure of the genetic impact) is high, then a smaller sample size can be used with a reasonable degree of statistical confidence. When the expected OR is low, the sample size needs to be greater to achieve the same level of statistical confidence. Power calculations are particularly important in relation to negative studies that fail to find any association. It has been suggested that the reason many studies have failed to replicate original results is due to the lack of statistical power in the replication cohort. This is true in many studies, but not in all. Once someone else has published their work it can be very difficult to publish results if they contradict the original findings. Indeed, there is likely to be a publication bias in favor of the initial studies. **False negatives** do occur, but so do **false positives**. Looking at the early history of association studies we can see that the most frequently replicated findings are those where the level of association is high, with an OR greater than 3.

Case selection can introduce bias into a study

Selection of both cases and controls can introduce bias into a study. To avoid **recruitment bias**, case selection needs to be considered at the beginning of the study during the planning process. Criteria for inclusion and exclusion should be as precise as possible. In syndromic diseases, such as inflammatory bowel disease (IBD), the subgroups should be identified clearly. The diagnostic criteria must be set out in order to create a homogeneous study population. Cases where the diagnosis is in doubt should be excluded to reduce the likelihood of bias. Post hoc changes should not be made. Where there are multiple cases in a family, it is standard practice to consider only the index case. Healthy relatives from patient's families should also be excluded. The methods for detecting sample relatedness are discussed in Chapter 5. Where studies are being designed for replication of previous research, the plan should if possible meet the same criteria and employ the same clinical parameters.

It is important to consider whether we are studying a disease, a syndrome, or a trait within a disease subgroup

We could consider allergy to be a syndrome if we are not specific about the allergen. IBD is a syndrome even though it carries the label disease. Some disorders are syndromes. A syndrome can be best described as a collection of diseases with similar signs and symptoms, but in this case different and distinct pathogenesis. The *Oxford English Dictionary* defines a syndrome as a disease with concurrent symptoms. A syndrome usually involves more than one disease, e.g. IBD, which can describe Crohn's disease, ulcerative colitis, or a less severe and less well defined disorder simply referred to as "inflammatory bowel disease." There is some overlap between different forms of IBD (Crohn's disease and ulcerative colitis), but the overlap is not complete. This introduces a greater variety into the sample being tested. At the other end of the scale we may want to test patients in subgroups looking at disease severity, progression, and response to treatment. These are all qualitative traits, whereas clinical expression of a disease is a quantitative trait. Some consideration of these different parameters will be given in later chapters of the book. At this stage it is important to state clearly that we are not confined to quantitative traits, we can also consider qualitative traits.

Selection of appropriate controls is equally important in any study

Recruitment bias discussed above can also be a problem with controls. Selection of appropriate controls for comparison is a key requirement of association studies. However, few studies look for completely matched controls. Complete matching is not essential in most cases, but it is desirable and should be seen as a gold standard in good practice.

Healthy controls

Most studies refer to their control group as healthy controls. This is at best a vague description and is mostly incorrect. What are healthy controls? In general terms, the authors of such studies mean individuals who have not been diagnosed as positive for the disease under study. It is arguable that comparator populations should contain the normal number of cases of other diseases that are not included in the study. Exclusion of all diseases could create a genetic bias in the controls. Therefore a good comparator population will not all be healthy.

Matched controls

Matched implies a degree of similarity between the patient group and the control group. Most studies use racial and ethnically matched controls. This makes sense because there are racial and ethnic differences in the distribution of different gene polymorphisms. In addition, there are racial and ethnic differences in environmental factors, examples include smoking, diet, and alcohol consumption, that may impact on disease susceptibility or severity. This is the definition of matched for most studies and no further matching applies. However, further matching may be appropriate where there are subpopulations (**population stratification**) within the study group, or where there is a preponderance of male or females within the disease group. For example, many forms of autoimmune disease are more common in women than in men and we could argue that this difference should be reflected in the control population. Body mass index may also be a factor. Similar caveats apply when selecting controls for studies of diseases in children or diseases in the elderly population. However, though matching controls and taking the age of onset into account is desirable, it is not always practical, particularly in children.

Some studies, such as the WTCCC1 study, which is referred to throughout this book, used a common control set. However, even this can be criticized because there is no consideration of latent onset of disease in the control cohort. Indeed, we may question the whole idea of a healthy control. It is quite common to find studies that have used a workplace population as controls (**Table 3.4**). In these circumstances this population may be very different from the general population, reflecting bias in terms of the socioeconomic group from which the controls come. Finally, on some occasions there will be no healthy controls; instead, there may be patient subgroups where each subgroup has a different disease phenotype.

Number of controls

General advice is that the number of controls should be at least the same as the number of patients. If possible, the number of controls may exceed the number of patients provided

Table 3.4 Characteristics of matched controls.

Health	Negative for the disease under study at time of recruitment; cannot control for latent onset
Race/ethnicity	Important to reduce effects of population stratification
Age of onset	Age of onset is especially important if the illness is likely to have a high rate of mortality and is common
Sex	Male/female differences are common in many common diseases
Socioeconomic factors	In some populations this is more important than others and populations are stratified on this basis

the number is not too excessive. As a rough guide, a control sample of up to but not exceeding five times the patient sample size is considered acceptable. The upper limit is important because the greater this number becomes, the greater the possibility of introducing unmatched controls and bias into the control sample.

Errors in the laboratory and in sample handling can also introduce bias into a study

It is important to ensure that no handling errors occur during sample collection and processing. There are ways of checking for errors through haplotype analysis. Where there is an unexpectedly high level of genetic mixing, samples may be excluded and replaced with fresh samples. Sample checking is performed as part of standard quality assessment in some studies.

Statistical analysis is the key in any study of complex disease

Statistics are considered separately in Chapter 5. However, there is one important concept that relates to study design to report here. If we test multiple markers, we must make some allowance for this in the analysis. In the past we have applied **Bonferroni's correction** by multiplying the probability value by the number of tests performed. This is a crude form of correction that reduces the likelihood of individuals reporting weak false-positive associations. However, this correction is much criticized and may cause many weak associations to be overlooked. As an alternative to this correction, the level at which a probability value is accepted as significant may be reduced, taking into account the number of variables tested. Even for students with no interest in statistics it is essential that there is an awareness of this practice.

SNP chip selection is an important factor to consider in study design

In science, as everywhere, new methods and ideas are greeted with great enthusiasm. GWAS has been, and still is, the new boy in town, but we need to be cautious because GWAS have limitations. All of the above apply to all methods discussed so far, but SNP chip selection applies to GWAS and subsequent studies only. The selection of the genotyping arrays is very important. There are a large number of different commercially available genotyping arrays. It is essential to ensure that the array used is appropriate for the disease being studied. It is also helpful when comparing studies that have used the same or similar arrays. This makes it possible to confirm initial findings from other research groups and it facilitates **meta-analysis**. However, evolving technologies such as direct sequencing and cheaper SNP chips means that this is no longer the major issue in these studies. This is discussed further in Chapter 12.

Allele frequency

When selecting arrays it is important to consider the frequency of the alleles under test. Alleles that are present at low frequencies (minor alleles) are not informative in association analysis and it is important to filter out any markers that are unlikely to be informative. The reason for this is that the statistical power of a study depends partly on the allele frequency and is lower when allele frequencies are low. Previously, the threshold chosen to exclude SNPs and other markers was 5%, but in 2015 as much larger sample collections have become available, SNP frequencies of 1% or lower are being included in studies. This addresses some of the concerns that have been raised about excluding rare polymorphisms, which suggested some associations may be missed and therefore exclusion of rare SNPs may be counter-productive.

Unfortunately publication bias does occur

All research work needs to be published and pass through peer review. The process can be quite punishing and often we find our favorite manuscript bouncing around journals while, for no apparent reason, manuscripts of lesser quality (or so we think) fly through the process. However, time heals all and over time as our list of publications grows we find there is a leveling out process. It is important to be realistic about our studies and where possible publication should be considered at the outset of a study and not post hoc. In exceptional circumstances results that exceed the expectations of the research group may be sent to a higher-level journal than initially planned, but it is important to be realistic to avoid disappointment.

Publication bias applies less often to papers with positive associations and more often to those with negative associations. Essentially, failure to find statistically significant associations can mean that a manuscript will remain unpublished for some time. This then means there is a bias towards papers with more positive associations. However, the production of papers with positive results will eventually create openings for papers with negative results.

The problem with publication bias where it exists is that it induces the authors to over-analyze data until a statistically positive observation is found. Good practice is to set the aims of a study at the beginning with clear indications of the planned analysis (**Box 3.2**). Post hoc subgroup analysis is considered bad practice and should be avoided. There has

BOX 3.2 BASIC GUIDELINES FOR GOOD PRACTICE IN ASSOCIATION STUDIES

- Replication of associations with alleles should be normal practice prior to publication.
- Case selection should be based on well-defined criteria and unusual or uncertain cases should be excluded from the main analysis so as to reduce bias. Variation between study groups must be taken into account when performing a meta-analysis.
- The appropriate SNP chip should be selected in GWAS.
- The minimum threshold set for tagged SNPs and other markers should be based on the number of SNPs being tested and size of the study population. Generally, early GWAS used 5% as the lowest level for the least frequent variant, but threshold limits of 1% or even lower are being accepted more recently due to the availability of a greater number of SNPs (many with low frequencies) and larger sample sizes.
- Control groups should be selected from an appropriate population.
- Consideration should be given to race, ethnicity, and other factors.
- Close matching of controls to the disease group, with age of onset and sex matching if possible.
- Sample size should be the maximum achievable within reasonable boundaries. Consideration of collaboration to achieve appropriate numbers should be encouraged. Sample sizes should reflect any expectation of the likely size of the anticipated genetic effect.
- The statistical analysis applied to the study should be clearly represented in the manuscript. This should include initial power calculations performed prior to the study.

BOX 3.2 BASIC GUIDELINES FOR GOOD PRACTICE IN ASSOCIATION STUDIES *(Continued)*

Where there are multiple alleles tested, appropriate statistical correction (thresholds) must be applied and the corrected probabilities (P_c) shown. All of the tests performed should be included in the consideration of the appropriate correction factor. Confidence intervals (CIs) should be shown as standard.

- If possible, closely linked loci should be investigated to provide haplotype data and define regions of linkage disequilibrium. Linkage disequilibrium is a very useful tool in confirming genotypes and haplotypes for genes with multiple polymorphisms, such as within the MHC.

Guidance on strategy is not included here as a specific point. The choice we make when planning these studies is made up of a cocktail of information. The plan (phase 2) is also based on that information, but guided by some of the considerations in this box.

been considerable discussion of publication bias and some effort to encourage journals to be more accepting of good quality studies that carry negative data. In addition, online publication of such studies is becoming increasingly popular, creating a pathway through which this bias can be reduced.

Replication in an independent sample is crucial for all association studies, especially GWAS

Replication or repeating a study is primarily used to validate the initial findings and it is essential in all association studies, but particularly in GWAS. However, because GWAS have very stringent statistical parameters for acceptance of identified associations, some studies have used the replication cohort as a tool to allow them to apply a lower cutoff point for statistical significance (**significance threshold**) in their first round. Using this plan, the study group is divided into two: the first group is genotyped for all of the SNPs and markers, and the second group is genotyped only for those SNPs and markers that are found to be significantly associated with the disease in the first round. By applying a less stringent cutoff for significance in the first round, it is hoped to reduce the likelihood that weak genetic associations will be missed (fewer false negatives). In practice, this may mean testing more SNPs and markers in the second round than would be tested if a higher significance threshold were set in the first round, but it does ensure that significant associations are not overlooked.

This has become one model for GWAS involving multiple rounds of analysis. **Figure 3.11** shows a study design that employs this multistage approach for replication. In the first phase, the entire genome is scanned for potential associations using approximately 500,000 SNPs. Two alternatives are offered for the second phase. Replication of initial findings at a significance threshold of $P < 10^{-7}$ or replication using a lower significance threshold in the order of $P < 10^{-3}$ to 10^{-5} focused on the identified alleles only. Whichever is used replication of the initial findings is essential to validate them. In many cases a third-step validation may be applied, focusing on either the same geographic or other geographic populations. Follow-up studies and fine mapping of associations will also be performed on specific chromosomal regions, haplotypes, and genes with potential causal links with the disease. Causality can only be truly proven once functional studies linking phenotype to genotype have been performed to consider the relationship between phenotypes and specific systems that may explain the disease.

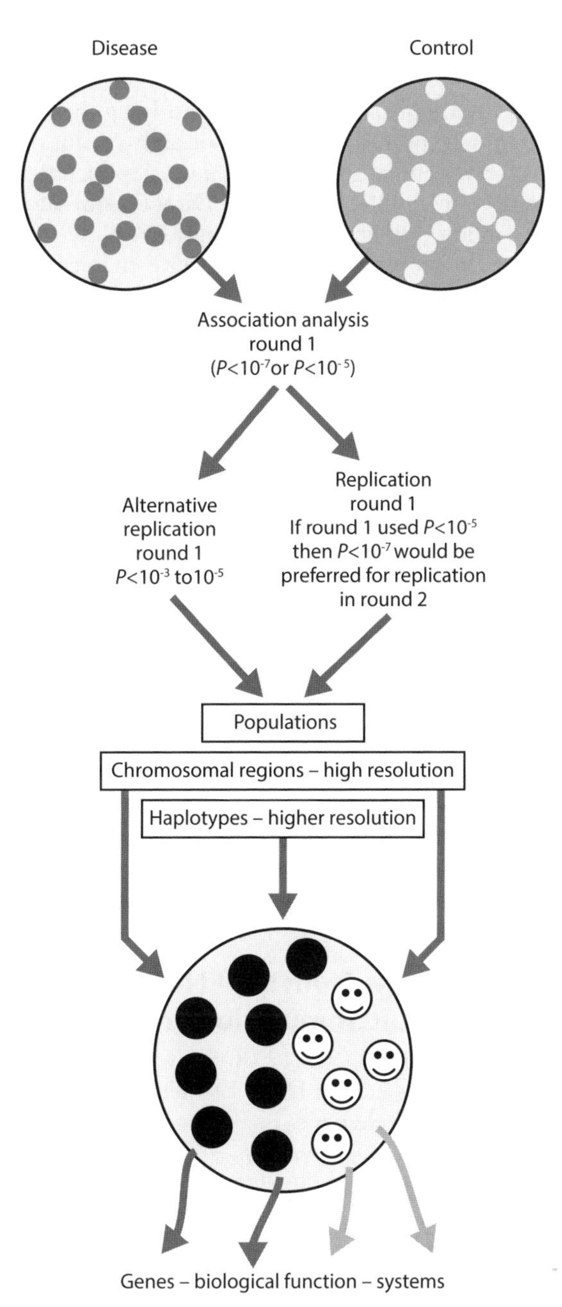

Figure 3.11: Schematic illustration of the study design of a GWAS and the follow-up studies, including fine mapping and functional investigations. The study employs a multistage approach for replication. First, the entire genome is scanned for potential associations using approximately 500,000 SNPs and selected significance thresholds ($P < 10^{-5}$ or $P < 10^{-7}$). Then either replication at a significance threshold of $P < 10^{-7}$ or sometimes between $P < 10^{-3}$ and $P < 10^{-5}$ can be applied. After re-testing, a second case control cohort focusing on the alleles identified in the first analysis provides confirmation that the associations are valid or not. In most cases, third-step validation is then applied focusing on other populations if possible. Follow-up studies include fine mapping, focusing on chromosomal regions, haplotypes, and genes with potential causal links with the disease. The final step once genes and regions have been identified is to perform functional studies linking phenotype to genotype, and to determine the relationships between phenotypes and specific systems that may explain the disease.

3.4 NEW TECHNOLOGIES AND THE FUTURE

GWAS is currently the primary phase 1 method for studying the genetics of any complex disease. Even with the promise of multiple rounds of high-resolution genotyping we cannot carry on doing this forever and at some point we need to focus on the individual genes and haplotypes to identify the specific risk alleles. **Direct sequencing** is rapidly catching up and will soon be the method of choice, replacing GWAS (see Chapter 12). The technology is not entirely new, but recent developments have enabled much longer sequences to be determined with a higher degree of accuracy and at significantly lower cost. As a consequence, it may be easier to abandon the use of tagged SNPs and high-resolution genotyping for direct sequencing of identified candidate genes. More laboratories will be encouraged to use this method as costs fall.

The technological advances of the past decade have had a major impact on research into the genetics of complex disease and the rate of change is going to increase

New information is constantly appearing. Recent advances in genotyping technology have made it possible to genotype millions of SNPs in a high-throughput fashion with lower and lower costs per genotype. An increasing number of databases and information have been released on the web. In 2012 the 1000 Genomes Project published information stating that they had identified and genotyped 38 million SNPs. HapMap (http://hapmap.ncbi.nlm.nih.gov/) describes the haplotype map of the human genome based on DNA sampled from eight different populations. HapMap phase 3 (2012) described the human genetic variants in 1301 DNA samples. The project included both SNPs as well as large differences from the loss, gain, or duplication of regions (i.e. CNVs). These technological advances will add even more power to GWAS in complex disease, but this is only one part of the story.

New developments will come from the ENCODE project, and will also involve more epigenetics and imputation analysis

The ENCODE project was published in 2012 (http://www.genome.gov/10005107). The project name refers to the Encyclopedia of DNA Elements and was started in 2003. It was designed to identify all functional elements in the human genome. Most of the human genome has been labeled as junk DNA, but this is not quite correct. ENCODE set out to look at the functional elements only, thereby restricting the workload to manageable proportions.

In today's world, genotyping for all but the most complicated genetic systems is mostly being outsourced. Few laboratories now perform all of their genotyping *in situ* and the commercial cost of genotyping is falling. The 1000 Genomes Project is coming to a close (with much of the preliminary data available now) and many centers will use whole-genome sequencing in the future rather than SNP genotyping by GWAS, especially as the cost is reduced. In addition, epigenetics offers a different look at complex disease and **imputation analysis** offers an easier way to look at the genome at high resolution. Currently, few studies have looked at systems, but the indication from some of the more studied diseases is that systems analysis offers a way forward.

It is important to keep our feet on the ground in the face of these massive developments. Studies still depend on good quality collections of patients and appropriate controls.

However, everything else has changed or is changing. Increasingly, studies are collaborative rather than being based on a single center. This has obvious benefits in terms of quality, but it can introduce problems too.

The real debate about the future of complex disease research lies not in the genetics itself, but downstream from the genetics

Genetics only provides one piece of the jigsaw. The whole picture can only be achieved by considering the genetics, the systems that risk alleles map to, and the environment. Systems biology provides the basic science for the biological approach. The environment is more difficult to tackle. Genes and the environment interact through systems. Consider, for example, the impact of lipids in the diet and the way in which lipid genes may play a role in obesity through regulating lipid metabolism. However, it is not so simple. Genes interact, often in different directions. There is also **redundancy** in biology to allow for biological failure or breakdown and also for the impact of the environment. Redundancy exists in the environment too, for example there is more than one source of most essential food elements.

A **reductionist** approach will not be sufficient to understand human disease, because human biology is complex. There is interaction and redundancy within systems (as stated above). Thus, deconstructing a genetic model can be interesting, but it does not necessarily provide a clear indication of the relationship between genotype and phenotype.

Epistasis or gene–gene interaction is essential in biology and is increasingly spoken of in the genetics of complex disease. Redundancy is one example of interaction (albeit negative) that illustrates this point. Redundancy can be seen in many systems through **gene duplication** where multiple copies of genes exist or through more subtle dual coverage, whereby there are two expressed isoforms of a gene with overlapping functions. It is possible that this redundancy or genetic compensation allows for some, but not all situations. For example, the functionally expressed complement gene *C4B* may be able to compensate for the deleted *C4A* gene except when the system is stressed, at which point the absence of *C4A* may create a dysfunctional environment, permissive for the accumulation and deposition of immune complexes in the small blood vessels, kidneys, and joints leading to intense outbreaks of inflammation as seen in SLE.

CONCLUSIONS

Following the completion of the Human Genome Mapping Project, HapMap, and SNP Map, designing studies in complex disease has become relatively easy. However, despite all of the technological advances, the current options in study design in complex disease are still linkage and association analysis. The real difference between the past and the present is one of scale. At one end, it is possible to concentrate on a single gene; at the other, it is possible to search the whole genome. There are pros and cons for both of these options, and a number of factors influence the choice of which to apply.

There are fewer restrictions now than there were in the past. In the post-genome era there is a wealth of new information and technology to latch on to. Genome-wide association analysis has become the most favored form of study for the past few years. Unrestricted by prior knowledge and more-or-less hypothesis free, it has given researchers the freedom to roam across the genome screening for polymorphisms with relatively small effects.

Multiple GWAS has led to accumulation of huge DNA banks and collaboration on a global much greater scale as well as meta-analyses. This has produced some very significant reproducible, if small associations. With the right information at the beginning, a good study plan, and selection of the most appropriate method, which can be a low-technology candidate gene association study or a full-blown GWAS, the results can be very promising. However, caution needs to be applied when designing studies. Good planning will produce good quality results; poor planning leads to irreproducible results and wastes resources. The lessons of the past are well worth consideration. The literature from the 1960s and 1970s onwards is full of unconfirmed associations. However, we should not throw the baby out with the bathwater, as the saying goes. Some of the most important genetic associations in complex disease also date back to 1960s and 1970s. Some would even go so far as to argue that interest in the genetics of complex diseases, especially those where there are no multiplex families, would never have been considered without these early studies.

Careful consideration at the early stages of any research project will help to identify potential pitfalls and also identify potential new avenues for research. Resources, both financial and material, are limited. It is therefore essential to make the most out of genetic studies. The aim should be for high-quality, reproducible results. Choice of options depends on indicators; of a genetic component in the disease (such as concordance in twins, sibling relative risk), the available materials (mostly DNA banks), access to patients and their collaboration, prior knowledge of the disease pathology, and any results of previous studies.

FURTHER READING

Books

Frank L (2011) My Beautiful Genome: Exposing our Genetic Future, One Quirk at a Time. Oneworld.

Spector T (2012) Identically Different: Why You Can Change Your Genes. Weidenfeld & Nicholson.

Strachan T & Read AP (2011) Human Molecular Genetics, 4th ed. Garland. Chapter 16 deals with identifying human genes and susceptibility factors and provides excellent additional data for students wishing to delve deeper into medical genetics. There is also additional relevant information in the other chapters.

Articles

Brewerton DA, Hart FD, Nicholls A et al. (1973) Ankylosing spondylitis and HL-A27. *Lancet* 301:904–907. One of the earliest reported genetic associations with HLA—an association that has stood the test of time despite changes in HLA phenotyping and genotyping methods and nomenclature (note HL-A27 is now correctly referred to as *HLA-B*27*). This illustrates the point that not all genetic associations are recent and not all early reports failed to be replicated.

Bhugra D (2005) The global prevalence of schizophrenia. *PLoS Med* 2:e151.

Cardon LR & Bell JI (2001) Association study designs for complex diseases. *Nat Rev Genet* 2:91–99. This is an excellent review of association studies of complex disease. The timing of this review reflects the immediate post-genome era and pre-GWAS era, providing an interesting historical insight into the genetics of complex disease.

Colhoun HM, McKeigue PM & Smith GD (2003) Problems of reporting genetic associations with complex outcomes. *Lancet* 361:865–872. This work provides an excellent guide and critique of the common problems in studies on the genetics of complex disease. It is an informative and well-scripted paper, and recommended for students wishing to develop their critical skills.

Collins FS & McKusick VA (2001) Implications of the human genome project for medical science. *JAMA* 285:540–544. This is a very important review that provides the reader with a guide to and an illustration of the potential for understanding complex disease in the immediate post-genome period.

Davies JL, Kawaguchi Y, Bennett ST et al. (1994) A genome-wide search for human type 1 diabetes susceptibility genes. *Nature* 371:130–136. This is possibly the first genome-wide search in complex disease and one which used very different techniques compared with the large-scale GWAS of today.

Hallmayer J, Cleveland S, Torres A et al. (2011) Genetic heritability and shared environmental factors among twin pairs with autism. *Arch Gen Psychiatry* 68:1095–1102.

Hirschhorn JN & Daly MJ (2005) Genome-wide association studies for common diseases and complex traits. *Nat Rev Genet* 6:95–108. This review puts the idea of GWAS into context, and provides useful insight into how such studies are planned and executed.

Johnson GCL & Todd JA (2000) Strategies in complex disease mapping. *Curr Opin Genet Dev* 10:330–334. This is an alternative review of the issues in planning studies of genetic associations in complex disease. The paper provides interesting insight into the future development of GWAS, unachievable in 2002, but since realized.

Kieseppa T, Partonen T, Haukka J et al. (2004) High concordance of bipolar I disorder in a nationwide sample of twins. *Am J Psychiatry* 161:1814–1821.

Satsangi J, Parkes M, Louis E et al. (1996) Two-stage genome-wide search for inflammatory bowel disease provides evidence of susceptibility loci on chromosomes 3, 7 and 12. *Nat Genet* 14:199–202. This paper describes one of the first genome-wide studies in complex disease.

The 1000 Genomes Project Consortium (2010) A map of human genome variation from population-scale sequencing. *Nature* 467:1061–1073.

The 1000 Genomes Project Consortium (2012) An integrated map of genetic variation from 1,092 human genomes. *Nature* 491:56–65. The Human Genome (2001) *Nature* 409:813–958. The complete issue of *Nature* mostly dedicated to the HGM with multiple papers, letters, and editorial commentary from multiple authors. It is full of useful insight and critical discussion. Any student of human genetics should read this issue.

The International HapMap Consortium (2005) A haplotype map of the human genome. *Nature* 437:1299–1320.

The International HapMap 3 Consortium (2010) Integrating common and rare genetic variation in diverse human populations. *Nature* 467:52–58.

The Wellcome Trust Case Control Consortium (2007) Genome-wide association study of 14,000 cases of seven common diseases and 3,000 shared controls. *Nature* 447:661–678. This is another landmark paper highlighting the development and application of new technologies (GWAS) in complex disease research, and it provides a mass of useful information for those studying complex disease. This study provides some of the basic information for students studying Crohn's disease as well as a further six diseases not discussed in this chapter.

Wang WY, Barratt BJ, Clayton DG & Todd JA (2005) Genome-wide association studies: theoretical and practical concerns. *Nat Rev Genet* 6:109–118. This early discussion of the practicalities of GWAS indicates the way in which studies of genetics of complex diseases were about to develop and provides some critical insight into the thinking behind this now widely adopted strategy.

Willer CJ, Dyment DA, Risch NJ et al. (2003) Twin concordance and sibling recurrence rates in multiple sclerosis. *Proc Natl Acad Sci USA* 100:12877–12882.

Online sources

http://hapmap.ncbi.nlm.nih.gov
This site is a useful source for the results of the International HapMap Project.

http://www.ncbi.nlm.nih.gov/omim
This is an essential site for updates and information on genetic disease of all types.

http://www.ncbi.nlm.nih.gov/projects/SNP/snp_summary.cgi
This site provides information about reference SNP clusters in the genome.

http://www.ncbi.nlm.nih.gov/SNP
This is an essential site for updates on SNPs.

http://www.genome.gov/10005107
The ENCODE project – a useful site from which information is freely available.

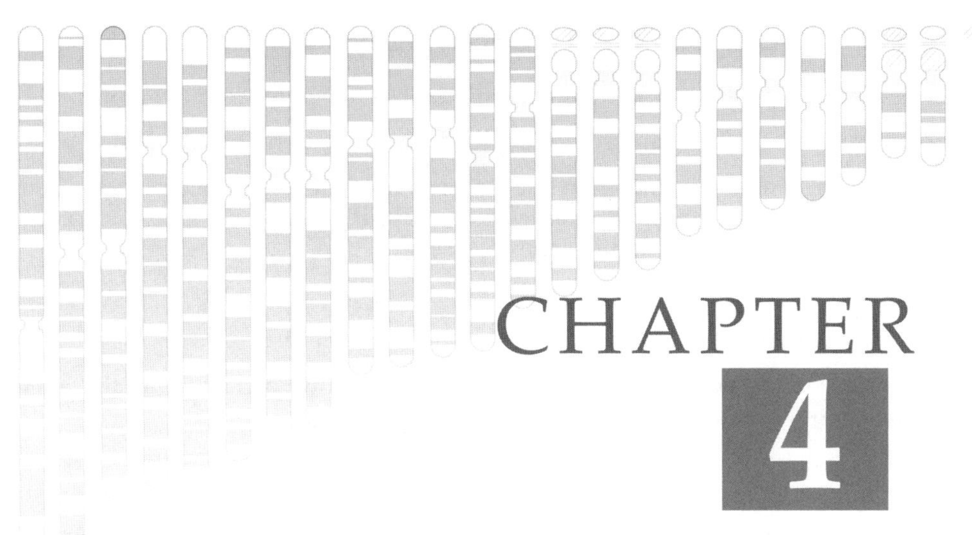

CHAPTER 4

Why Investigate Complex Disease Genetics?

We have already seen that complex diseases do not lend themselves to simple analysis, and by definition we are dealing with mutations and polymorphisms that increase or reduce the risk of a trait. It is also clear that there may be complex interactions between groups of risk alleles (epistasis), and between risk alleles and the environment. All of the above suggest that identifying genetic components in complex disease is and will be a difficult task, and yet it is a task that has attracted a considerable amount of attention in biomedical research. The potential rewards from this research are great and the technological advances made over the past 20 years have been astounding.

In this chapter, we will consider the question of why we invest so much time and so much of our resources to identify risk alleles in complex disease. We will use five key examples of complex disease to illustrate these: ankylosing spondylitis, cardiovascular disease, hepatitis C virus (HCV), rheumatoid arthritis, and bipolar syndrome or bipolar disorder. These five examples illustrate a host of key points, but they are selected here specifically to illustrate the potential to use genetics in the differential diagnosis in disease (ankylosing spondylitis), in understanding the response to infection and potentially planning for patient management (HCV), and also to increase our knowledge of disease pathogenesis (cardiovascular disease, rheumatoid arthritis, and bipolar disorder). These are complex issues as the text will show, but the central principals are simple.

4.1 WHY DO WE INVESTIGATE COMPLEX DISEASE?

In considering the question we need to remind ourselves of a number of key points.

- Complex diseases do not conform to simple Mendelian patterns of inheritance.
- The potential clinical outcome from research in complex disease, sometimes referred to as the promises of the **Human Genome Project (HGP)**, include the ideas that knowledge of the genetic background in complex disease:
  - may be used to aid disease diagnosis.
  - may be used to aid treatment/management and patient care.
  - may help us to understand the pathogenesis of the disease, and help with the development of novel therapeutic agents and strategies.

Each of these will be discussed in turn in this chapter.

Complex diseases do not conform to simple patterns of inheritance

By this stage in this book, the first point above is obvious, but it is worth reminding ourselves that we are dealing with complex issues. When things become complex, we are often forced to ask whether we should be spending vast amounts of time and effort to investigate them. It should not surprise us that the questions we are trying to answer are difficult. Science is full of questions that are difficult to answer. Yet we have to be cautious. To paraphrase Hugo Mencken, the author and satirist, "there is always an easy solution for every human problem (that is) neat, plausible, and wrong." We are inclined, as Nobel Laureate Herbert Simon suggests in his statement about the aims of science, to "find meaningful simplicity in the midst of disordered complexity." In other words, it is part of our humanity that we should seek to find simple solutions to complex problems (**Figure 4.1**). Mencken is telling us to beware, whereas Simon is encouraging us to tackle the issue and not be put off. However, complex diseases are complex by nature, and therefore we should not expect a neat and (simple) plausible answer. Once we agree with this principle then we

Figure 4.1: Human desire to find simple solutions to complex problems. It is our nature to seek simple solutions to complex problems. However, we need to be aware of simple solutions, especially in the genetics of complex disease. Are we looking for a neat and simple answer that is wrong or are we over-complicating science? It is important to ask these questions in science and reflect on the broad issues as well as the pure facts.

can proceed to answer the basic question. Why investigate the genetics of complex disease? This is not pessimism, it is realism and that "realism" is essential in science.

Part of understanding the reasons for the mass of research in complex disease is born out of an appreciation of the diseases themselves. Complex diseases are common, if we include cancer in the count they may account for as much as 95% of all forms of human disease. If we exclude cancer then the figure is likely to be closer to 75%. Not all cancers are complex; some are Mendelian and some are mosaics that develop during the life of an individual, that is they are not inherited from the parents. However, some mosaics can be stimulated into production by inherited genetic variations. Cancer as we shall see later in the book is a difficult area in complex disease. However, overall complex diseases whether 75% or 95% of all forms of human disease still have a major social and economic impact. Understanding genetics helps us to understand disease. For example, we currently know very little about the pathogenesis of many common diseases. This is because patients present relatively late in the disease process, sometimes long after the disease-initiating processes have done their work. Trying to understand disease pathogenesis by looking at late stages of a disease can be like looking for fingerprints in a room where the criminal wore gloves. However, genetics can help. As individuals, our whole genome does not change, while the clinical symptoms of diseases do. There are some changes in specific cells due to mutations during aging and in some cancers, but these are not present in all cells.

The HGP in research into genetically complex disease

Identifying risk alleles for common genetically complex disease has been set as one of the major challenges of the post-genome era and indeed it was the potential to identify such risk alleles that made the Human Genome Mapping Project (HGMP) such an attractive proposal. The idea of finding risk alleles in complex disease is not new. However, early studies were confined to selected candidate genes, the locations of which were mostly unknown. The publication of the HGM, SNP Map (http://www.ncbi.nlm.nih.gov/projects/SNP/snp_summary.cgi), and HapMap (http://hapmap.ncbi.nlm.nih.gov/) enabled the door to understanding the genetics of complex disease to be fully opened for the first time and now the race is on.

The promises of the HGP listed above and illustrated in **Figure 4.2** will be used as the basis for the discussion of the potential clinical outcomes and benefits of investigating

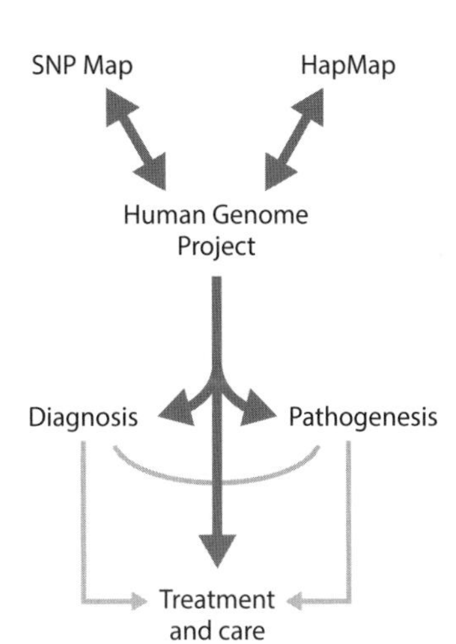

Figure 4.2: Clinical outcomes of the HGP. The figure illustrates the potential clinical outcomes (promises) of the HGP in medical sciences and the way in which the three main promises are interrelated. The outcomes are: to aid diagnosis, to help with patient management and care (including development of new therapies), and to improve the understanding of disease pathology. All are achievable to some extent, though some will take longer than others.

genetics in complex disease (below). The idea of individualized therapy and identifying key pathways in disease pathogenesis are those that have been most achievable so far, but ultimately all of these promises are linked. It is important to consider not only the data that studies produce, but also the application of the data in terms of meeting these expectations or promises.

4.2 DISEASE DIAGNOSIS

It was suggested in 2001 when the draft HGP was published that genetic tests would increasingly be used to predict disease and initiate preventative action. This is pre-diagnostic testing. Francis Bacon reminds us to be cautious; he said "the human understanding ... readily supposes a greater order and uniformity in things than it finds, (it) is infused by desire and emotion, which gives rise to wishful science." When we look at this promise in particular it is important to consider whether the promise constitutes wishful science. The extent to which identified risk alleles are currently used in disease diagnosis will provide a clue. However, the best examples that are currently available are those that were identified before the HGP began. This is by no means negative, as it shows what can be achieved.

Early studies on the genetics of ankylosing spondylitis indicated what could be achieved in terms of differential diagnosis in the post-genome era

One of the best examples that has received a great amount of attention because of the potential clinical value associated with it is the association between the rheumatic disease ankylosing spondylitis and the *HLA-B*27* allele family. The main ankylosing spondylitis risk allele is located in the major histocompatibility complex (MHC). The MHC encodes the genes for the human leukocyte antigens (HLA genes) and it is one of the most complex genetic systems in the human genome. The MHC is discussed in detail in Chapter 6 and further information for this chapter is given in **Box 4.1**.

The pathogenesis of ankylosing spondylitis is unknown; however, it is a relatively common rheumatic disorder that occurs in up to 0.5% of the UK population. It is a debilitating disease of the lower spine that develops in early adulthood. Though the disease usually involves the spine, it can cause other symptoms, such as abnormalities in the aortic valve and **pulmonary fibrosis**. The genetic association with *HLA-B*27* was first described in 1973 by Brewerton et al. and confirmed by Schlosstein et al. in the same year. Other clinical conditions that have similar symptoms have also been associated with *HLA-B*27*, including **reactive arthritis**, uveitis, and sacroileitis in inflammatory bowel disease, and psoriasis. However, none of these other conditions is as strongly associated with *HLA-B*27* as is ankylosing spondylitis. Estimates for the strength of this association between *HLA-B*27* and ankylosing spondylitis vary. The gold standard study of ankylosing spondylitis by Brown et al. published in 1996, suggested the odds ratio (OR) for ankylosing spondylitis in those with *HLA-B*27* may be as high as 171 (with 95% confidence intervals of 135–218). This is one of the strongest genetic associations ever described for an HLA allele family and clinically these data are of great potential significance. In terms of diagnosis however simply being *HLA-B*27*-positive is not an indication that a patient has ankylosing spondylitis or will develop the disease. Even in Brown's study, 4% of patients with ankylosing spondylitis were *HLA-B*27*-negative and approximately 9% of the UK population are *HLA-B*27*-positive. If possession of *HLA-B*27* were sufficient

BOX 4.1 A SHORT HISTORY OF THE MHC

The problem with HLA genes is that they reside within the **major histocompatibility complex (MHC)**, which is so-called because of the role many of the genes encoded there play in determining compatibility of tissues. A huge interest in HLA grew out of the practical need to find matched organs for transplantation, particularly kidney transplantation. In studying complex disease, the HLA genes are the most problematic. There are two reasons for this: they are the most variable genes in the human genome and they have the highest levels of linkage disequilibrium in the human genome. HLA antigens are not expressed on red blood cells. HLA typing was initially performed using serum to detect antigens on human white cells (leukocytes—hence the name "human leukocyte antigens" for the genes). The HLA phenotypes, as they were correctly called, identified a range of different antigens on four different HLA loci: *HLA-A*, *HLA-B*, and *HLA-C* (all class I loci), and *HLA-DR* (a class II HLA gene).

The quality of early studies was based on the availability of banks of sera to identify a range of antigens. Rare antigens were often missed or not tested and many remained undiscovered until the 1990s. Later studies were able to capitalize on the introduction of molecular genetics, first **restriction fragment length polymorphism (RFLP)** and then **polymerase chain reaction (PCR)**, to genotype the HLA genes. The introduction of molecular genetics led to the discovery of other HLA genes, in particular the *DQ* and *DP* family.

As methods changed, so the names changed. In the 1970s and 1980s, the genes were called HLA A, B, Cw, and DR, with the antigens identified in numbers without a space (e.g. A1) and roman script used. In the 1990s, the names were changed again, e.g. *HLA-A*, *HLA-B*, and *HLA-C* (the "w" was dropped from the name for all *HLA-C* antigens and genes). For each allele, the gene name is followed by an asterisk (*) and then a number starting with 01, e.g. *HLA-A*01*. If a variant of this allele family is known, then this is labeled with an additional colon (:) and two more numbers, e.g. *HLA-A*01:01*. The alleles are all given in italics.

Two additional essential facts are that unlike *HLA-A*, *-B*, and *-C*, which encode only a single α-polypeptide, the HLA class II antigens are the product of two expressed HLA genes: an A gene and a B gene encoding an α-peptide and a β-peptide, respectively. Both are found close together on chromosome 6p21.3. In all cases, these genes are polymorphic, but this polymorphism is limited in the case of the *DRA* gene.

One of the most difficult issues relating to the MHC is the extreme linkage disequilibrium between the alleles encoded within the region. This can make dissecting the details about which candidate gene may be responsible for increasing the risk difficult. It is also possible that the conserved combinations indicate multiple risk alleles on the same haplotype. The MHC plays a central role in complex disease and cannot be dismissed. More details about the MHC are given in Chapter 6 and associations with disease appear throughout this book.

for ankylosing spondylitis to occur, then the disease would be found in a much greater proportion of the population and all ankylosing spondylitis patients would be expected to have *HLA-B*27*. In practice, only a small proportion of *HLA-B*27*-positives develop ankylosing spondylitis; this is an example of incomplete penetrance (see Chapter 2). This

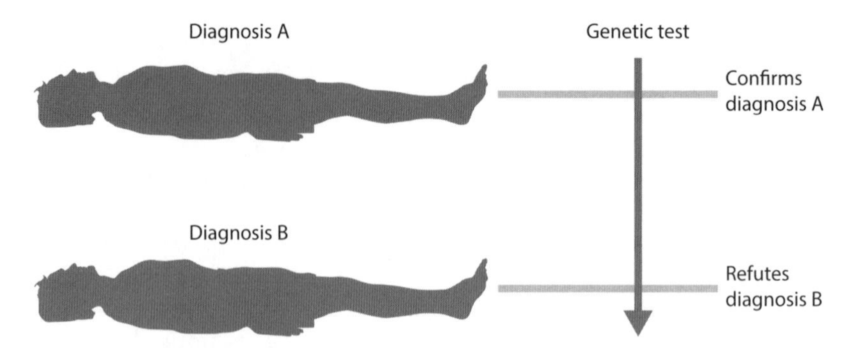

Figure 4.3: Use of genetic tests in differential diagnosis in complex disease. The patient illustrated has two potential diagnoses. For each diagnosis, genetic tests can help to determine which diagnosis is more likely to be correct. This is called differential diagnosis and is widely used, and genetics is likely to play an increased role in clinical practice. However, the genetic test is not being used here to diagnose the disease per se, but to add weight to the likelihood of one or other diagnoses. Note this example is not a Mendelian disease, but a complex disease.

departure from the expected Mendelian norm is exactly what we expect to find in genetically complex disease, because complex diseases do not conform to Mendelian patterns of inheritance. However, this does not mean that it is impossible to achieve the goal of using risk alleles as a diagnostic tool. Associated alleles can be used provided the limitations are correctly understood. In ankylosing spondylitis, *HLA-B*27* testing can be helpful in making a differential diagnosis as the simple illustration in **Figure 4.3** shows.

Genetic associations in complex disease confer small risks

Most genetic associations found in complex diseases are with alleles that confer relatively small risks. However, there are some exceptions. These exceptions include strong associations with mutations in the *PTPN22* gene in rheumatoid arthritis (see below), and with certain adverse drug reactions, such as the anti-human immunodeficiency virus (HIV) drug abacavir and *HLA-B*57:01* and the commonly used antibiotic flucloxacillin that is also associated with *HLA-B*57:01* (see Chapter 8).

Of course we cannot know what may arise in the future, just as the HGP could not predict the success in each area. Current evidence suggests that most common diseases are likely to be more complex than we have imagined and involve more risk alleles than we have thought likely. For example, Crohn's disease is associated with more than 163 risk alleles. Without the complete picture of how risk alleles interact we cannot know whether individuals have to have a group of risk alleles (risk portfolio) for a specific disease (**Figure 4.4**) or just one risk allele. If there are multiple interacting risk alleles, even though the individual alleles may have a small impact in terms of risk, the combinations may have much larger effects and identifying these risk portfolios may turn out to be useful in diagnosis.

4.3 PATIENT TREATMENT/MANAGEMENT AND CARE

As we search for alleles with small risk effects it will become increasingly difficult to use the observations to support diagnosis in the classical sense, and yet it is possible that these developments will help with planning patient treatment and care. Characteristics such as disease progression, severity, and response to treatment are traits that may have genetic associations.

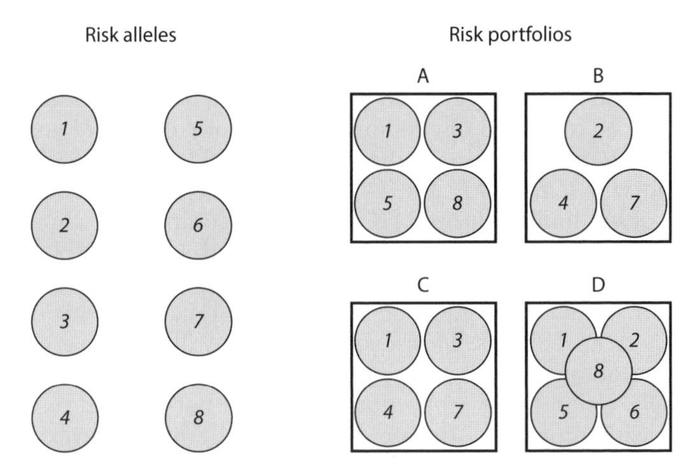

Figure 4.4: Risk portfolios in complex disease. The figure illustrates risk alleles numbered *1–8* in the circles on the left and four cases in square boxes labeled A–D. Not all cases may have the same risk alleles. This figure illustrates the idea that individual cases may have different risk portfolios, i.e. groups of risk alleles (shown in the black boxes). Here, case A has *1-3-5-8*, case B has *2-4-7*, case C has *1-3-4-7*, and case D has *1-2-5-6-8*. Note that each case has a different set of risk alleles and some cases have more risk alleles than others, e.g. B has three alleles, D has five alleles, while A and C each have four alleles. Note that whatever the risk portfolio, the sum of the risk is unlikely to be 1. These are complex diseases, thus being a carrier of a group of risk alleles will be neither necessary nor sufficient for the disease to occur.

It is very easy in a research laboratory to forget that patients are individuals. In a laboratory, patients can become numbers within a population. We tend to lose sight of the fact that each individual patient is different and we tend to assume that all cases with the same disease are exactly alike. This is not true. A disease group is a collection of patients with the same signs and symptoms. However, there is usually a great deal of variability in terms of age of onset, severity, response to treatment, and prognosis.

Identifying risk alleles that predict onset of complex diseases may enable patients to make beneficial lifestyle changes

In Alzheimer's disease, there are early- and late-onset models (EOAD and LOAD; discussed in Chapter 2). The genetics of these two forms are different, though the signs and symptoms are very similar. Identifying markers that will predict onset of complex disease could be helpful if changes in lifestyle, for example diet, would prevent the disease or at least extend the time to onset. In addition, genetic prediction of onset may be helpful if there are preventative treatment options that can be applied. This may be particularly useful in familial cases, but the same limitations apply to the use of identified risk alleles in disease prevention as apply to their use in diagnosis. Possession of the risk allele does not guarantee that the disease will occur. There is also no guarantee that lifestyle changes will prevent the disease. It is all about balancing the risk.

Predicting disease severity through genetic analysis may have clinical significance in terms of patient management

More promising are the possibilities to use identified risk alleles as markers of severity and also response to treatment and prognosis. There have been several associations with infectious disease and genetic variation in the human genome. Many of these concern the outcome following HIV-1 infection and are discussed in Chapter 7. Response to HCV has also been found to be strongly associated with polymorphism in the *HLA-DQB1* gene. This association is thought to be primarily with the *DQB1*03:01* allele. The *DQB1*03:01* allele is in linkage disequilibrium with both *DRB1*04* and *DRB1*11*. In the UK *DQB1*03:01* is most often found with *DRB1*04* and in Southern France it is most often found with *DRB1*11*. The *DQB1*03:01* allele has been associated with a higher incidence of self-limiting (acute) HCV infection compared to chronic HCV infection (**Figure 4.5**). Further studies have also shown that this polymorphism is associated with a stronger host T cell response to synthetic HCV peptides. The clinical significance of this observation is that HCV infection is very common and the majority of cases do not spontaneously clear the virus. Most cases (50–80%) will develop a chronic infection that over a long period of time can progress to liver fibrosis, liver failure, liver cancer, and death. Interestingly, a significant proportion of cases do not respond to treatment when treated. Patients with progressive disease may eventually require liver transplantation. From a diagnostic perspective this association is of limited value. *DQB1*03:01* is common (about 30% of the UK population have a *DQB1*03:01* allele), but from a biological perspective it is interesting. What is it about *DQB1*03:01* that enables the majority of those who carry this allele to clear the virus? In other words, what is *DQB1*03:01* doing or not doing? Information about this allele and how it works may be used in clinical practice to develop a vaccine or to manipulate the host response to the virus in such a way that it is spontaneously eliminated. At the time of writing, new treatments in HCV are having very significant effects so the potential use of host genetics may change. More recently, there have been quite a number of studies identifying genetic variations that are associated with the outcome following HCV infection and response to treatment. These have identified a number of interferon (IFN) genes, among others, that may prove important in developing new personalized therapy plans for patients with HCV and other infectious diseases.

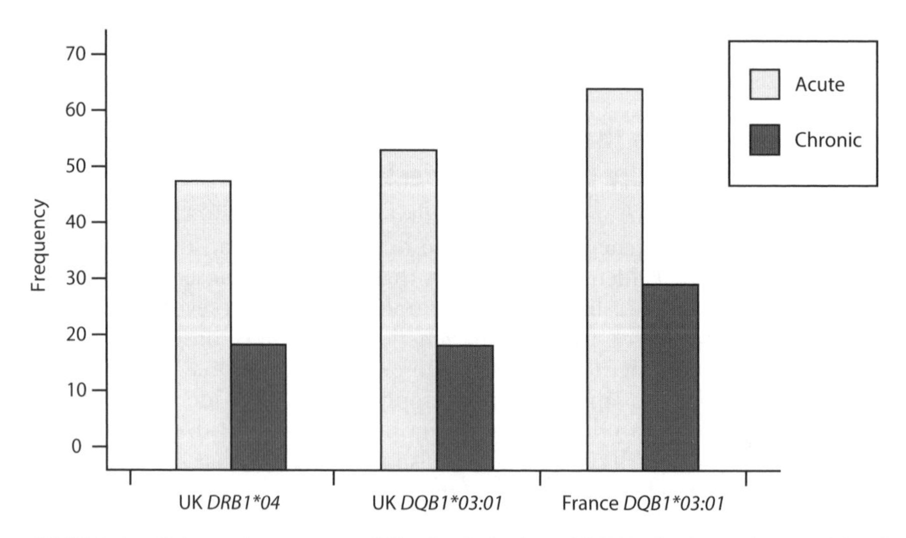

Figure 4.5: HLA class II determines outcome following infection with HCV. The figure shows real data from two different studies. Both studies reported genetic associations with HLA in two different patient groups: those who developed chronic infection and those who had an acute (self-limiting) infection following HCV infection. A group from the UK and one from France each reported a lower frequency of the *DQB1*03:01* allele in patients with chronic HCV infection. In the UK population, this allele is most often carried on the *DRB1*04* haplotype (data for *DRB1*04* shown); in France, this allele is found more frequently with *DRB1*11:01* (data not shown). The important observation is that acute infection is associated with *DQB1*03:01* in both populations.

Common genetic variation may predict response to treatment and be critical in patient care

Response to treatment is considered more extensively under the heading of pharmacogenetics in Chapter 8. This is a hot spot for research at present. Recently, there have been major discoveries associating MHC genes with adverse reactions to commonly used pharmacological agents. However, pharmacogenetics has a long history, with the first direct study of pharmacogenetics beginning reported in 1932. Most of this data has yet to be applied in a clinical setting. However, that is not to say the data could not be used. Genetic variations in the response to the antiviral abacavir, are one such example.

Onset, severity, and response to treatment are all part of patient management

All of the above are important in patient management. Deciding which drugs to use and the correct dose is essential. Close monitoring of patient progression, and taking timely and decisive action is essential. This does not only apply to pharmacogenetics, risk alleles may also identify which patients within a group will progress more rapidly and require more radical treatment. In the future, genetic risk assessment may be one means of ensuring healthcare resources are used appropriately and with maximum efficiency.

Consider, for example, transplantation. At present, demand for transplants outweighs the availability of donor organs (**Figure 4.6**). In the future, genetics could be used to ensure resources, in this case donor organs, are used efficiently and appropriately. Consider the example of two patients (A and B) with the same autoimmune liver disease (**Figure 4.7**). Imagine that the disease is a progressive disorder for which there is no effective treatment. Both patients will each eventually require a liver transplant. Both patients A and B present clinically at the same age and the same stage, but only one can be listed for transplantation due to a lack of donors. Presently this is a dilemma, but in the future it may be possible

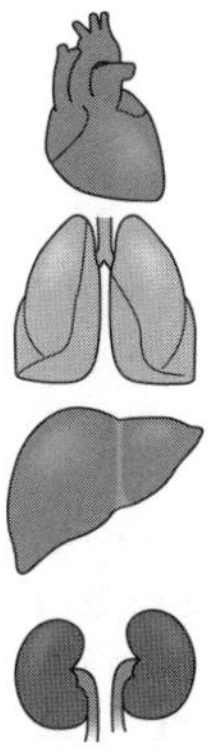

	UK waiting list	Approximate number of donors
Heart	196	142
Lungs	227	188
Liver	462	775
Kidneys	6111	1750

Figure 4.6: Estimated donor organ availability versus demand. The figure illustrates the difference between donor organ availability in the UK and need for four major organs in 2013. Note that the available figures are constantly updated. These figures illustrate a major clinical issue, which is that in most years only the donor numbers for livers seem to match their target. In some years there is even a deficit in liver donors. A large number of the transplanted kidneys are from matched, related donors. Approximate donor figures are based on the previous year's transplant figures as there is no way to know the actual number of donors for the year ahead. On a global scale, the donor deficits are much more pronounced and there is a serious debate over the advantages of opt-in versus opt-out systems.

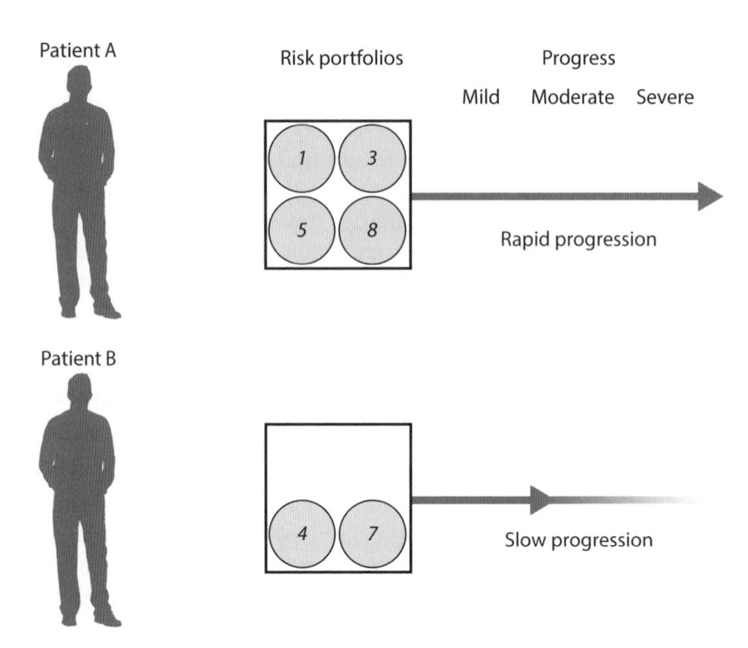

Figure 4.7: Disease progression and risk portfolios. This figure shows the potential to apply genetic testing to identify patients at risk of rapid progression versus those with slow or moderate progression. Such information could be used to determine the best course of treatment and management for individual cases. The use of genetics in this way may also maximize the use of key resources, such as donor organs for transplantation.

to use genetic profiling to determine the risk portfolio for the two patients. This genetic risk portfolio could then be used to determine the likely rate of progression for the two patients and this information could be used to inform the selection about who (A or B) to list for transplantation. This would lead to a more rational use of a precious resource (transplant donors) and be of benefit to both patients as the unlisted patient could be listed for transplantation later.

4.4 DISEASE PATHOGENESIS

One of the ultimate goals for the HGP is to map and understand the genetics of all human disease. Understanding disease pathogenesis is the key element within this. The sad fact is that we know too little about the pathogenesis of many common diseases. This means that details of the early pathogenesis are lost to investigators and complicated by downstream pathology. To understand what causes a disease we need to see it at the outset. We need to monitor the processes and mechanisms that lead to the symptoms as pathology changes with time. Systems fail due to the slow breakdown of intricate, interacting networks. This breakdown almost certainly takes place over a considerable length of time, but in chronic disease we most often see the end of the process only, making the process of understanding disease pathology difficult to say the least. Fortunately, the genome of individuals does not change, though there may be mutations in individual cells, and **population genetics** evolves. Therefore, we can look at commonly inherited genetic variation to seek risk alleles that may act as markers to outline disease pathology.

Unlike using risk alleles for diagnostic purposes, we are not restricted to the consideration of alleles with large effects when studying pathology – we can use data on any allele with a significant effect. This means that even markers for risk alleles associated with small effects are of value. Consequently, it is in this arena that the majority of studies have been focused and there have been some notable successes, and even though there is still much, much more to come, we do not need to look back to the pre-genome era to find good examples.

The Wellcome Trust Case Control Consortium (WTCCC1) study published in 2007 included seven diseases: Crohn's disease (considered in Chapter 2), type 1 diabetes (T1D) and type 2 diabetes (T2D) (both included in Chapter 10), rheumatoid arthritis, bipolar disorder, coronary artery disease, and hypertension. There had been considerable work prior to 2007 in each of these diseases but in each case the WTCCC1 study represented a milestone in genetic investigation, though not necessarily the first genome-wide association study (GWAS) in each case.

The WTCCC1 study confirmed strong associations, i.e. with a significance threshold of $P < 10^{-7}$ with the MHC region and *PTPN22* in rheumatoid arthritis, and identified novel significant associations with rs420259 on chromosome 16p12 in bipolar disorder and with rs1333049 on chromosome 9p21 in cardiovascular disease. In addition, the study identified nine moderate (i.e. significance threshold of $P < 10^{-5}$) associations with rheumatoid arthritis, 13 with bipolar disorder, and six with hypertension.

In the remainder of the present chapter we will consider four diseases: ankylosing spondylitis, rheumatoid arthritis, bipolar disorder, and cardiovascular disease. Three of these are from this list of seven examples in the WTCCC1 study above, and will be considered in detail to illustrate the concept that genetics can help us understand disease pathology. Ankylosing spondylitis was not considered in the WTCCC1 study, but has been considered more recently in the WTCCC2 study. We will look at early pre-GWAS, early GWAS (2007), and more recent studies for each disease. Regarding the other diseases in the WTCCC1 study, Crohn's disease is discussed in Chapter 2, T1D and T2D are discussed in Chapter 10, and hypertension is not discussed, though many other diseases and traits are discussed throughout the book.

Early studies offered potential insight into the biology of ankylosing spondylitis

The *HLA-B*27* association in ankylosing spondylitis has also been considered as having a role in disease pathology. *HLA-B*27* encodes a molecule with an unpaired cysteine residue at position 67 that forms a disulfide bond with thiol-containing agents. It has been suggested this results in an immune response to altered self-antigens, either by antibodies or by **natural killer (NK)** cells (**Figure 4.8**). An alternative hypothesis is that HLA-B molecules with cysteine 67 may have a different preference for certain athritogenic peptides leading to an increased likelihood of ankylosing spondylitis in those with *HLA-B*27*. Crystallographic studies of B27 molecules with a variety of bound peptides in the **antigen-binding groove** have shown arginine in the second position of all bound peptides. Arginine carries a long side chain that engages a specific pocket in the antigen-binding groove where cysteine 67 is also located. Subtyping *HLA-B*27* suggests not all *HLA-B*27* alleles are associated with ankylosing spondylitis. In particular, *HLA-B*27:02*, *-B*27:04*, and *-B*27:05* are associated with ankylosing spondylitis, but the association with *HLA-B*27:03* is uncertain, and *HLA-B*27:06* and *-B*27:09* are not associated. The hope for the future in ankylosing spondylitis is that such studies will narrow the search for athritogenic peptides, leading eventually to novel and more effective therapies.

Figure 4.8: Molecular mimicry in ankylosing spondylitis. The figure illustrates one of the major theories behind the genetic association of ankylosing spondylitis with *HLA-B*27*. The pathogenic peptides (represented by the dark gray smiley face with a hat) have sequences that are very similar to the host (represented by the smiley white face without a hat).

Later GWAS have offered even further insight into the biology of ankylosing spondylitis

The WTCCC2 study and others, including the International Genetics of Ankylosing Spondylitis Consortium (IGAS), have identified at least 25 risk alleles for ankylosing spondylitis, all of which are linked to immune function. The IGAS study alone found 25 risk loci at a significance threshold of $P < 5 \times 10^{-7}$. Twelve of these had been previously identified between 2007 and 2011 (including *ERAP1*, *IL12B*, and *IL23R*). *ERAP1* (also known as ARTS) stands for endoplasmic reticulum aminopeptidase 1. In addition, in the Han Chinese population, two other loci have been reported: *HAPLN1–EDIL3* and *ANO6*. A large number of the 25 identified genes produce proteins involved in immune inflammatory activity, e.g. *IL6R* and *IL23R*. The impact of the cytokines and their receptors has received considerable attention with interest in interleukin (IL)-23 and IL-17 interaction in pathogenesis, and the development of potential anti-IL-17 therapy. It is possible that in ankylosing spondylitis we are now quite close to reaching the stage where understanding the genetic basis of disease pathology will lead to the development of individualized treatment and better diagnosis based on knowledge of each individual's genetic risk portfolio.

Rheumatoid arthritis has many strong genetic associations, some of which can be used to help us unravel the pathogenesis of this disease

Rheumatoid arthritis is a common disease affecting approximately 400,000 people in the UK. Globally, the disease may affect as many as 1% of the adult population, it is more common in women and in the aging population (**Figure 4.9**), and is characterized by inflammation of the joints and by increased or abnormal levels of inflammatory activity in the synovial fluid, leading to destruction of the synovial joints. Two-thirds of cases are rheumatoid factor-positive, carrying antibodies against cyclic-citrullinated peptide (anti-CCP). Generally, rheumatoid arthritis is thought to be an autoimmune disease and it has many of the classical markers of such a disease, i.e. a female preponderance, autoantibodies, and genetic associations with risk markers in the MHC. There is an imbalance in the normal immune regulatory procedures in rheumatoid arthritis, such that though

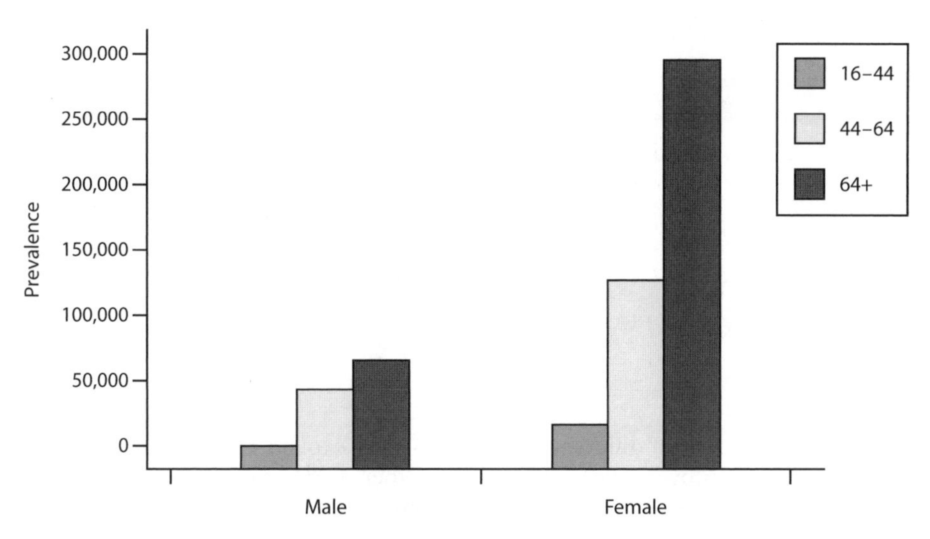

Figure 4.9: Prevalence of rheumatoid arthritis in the UK. The figure shows the prevalence (total number of cases) in each age category for male and female patients from Symmons et al. (2002). Note that the prevalence rises in each age group compared with the previous group and the greater prevalence in females compared with males for all three age groups. (Data from Symmons D, Turner G, Webb R et al. [2002] *Rheumatology* 41:793–800.)

anti-inflammatory mediators are present and activated they fail to adequately downregulate the immune response. Genetic studies in seropositive European cases have identified a large number of risk alleles for rheumatoid arthritis and these alleles are mostly associated with immune or immune-related function. In the sections below we will briefly consider some of the early studies of rheumatoid arthritis and then concentrate on three recent studies, all based on GWAS and meta-analysis of GWAS data.

The early history of rheumatoid arthritis, the MHC, and *PTPN22*

Evidence of a genetic component in rheumatoid arthritis is based on familial occurrence and elevated sibling relative risks (λs) of approximately 10. Early studies identified strong reproducible associations with the HLA genes on chromosome 6p21.3 and also with the *PTPN22* gene on chromosome 1p13, which encodes the gene for the lymphoid-tyrosine phosphatase (Lyp) protein tyrosine phosphatase non-receptor type 22.

The MHC in rheumatoid arthritis

The WTCCC1 study identified HLA as the major risk region for rheumatoid arthritis. This finding was not a surprise as the association had been described long before this. The genetic association between rheumatoid arthritis and HLA is very neatly summarized by Gerald Nepom in his chapter in the book *HLA in Health and Disease* (Lechler and Warrens, 2000) (summarized here in **Table 4.1**). The primary associations in each case are with alleles from the *DRB1*04* family, with the exception of Native Americans, where the primary association is with the *DRB1*14:02* allele. All of these susceptibility alleles either encode the amino acid sequence QKRAA or QRRAA (Q = glutamine; K = lysine; R = arginine; A = alanine) at positions 70–74 of the DRβ polypeptide of the HLA DR molecule. This shared epitope is within the **HLA-binding groove** and a site where antigenic side chains engage with the HLA molecule. At the molecular level, the crucial consideration is the **electrostatic charge** in and around the binding pocket. The electrostatic charge at each pocket and within the groove will determine which peptide antigens are preferentially bound and presented to the T cell receptor (see Chapter 6). Allelic variation in the molecular structure of the expressed

Table 4.1 HLA and shared epitopes associated with rheumatoid arthritis.

HLA class II allele	Relative risk	Shared epitope (DRβ-70–74)
*DRB1*04:01*	6–11	QKRAA
*DRB1*04:04*	5–14	QRRAA
*DRB1*04:05*	6–10	QRRAA
*DRB1*14:02*	1–2	QRRAA

Lechler R & Warrens A (eds) (2000) HLA in Health and Disease, 2nd ed. Elsevier. With permission from Elsevier.

HLA molecule encoded by different HLA alleles may explain the genetic association in rheumatoid arthritis, providing a platform for further work to identify athritogenic peptides and potentially develop novel therapies for the disease.

In addition, further work in rheumatoid arthritis suggests that there may be a gene dose effect. Not all *DRB1*04* family members carry the QKRAA or QRRAA sequence. In those that do carry this sequence at the *DRB1* 70-74 position, especially: *DRB1*04:01*, *DRB1*04:04*, and *DRB1*04:05*, all have a greater risk of RA. This is higher in those who carry two copies of this sequence (that is they are homozygous), than in those who carry only one copy (that is they are heterozygous). Despite the strong association between HLA and rheumatoid arthritis, the risk ratios listed indicate that *HLA-DR* genotyping would be of very little value in the diagnosis of rheumatoid arthritis. However, it may be of some use as a prognostic indicator. The *DRB1*04* alleles are associated with more progressive severe erosive disease.

The story of rheumatoid arthritis is more complex than first thought

It is always nice to see a clear explanation for HLA associations with disease, such as the shared epitope hypothesis above for rheumatoid arthritis. However, recent GWAS indicate a more complex situation in rheumatoid arthritis with at least three different HLA genes involved. *DRB1* remains the strongest, but independent associations with *HLA-B*08* and *DPB1* have also been identified. In this analysis some elements of the shared epitope hypothesis for *DRB1* still stand, but the amino acids at positions 11, 71, and 74 are highlighted as the most important, and potentially the only *DRB1* amino acids with causal effects in rheumatoid arthritis. This idea is supported by observations of significant, if weaker associations with amino acids at position 9 in the binding grooves of both *HLA-B*08* and *DPB1*. This is a departure from the idea of a string of amino acids and focuses more on specific isolated amino acids only. These studies involved a complex analysis with a mix of GWAS using 3117 single nucleotide polymorphisms (SNPs) across the MHC in over 5000 cases with seropositive rheumatoid arthritis and close to 15,000 controls. Imputation analysis was used to identify the HLA alleles and haplotypes. Rheumatoid arthritis is not a rare illness and therefore these findings are likely to be either confirmed or challenged by follow-up studies. It will also be interesting to see how direct sequencing impacts on these findings and how imputation analysis compares with direct sequencing.

PTPN22 in rheumatoid arthritis

In mouse models, mutations in the *PTPN22* gene have been shown to be associated with an enlarged thymus and spleen, and with increased T cell, B cell, and dendritic cell activity. The exact mechanism for this was until recently unknown. Recent data suggests *PTPN22* is involved in calpain-mediated degradation of Lyp. In mice, polymorphisms in the *PTPN22* gene leads to lower Lyp expression, and thus to T, B, and dendritic cell hyper-responsiveness, such as that seen in rheumatoid arthritis and T1D.

The genetic associations with the MHC and *PTPN22* in rheumatoid arthritis are not surprising as both are associated with a wide variety of other autoimmune diseases, including T1D, systemic lupus erythematosus, and multiple sclerosis. In rheumatoid arthritis, estimates suggest that together the MHC and *PTPN22* account for as much as 50% of the familial genetic risk of the disease.

A sample of recent genome-wide studies in rheumatoid arthritis

The 2007 WTCCC1 GWAS identified both the MHC and *PTPN22* as major susceptibility loci in rheumatoid arthritis with probability values $P < 7.5 \times 10^{-27}$ and $P < 5.6 \times 10^{-25}$ respectively (**Table 4.2**). In addition, a further nine SNPs with moderate *P* values mapping to loci not previously associated with rheumatoid arthritis were identified. Among these are SNPs that map close to the α- and β-chains of the IL-2 receptor (*IL2RA* and *IL2RB*). The IL-2 receptor is very important in T cell stimulation and a number of SNPs in this region have also been associated with T1D.

Seven new associations were reported in a more recent GWAS meta-analysis by Stahl et al. (2010) involving 5539 cases and 20,169 controls, and a replication set of 6768 cases and 8806 controls. The new associations were with SNPs close to the genes encoding *IL6ST* (IL-6 signal transduction), *SPRED2* (sprout-related EVA domain containing 2), *RBPJ* (regulation signal-binding protein for immunoglobulin κJ region), *CCR6* (C-chemokine receptor 6), *IRF5* (IFN regulating factor 5), *C5orf30* (chromosome 5 open reading frame 30), and *PXK* (PX domain containing serine/threonine kinase). The study also confirmed the previously identified associations at 27 different loci including *IL2RA* (IL-2 receptor A), *CCL21* (C-chemokine ligand 21) and *AFF3* (AF4/FMR2 family number 2). Interestingly, the probability values associated with many of the risk loci in this study were higher than those usually found in many GWAS (i.e. $P > 10^{-7}$ or 10^{-8}). For example the *P* values for *HLA-DRB1* and *PTPN22* were $P < 10^{-299}$ and $P < 9.1 \times 10^{-74}$ respectively. These higher *P* values for these well known, strong associations reflect the larger numbers in the Stahl et al. study. All of the associations were confirmed in the replication cohort. The most significant associations from the WTCCC1 study and the Stahl study are also shown in Table 4.2.

Stahl et al. also listed a further 10 moderate risk alleles for rheumatoid arthritis (**Table 4.3**). This latter group all have $P > 5 \times 10^{-7}$, i.e. they are outside the normal significance threshold (hence the term moderate associations). The major gene groups identified in this study are all associated with immune regulation. In addition, many of the identified SNPs mapped to loci associated with other autoimmune disorders, suggesting common pathways in the genesis of rheumatoid arthritis (**Figure 4.10**).

New susceptibility alleles are continually being added to the list for rheumatoid arthritis. A meta-analysis of six GWAS in 2011 identified seven new rheumatoid arthritis loci and listed a total of more than 37 risk loci for rheumatoid arthritis. With at a least a further nine risk loci being suggestive (moderate associations) from the three studies above, and a very large number of confirmed and unconfirmed loci listed on the OMIM (Online Mendelian Inheritance in Man) database, this brings the current total of risk alleles for rheumatoid arthritis to more than 48.

The problem with information supplied in Tables 4.2 and 4.3 is the OR (risk) values for the highly significant associations are in all but two obvious exceptions less than 1.5 (see **Figure 4.11**). The critical question is how are these data to be used? Or perhaps even how useful are these data? The answer is quite simple – they cannot at this stage be used to aid

Table 4.2 A short list of selected major risk genes for rheumatoid arthritis.

Gene/allele	Location	*P* value
HLA-*DRB1*	6p21	7.5×10^{-27} *Increases to maximum* 10^{-299} *(OR >2.5)*
PTPN22	1p13	5.6×10^{-25} *Increases to maximum* 10^{-74} *(OR > 1.5)*
IRF5	7q32	2.65×10^{-6} 4.2×10^{-11} *(OR = 1.25)*
IL2RA	10p15	5.92×10^{-8} *Increases to* 1.4×10^{-11} *(OR = 1.11)*
TNFA1P3 (rs6920220)	6q23	1.58×10^{-5} *Increases to* 8.9×10^{-13} *(OR = 1.4)*
IL2RB	22q13	1.15×10^{-6}
CTLA4	2q33	6.25×10^{-9} (OR = 0.85) *Lower in Stahl* 1.2×10^{-8} *(OR = 0.87)*
STAT4	2q32	2.9×10^{-7} *(OR = 1.16)*
TRAF1/C5	9q33	2.1×10^{-7} *(OR = 1.13)*
CD40	20q13	2.8×10^{-9} *(OR = 0,85)*
CCL21	9p13	3.9×10^{-10} *(OR = 0.87)*
AFF3	2q11	1.0×10^{-14}
REL	2p13	6.01×10^{-10} *Lower in Stahl* 7.9×10^{-7} *(OR = 1.13)*
BKL	8p23	5.69×10^{-9} *Lower in Stahl* 1.5×10^{-5} *(OR = 1.12)*
SPRED2	2p14	5.3×10^{-10} *(OR = 1.13)*
IL6ST/ANKRD55	5q11	9.6×10^{-12} *(OR = 0.85)*
C5orf30	5q21	4.1×10^{-8} *(OR = 0.93)*
PXK	3p14	4.6×10^{-8} *(OR = 1.13)*
RBPJ	4p15	1.06×10^{-16} *(OR = 1.18)*
CCR6	6q27	1.5×10^{-11} *(OR = 1.11)*

Where two sets of figures are presented, the top figures are from the WTCCC1 (2007) study (not italic) and the lower figures are from Stahl et al. (2010) (italic). Where only one set is represented for a gene, not italic or italic indicates the source as WTCCC1 or Stahl et al. respectively.

disease diagnosis and patient management, but they do point at the disease pathology. As almost all of the risk alleles point to the immune system, this suggests that using genetic associations to target and unpick specific immune pathways and/or correct defects in specific pathways may be the way forward. There are two elements to this: (1) identifying pathways helps us understand the pathogenesis of this disease, and (2) by improving our understanding of the disease, we may be able to produce better treatment and management plans for patients. For example, we could consider the *IL4–STAT6* pathway. These

Table 4.3 Weak (suggestive) genetic associations with rheumatoid arthritis from Stahl (2010) and listed on OMIM.

Gene	Location	Probability (OR)
Other suggestive risk alleles (from Stahl et al. 2010)		
IL6R	1q21	7.9×10^{-5} (OR = 1.13)
CD247	1q24	3.6×10^{-5} (OR = 0.9)
IL2/IL21	4q26	1.0×10^{-3} (OR = 0.89)
ZEB1	10p11	2.0×10^{-3} (OR = 0.93)
SH2B3	12q24	4.0×10^{-3} (OR = 0.93)
BATF	14q24	1.0×10^{-5} (OR = 1.16)
CD19/NFATC21P	16p11	5.3×10^{-5} (OR = 1.14)
IKZF3	17q12	4.7×10^{-5} (OR = 1.1)
UBE2L3	22q11	7.0×10^{-4} (OR = 1.1)
Listed on OMIM		
CCP		Not given
IL1A	2q13	Not given
NFKBIL1	6p21.33	Not given

two genes interact. IL-4 activates STAT-6, a signal transducer and activator of transcription (type 6). This pathway has been found to be activated in patients with both short- and long-term rheumatoid arthritis. If those carrying risk alleles could be given a dietary supplement to enhance or reduce the activity in this pathway, then maybe it would be possible to reduce or even eradicate rheumatoid arthritis. This may seem overly optimistic considering the statement that Hugo Mencken applied to seeking solutions to complex problems, and maybe we are over-inclined to seek simple solutions, but equally the best solutions often turn out to be simple.

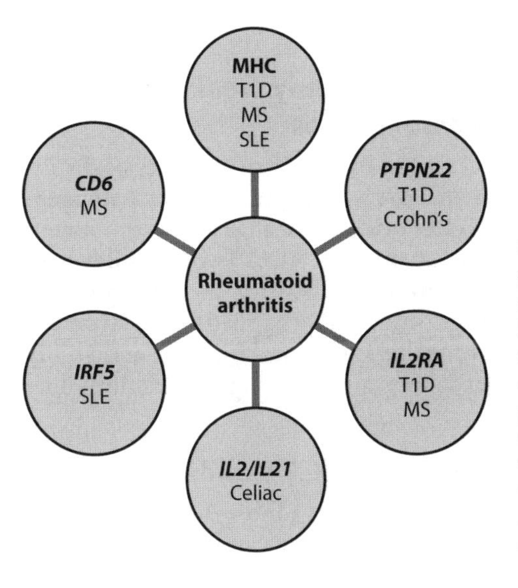

Figure 4.10: Overlap between risk alleles of specific genes and different diseases focusing on rheumatoid arthritis. This diagram illustrates some of the overlap between associated genetic markers in rheumatoid arthritis and other diseases. Overlap is common among complex diseases and it is helpful in identifying common pathways in pathogenesis. Each of the major risk alleles or genetic regions identified in the circles is associated with rheumatoid arthritis. In each case, other diseases that share some degree of overlap are illustrated, e.g. *IRF5* and systemic lupus erythematosus (SLE), and *CD6* and multiple sclerosis (MS).

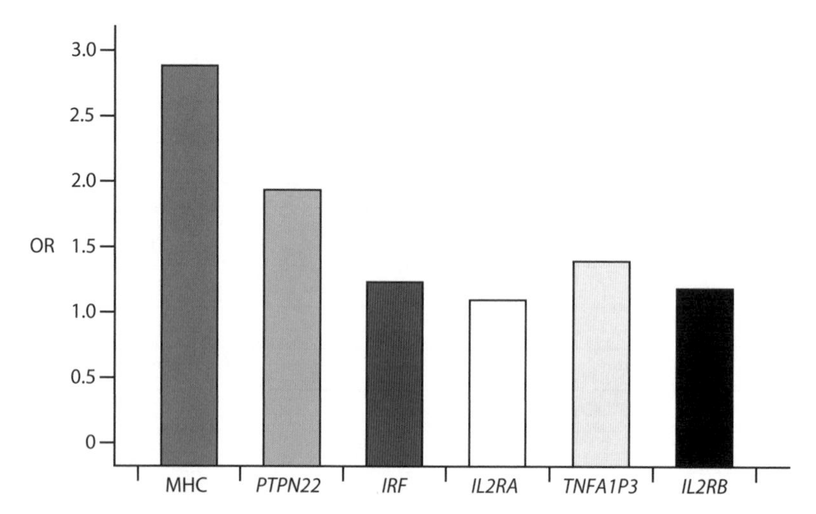

Figure 4.11: ORs for six of the most strongly associated genetic risk markers for rheumatoid arthritis. The figure shows the small size of the effect for most risk alleles as determined by the OR. Figures vary from study to study, and indeed even within studies, and therefore it is advisable to consider these as approximations or estimates rather than absolute values, see Box 2.2.

The story of rheumatoid arthritis so far tells us that there are difficulties in reaching our goals, but we should be optimistic about the possibilities of achieving at least one of the aims of the HGP with regard to this disease, that of understanding pathways that lead to pathogenesis. The problem is that we have more data than we can handle at present. In one respect, it is reassuring that there are so many identified alleles because this will enable us to explore multiple pathways and interactive networks. However, this takes time and resources, and pathways and networks are complex. We need time to work on the data, but more genetic data is piling up all around us as we work and it can be overwhelming. The answer is currently out of our reach, but as we shall see in some of the other chapters in this book, it is getting nearer by the day. It will take time to complete this exercise in rheumatoid arthritis, but hopefully the final outcomes will be well worth the wait and the investment in time and money.

Bipolar disease is a disease for which there are many weak genetic associations, but few strong consistent associations

Bipolar disease is a manic depressive illness characterized by recurrent episodes of disturbance in mood, ranging from extreme elation or mania to severe depression. The pathogenesis of this disorder is poorly understood. There is clear evidence of a heritable component with a sibling relative risk of 7–10 and heritability rated at 80–90%. **Twins studies** have shown variable rates of concordance with values from 33–75% for monozygotic and 0–13% for dizygotic twins. Differences in the diagnostic criteria applied, **ascertainment bias**, and selection of controls all help to account for this variation. However, heritability is not just about genetics, it involves many other factors and thus high heritability does not mean strong genetic associations.

Regions identified in early linkage and the 2007 WTCCC1 study

Several genome regions have been implicated in linkage studies (**Table 4.4**), and there is clear evidence of overlap with genetic susceptibility to schizophrenia and bipolar disorder.

Table 4.4 Linkage studies for risk loci in bipolar disorder.

Gene/allele	Location
MAFD1	18p
MAFD5	2q22–q24
MHW1	4p
MHW2	4q
MADF4 (region includes *PALB2*, *NDUFABI*, and *PCTN5*)	16p12

MHW indicates mental health and welfare. *MHW1* and *MHW2* are based on linkage studies in the Amish. MAFD is the name for bipolar disorder, based on manic affective disorder, then a number. Interestingly, OMIM reports linkage on chromosomes 5, 11, and 18, which have not been identified in association studies.

Associations with *DAOA* (D-amino acid oxidase activator), *DISC1* (disrupted in schizophrenia 1), *NRG* (neuregulin 1), and *DTNBP1* (dystrobrevin binding protein 1) have all been reported. In the 2007 WTCCC1 study, the strongest signal was on chromosome 16p12 (**Table 4.5**). Genes close to this region with potential clinical relevance include:

- *PALB2* (partner and localizer of *BRCA2*), which is involved in key structures within cells including chromatin.
- *NDUFAB1* [NADH dehydrogenase (ubiquinone) 1, α/β complex 1], which encodes a mitochondrial respiratory chain subunit.
- *DCTN5* (dynactin 5), which encodes a protein involved in intracellular transport that is known to interact with the DISC1 protein, the gene for which is already implicated in susceptibility to both bipolar disorder and schizophrenia.

Expanded analysis in the WTCCC1 study found strong associations with four regions of the genome, one of which is close to the *KCNC2* gene that encodes the Shaw-related voltage-gated potassium channel. Changes in the dynamics of ion channel activity are known to cause episodes of central nervous system disorders, including seizures, ataxia, and paralysis. This may include the extreme mood changes seen in bipolar disorder. Other highly ranked SNP signals indicated the importance of the γ-aminobutyric acid (GABA) neurotransmission pathway identifying three signals in particular rs7680321, rs1485171, and rs11089599, which are close to or within the genes for *GABRB1* (encoding the GABA A receptor β1), *GRM7* (glutamate receptor, metabotropic 7), and *SYN3* (synapsin III).

More recent GWAS

Following on from the 2007 WTCCC1 GWAS, additional studies of bipolar disorder have identified novel/different SNP associations that include:

- *DGKH* (diacyclglycerol kinase).
- *MYO5B* (myosin 5B).
- *NCAN* (neurocan), which is an extracellular matrix glycoprotein involved in cell adhesion and migration.
- *MAD1L1* (meiotic arrest deficient like 1), which is important in cell division.

Table 4.5 The main findings from GWAS in bipolar disorder.

Gene(s)	Location(s)	*P* value (if available)
Major observations of bipolar disorder genes from 2007 WTCCC1 GWAS		
MADF4 (region includes *PALB2*, *NDUFABI*, and *PCTN5*)	16p12	$<5.14 \times 10^{-8}$
KCNC2	12q21	$<2.18 \times 10^{-7}$
Most likely *SYN3*	22q12	$<1.61 \times 10^{-7}$
GABRB1	4p12	Not listed
GRM7	3p25–26	Possibly $<7 \times 10^{-6}$ in WTCCC study
NRG1	8p12	$<6.86 \times 10^{-6}$
Genes identified in other GWAS of bipolar disorder		
DGKH	14q11	Not listed
MYOB	2q32.3	Not listed – this could be due to linkage disequilibrium with risk allele(s) in 2q33 – WTCCC1 study above identified tagged SNP $<5.4 \times 10^{-5}$ as closest.
*HLA C*01:02*	6p21.3	Not listed
TRIM26	6p22.1	$<5 \times 10^{-16}$ in schizophrenia
Identified from meta-analysis of GWAS		
CACNAIC	12p13.3	$<7 \times 10^{-8}$
ANK3	10q21.2	$<9.1 \times 10^{-9}$
JAM3	11q25	$<1 \times 10^{-6}$
SLC39A3	19p13.3	$<5 \times 10^{-6}$
NCAN	19p13.11	$<2.14 \times 10^{-9}$ (OR = 1.17)
MAD1L1	7p22.3	$<1.28 \times 10^{-7}$ (OR = 1.6)
PBRM1	3p21.3	$<2.68 \times 10^{-9}$ (OR = 0.875)
ODZ4	11q14.1	$<4.4 \times 10^{-8}$ (OR = 0.89)

OMIM reports linkage on chromosomes 5, 11, and 18 which have not been identified in association studies. The WTCCC1 study data can be read in several ways; there are major associations and moderate associations, and associations only reported in the supplementary texts. In this chapter, the major associations are listed together with some of those from the supplementary text that could be identified by gene name, this leaves 13 listed as moderate unaccounted for in Table 4.4 (these are on chromosomes 1p31, 2p25, 2q12, 2q14, 2q31, 2q37, 3q37, 6p21, 6q22, 9q32, 14q22, 14q32, and 20p13). It is assumed that the tagged SNPs for 3p23, 8p12, and 16p12 are markers for *GRM3*, *NRG1*, and *PALB2*, respectively. *DAOA*, *DTNBP1*, and *DISC* are also not listed in Table 4.3 as they fail to reach minimum values for moderate association $P < 10^{-5}$.

The *DGKH* and *MY05B* genes are not currently listed as potential candidates in bipolar disorder, and interestingly these post-WTCCC1 studies did not immediately replicate the findings of the WTCCC1 study itself. Altogether, it is quite challenging trying to find a consistent message when reading through all of the papers on genetic association studies in bipolar disorder. On first review of the data there appears to be no consistency between them. However, further analysis, including meta-analysis, has revealed some important new and consistent associations. These include novel associations with the *CACNA1C* (calcium channel, voltage dependent, L-type α 1C subunit), and *ANK3* [ankryin 3, node of Ranvier (ankryin G)] genes, and confirmation of the *NCAN* association.

There are many reasons for these reported differences between initial studies and the meta-analysis studies. Bipolar disorder is not easily diagnosed, the genetic associations with the disease are quite weak, and some of the studies have been based on quite small case cohorts. Meta-analysis can iron out some of these wrinkles. One particular study that stands out and illustrates this well is that of Ferreira et al. (2008). In a GWAS of 1098 cases, Ferreira et al. first reported no major associations with bipolar disorder. The authors decided that this inconsistency may have been due to the small number in their own study cohort or the use of different tagged SNPs for their analysis. However, the authors were not discouraged. They decided to perform further analysis focusing on areas of known association from other studies, especially the recently identified weak association with the *CACNA1C* gene that had been identified by combining two data sets in a meta-analysis. When the authors fed their own data into a meta-analysis with the two previous studies, there was a weak association with the *CACNA1C* gene. In addition, by referencing the existing data to HapMap they were able to impute the genotypes for each case using PLINK software. This software uses the expected pattern of linkage disequilibrium to assign haplotypes and increases the genome coverage—in this case from approximately 350,000 to over 1.5 million SNPs (see Chapter 12). Imputation is increasingly used on GWAS, provided there is an adequate database of fully genotyped individuals on which to make the haplotype assumptions for the test sample. In this case, assigning non-genotyped alleles to the test samples on the basis of known linkage disequilibrium is a very profitable exercise, reducing cost and increasing efficiency. However, the quality of the study relies on the control data bank. This analysis identified the *ANK3* gene on chromosome 10q21 as the major susceptibility gene in bipolar disorder, with *CACNA1C* as the second strongest.

It must be stressed that many of these associations with bipolar disorder did not reach the significance threshold of $P < 10^{-7}$ in any of the initial analyses or for any individual group. Only the combination of the data provided P values of the required acceptable magnitude.

The future of genetic studies in bipolar disorder

Clearly, much more needs to be done in bipolar disorder, but taken together the observations identify a number of potentially important pathogenic pathways in bipolar disorder, particularly the GABA neurotransmission pathway. These studies also indicate the potential for polymorphisms in the genes involved in calcium channel activity essential for neurotransmitter release and genes from the solute carrier family (*CACNA1C* and *SLC39A3*, respectively) to be functionally associated with bipolar disorder. The *SLC39A3* gene encodes a solute carrier family 39 (zinc transporter) member 3. The association with *GRM7* illustrates the possibilities of unraveling the complexities of genetic links in bipolar disorder. *GRM7* encodes a member of a family of glutamate receptors. L-Glutamate is a major excitatory neurotransmitter in the central nervous system. L-Glutamate activates glutamate receptors such as *GRM7* and is active during most normal brain activities. *GRM7* is one of a family of metabotropic G-protein-coupled receptors that has been linked to the inhibition of the cyclic AMP cascade. It is quite easy to consider how polymorphisms that regulate this receptor could have functional effects in terms of neurotransmission and how the downstream consequences of such interpersonal variation may lead to a trait such as bipolar disorder (**Table 4.6**).

Table 4.6 The function of selected genes associated with increased risk of bipolar disorder.

Gene/allele	Function
PALB2	Partner and localizer for stabilization of *BRCA2*. Acts as a molecular scaffold attracting *BRCA1* and *RAD51* ensuring stabilization of the *BRCA1–PALB2–BRCA2* complex, which is required for homologous recombination.
KCNC2	Potassium-gated voltage channel Shaw-related family member 2. As the name implies, this mediates potassium ion permeability across the cell membrane.
NRG1	Glial cell growth factor interacting with tyrosine kinase receptors to recruit ERBB1 and ERBB2 co-receptors resulting in ligand-stimulated tyrosine phosphorylation and activation of ERBB receptors. Involved in induction of growth of glial and neuronal cells, among other cells Hence, early identification as a glial cell growth factor.
GABRB1	γ-Aminobutyric acid A receptor β1. Encodes a subunit of a chloride channel that mediates rapid inhibition of synaptic transmission in the central nervous system. This gene is strongly associated with schizophrenia.
GRM7	Metabotrophic G-coupled-protein receptor for glutamate associated with inhibition of the cyclic AMP cascade.
SYN3	May be involved in the regulation of neurotransmitter release and genesis of synapse.
CACNAIC	Calcium channel voltage-dependent type 1 α1 C subunit gene. Mediates influx of calcium into cells and is involved in neurotransmission.
ANK3	Ankyrin G is specifically found at a neuronal junction or axonal segment known as the node of Ranvier in the central and peripheral nervous systems. It is one of a group of proteins thought to be associated with various functions in the cytoskeleton.
JAM3	Junctional adhesion molecule 3. Promotes cell–cell adhesion.
SLC39A3	Solute carrier family member 39 family 3. Zinc transporter. Another member of the SLC family, *SLC6A4*, may be important in serotonin release.
NCAN	Neurocan. A chondroitin sulfate proteoglycan that is thought to be involved in control of cell migration and adhesion. Studies show that it is expressed in the hippocampus of mice and humans.
MAD1L1	Meiotic arrest deficient-like 1. Important in cell division and regulation of processes in meiosis.
PBRM1	Polybromo-1. Involved in transcriptional activation and repression of select genes by chromatin remodeling. It is thought to act as a negative regulator of cell proliferation.
ODZ4	The *ODZ4* gene encodes a human homolog of the *Drosophila* pair-rule gene Tenm (*odz*). The gene product may function as a signal transducer.

Functional studies on human tissue samples have shown that the *NCAN* and *MAD1L1* genes are both expressed in the hippocampus, and studies on mice show *NCAN* to be present in the cortical and hippocampal areas of the brain that are involved in cognition and regulation of emotions.

The study examples above also indicate that similar diseases can share risk alleles, but they also illustrate some of the difficulties of working with disorders that are not easily defined, such as bipolar disorder, compared with working with a disease with more physical manifestations, such as rheumatoid arthritis. Altogether, these studies serve as reminders about study planning, design, and the limitations of association studies (discussed in Chapter 3 and 5). Finally, these findings also illustrate the importance of taking a holistic view and not dismissing non-genetic factors. Many factors are known to contribute to bipolar disorder, not all of them are genetic (**Figure 4.12**).

Coronary artery disease is the most common cause of death in the developed world

The environment is known to play a major role in coronary artery disease, but genetic variation is also thought to have an impact. It is the most common cause of death in the developed world and second most common cause in medium- to low-income countries. Most patients develop atheromatous plaques on the walls of the coronary arteries, and this causes angina and myocardial infarction (heart attack). The plaques are made up of a lipid and fibrous matrix. The pathogenesis of the disease is complex with several known contributors, including endothelial cell dysfunction, oxidative stress, and inflammation, as well as environmental factors such as smoking and diet. The sibling relative risk for cardiovascular disease is quite low, between 2 and 7, reflecting the strong environmental input into the disease pathogenesis. In contrast to this, data from a study of 20,000 twins in the Swedish Twin Registry suggest that heritability for angina in men may be as high as 39% and death from coronary disease as high as 57%, with figures of 43 and 38% for women. This suggests that heritability is important and contributes to cardiovascular disease, but it is not as clear-cut as seen in other diseases. Once again this reminds us that the term heritability has multiple uses and does not simply refer to our genes. Heritability can include the

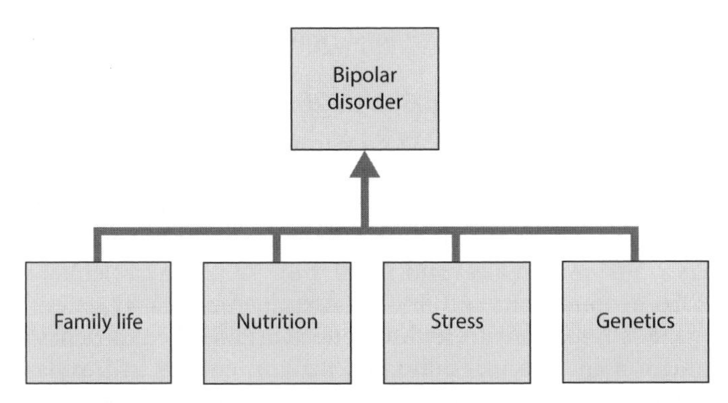

Figure 4.12: Different potential causes of bipolar disorder. The figure illustrates probable contributing factors to bipolar disorder. They may act individually or in combination. There is no restriction on the number of elements active. For example, stress in a genetically susceptible individual may trigger bipolar disorder, but nutritional status may also be a factor. However, it must be noted these are only possible relationships and not definite relationships.

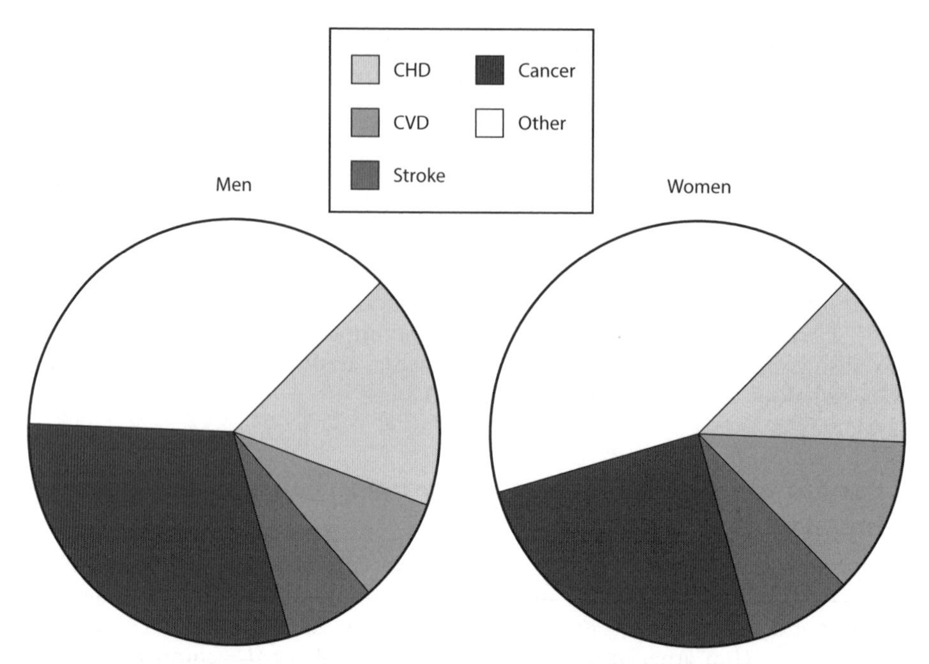

Figure 4.13: Impact of coronary artery disease based on figures from the British Heart Foundation. These figures for the UK population in 2008 show morbidity for coronary heart disease, cardiovascular disease, stroke, and cancer against other diseases and pathologies in men and women. It is quite clear that coronary artery disease-related conditions [coronary heart disease (CHD), cardiovascular disease (CVD), and stroke] account for a significant portion of deaths considering that coronary heart disease, cardiovascular disease, and stroke may all be the consequences of coronary artery disease. Compare the figures given for these three problems with diabetes, which is considered a major scourge and accounted for approximately 1% of deaths in 2008 in the UK. In comparison, the figures for coronary heart disease, cardiovascular disease, and stroke were 18, 8, and 7% in men (total 33%) and 13, 12, and 8% in women (total also 33%) that same year.

environment. As cardiovascular disease is a common cause of mortality (**Figure 4.13**), the search for risk alleles remains as intense as that for diseases with a greater and more obvious genetic component.

Genome-wide studies in cardiovascular disease can be divided into those produced prior to the publication of the WTCCC1 GWAS in 2007 and those that followed.

Studies of cardiovascular disease prior to 2007

Though the potential promise of GWAS in cardiovascular disease was well established before 2007, there were only two GWAS performed prior to 2007 that identified putative candidates. These two candidate genes were *ALOX5AP* on 13q12 [arachidonate 5-lipoxygenase-activating protein (also known as FLAP), with probability values around 2×10^{-6} for males and 24×10^{-4} for females] and *LTA4H* on 12q22–q23 (leukotriene A4 hydrolase, with OR 1.16 in Europeans and 3.57 in African-Americans) genes. The product of the *ALOX5AP* gene is important in **leukotriene** synthesis. Leukotrienes are involved in immune inflammation and vascular adhesion. Leukotriene A4 hydroxylase is an epoxide hydrolase that catalyses the final step in the biosynthesis of the pro-inflammatory mediator leukotriene B4—a leukotriene involved in cell adhesion via the STB4 biosynthesis pathway. Both of these genes are plausible candidates as risk markers for cardiovascular disease. However, the WTCCC1 report did not find any strong associations with these genes and nor did any other later GWAS report associations at these sites (**Table 4.7**).

Table 4.7 Risk loci for coronary artery disease identified in the 2007 WTCCC1 GWAS.

Gene	Location	*P* value and Odds Ratio (OR)
CDKNA2I CDKN2B (close to *MTAP*) (rs1333049)	9p21.3 (9p21.3)	1.79×10^{-14} (OR = 1.47) (1.75×10^{-16})
ADAMTS17 (close to *ADAMTS7* at 15q25.1 rs19994016)	15q24–q26	1.1×10^{-4} (1.6×10^{-7})
MTHFD1L (rs6922269)	6q25.1	6.33×10^{-6} (OR = 1.17)
(rs17672135) (close to *MIA3* at 1q41 rs17465637)	1q43	2.35×10^{-6}
(rs383830) (note association with 5q31 C4D group)	5q23	1.34×10^{-5}
(rs8055236)	16q23	5.6×10^{-6}
(rs7250581) (note *APOE* at 19q13)	19q12	2.5×10^{-5}
(rs688034)	22q12	3.75×10^{-6}

The lower case "rs" denotes the tagged SNP for the region. Not all potential regions identified were associated with specific genes in the WTCCC1 study. Many potential candidates were suggested following further analysis. Note association with *ALOX5AP* and *LTA4H* identified as candidates prior to 2007 were not confirmed in the WTCCC1 study. The values quoted depend on the reading of the paper and which table they were extracted from, as a range of possibilities are given in the different analyses. Note that not all associations were replicated in later studies.

The WTCCC1 study was also unable to replicate previously reported associations with the *APOE* gene. This problem may have been due to the fact that the tagged SNPs on the **Affymetrix gene chip** that was used in the study were not good markers for identifying *APOE* compared with direct genotyping. Interestingly, this explanation appears to be sound because studies in 2011 confirmed the association with *APOE* in cardiovascular disease. Considering the function of *APOE*, which mediates the binding, internalization, and catabolism of lipoprotein particles, this gene makes a very likely candidate. The APOE protein can serve as a ligand for the low-density lipoprotein receptor as well as interacting with the APOE receptor.

The 2007 WTCCC1 study and cardiovascular disease

The most important single observation on cardiovascular disease in the WTCCC1 study was a new and strong ($P < 10^{-14}$) association with a SNP on chromosome 9p21.3. The region contains genes for two cyclin-dependent kinase inhibitors (*CDKN2A* and *CDKN2B*). Both genes are widely expressed and are involved in the regulation of the cell cycle. *CDKN2B* is expressed in **macrophages** but not smooth muscle cells with fibrous lipid lesions. *CDNK2B* expression is induced by transforming growth factor-β, a signaling system that has been associated with cardiovascular disease. In addition to the two *CDNK* genes, the SNP on 9p21.3 maps close to the *MTAP* gene, coding for a methylthioadenosine phosphorylase enzyme that contributes to processes involved in adenine and methionine salvage. This is a ubiquitously expressed gene present in the cardiovascular system and it is also listed as a candidate for cardiovascular disease. Genetic variation in any one or all of these genes could contribute to the risk of cardiovascular disease. The association with *CDNK2A–CDNK2B* remains the strongest reported and most consistent association for cardiovascular disease.

The other associations with cardiovascular disease reported in the 2007 WTCCC1 study include modest associations with a SNP (rs6922269) within the *MTHFD1L* gene [methylene tetrahydrofolate dehydrogenase (NADP$^+$-dependent) 1-like]. This gene encodes the mitochondrial isozyme of C1 tetrahydrofolate (THF) synthase, which converts the single carbon units carried in folic acid. C1 THF synthase is used in several biological processes, including purine synthesis. Cardiovascular disease has also been associated with mutations in another THF enzyme, i.e. THF reductase (encoded by *MTHFR*). Interestingly, both *MTHFD1L* and *MTHFR* activity can influence plasma levels of homocysteine, indicating a common pathway for cardiovascular disease. One other gene identified in the expanded analysis as having a major risk for cardiovascular disease is the *ADAMTS17* (a disintegrin-like and metalloproteinase with thrombospondin type 1 motif 17). This gene encodes a protein involved in vascular extracellular matrix degradation and remodeling—a process central to atherosclerosis.

GWAS in cardiovascular disease beyond 2007

Since the publication of the WTCCC1 GWAS there have been many other studies of cardiovascular disease using GWAS. Two studies published in 2011 stand out. The first study was a meta-analysis of data from 14 GWAS comprising 22,233 cases and 64,762 controls with a replication cohort of 56,682 cases performed by the CARDIoGRAM Consortium.

In this analysis, 13 loci were identified for the first time as risk loci for cardiovascular disease and 10 of 12 previously reported risk loci were confirmed. The data from this study are summarized in **Table 4.8**. The authors of this study found that only three of the new loci were associated with traditional markers for cardiovascular disease, with the remainder being in gene regions not previously implicated in the pathogenesis of cardiovascular disease. The report also highlights the overlap between risk markers as five of the new loci have strong associations with other diseases or traits.

The second study in 2011 (also summarized in Table 4.8) used the data from the study above and a second data set produced by the C4D Consortium based on approximately 40,000 cases. This combined report identified 35 common gene variants at 34 loci as risk factors for cardiovascular disease and suggested that these common variants account for only 13% of the overall genetic risk of cardiovascular disease (**Table 4.9**).

The reports themselves are too detailed to reiterate every fact in this chapter, but they raise a number of important issues for anyone looking at published data or about to set out on a GWAS.

Many of the associations were not replicated

A review of the report of the 2007 WTCCC1 study in the light of the 2011 studies shows that only one of the associations for cardiovascular disease reported in 2007 has been replicated and all of the others are not confirmed (*ADAMTS7*, *MTHFD1L*, and *MTHFR* in particular) at a significance threshold of $P > 5 \times 10^{-8}$. One explanation for these failures in replication is the application of different gene chips or the closer association with tagged SNPs in a different location which may be identified in subsequent studies, but not identified (or perhaps even tested) in the initial study. However, all is not lost as some of this discordance may be explained through linkage disequilibrium whereby SNPs associated with CAD in original studies are in close linkage with other loci that have themselves been claimed as risk genes for cardiovascular disease. For example: *MIA3* (melanoma inhibitory activity family 3) on 1q41, *APOE* on 19q13, and *ADAMTS17* (ADAM

Table 4.8 Risk loci identified in the 2011 CARDIoGRAM study.

Gene(s)	Location	P value (OR)
Confirmed		
PCSK9	1p32.3	9.1×10^{-8} (OR = 1.15)
SORT1	1p13.3	2.89×10^{-10} (OR = 1.29)
MIA3	1q41	1.36×10^{-8} (OR = 1.2)
WDR12	2q33.1	1.12×10^{-9} (OR = 1.16)
MRAS	3q22.3	3.34×10^{-8} (OR = 1.15)
PHACTR1	6p24.1	1.15×10^{-9} (OR = 1.13)
LPA	6q25.3	3.0×10^{-11} (OR = 1.92)
CDKN2A/CDKN2B	9p21.3	1.35×10^{-22} (OR = 1.25/1.37) Identified in WTCCC
CXCL12	10q11.21	2.93×10^{-10} (OR = 1.33) Identified in MIGen Consortium
SH2B3	12q24.12	6.35×10^{-6} (OR = 1.13)
LDLR	19p13.2	9.73×10^{-10} (OR = 1.14) Identified in MIGen Consortium
MRPS6	21q22.11	4.22×10^{-10} (OR = 1.19)
Novel associations		
PPAP2B	1p32.2	3.81×10^{-19} (OR = 1.17)
ANKS1A	6p21.31	1.36×10^{-8} (OR = 1.07)
TCF21	6q23.2	1.07×10^{-12} (OR = 1.08)
ZC3HC1	7q32.2	9.18×10^{-18} (OR = 1.09)
ABO	9q34.2	4.08×10^{-14} (OR = 1.10)
CYP17A1/CNNM2/NT5C2	10q24.32	1.03×10^{-9} (OR = 1.12)
ZNF259/APOA5-A4-C3-A1	11q23.3	1.02×10^{-17} (OR = 1.13)
COL4A1/COL4A2	13q34	3.84×10^{-9} (OR = 1.07)
HHIPL1	14q32.2	1.14×10^{-10} (OR = 1.07)
ADAMTS7	15q25.1	1.07×10^{-12} (OR = 1.08)
SMG6/SSR	17p13.3	1.15×10^{-9} (OR = 1.07)
RASD1/SMCR3/PEMT	17p11.2	4.45×10^{-10} (OR = 1.07)
UBESZ/GIP/ATP5G1/SNF8	17q21.32	1.81×10^{-8} (OR = 1.06)

The CARDIoGRAM study data presented here indicates confirmation of 12 locations and 13 novel genes with $P < 10^{-8}$, with one exception (*SH2B3* at 6.35×10^{-6}).

Table 4.9 Risk loci identified in other GWAS (excluding WTCCC and CARDIoGRAM reports shown in Tables 4.7 and 4.8 respectively), but included in the list of 35 loci reported by Peden and Farrall (2011).

Locus	Location	OR (Source)
ABCG8	2p21	OR = 1.09 (MIGen Consortium)
IL5	5q31	OR = 1.02 (C4D Consortium 2011)
7q22	7q22	OR = 1.08 (C4D Consortium 2011)
TRIB1	8q24.13	OR = 1.04 (C4D Consortium 2011)
KIAA1462	10p11.23	OR = 1.07 (C4D Consortium 2011)
LIPA	10q23.31	OR = 1.08 (C4D Consortium 2011)
PDGFD	11q22.3	OR = 1.08 (C4D Consortium 2011)
APOE	19q13	OR = 1.14 (C4D Consortium 2011)
MRPS6	21q22.11	OR = 1.18 (MIGen Consortium)

The data presented here represent those loci not listed in Tables 4.7 and 4.8. Interestingly, the list includes *APOE*, which was not reported by the WTCCC1 2007 study, and *IL5*. Both of these gene loci have been associated with susceptibility to other diseases.

metalloproteinase with thrombospondin type 1 motif 17) on 15q25. Interestingly, *ADAMTS7* and *ADAMTS17* map close to each other, and it would be easy to make a false assumption that an association with a SNP on chromosome 15q25.1 was due to linkage with *ADAMTS7* rather than with *ADAMTS17*. This illustrates the importance of searching an area thoroughly and in fine detail. Weak or moderate associations are still at a risk of being overlooked, but it also is possible to make false assumptions through accepting weak associations.

There is at least one very strong and reproducible genetic association with cardiovascular disease—that with chromosome 9p21.3 and the *CDKN2A–CDKN2B* region. Here, the probability may be as high as 1.35×10^{-22}. However, the quoted probability values vary from study to study and even within studies, especially where numerous analyses are performed.

In summary:

- The impact of these genetic polymorphisms is small. All of the OR values indicate small effects. It is important to remember the role of environmental factors in cardiovascular disease and consider OR values for gene polymorphisms in that context.
- There is now a mass of data available and not all of the genes identified indicate obvious candidates for cardiovascular disease. This is important because identifying novel candidates may help to unravel the pathogenesis of cardiovascular disease.
- All of the studies above relate to European patients. However, it is important to consider other populations.

Genetic studies of cardiovascular disease in different populations

As with the European population, there have been many studies of the genetics of cardiovascular disease in populations outside Europe. One study in particular stands out. Published in 2012, a study of the Han Chinese population based on 1515 cases and 5159 controls with a replication cohort of 15,460 cases and 11,472 controls identified four new loci at the 5×10^{-8} threshold of significance and confirmed the associations with *CDKN2A–CDKN2B* reported in the European population, as well as the associations with *PHATCTR1* (phosphatase and actin regulator 1), *TCF21* (transcription factor 21), and *C12orf51* (chromosome 12 open reading frame 51).

On close reading, the Chinese study illustrates a number of important issues in this area of research. Most important among these is the observation that not all of the associations reported in Europeans are replicated in the Chinese population (**Table 4.10**). This population variation is well demonstrated by the SNPs that identify common genetic variations in the 12q24 region. The variations in this region that are common in Europeans are not

Table 4.10 Risk loci for coronary artery disease reported in the Han Chinese population.

Gene(s) and tagged SNP	Location	*P* value (OR)
Han Chinese 2012		
TTC32–WDR35 (rs2123536)	2p24.1	6.83×1^{-11} (OR = 1.12)
GUY1A3 (re1842896)	4q32.1	1.26×10^{-11} (OR = 1.14)
6orf10–BTNL2 (rs9268402)	6p23.32	2.77×10^{-15} (OR = 1.16)
APT2B1 (rs7136259)	12q21.33	5.68×10^{-10} (OR = 1.11)
Other Chinese studies		
C6orf105 (rs6903956)	6p24.1	Androgen-dependent TFPI-regulating protein
Strong association reported in Europeans and confirmed in the 2012 Han Chinese Study		
PHACRT1	6p24.1	
TCF21	6q23	
CDKN2A–CDKN2B	9p21.3	
C12orf51	12q24.13	
Weak but consistent other European (confirmed in the 2012 Han Chinese study)		
PPAP2B	1p32.2	
MIA3	1q41	
LIPA	10q21.31	
CYP17A1/CNNM2/NT5C2	10q24.32	
ZNF259/APOA5-A4-C3-A1	11q22.3	
ADAMTS7	15q25.1	
SMG6/SSR	17p13.3	

This table illustrates the importance of studying several different populations. The listed risk loci for coronary artery disease in the Han Chinese population includes some loci also identified as risk loci in Europeans and some only found in the Han Chinese. Not all the loci identified in Europeans are confirmed in the Han Chinese.

found in the Chinese (i.e. the Han population appears to be monomorphic with respect to these particular variants), whereas all of the variants at 12q24 reported in the Han Chinese appear to be monomorphic in the European population. Population variation is a major factor in genetic association studies. Even where there is general agreement about associations, there are often subtle differences between populations. This can be seen with respect to the *CDNK2A–CDNK2B* loci, where the pattern of linkage disequilibrium varies between the Han Chinese and the European populations even though both populations carry genetic associations with the *CDNK2A–CDNK2B* haplotype. However, not all associations are restricted to single populations and there is considerable overlap here. In addition to the four confirmed associations reported, the Chinese study investigated 29 different cardiovascular disease associated loci identified from European studies and found a further seven with consistent nominal associations (1p32.2, 1q41, 10q23.31, 10q24.32, 11q22.3, 15q25.1, and 17p13.3). Furthermore, 11 of the SNPs (at 10 loci) associated with cardiovascular disease in Europeans were monomorphic in the Han Chinese population, and use of proxy SNPs as substitutes for the monomorphic SNPs identified three further associations at 3q22.3, 6p26, and 17p11.2. In their conclusion, the authors stated that both shared and unique genetic associations occur in this complex disease. This idea of a mixture of shared and unique susceptibility alleles in different populations is fundamental in complex disease. Comparison of populations may be very fruitful in helping us to understand diseases pathogenesis. In this **compare and contrast model**, comparing and contrasting data may pull out the real plums.

Genetic polymorphism in cardiovascular disease and disease pathogenesis

What does all this information (so far) tell us about the pathogenesis of cardiovascular disease and how does it direct functional research? It is quite interesting to create pictorial views of how these genes may interact to produce a disease portfolio for cardiovascular disease. We can do this by looking at the function of the different genes associated with cardiovascular disease listed in Tables 4.7–4.10. There are obvious groups of genes (**Figure 4.14**):

- Genes associated with the immune response through inflammation and vascular adhesion and cell migration, e.g. *CXCL12, SH2B3, IL5, PDGFD*, and *6orf10–BTNL2*.
- Genes associated with lipid metabolism and cholesterol homeostasis, e.g. *PCSK9, LIPA, LDLR, APOA5-A4-C3-A1, ABCG8, APOE*, and *APT2B1*.
- Genes associated with cell growth, e.g.*CDKN2A* and *CDKN2B*.
- Genes associated with fibrosis and regeneration, e.g. *MIA3, COL4A1, COL4A2*, and *ADAMTS7*.
- Genes associated with coagulation and anticoagulation activity, e.g. *C6orf105* and *ABO*.
- Genes associated with other activities, including formation of cellular particles (*WDR12*), different kinase pathways (*MRAS*), phosphatase activity (*PHACTR1*), mitochondria (*MRPS6*), cell differentiation (*TCF21*), a number of zinc finger genes (*ZC3HC1* and *ZNF259*), a number of cytochromes (*CYP17A1*), ATP synthatase (*ATP5G1*), and nitric oxide signaling (*GUY1A3*).

The associations that have been reported in cardiovascular disease are mostly weak. On a good day it is easy to look at these lists and identify obvious targets for further genetic studies, and in some cases functional studies, links with lipid metabolism, and cholesterol homeostasis, for example. However, it is important to be mindful of seeking simple solutions to complex problems. There is no doubt that strong associations that have been

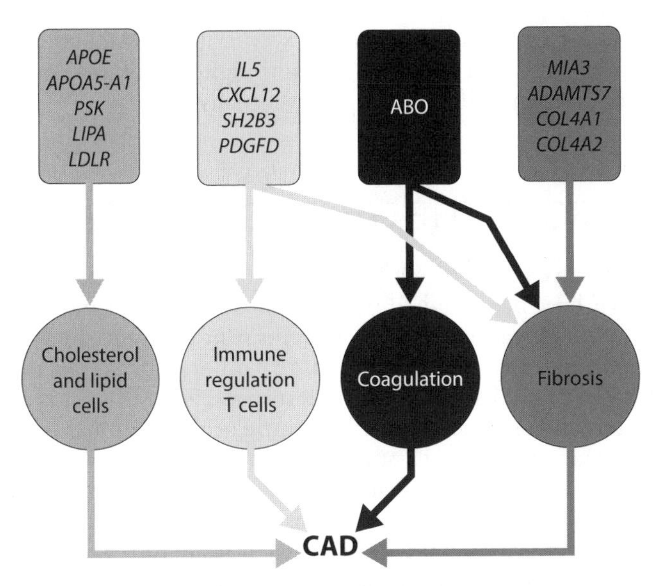

Figure 4.14: Systems and pathways in coronary artery disease (CAD). This simple picture illustrates the broad concept that identifying risk genes may help us understand disease pathology. Some of the candidate alleles identified (e.g. cholesterol and lipid) are quite obvious candidates for coronary artery disease. It is not possible to list all of the susceptibility alleles nor is it yet possible to make a clear link between some of the identified associations and disease pathology. Genes that are associated with cell growth (e.g. *CDKN2A* and *CDKN2B*) are not included here nor are genes that are associated with other activities, including formation of cellular particles (*WDR12*), different kinase pathways (*MRAS*), phosphatase activity (*PHACTR1*), mitochondria (*MRPS6*), cell differentiation (*TCF21*), a number of zinc finger genes (*ZC3HC1* and *ZNF259*), a number of cytochromes (*CYP17A1*), ATP synthetase (*ATP5G1*), and nitric oxide signaling (*GUY1A3*). Finally, this illustration does not take account of the strong environmental impact in coronary artery disease.

confirmed are the best place to start when we begin to construct a picture of the disease pathology. These genetic associations are after all the most reliable and most frequently confirmed, and they also tend to be those which are seen in multiple populations. It is more difficult to link function to other genes (e.g. the zinc finger genes) and cardiovascular disease, but these too are also associated with an increased risk of cardiovascular disease and therefore a better understanding of genotype–phenotype relationships is required.

Shared risk alleles between cardiovascular disease and other diseases

We should also not overlook the genetic similarities between diseases. Autoimmune diseases share risk alleles. Genetic variants associated with the immune response play a major role in many diseases and often the same allele is listed in two or more. With reference to cardiovascular disease, it is tempting to look for links with T2D (Chapter 10). However, despite there being considerable overlap between cardiovascular disease and T2D, studies so far have identified very few shared risk alleles. The two genes that have been identified in both T2D and cardiovascular disease are *CDKN2A*, which encodes a cyclin-dependent kinase inhibitor on chromosome 9p21.3, and *SH2B3* on chromosome 12q13. *CDKN2A* is one of two cyclin-dependent kinase inhibitors associated with CAD with cardiovascular disease. *SH2B3* maps close to a tyrosine kinase precursor gene *ERBB3*. A functional link is yet to be established between these genes and either of the two diseases, but associations shared between diseases are always interesting because they may indicate a common functional link in the disease pathology. Looking at cardiovascular disease and T2D, it is surprising to see so few shared associations. However, it is possible that there are shared associations out there, and with larger studies and higher-resolution genotyping or even direct sequence analysis they will be cataloged eventually.

4.5 WHAT ABOUT THE OTHER DISEASES?

It would be wrong to give the impression that the five disease examples in this chapter are the only diseases that have been studied in this way. As the book shows us, there are many more diseases under investigation, though even this is not enough. In 2012, 1350 GWAS had been published reporting data on at least 421 different traits. In the last quarter of 2013 the total number of published GWAS reports was 1960. This list is updated regularly by the National Human Genome Research Institute at the National Institutes of Health website (http://www.genome.gov/gwasstudies). Some of the more interesting associations described include those we have discussed already (Alzheimer's disease and Crohn's disease, Chapter 2) or will be discussing later in this book (cancer in Chapter 9 and diabetes in Chapter 10). However, we cannot consider 421 or more different diseases here or even in the whole book. That is not our purpose and therefore our consideration will be limited to selected examples. These will be used to illustrate concepts and ideas as well as provide knowledge on specific diseases. This discussion and the lists that accompany the examples are intended to be a starting point for further exploration by students of this topic and are by no means comprehensive. The intention is to provide examples that illustrate concepts that are transferable from one disease to another. It is also important to consider the overlap between disease and between populations, where possible (**Figure 4.15**). Associations that persist in different populations can be more informative than those that are population specific, but not always.

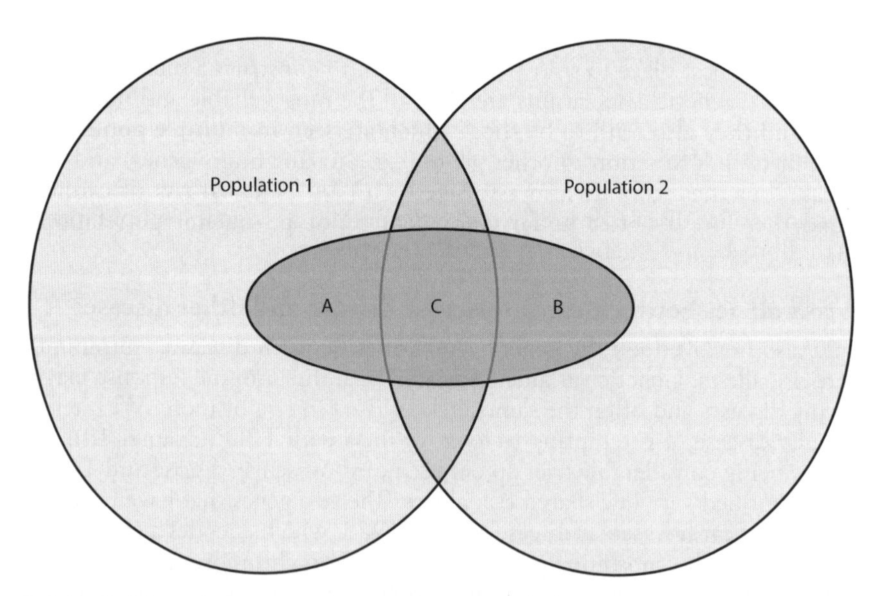

Figure 4.15: Similarities and differences in risk alleles in different populations with the same diseases. The figure illustrates the genetic risk portfolio for a disease in two populations: population 1 and population 2. The smaller oval shape illustrates the disease population for both populations 1 and 2. The oval is subdivided into three regions. Those in areas A and B share no risk alleles between the two populations, while those in areas C share risk alleles.

CONCLUSIONS

Considerable scientific resources, both financial and intellectual, have been ploughed into studying the genetics of complex disease. One may ask the question why? After all, complex diseases are not simple, they do not lend themselves to simple analysis, and they do not conform to Mendelian patterns of inheritance. The answer lies with the promises of the HGP. Under this banner, three promises can be considered: (1) identifying risk alleles will aid disease diagnosis, (2) identifying risk alleles will aid patient treatment/management and care, and (3) identifying risk alleles will inform the debate on disease pathogenesis and may ultimately lead to the development of new therapies. Clearly these outcomes are all linked and though at present there is less potential to use risk alleles in disease diagnosis, there is great potential to use the new genetics in patient management, including the potential for developing personalized therapy. However, so far, the greatest success has been with the third promise.

This is illustrated through the selected disease examples discussed above: ankylosing spondylitis, HCV, rheumatoid arthritis, bipolar disorder, and cardiovascular disease. In ankylosing spondylitis, the idea of molecular mimicry based on the shared sequence of self-peptides and those of pathogens was proposed long before the publication of the map of the human genome, but recent GWAS have led to the identification of multiple immunoregulatory genes, all of which helps to support the concept of this being an immune-mediated or autoimmune disease and will enable us to dissect the pathogenesis of this disease. In fact, significant progress is being made in ankylosing spondylitis following recent GWAS. As with ankylosing spondylitis, in HCV the association with HLA *DQB1*03:01* has been linked to T cell function, though this was also identified prior to the publication of the first draft of the map of the human genome. However, once again, subsequent genetic studies have identified several other risk immune alleles for HCV, some of which may interact with *DQB1*03:01* and others that may influence viral clearance independently. These include *IFNL3* (previously *IL28*) and *IFNL4*. In rheumatoid arthritis, multiple risk alleles mostly associated with immunity have been identified, including some cytokines and some chemokine receptors (e.g. *IL6* and *CCR6*). In bipolar disorder, the two major genes for focus are *ANK3* and *CACNA1C*. It is easy to see the potential for polymorphisms in the genes involved in calcium channel activity (such as *CACNA1C*) that is essential for neurotransmitter release being important in bipolar disorder. In cardiovascular disease, it is easy to see the potential for genes in lipid metabolism, cell growth, fibrosis, and regeneration as potential risk genes.

However, this will all mean nothing unless we can link genotype to phenotype—the next piece in the jigsaw. It is OK to be selective in terms of deciding which genes to look at as long as all of the potential candidates are considered at some point. Otherwise, the genotyping will have been a waste of time and the phenotyping work will fall into the same trap that applied to early genetic association studies, i.e. rejection of positive associations, but not based on numbers or study design (as in the past), but this time based on absence of knowledge or absence of an obvious (simple) link.

Together, these examples illustrate the potential of this work. There is clear progress and there is a great deal to be gained from this work. However, science is a slow process and not all studies are easily replicated. This is most strongly illustrated in studies of bipolar disorder. Overall, we must be sensible and proceed with caution because there is still a great deal of work to be done.

FURTHER READING

Books

Lechler R & Warrens A (eds) (2000) HLA in Health and Disease, 2nd ed. Elsevier.

Strachan T & Read AP (2011) Human Molecular Genetics, 4th ed. Garland Science. Chapter 16 deals with identifying human genes and susceptibility factors, and provides excellent additional data for students wishing to delve deeper into medical genetics. There is also additional relevant information in the other chapters.

Tiwari JL & Terasaki PI (1985) HLA and Disease Associations. Springer. Chapter 8 on Neurology covers early studies of multiple sclerosis and narcolepsy plus other neurological conditions. One of the authors of the 1985 book, Paul Terasaki, is considered as one of the founding fathers of HLA typing and immunogenetics.

Articles

Barrett JC (2012) From HLA association to function. *Nat Genet* 44:235–236. Review of the paper by Raychaudhuri et al. (2012).

Baum AE, Akula N, Cabanero M et al. (2008) A genome-wide association study implicates diacylglycerol kinase eta (DGKH) and several other genes in the etiology of bipolar disorder. *Mol Psychiatry* 13:197–207.

Brewerton DA, Hart FD, Nicholls A et al. (1973) Ankylosing spondylitis and HL-A 27. *Lancet* 301:904–907. This is the original identification of association between ankylosing spondylitis and *HLA-B*27*; see also Schlosstein et al. (1973).

Brown MA, Pile KD, Kennedy LG et al. (1996) HLA class I associations of ankylosing spondylitis in the white population in the United Kingdom. *Ann Rheum Dis* 55:268–270. This paper is the gold standard paper for ankylosing spondylitis and *HLA-B*27*.

Cichon S, Muhleisen TW, Degenhardt FA et al. (2011) Genome-wide association study identifies genetic variation in neurocan as a susceptibility factor for bipolar disorder. *Am J Hum Genet* 88:372–381. This article is a very good overview of the genetics of bipolar disorder with original data and some good examples on how to plan GWAS.

Collins FS & McKusick VA (2001) Implications of the human genome project for medical science. *JAMA* 285:540–544. This is a very important review that provides the reader with a guide to and an illustration of the potential for understanding complex disease in the immediate post-genome period.

Ferreira MA, O'Donovan MC, Meng YA et al. (2008) Collaborative genome-wide association analysis supports a role for *ANK3* and *CACNA1C* in bipolar disorder. *Nat Genet* 40:1056–1058.

Helgadottir A, Manolescu A, Thorleisfsson G et al. (2004) The gene encoding 5-lipoxygenase activating protein confers risk of myocardial infarction and stroke. *Nat Genet* 36:233–239.

Helgadottir A, Manolescu A, Helgason A et al. (2006) A variant of the gene encoding leukotriene A4 hydrolase confers ethnicity-specific risk of myocardial infarction. *Nat Genet* 38:68–74.

Hirschhorn JN, Lohmueller K, Byrne E & Hirschhorn K (2002) A comprehensive review of genetic association studies. *Genet Med* 4:45–61. This paper provides an interesting pre-GWAS view of association studies.

Lees CW, Barrett JC, Parkes M & Satsangi J (2011) New IBD genetics: common pathways with other diseases. *Gut* 60:1739–1753. This paper provides a current summary of the genetics of Crohn's disease and also highlights the possibility of shared pathways.

Lu X, Wang L, Chen S et al. (2012) Genome-wide association study in Han Chinese identifies four new susceptibility loci for coronary artery disease. *Nat Genet* 44:890–894. This study not only identifies association with cardiovascular disease specific to the Han Chinese population, but also confirms some shared associations with the European population. Data from this study are summarized in Table 4.10.

Peden JF & Farrall M (2011) Thirty-five common variants for coronary artery disease: the fruits of much collaborative labour. *Hum Mol Genet* 20:R198–R205. This review paper summarizes the findings of two major groups (the CARDIoGRAM Consortium and the C4D Consortium) and identifies a total of 35 associated common variants. The data from this paper are summarized in Tables 4.8 and 4.9. The paper provides a useful discussion of the differences between the two studies.

Raychaudhuri S, Sandor C, Stahl EA et al. (2012) Five amino acids in three HLA proteins explain most of the association between MHC and seropositive rheumatoid arthritis. *Nat Genet* 44:291–296. This original paper is reviewed by Barrett (2012)—reading both is useful to give a full understanding of the paper and the meaning of the results.

Schlosstein L, Terasaki PI, Bluestone R & Pearson CM (1973) High association of an HL-A antigen W27 with ankylosing spondylitis. *N Engl J Med* 288:704–706. One of two original reports (see also Brewerton et al. 1973) of the genetic association of *HLA-B*27* with ankylosing spondylitis. This example illustrates how not all findings from pre-genome and pre-GWAS studies failed to be replicated. Notice in both cases the HLA nomenclature differs from present. Standard nomenclature was only just coming into use at this time and in retrospect these differences can cause some confusion when looking at early texts. However, this allele family is now labelled *HLA-B*27*. It is important to note the quality of phenotype for B antigens in 1973 was reasonable for common antigens such as B27, but poor for rare antigens and variants.

Schunkert H, Konig IR, Kathiresan S et al. (2011) Large scale association analysis identifies 13 new susceptibility loci for coronary heart disease. *Nat Genet* 43:333–340. This paper describes a large-scale study on cardiovascular disease based on a meta-analysis of 14 GWAS. The data are presented in Table 4.8.

Sklar P, Smoller JW, Fan J et al. (2008) Whole genome association study of bipolar disorder. *Mol Psychiatry* 13:558–569.

Stahl EA, Raychaudhuri S, Remmers EF et al. (2010) Genome-wide association study meta-analysis identifies seven new rheumatoid arthritis risk loci. *Nat Genet* 42:508–515. This is an excellent review of genetics of rheumatoid arthritis and provides a good starting point for any student looking at this disease.

Stranger BE, Stahl EA & Raj T (2011) Progress and promise of genome-wide association studies for human complex trait genetics. *Genetics* 187:367–383. This is an excellent review of the planning and how to do GWAS with data on rheumatoid arthritis to illustrate the ideas. There is fresh data in this paper to add to the paper by Stahl et al. (2010).

Symmons D, Turner G, Webb R et al. (2002) The prevalence of rheumatoid arthritis in the United Kingdom: new estimates for a new century. *Rheumatology* 41:793–800.

The 1000 Genomes Project Consortium (2010) A map of human genome variation from population-scale sequencing. *Nature* 467:1061–1073.

The Human Genome (2001) *Nature* 409:813–958. The complete issue of *Nature* mostly dedicated to the HGM with multiple papers, letters, and editorial commentary from multiple authors. It is full of useful insight and critical discussion. Any student of human genetics should read this issue.

The International HapMap Consortium (2005) A haplotype map of the human genome. *Nature* 437:1299–1320.

The International HapMap Consortium (2010) Integrating common and rare genetic variation in diverse human populations. *Nature* 467:52–58.

The Wellcome Trust Case Control Consortium (2007). Genome-wide association study of 14,000 cases of seven common diseases and 3,000 shared controls. *Nature* 447:661–678. This is another landmark paper highlighting the development and application of new technologies (GWAS) in complex disease research, and it provides a mass of useful information for those studying complex disease.

Vassos E, Steinberg S, Cichon S et al. (2012) Replication study and meta-analysis in European samples supports association of the 3p21.1 locus with bipolar disorder. *Biol Psychiatry* 72:645–650. This paper describes a recent GWAS and meta-analysis study of bipolar disorder that illustrates some of the problems and uses of this type of analysis.

Zhang J, Zahir N, Jiang Q et al. (2011) The autoimmune disease-associated *PTPN22* variant promotes calpain-mediated Lyp/Pep degradation associated with lymphocyte and dendritic cell hyperresponsiveness. *Nat Genet* 43:902–907.

Online sources

http://www.wtccc.org.uk/ccc2

http://www.ncbi.nlm.nih.gov/SNP
This an essential site for updates on SNPs.

http://www.ncbi.nlm.nih.gov/omim
This is an essential site for updates and information on genetic disease of all types.

http://www.ncbi.nlm.nih.gov/projects/SNP/snp_summary.cgi
This site provides information on reference SNP clusters in the genome.

http://hapmap.ncbi.nlm.nih.gov
This is the site of the results of the International HapMap project.

http://pngu.mgh.harvard.edu/~purcell/plink
PLINK is a freely available, open source, whole genome association analysis toolset designed for quality control and analysis of GWAS data.

http://www.r-project.org
This is another freely available and excellent package for statistical computing that can be used with existing commercial packages such as SPSS or SAS.

http://www.genecards.org
This site is sponsored by the Weizmann Institute of Science and is free for use by academic and non-profit institutions. Good for gene identification, location, and function.

http://www.genome.gov/gwastudies
A catalog of published GWAS. This website provides a catalog of listed associations for GWAS with SNP numbers greater than 100,000 and associations with a significance threshold cut of 1.0×10^{-5}.

Statistical Analysis in Complex Disease: Study Planning and Data Handling

Now that we have considered why we are interested in the genetics of complex disease and how to design studies, it is important to consider how to handle the data from these studies. This is done through the application of complex statistical analyses. No longer is statistics a minor consideration in genetics; statistical genetics has become a science in itself and statistical geneticists have become major players in complex disease. The reasons behind this are clearly illustrated in Chapter 3. In the past, without good statistical advice, studies were often badly designed and this led to the production of numerous false-positive and false-negative genetic associations. However, this is not a statistics textbook, but given the importance of this subject it is essential to consider the basic principles that are currently applied in complex disease. Therefore, this chapter will consider some of the different methods available to test statistical significance in both linkage and association analysis. It will start with some basic methods applied to early studies, but will concentrate mostly on statistics in genome-wide association analysis.

As stated in Chapter 3, there are two types of study in complex disease: linkage and association. These have been discussed at some length in previous chapters and they will only be considered here from a statistical viewpoint.

5.1 LINKAGE ANALYSIS

Linkage analysis is based on the detection of a physical link between a genetic marker and a disease or disease subgroup. Linkage analysis can only be performed in families and sibling pairs. Linkage analysis is most often performed in Mendelian diseases where familial occurrence is common. Linkage analysis is less useful in complex disease. To illustrate linkage we can consider the example of nail–patella syndrome shown in **Figure 5.1**. Nail–patella syndrome is an **autosomal dominant disorder** characterized by small poorly developed nails and primitive knee caps. There is close linkage between this syndrome and the genes encoding the ABO blood group on chromosome 9. In this illustration the ABO alleles are I^A for blood group A, I^B for blood group B, and *i* for blood group O. The A and B phenotypes are co-dominant, and the alleles I^A and I^B are dominant over *i*.

In Figure 5.1, the ABO genotypes can be inferred from the phenotypes of the offspring produced. Female I-2 (i.e. generation I, person 2), for example, has blood group B, which has two possible genotypes: I^BI^B or I^Bi. Looking at her offspring, one is blood group O and must therefore have inherited an *i* allele from each parent. Consequently, the genotype of female I-2 must be I^Bi. Similarly, because O requires *i* from both parents, then the genotype of male I-1 (who displays the A blood group) must be I^Ai. Otherwise his offspring would have blood group A because A is dominant over O. Nail–patella

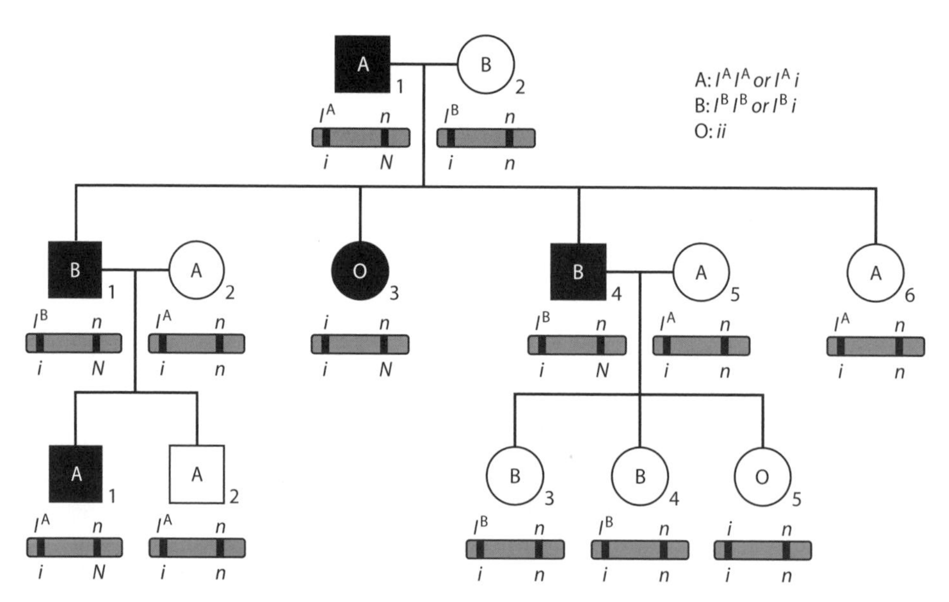

Figure 5.1: Linkage between ABO blood group and nail-patella syndrome. The figure illustrates three generations of a family, and shows the blood group and phenotype for nail-patella syndrome (a Mendelian autosomal dominant disorder). The ABO blood type is indicated in each circle or square (I^AI^A, I^BI^B, and *ii*). The genotype inferred from the phenotype is given below each circle or square. There are two possibilities for the nail-patella genes: *n* for wild-type and *N* for mutant. Circles indicate female family members and squares indicate male family members. In this figure the grandfather, generation I, person 1 (I-1), is blood group A heterozygous I^Ai and is heterozygous for the nail-patella trait (*nN*), and will display the trait (indicated here by black symbols). The grandmother, generation 1 person 2 (I-2), is blood group B heterozygous and carries two copies of the wild-type nail-patella gene, and is therefore unaffected (indicated in white). The grandfather's haplotypes are I^An and *iN*. Those who inherit the paternal *i* allele in this family also inherit the *N* nail-patella gene mutation (all shown as black squares or circles). Not all O blood group individuals have nail-patella syndrome (it is quite a rare trait); however, this example indicates extreme linkage disequilibrium and suggests that the two genes are not independently assorted.

syndrome is illustrated in Figure 5.1 by either n (wild-type) or N (mutant). The N allele is the autosomal dominant disease-causing allele. In this model we assume male I-1 is heterozygous for the nail–patella trait (nN) and therefore affected. In contrast, female I-2 is homozygous wild-type (nn) and therefore unaffected by this autosomal dominant disease. The genotype of the generation I is:

$$I^{A}i\ Nn \times I^{B}i\ nn$$

In the second generation we find that only those offspring who inherit the paternal i allele have the nail–patella trait and offspring with blood group A do not. This suggests that the genes for the nail–patella syndrome are not **independently assorted**. If these two genes were independently assorted, then some of the children with the paternal A blood group could inherit the nail–patella syndrome from their father. This outcome is really important as we need this information to determine whether the nail–patella gene in the father is on the same chromosome as the marker allele (I^{A}) or not. This is called determining the **phase** of these two loci. We can establish the phase of the above loci in the parents as:

$$\frac{I^{A}n}{iN} \times \frac{I^{B}n}{in}$$

Further examination of the pedigree in Figure 5.1 shows that while there is no recombination in the offspring of generation I, there are two instances of recombination in the offspring of generation II. If we focus on the couple II-1 and II-2 with the following genotypes:

$$\frac{I^{B}n}{iN} \times \frac{I^{A}n}{in}$$

we can see that one of the children (male III-2) is unaffected and has blood group A with the following genotype:

$$\frac{I^{A}n}{in}$$

This child must have inherited the i and n alleles from his father and not the i and N combination that were expected, and because these alleles are on different chromosomes in the father a crossover must have taken place. Similarly, crossover must have occurred in child III-3.

Individuals III-2 and III-3 are examples of recombination between the two loci for blood type (in this case the marker locus) and disease, and from this information we can calculate the **recombination fraction** (θ) for this family. The distance between two loci is defined in terms in units of **centimorgans** (named after geneticist Thomas Hunt Morgan and denoted cM). If two loci are 1 cM apart, then there is a 1% chance of recombination between the two loci as the chromosome is passed from one generation to the next.

It is essential to have informative families in linkage analysis, preferably with a large number of offspring and several generations with both affected and unaffected members in each. When small numbers are included there is a greater risk of false-positive and false-negative results. Recombination is more frequent in cases where the trait and **marker locus** are not close together, and the evidence for linkage can be quite low. Linkage analysis needs to take into account these possibilities. The LOD score (logarithm of odds) is a measure of the significance of linkage.

The LOD score is a measure of significance of linkage between a trait and a marker allele

The LOD score compares the likelihood of obtaining the results if the two loci are linked versus the likelihood of observing the same data purely by chance. The likelihood of linkage depends on the frequency of recombination and recombination depends on the genetic distance between the genes. In 1956, Morton published a model for the calculation of the LOD score symbolized at the time as the ***Z* score**. The calculation is based on standard pedigrees and is calculated for a range of recombination fraction values (θ). In this model, the highest LOD score identifies the recombination fraction for the marker and the trait, and is presented as a logarithm in base 10. A positive LOD score indicates the presence of linkage, whereas negative LOD scores indicate that linkage is less likely. By convention, a LOD score of 3 or greater is considered as strong evidence of linkage between two loci. A LOD score of −2 or less is taken as evidence of no linkage. A LOD score between −2 and 3 is a gray area, with some statisticians arguing the upper limit should be lowered to 2 from 3 and others suggesting that even 3 is too low. Currently, it is reasonable to say a value in-between the two above values is inconclusive. It is important to keep in mind that when using a logarithm to base 10, a value of 3 indicates 1000:1 odds for linkage, suggesting that the likelihood (L) that the observed linkage occurs by chance is less than 0.001. An example of how LOD score is calculated is shown in **Box 5.1**. Finally, linkage analysis can be either two-point mapping or multipoint (locus) mapping. **Multipoint mapping** is considered more useful and is most frequently used in the present era.

BOX 5.1 THE LOD SCORE METHOD FOR LINKAGE

The **LOD score** indicates the logarithm of the odds to base 10 and it is calculated using the recombination fraction (distance) between two loci on the same chromosome, which is denoted by the Greek letter θ. θ is the likelihood (L) of a recombination event between two loci and it is a function of the distance between the two loci. Two unlinked loci have $\theta = 0.5$; the closer they are, the lower the recombination fraction becomes. We use the likelihood ratio or OR (the ratio between two probabilities) to estimate the probability that two loci are linked such as:

$$\frac{L(\theta)}{L(\theta = 0.5)} = \frac{\text{likelihood of an offspring's sequence with recombination frequency } \theta}{\text{likelihood of an offspring's sequence if loci are unlinked}}$$

If loci were unlinked, the most likely recombination frequency would be 1/2 or 0.5, and in this case the numerator and the denominator of the ratio would be the same, and thus the OR will be 1. The easiest way to calculate the above likelihood ratio is to employ the logarithms (logarithm of odds to the base 10 LOD score) such as:

$$\text{LOD score} = Z(\theta) = \log_{10} \frac{L(\theta = \hat{\theta})}{L(\theta = 0.5)}$$

BOX 5.1 THE LOD SCORE METHOD FOR LINKAGE (*Continued*)

where $\hat{\theta}$ is the calculated recombination rate expressed as the distance between two loci. A likelihood ratio of 1:1 equals to a LOD score of 0, a ratio of 10:1 equals of LOD score of 1, a ratio of 100:1 equals a LOD score of 2, etc. A way of calculating LOD scores using the above equation is to consider the number of births in a family (offspring) and the number of recombinants as:

$$\text{LOD score} = Z(\theta) = \log_{10}\frac{(1-\theta)^{n-r}\theta^{r}}{(0.5)^{n}}$$

where n is the total number of births and r is the number of recombinant types. If we assume a test family with 18 non-recombinant offspring and two recombinant offspring between two loci A and B, we need first to know the distance between the two loci. Distance is measured in centimorgans (cM) and in this example is 10 cM, describing a recombination frequency of 1% and estimates the value of θ. We will have:

$$n = 18 + 2 = 20$$

$$r = 2$$

$$\theta = (10/100) = 0.1$$

$$Z(0.1) = \log_{10}\frac{(1-0.1)^{20-2}0.1^{2}}{(0.5)^{20}} = 3.2$$

The resulting LOD score of 3.2 indicates that the two tested loci A and B show a strong linkage.

By convention a $Z(\theta)$ greater than or equal to 3 (likelihood ratio = 1000:1, in favour of linkage) is considered to be proof of linkage between two loci. A $Z(\theta)$ of −2 (likelihood ratio = 0.01:1) or less instead, is taken as proof of not linkage. A $Z(\theta)$ between −2.0 and 3.0 is considered inconclusive and warrants for further studies. Some authors, however, consider a $Z(\theta)$ of 2 (likelihood ratio = 100:1, in favour of linkage) as strong evidence of linkage between two loci.

5.2 THE BASIC STATISTICAL CONCEPTS OF ASSOCIATION ANALYSIS AND THEIR APPLICATION IN STUDY DESIGN

Association studies are not the same as linkage analyses. Different study plans have been discussed at some length in previous chapters, but it is important to remember this. Association studies seek to find association between alleles at one or more loci and a disease, but they do not establish a physical link. LOD scores do not apply in this type of study. Association studies are mostly performed as case control studies, but the transmission disequilibrium test (TDT) and FBAT (family-based association test) are also association-based methods (though they use families).

In statistical terms, there are two different hypotheses to consider in the analysis of genetic association studies: a null hypothesis and an alternative hypothesis

Differences between two groups in case control analyses are assayed using statistical tests. Two possibilities are considered: the **null hypothesis H_0** and the **alternative hypothesis H_1**. These two possibilities are tested to make decisions about differences between the two groups. The null hypothesis H_0 assumes that there is no difference between the two groups and that all samples are drawn at random from the same population. If we accept the null hypothesis, then any differences between the two groups are considered to be due to chance. The alternative hypothesis assumes the opposite and is incompatible with the null hypothesis.

When statistical testing is applied to a study, if the probability that the observed differences are due to chance variation is sufficiently small (usually less than 1:20, $P < 0.05$), then the null hypothesis is rejected and we conclude that there is a genuine difference between the groups. In other words the sample populations are genetically different. In this case we reject the null hypothesis and thus accept the alternative one. If we use regression analysis (as demonstrated below) we can conclude that the independent variable x has some effect on the dependent variable (y).

5.3 STATISTICAL ERROR, POWER, AND *P* VALUES

Making the right decision and avoiding errors in the hypothesis testing

Before we address the concept of probability values (P) we must first consider different types of statistical error and the meaning of statistical power. Though we think we will achieve the right conclusion after we run an experiment, we may in fact make the wrong conclusion. Whenever we decide—either to reject or accept a null hypothesis—we could make mistakes, which are described as type I and type II statistical errors (**Figure 5.2**).

We may either accept or reject the null hypothesis. If we reject the null hypothesis, we are saying there is a difference between the two groups and that the null hypothesis is false. If we accept (or fail to reject) the null hypothesis, we are saying there is no difference between the groups and that the null hypothesis is true. However, in both instances we can be wrong, making statistical errors as described below.

A type I statistical error is the incorrect rejection of the null hypothesis when it is in fact true. This generates false-positive associations. The probability of making a type I error is

	Null hypothesis (H_0) is true	Null hypothesis (H_0) is false
Reject null hypothesis (H_0)	Type I error α False positive	Correct outcome True positive
Accept null hypothesis (H_0)	Correct outcome True negative	Type II error β False negative

Figure 5.2: Type I and type II statistical errors. The figure shows a number of different possibilities arising from statistical errors. The null hypothesis (H_0) states that there is no statistical difference between two groups.

denoted α and is the same α as the significance level of a test; we reject the null hypothesis if the inferred *P* value is less than the significance level (or threshold) α. We need to establish the value of α before we run the analysis. A conventional *P* value of 0.05 is commonly assigned, though we may choose a more restrictive value such as 0.01, 0.001, or even 5×10^{-7} depending on the nature of the study and the number of tests being run. For a given null hypothesis H_0, the type I error rate *P* is shown as:

$$P(\text{reject } H_0 | H_0 \text{ is true}) \leq \alpha$$

where *P* is the probability of rejecting the null hypothesis when it should instead be accepted, i.e. a false positive.

Type II statistical errors arise when we accept the null hypothesis when it is in fact false, giving rise to false negative associations. A type II error is denoted by β and this corresponds to the probability of not rejecting the null hypothesis when it is false; $1 - \beta$ represents the power of the test. For a given null hypothesis H_0, the type II error rate is shown as:

$$P(\text{not reject } H_0 | H_0 \text{ is false}) \leq \beta$$

The way to avoid these problems is outlined further in this chapter, and includes good study planning, and large-scale studies with clear statistical goals and significance thresholds. In many clinical situations the type I error is potentially serious and a procedure able to minimize the probability of a type I error is essential. The significance level of α is crucial in reducing the likelihood of a type I error. If you set α at 0.05, it means that you want the probability of a type I error to be no more than 0.05 or no more than approximately 1:20 cases to be a false positive.

The likelihood of detecting a significant difference in an association study is directly related to sample size

The **power** of a test $(1 - \beta)$ is the probability of correctly rejecting the null hypothesis when it is false (i.e. H_1 is true) for a given significance threshold α. Statistical power is an important concept in association studies, especially because it is possible to miss important statistical differences if a study is under-powered. It is essential therefore that we have information on the statistical power of a proposed investigation before we embark on a study. A number of factors have a direct influence on statistical power, such as sample size, the expected size of the impact in terms of risk in the disease population (or disease subgroup), and the significance threshold (α), which is set prior to the study. The power increases if we enlarge the sample size and also with increases in the size of risk expected to be found. A study design aimed at decreasing the variability of the observations also produces more power. The power is higher with less stringent significance levels. However, this also results in an increased likelihood of a type I error. Thus, we are more likely to detect an effect if we decide to use a threshold of $\alpha = 0.05$ rather than $\alpha = 0.01$, but the effect detected is also more likely to be false.

One way to maximize the power is to increase the sample size, other methods are available related to study design, and sample and genotype selection, but sample size is the only method discussed in detail here. Differences in sample size may account for the differences in reported genetic associations in complex disease, especially in early pre-genome-wide association studies (GWAS) (see Chapter 3).

Probability (*P*) values are simply statements of the probability that the observed differences between two groups could have arisen by chance

As the *P* value falls, the evidence against the null hypothesis in favor of the alternative hypothesis becomes stronger. This also increases the likelihood of obtaining the same or similar results in a second (validation) cohort. The lower the *P* value, the greater the confidence one can have in the results obtained.

Large-scale studies are also performed to increase statistical confidence

In large-scale studies, especially GWAS, the threshold for significance (*P* value) varies from the above norms depending on the study size. The Wellcome Trust Case Control Consortium (WTCCC1) GWAS published in 2007 recommended a minimum value of 5×10^{-7}, whereas Dudbridge and Gusnanto (2008) suggested an even lower value of 7.2×10^{-8} for the UK population. The reason for this higher level of stringency in the acceptable threshold is that an average GWAS investigates at least 500,000 single nucleotide polymorphisms (SNPs) and in such a study we would expect approximately 25,000 false-positive associations if the conventional significance threshold of $P < 0.05$ was applied. As 500,000 SNPs are tested, 500,000 statistical tests are being performed (multiple testing) and therefore correction for multiple testing should be applied in order to avoid large numbers of false positives. One option is to use Bonferroni's correction, which calculates a new significance cut-off according to the number of SNPs or alleles tested.

Bonferroni's correction

Bonferroni's correction sets the new significance cut-off at α/n where α is the significance level (usually $\alpha = 0.05$) and n represents the number of SNPs assayed (or the number of tests carried out). For example, in the above investigation of 500,000 SNPs, Bonferroni's correction sets the new threshold at 5×10^{-7}.

Under Bonferroni's correction many authors prefer to adjust each single *P* value instead of the α significance level. Each *P* value is multiplied by the number of SNPs tested and called significant if the corrected *P* value is still under 0.05.

$$\text{Corrected } P \text{ value} = P \text{ value} \times n \text{ (number of SNPs in test)} < 0.05$$

Bonferroni's correction assumes there is independence between markers, and thus independence between each SNP (or other genetic variant) and the trait. In the context of genetic association studies, as many of the SNPs under investigation are located on the same chromosome and linked through varying degrees of linkage disequilibrium, this correction has been criticized as excessively conservative, which may lead, especially for tightly linked SNPs, to loss of power and thus to an increase in the likelihood of a type II error. For this reason many GWAS studies are performed in stages, with preliminary studies applying less stringent *P* values in the first round aimed at identifying potential loci of interest that can then be confirmed in subsequent rounds. This type of multistage planning is discussed in Chapter 3.

Alternative methods have been proposed in order to avoid the high penalty applied by using Bonferroni's correction and to correct for multiple testing. A practical alternative, for example, is to approximate the type I error rate using a **permutation procedure**. This procedure aims to calculate the approximate false-positive rate that can be then be used as the threshold for the *P* value in the data analysis. The method is conceptually simple, but is computationally demanding and is beyond the scope of this book, and therefore will not be discussed further. Other approaches have been proposed, including an estimation of the **false discovery rate** and a **Bayesian** approach. The false discovery rate is the proportion of positive tests that are false positive and often leads to low threshold levels, while the Bayesian approach incorporates the prior probability of association.

Sample relatedness

It is assumed in all case control association studies that cases and controls are unrelated. Alleles can be **identical by state (IBS)**, i.e. they are the same allele having the same phenotype. Alternatively, they can be **identical by descent (IBD)**, i.e. they are the same and share a common ancestor, and thus have a shared haplotype from one of the parents (**Figure 5.3**).

In order to evaluate the degree of relatedness of a sample, **pair-wise probability (of IBD)** between every individual in the study is calculated. As GWAS uses dense SNP arrays this makes it is easy to compute **pair-wise kinship estimates**. In practice, there is no need to perform analysis of the whole genome. A GWAS SNP data set from only 100,000 markers will yield stable estimates of **kinship coefficients**. On average, siblings share zero, one, and two alleles IBD at 25, 50, and 25% for each gene, respectively. Unrelated individuals do not share alleles IBD. The most commonly used approach to incorporate locus-specific

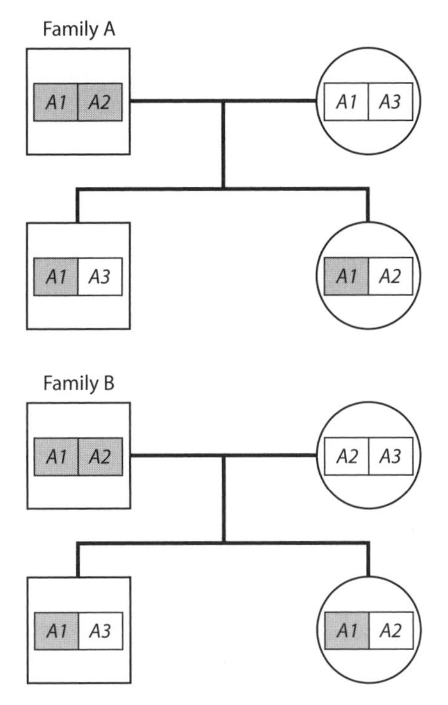

Figure 5.3: Identity by state (IBS) versus identity by descent (IBD). The figure illustrates two families: A and B. Consider a single gene (*A*) with three possible alleles *A1*, *A2*, and *A3*. The father's alleles are shown in gray boxes. In family 1, the paternal genotype is *A1/A2* and the maternal genotype is *A1/A3*. The children both inherit the *A1* allele. However, this is inherited from the father in the male child and the mother in the female child. The male child receives the maternal *A3* allele and the daughter receives the paternal *A2* allele. Therefore, in family 1 the male child has the genotype *A1/A3* and the female child has the genotype *A1/A2*. Though both children carry the *A1* allele, they are IBS for this allele and as a consequence may carry different alleles in linkage disequilibrium with the *A1* allele. In family 2, the pattern of inheritance is subtly different. The father has the same genotype as family 1, but the mother carries the *A2* and *A3* alleles only. Both children inherit *A1* from their father. Thus, with respect to the *A1* allele they are IBD. The two children do not, however, have the same genotype as they inherit different maternal alleles, creating *A1/A2* in the male child and *A1/A3* in the female child.

IBD sharing probabilities into analysis is the **pi-hat** approach, which is implemented in the PLINK whole-genome association analysis software. PLINK is a single-syllable term for a free, open-source whole-genome association analysis toolset, designed to perform a range of basic, large-scale analyses in a computationally efficient manner. PLINK can be used for a range of tests and analyses. Using the --genome option in PLINK, it is possible to estimate pair-wise IBD to detect pairs of individuals who look more or less different from each other than we would expect in a random sample. As a rule of thumb a pi-hat value of 0.95 or greater resulting from pair-wise IBD comparisons is taken as individual relatedness. Detailed instructions on producing this data can be found on the PLINK web page (http://pngu.mgh.harvard.edu/~purcell/plink/).

Genotyping efficiency and sample call rate

Genotyping efficiency per sample, or **sample call rate**, is an informative indicator of sample quality. The call rate per sample represents the percentage of SNPs genotyped in each individual. If a large proportion of SNPs assayed on an individual DNA sample fails to be genotyped it may be indicative of a poor quality DNA sample, which could lead to abnormal genotype calling. False positives will arise if DNA quality differs with phenotype, as the frequency of called genotypes will differ. False negatives will also arise by reducing the statistical power as a result of reduced sample size. Samples with low genotyping efficiency should be removed from further analysis. Acceptable call rate thresholds vary from study to study depending on the genotyping platform used, but they are usually set at 98–99%. Some manuscripts or books may refer to sample genotyping efficiency as **individual missingness**, which, for each subject, is the fraction of missing SNPs (not genotypes) and it ranges from 0 (all SNPs are genotyped) to 1 (all SNPs failed). Genotyping efficiency can be checked using the -missing option in PLINK. This will produce a file showing the genotype missingness rate (1 – efficiency) for each individual (proportion of SNPs that failed on each sample) and for each SNP (proportion of individuals for which no genotype was called). Samples below a desired threshold should be eliminated from any further (downstream) analyses.

SNP quality and SNP call rate

One of the determinants of failure or individual missingness is SNP quality. SNP assays that fail on a large number of samples are poor assays with a low call rate and are likely to result in spurious results. As with individual missingness, the suggested threshold for removing SNPs with a low call rate is approximately 98–99%, though this threshold may vary from study to study. The higher the threshold, the lower the proportion of miscalled data, but this increases the proportion of missing SNPs and leads to a loss of potentially useful information. This can generate a sort of false-negative result whereby potentially informative tagged SNPs are removed from the test set before the association analysis is run. Some studies have found that call rates may be lower where there are higher proportions of heterozygotes and this may introduce bias into a study. Differential rates of SNP missingness between cases and controls may also be a problem. Quality control of SNP genotyping should be performed before any quality control on sample genotyping because in this way fewer samples will be dropped from the analysis as they were genotyped with tagged SNP that had poor performance. In PLINK, markers can be removed based on call rate by using the --geno option, followed by a threshold for a lower limit of missingness (e.g. --geno 0.02 would remove SNPs with more than 2% missing, i.e. less than a 98% call rate).

Issues such as sample relatedness, sample call rate, and SNP call rate can all be built into the study design, but a certain amount of prior information is needed for each and therefore these are often dealt with at the analysis stage, especially in the first round of GWAS.

Data storage format and software: GWAS and other studies

The most commonly used format to store and analyze association data (including GWAS and genome-wide linkage analysis) is the **pedigree file** format. This file is matrix-like, where each individual is in a row, and phenotype and genotype information are stored in columns. The first six columns are identifying information (family ID, individual ID, father ID, mother ID, sex, phenotype), and the remaining columns (column 7 onwards) are genotypes (two columns per genotype; one for each allele, white-space delimited). There are several variations of this format, including transposed (long) formats (typed) and compressed (binary) formats. Descriptions of these file formats can be found on the PLINK homepage (see above). PLINK is frequently used for handling GWAS data and quality control analysis; it is specifically designed and optimized for GWAS and is computationally efficient. PLINK is only one package that allows quality control for genome-wide associations and there are other possibilities, including those writing custom codes using the freely available and excellent R package for statistical computing (www.r-project.org/).

5.4 THE BASIC STATISTICAL CONSIDERATIONS FOR ANALYSIS OF CASE CONTROL ASSOCIATION STUDIES AND THEIR APPLICATION TO DATA COLLECTION AND ANALYSIS

Checking for Hardy–Weinberg equilibrium (HWE) is one step in the quality control analysis of markers in any association analysis including GWAS data. HWE is discussed at some length in Chapter 1; however, as a focal principle, it is important to remind ourselves of the role and importance of HWE. Under HWE, alleles segregate randomly in the population, allowing expected genotype frequencies to be calculated from observed frequencies. Departure from this equilibrium can be indicative of potential genotyping errors, population stratification, or even actual association with the trait under study.

A simple comparison of the expected and observed genotype frequencies provides a test of HWE and departure from HWE is generally tested for by using Pearson's χ^2 test. This test evaluates goodness of fit between the observed genotype counts and expected genotypes under HWE, and is illustrated in **Table 5.1**.

Departures from HWE can have different causes

Departures from HWE may be caused by technical errors, e.g. in genotype assignment or detection (most frequently this is due to missing heterozygotes), but there may also be biological explanations, such as non-random mating, population stratification, and selection. In addition, some cases of departure from HWE may occur as a result of a genotyped SNP having a strong influence on disease susceptibility. Genotypes for a disease mutation will only be in HWE if the genetic model is multiplicative (see below). Occasionally, one allele is strongly associated with disease in a near-Mendelian pattern. In these cases, HWE may appear to have been broken down. Disease-free controls by contrast should follow HWE more closely. No standard guidelines for rejecting SNPs that depart from HWE have been developed. In practice, all SNPs for which the P values obtained by Pearson's χ^2 test are below a predetermined threshold should be checked manually for genotyping quality. In **Haploview**, a software program that is used extensively with HapMap data, the default significance threshold for departure from HWE is set at 0.001.

Table 5.1 Testing for departure from HWE.

	Genotype count				Estimated frequency of *A* allele	Estimated frequency of *C* allele
	AA	*AC*	*CC*	Total		
General						
Observed	a	b	c	N	$p = \frac{2a + b}{2N}$	$q = \frac{2c + b}{2N}$
Expected	p^2N	$2pqN$	q^2N			
Pearson's χ^2	$\chi^2 = \sum_{i=1}^{3} \frac{(O_i - E_i)^2}{E_i}$					
Data set						
Observed	50	30	20	100	$p = \frac{2 \times 50 + 30}{200} = 0.65$	$q = \frac{2 \times 20 + 30}{200} = 0.35$
Expected	42.2	45.5	12.2			
Pearson's χ^2	11.6					
P value	0.0008					

a, *b*, and *c* represent numbers of *AA*, *AC*, and *CC* genotypes, respectively. There is no "*b*" allele in this illustration; this represents the *AC* heterozygotes in the general section of this table only.

Analyzing the data

In a simple single-gene locus study with a single polymorphism under consideration, genotype data are usually illustrated in a 2 × 3 contingency table, where rows represent the disease status and columns represent the genotype counts. Allele frequencies are presented in a 2 × 2 contingency table with a similar structure. Different statistical analysis methods can be applied to these types of tables. In the section below we will consider a number of different statistical methods, starting with **Pearson's χ^2 test** then moving on to the **Fisher's exact probability test** (which in theory is simpler, but requires more computational capacity), and the more advanced **Cochran–Armitage** and logistic regression tests.

Pearson's χ^2 and Fisher's exact test are used to assess the departure from the null hypothesis

To consider Pearson's χ^2 test, let us use the data illustrated in the 2 × 3 table in **Table 5.2**, which represents data from a single SNP for a standard case control association study. The total number of genotyped individuals n is represented as n_{cases} and n_{controls} for cases and controls, respectively. The table represents a single SNP with alleles *A* and *C*, where n_{AA} is the total number of *AA* genotypes observed, n_{AC} is the total number of *AC* genotypes observed, and n_{CC} is the total number of *CC* genotypes observed.

Pearson's χ^2 test is used to assess departure from the null hypothesis, which states that cases and controls have the same distribution of alleles and of genotypes. In this example we will use this statistical test to determine the χ^2 distribution with 2 **degrees of freedom (d.f.)** from the 2 × 3 contingency table. In this instance, 2 d.f. is selected because there are two

Table 5.2 Simplified 2 × 3 for Pearson's χ^2 test.

	AA	*AC*	*CC*	Total
Cases	*a*	*b*	*c*	n_{cases}
Controls	*d*	*e*	*f*	n_{controls}
Total	n_{AA}	n_{AC}	n_{CC}	*n*

possibilities *A* or *C*. The null hypothesis being tested is that there is no significant difference in genotype distribution between cases and controls. In order to understand this test, consider **Figure 5.4** and focus on the observed value for the *AA* genotype ($O_{AA} = a$). When we apply Pearson's χ^2 test, this observed value is compared with its expected value E_{AA} calculated from an equation involving the total number of cases (n_{cases}), the total number of *AA* genotypes (n_{AA}), and the total number of individuals (n):

$$E_1 = \frac{n_{AA} \times n_{\text{cases}}}{n}$$

The above equation is used to calculate each expected genotype *i* as:

$$E_i = \frac{n_i \times n_{cc}}{n}$$

where $n_i = n_{i\ \text{cases}} + n_{i\ \text{controls}}$ is the observed number of each genotype given by the sum of *i* cases plus *i* controls, n_{cc} is the total number of cases (n_{cases}) or controls (n_{controls}), and n is the total number of individuals genotyped ($n_{\text{cases}} + n_{\text{controls}}$). Each observed genotype is then compared with each expected value and the full statistical test is applied:

$$\chi^2 = \sum_{i=1}^{6} \frac{(O_i - E_i)^2}{E_i}$$

where the summation is over all six cells in the table and O_i are the observed values (*a*, *b*, *c*, *d*, *e*, and *f* in Table 5.2) in each cell, while E_i is the expected value calculated according to the equation above for each genotype in cases and controls.

	Allele A	**Allele C**	**Totals**
Cases	*a*	*b*	*a* + *b*
Controls	*c*	*d*	*c* + *d*
Totals	*a* + *c*	*b* + *d*	*n*=*a* + *b* + *c* + *d*

Figure 5.4: A simplified form of the χ^2 test. This simplified version of the χ^2 test is often applied in association studies. χ^2 is calculated from a 2 × 2 table above using values for alleles in boxes for cases and controls allele *A* and allele C above (*a*, *b*, *c*, and *d* here) and the data from sum boxes (containing *a* + *b*, *c* + *d*, *a* + *c*, *b* + *d*, and *n* here). The 2 × 2 table can also be used to calculate the OR using the values in the boxes containing *a*, *b*, *c*, and *d*.

Pearson's χ^2 test compares the observed and expected genotypes in cases and controls assuming both cases and controls have the same frequency of genotypes. The test is asymptotic, implying that the analysis becomes more accurate with larger data sets. A low count in any of the cells of the table can violate the above assumption. In practice, an expected value of at least five observations in each cell is regarded as a minimum number needed in order to apply Pearson's χ^2 test. Fisher's exact test is recommended if any of the cells have values below five.

When using χ^2 it is important to use real numbers and not percentages or mean values as some authors have done in error. It should also be noted that some calculations use a 2×2 table in a simplified version of Pearson's χ^2 test (Figure 5.4). The principles are the same in this analysis; however, allele frequencies are counted rather than genotypes. In complex systems where genes have multiple alleles it is often easier to count individuals as either positive or negative for an allele of interest rather than count genotypes. Homozygosity is ignored in this model and no assumptions are made about homozygous or heterozygous advantage.

Data from 2×2 tables illustrated in Figure 5.4 can be used to calculate χ^2 using the simple formula:

$$\chi^2 = \frac{n(ad - bc)^2}{(a + c)(b + d)(a + b)(c + d)}$$

The value of this calculation is that it makes no assumptions about the impact of heterozygotes versus homozygotes—the calculation is simply based on the number of alleles in the population. Homozygotes are counted once only. This same formula can be used to calculate the odds ratio (OR) as:

$$\text{OR} = \frac{a \times d}{c \times b}$$

These two calculations (simplified χ^2 and OR) are among the most well-known and frequently used in association studies, especially in the early phases of data analysis.

Fisher's exact test calculates the exact probability (*P*) of observing the distribution seen in the contingency table

Fisher's exact test is computationally more demanding than Pearson's χ^2 test and requires factorial calculations. The problem of working with factorials is that they generate very large values quickly, consequently where it is necessary to use them most students will use computer statistical packages such as R for this type of analysis. However, it is important to understand how this test works. Therefore, for the sake of clarity, the formula to calculate Fisher's exact test from a standard 2×2 contingency table is calculated as:

$$P = \frac{r_1!r_2!c_1!c_2!}{n!a!b!c!d!}$$

where P is Fisher's exact probability. The symbol "!" refers to the **factorial** for the cells and the marginal values that are identified in the 2×2 Punnett Square in Figure 5.4. In this table, a, b, c, and d are the observed values and r_1, r_2, c_1, and c_2 are the marginal values (computed as the sum of each column or row depending on position). The letter n denotes the total number.

An exhaustive explanation of this test is beyond the scope of this book and we advise the reader to refer to a good statistics book.

Figure 5.4 shows a 2 × 2 contingency table with two alleles *A* and *C* in two groups: cases and controls. Allele frequencies for the two groups are represented by the values *a*, *b*, *c*, and *d*. There is no assumption about the status of homozygous genotypes in this calculation. The values $a + c$ and $b + d$ are the sum values for the two columns (c_1 and c_2) and $a + b$ and $c + d$ are the sum values for the two rows (r_1 and r_2). The value n is the total for the whole cohort. The marginal values used in Fisher's test are c_1, c_2, r_1, and r_2 corresponding to $a + c$, $b + d$, $a + b$, and $c + d$ above, respectively.

	Allele *A*	Allele *C*	Totals
Cases	a	b	$a + b = r_1$
Controls	c	d	$c + d = r_2$
Totals	$a + c = c_1$	$b + d = c_2$	n

The Cochran–Armitage test looks for a trend for a difference between cases and controls across the ordered genotypes in the table

The Cochran–Armitage test was developed specifically to detect a linear trend in proportions over levels of exposure for a risk variable; when applied to genetic association analyses it tests whether the probability of disease increases in a linear pattern with the three corresponding genotypes and the binary trait (the latter meaning cases or controls in this instance). The Cochran–Armitage test begins by taking into account the ordering of the columns of the contingency table (shown in **Figure 5.5**). In **Figure 5.6**, the column order is *AA*, *AC*, and *CC*. Cochran–Armitage tests the null hypothesis of zero slope for a line of best fit for the three genotypic risk estimates. We can consider the following example to illustrate the Cochran–Armitage equation for genotypes.

The hypothetical counts of genotypes in a 3 × 3 contingency table for each marker of hypothetical cases and controls are $N_{1\bullet} = n_{11} + n_{12} + n_{13}$ and $n_{\bullet 1} = n_{11} + n_{12}$, etc..

	Genotype			Totals
	AA	*Aa*	*aa*	
Cases	n_{11}	n_{12}	n_{13}	$N_{1\bullet}$
Controls	n_{21}	n_{22}	n_{23}	$N_{2\bullet}$
Totals	$n_{\bullet 1}$	$n_{\bullet 2}$	$n_{\bullet 3}$	N

The Cochran–Armitage statistic is calculated as:

$$T^2 = \frac{\left[\sum_{i=1}^{3} w_i (n_{1i} n_{2\bullet} - n_{2i} n_{1\bullet}\right]^2}{\frac{n_{1\bullet} n_{2\bullet}}{n}\left[\sum_{i=1}^{3} w_i^2 n_{\bullet i}(n - n_{\bullet i}) - 2\sum_{i=1}^{2}\sum_{j=i+1}^{3} w_i w_j n_{\bullet i} n_{\bullet j}\right]}$$

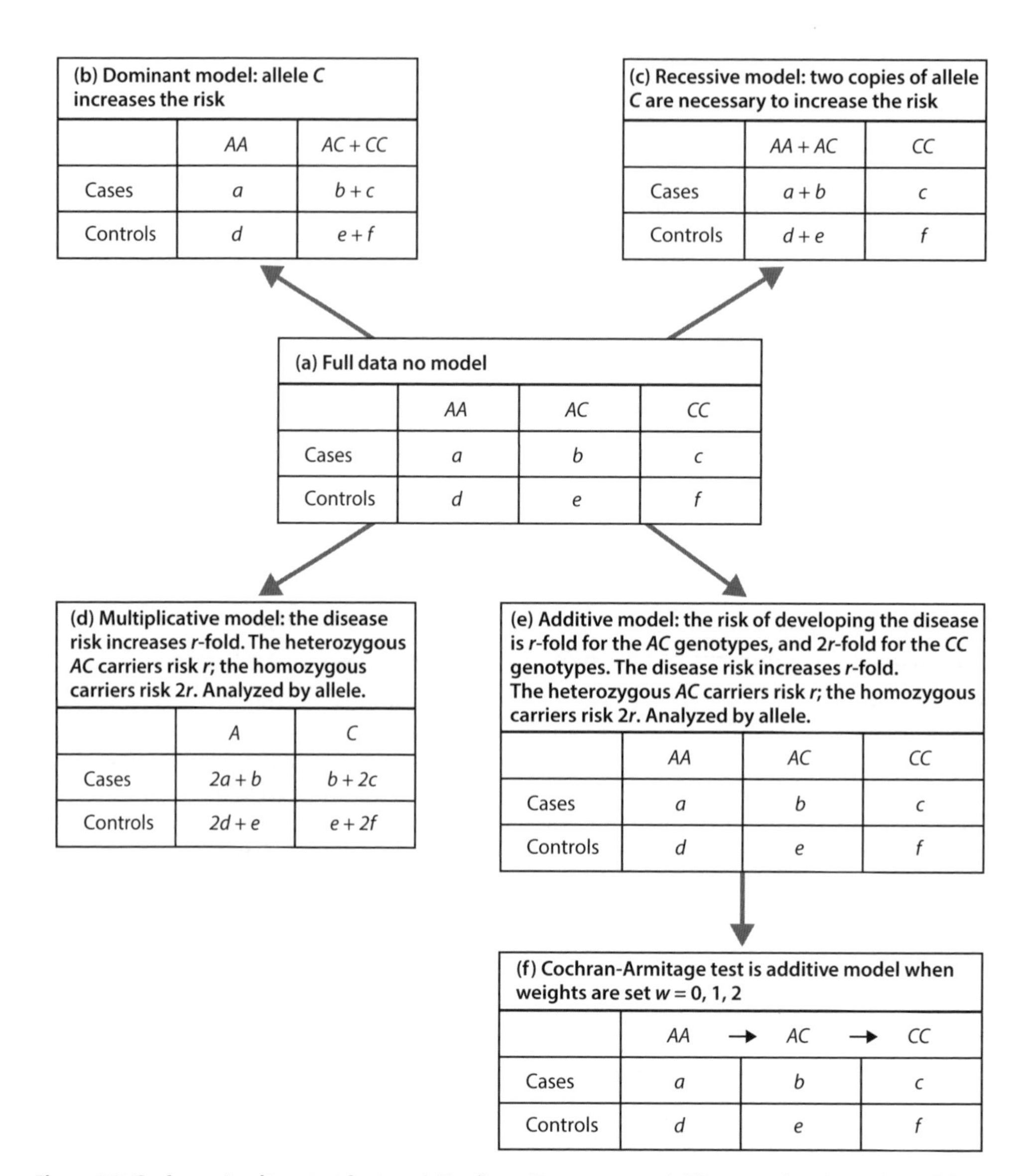

Figure 5.5: Cochran-Armitage test for trend. The figure illustrates several different options for analysis of data from single SNP association studies. Several different models are illustrated. Here, allele *C* is assumed to increase the risk for the disease. The models applied to (a) are: dominant (b), recessive (c), multiplicative (d), and additive (e). In the additive model, the risk of the trait is greater than two-fold in homozygotes, whereas in the multiplicative model the risk is two-fold in homozygotes compared with heterozygotes. (Adapted from Lewis CM & Knight J [2012] *Cold Spring Harbor Protoc* 2012:297–306. With permission from Cold Spring Harbor Laboratory Press.)

where $w_i = (w_1, w_2, w_3)$ weights are set to detect particular types of association. In genetic association studies and GWAS, the Cochran–Armitage test is mostly used to examine the potential additive effect of the alleles. To do this the weights are set as $w = (0, 1, 2)$, which correspond to Figure 5.5, for the additive effect of the allele *C* in developing the risk of the disease.

In addition, the Cochran–Armitage test can also be used to test for a potential dominant ($w = 0, 1, 1$) or recessive effect ($w = 0, 0, 1$) effect of the *C* allele in Figure 5.5.

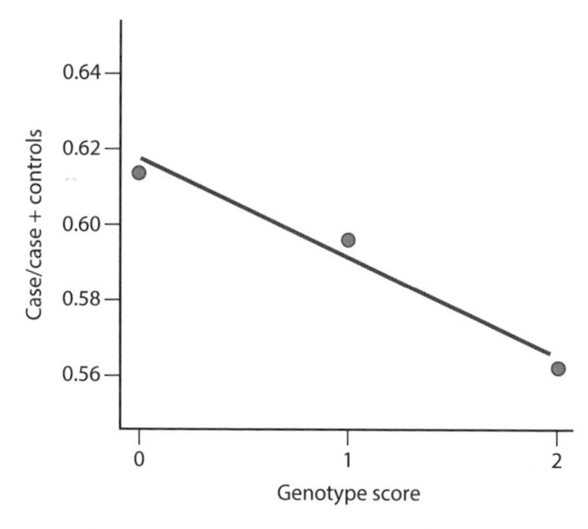

Figure 5.6: Cochran-Armitage test for trend in a single SNP case control association study. In this case we are looking at a single gene with two alleles (alleles *1* and *2*). The genotypes are scored as 0, 1, and 2 for homozygotes allele *1*, heterozygotes alleles *1* and *2*, and homozygotes allele *2*, respectively. The circles indicate the genotype score for the cases and controls combined together with their least-squares line. Here, the line fits the data reasonably well as the heterozygote risk estimate is intermediate between the two homozygote risk estimates, corresponding to an additive genotype risk. (Adapted from Balding DJ [2006] *Nat Rev Genet* 7:781-791. With permission from Macmillan Publishers Ltd.)

T^2 has a χ^2 distribution with 1 d.f. under the null hypothesis of no association. The Cochran–Armitage test for trend performs with robust power in the case of an additive risk of disease developing. Conversely, the power is reduced by deviations from an additive model. Cochran–Armitage is a more conservative test, minimizing the probability of incorrectly rejecting the null hypothesis and thus reducing the risk of a type I statistical error. Cochran–Armitage is also more robust when it comes to departures from HWE than either Pearson's χ^2 or Fisher's exact probability tests.

There is no simple answer to the question of which test to choose

There is no easy answer about which statistical test to use for case control association studies. Adopting the Cochran–Armitage test can mean sacrificing statistical power if the risks for each genotype do not conform to an additive model. Using Pearson's χ^2 or Fisher's test spreads the analyses over the full range of risk models, and this inevitably means investing less effort in the detection of additive risks. Pearson's and Fisher's tests have reasonable power regardless of the underlying risks, but when the genotype risks become additive, these tests lose their power. It is also worth remembering that Sasieni (1997) has shown that when HWE is violated, an allele-based test is inappropriate and the frequency of genotypes rather than alleles should be compared using the Cochran–Armitage test for trend.

Data may also be analyzed assuming a predefined genetic model

Genetic contributions to disease risk for complex traits are complex and do not fit easily into such simple models. However, it is possible to identify patterns looking at individual loci.

Dominant, recessive, and additive (or over-dominant) models

In Figure 5.5, if we assume that allele *C* is a high-risk allele for the disease being tested, then we can test to see if the effect is dominant (Figure 5.5b) or recessive (Figure 5.5c). The model assumes that carrying allele *C* increases disease risk, and to test it *AC* and *CC* genotypes are pooled together to give a 2 × 2 table. This is particularly relevant when allele

C is rare, with few *CC* observations in cases and controls. Alternatively, under a recessive model for allele *C*, genotypes *AA* and *AC* are pooled together (Figure 5.5c). This model assumes that two copies of allele *C* are required for increased risk. However, in some cases risk is additive, meaning that the risk in homozygotes is higher than that for heterozygotes. In an additive model if we assume allele *C* is the risk allele, the risk of developing the disease is r for the *AC* genotypes and $2r$ for the *CC* genotypes (Figure 5.5e).

Though the described models and tests are valid methods for analysis of association studies, these methods should be applied according to a predefined study plan because a random application of such tests increases the probability of a false-positive result. These models are rarely used in the primary analysis of complex genetic disorders, but they may be employed as a secondary tool to explore the potential mode of inheritance of an associated SNP or to test a predefined hypothesis.

Multiplicative risk

Under this model, if *C* is the risk allele, the risk of developing a disease is modeled by increasing risk factor r from r for heterozygotes (*AC*) to r^2 for homozygotes (*CC*). Under the multiplicative model the allele count is then the most powerful method of testing. The total number of *A* and *C* alleles in cases and controls are compared, regardless of the genotypes from which these alleles are constructed (Figure 5.5d).

Finally, we must remember that in complex diseases we anticipate the involvement of several genetic risk factors that may interact in complex models. We must be cautious about simple patterns. In fact, most single-gene studies count alleles and make no assumptions about homozygotes or models. This is especially true for human leukocyte antigen (HLA) studies where the number of alleles at each locus is extreme.

Logistic regression

Regression is a statistical method employed to determine the dependence between a response variable (y), known as the dependent variable, and one or several predictors (x), called the independent variables. In the simplest case, y (the dependent variable) is predicted by only one x (independent variable) by a linear equation or the equation of a straight line:

$$y = \alpha + \beta x + \varepsilon$$

where α is the intercept, β is the regression coefficient (i.e. the slope of the line), and ε represents the error (i.e. the differences between predicted and observed y values). The above equation illustrates a simple **linear regression model** where a line fits all values of x with y. Where there are infinite lines we need to find the linear equation able to best fit all observed values of x. To achieve this aim we use the **least-squares criterion**, which states that the line of best fit for the data is the line where the sum of the squares of the vertical distances from the observed points to the line are as small as possible. This criterion is illustrated in **Figure 5.7** with an example taken from population ecology regarding the distribution of a predator population (e.g. wolves) and their prey (e.g. hares), where each dot represents an observed value and x is the number of prey observed within an area and y the corresponding number of predators. In this equation, d represents the difference between the y coordinate of the data point and the corresponding y coordinate on the line. The best line of best fit ($S_{\min}$) is the line which minimizes the sum of the d squares (d^2):

$$S_{\min} = \sum_{i}^{n} d_i^2$$

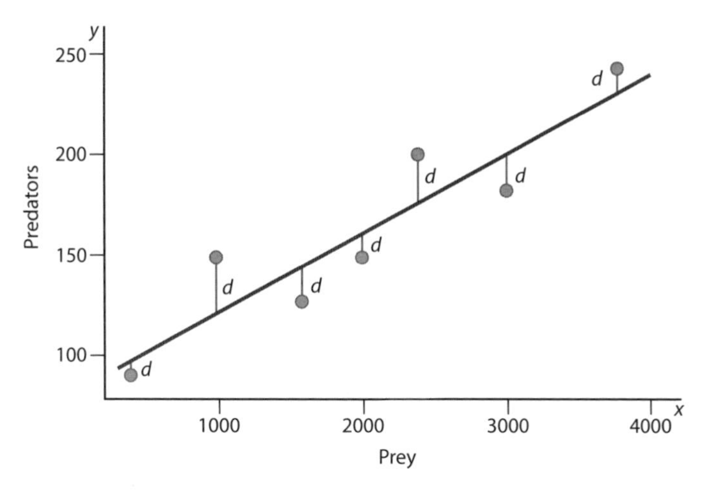

Figure 5.7: Linear regression using a population ecology model. An example from population ecology that represents the relationship between prey and predator numbers within a studied area, where each dot represents an observed value; *x* is the number of prey observed within an area and *y* is the number of predators, *d* represents the difference between the *y* coordinate of the data point and the corresponding *y* coordinate on the line. In this case, an increase in the prey population corresponds with an increase in the number of predators.

The distance of each point in the scatter from the regression line is known as the residual or error, denoted *d*. When all of these residuals are squared and then added together they are given the term $\sum d^2$. The "best" straight line is the one with the lowest value for $\sum d^2$ hence the name ordinary "least squares." Another example is shown within **Figure 5.8**, where a scatter plot is plotted to illustrate the relationship between body mass index versus hip circumference for a sample of 1500 women in a diet and health cohort study.

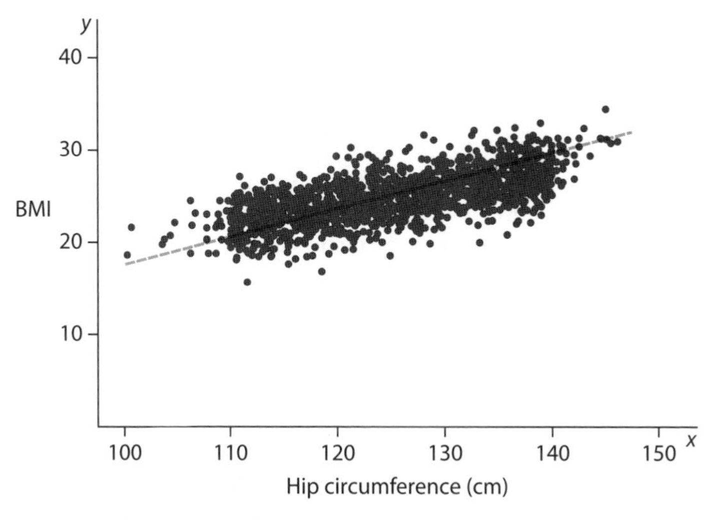

Figure 5.8: A scatter plot of body mass index (BMI) versus hip circumference. The figure is a scatter plot based on data from a sample of 1500 women in a diet and health cohort study. The scatter illustrates the relationship between BMI and hip circumference. The scatter of values appears to be distributed around a straight line, i.e. the relationship between these two variables appears to be almost linear.

A simple linear regression model is one with only one independent variable. When we have more than one independent variable the regression model is called a **multiple linear regression model**. This model is an extension of simple regression where the dependent variable y is predicted by several independent variables (x_i) simultaneously:

$$y = \alpha + \beta_1 x_1 + \beta_2 x_2 + \cdots + \varepsilon$$

This multiple linear regression allows us to include more predictors or covariates in the analysis. This will reduce the residual variance (d) of y.

In genetic association studies, however, a **logistic regression model** is preferred compared with a linear regression model.

Logistic regression is frequently used in association studies

There are two regression models: linear regression, which may be either simple or multiple, and logistic regression. A logistic regression model is preferred in case control genetic association studies where we have a binary outcome of interest (e.g. presence/absence of a disease) and a number of explanatory variables. In studies of quantitative traits logistic regression would not be appropriate. In order to have a valid analysis we need to satisfy the assumptions of logistic regression or we may derive invalid statistical inferences. Before we can use our model and make any statistical inference, we need to check that our logistic model fits the assumptions underlying logistic regression:

- The observations are independent. Observations are independent if there is one pair of observations on each individual.
- The dependent variable must be a dichotomy (i.e. there are two categories).
- No important variables are omitted.
- No extraneous variables are included.
- The independent variables are measured without error.

Furthermore, a logistic regression does not require assumptions of normality, linearity, and homoscedasticity (homogeneity of variance) usually needed in linear regression. In summary, in logistic regression:

- The independent variables do not need to be normally distributed, though normality yields a more stable solution. Also, the residuals do not need to be normally distributed.
- Assumption of a linear relationship between dependent and independent variables is not needed. As logistic regression applies a non-linear log transformation to the predicted OR (see below) it can deal with all sorts of relationships.
- Assumption of homoscedasticity (equal variance) of the residuals is also not needed.

We can start to make a logistic regression model by creating a binary variable to represent two possible outcomes of the dependent variable, such as 1 for having the disease and 0 for not having the disease. This binary variable cannot be used as the dependent variable because we cannot explain predicted values that are not 0 or 1 and after all such a variable is also not normally distributed. However, if we consider the probability P of an individual being classified into the highest coded category (i.e. having the disease) as the dependent variable we can overcome the above problem. In order to predict this probability, logistic regression employs **maximum-likelihood estimation** where the outcome is the probability of belonging to a condition of y (i.e. having the disease), which can take any value between 0 and 1. To overcome mathematical difficulties we need to transform the probability P to a logarithmic distribution. This transformation normalizes the distribution and creates a link with the linear regression equation. The logarithmic distribution (or logistic transformation of P) is also called the logit of P or **logit(P)** or log odds. Logit(P) is the natural logarithm (log to base e) of the odds ratio and it is given by the formula:

$$\text{logit}(P) = \log_e\left(\frac{P}{1-P}\right)$$

According to this formula, $\log_e$ is a natural logarithm and the ratio ($P/1 - P$) is an odd (odds ratio) of the event occurring and can range from 0 to infinity. Odds values tell you how much more likely it is that an observation is a member of the target group rather than a member of the other group. For example, if the probability P is 0.80, the odds are 0.80/(1 – 0.80) or 0.8/0.2 or 4 to 1. The probability P can only range from 0 to 1, but logit(P) can range from negative infinity to positive infinity. We can also express the above equation using a linear regression equation in which the logit of P or y (dependent variable) is determined by a linear function of the independent variables x_i (where P is a function of x corresponding to the dependent variable y):

$$\begin{aligned}\text{logit}(y) &= \log_e\left(\frac{P}{1-P}\right)\\ &= \beta_0 + \beta_1 x_1 + \beta_2 x_2 + \cdots + \beta_n x_n + \varepsilon\end{aligned}$$

The terms y and x represent dependent and independent variables, respectively. Each β_i coefficient measures each x_i partial contribution to variation in y. Logistic regression uses a maximum likelihood method, which maximizes the probability of getting the observed results given the fitted regression coefficients.

We can use a more complicated logistic regression model when additional covariates may affect the onset of a disease. Examples of this are situations in which environmental effects such as epidemiological risk factors (e.g. smoking and gender), clinical variables (e.g. disease severity and age at onset), population stratification, and other marker loci have interactive effects on disease risk (and gene–gene interaction or epistasis).

Fortunately, we do not have to calculate any of these mathematical equations by hand. Computer software is easily available, but, once again, it is important to understand the principles and how we get from a regression formula to a line to the logistic analysis.

Odds Ratio (OR)

In order to interpret the logit coefficients, however, we need to introduce the concept of the OR. The OR is a measure of the strength of an association between two variables and is widely used in case control studies. In a genetic association study, OR values give us useful information for understanding the relationship between alleles and disease risk, and we can define the OR as the ratio of the odds of bearing an allele in the group with the disease (cases) to the odds of carrying the same allele in the group without the disease (controls). In other words, OR estimates the risk of a disease or trait for a specific allele.

In a standard 2 × 2 table, the OR can be expressed as the cross-products ratio (ad/bc), as seen in **Table 5.3** (also discussed in Chapter 3). In this calculation a, b, c, and d are the frequencies of the alleles in cases and controls. OR values range from 0 to infinity. OR values close to 1 indicate no relationship between the allele and the trait, while OR values of less than 1 suggest a protective effect and values greater than 1 suggest an increased risk.

Table 5.3 A standard 2 × 2 plot with allele frequencies shown as *a*, *b*, *c*, and *d* (this table would be used for calculation of simple χ^2 or OR).

	Cases	Controls
Risk allele	*a*	*b*
Other allele	*c*	*d*

However, causality cannot be established from an association because associations may arise as a result of linkage with a true disease-causing mutation. In other words, association analysis detects associations only and these associations may reflect linkage with hitherto uninvestigated disease-causing genes or variants.

The OR is closely connected to logistic regression by modeling the natural logarithm of the OR as a linear function of independent variables. This is a powerful and very common method, and it is calculated using the exponential function e^{β} and the regression coefficients from one of the two following equations:

$$OR = e^{\beta_0 + \beta_1 x_1 + \beta_2 x_2 + \cdots + \beta_n xn + \varepsilon}$$

or

$$OR = e^{\beta_0} e^{\beta_1 x_1} e^{\beta_2 x_2} \cdots e^{\beta_n xn} e^{\varepsilon}$$

In a simple example ($\text{logit}(y) = \beta_0 + \beta x + \varepsilon$), a value of $\beta = 2$ means that $e^2 \sim 7.4$, which means that when x increases by one unit, the odds of being affected ($y = 1$) increase by a factor of 7.4 when other variables are not changed.

The pitfalls and problems of GWAS

A major challenge in GWAS is false-positive associations typically resulting from type I statistical errors and/or population structures. In consideration of the fact that GWAS may have clinical implications, false-positive results are an important issue and it is vital to reduce the incidence of them. We have already discussed quality control as an essential step to reduce the number of false-positive findings and we have introduced the importance of correcting for multiple testing across the number of SNPs. In this section, we will discuss the impact of some of problems associated with study design and **population structure**, its implications, and how to deal with this phenomenon.

Study heterogeneity

One problem in GWAS is that heterogeneity between and across the studies may lead to equally valid but different conclusions about the role of an allele in disease risk.

Different studies of the same disease may use different clinical criteria reflecting subtle differences in disease severity, disease subtype, age of diagnosis, or duration of the disease—all of which can influence the outcome of genetic studies.

Population heterogeneity

In addition, there are variations between human populations that may lead to different associations in different populations. Evolutionary forces such as random drift, novel mutations, and natural selection (less common) act on human populations, and thus variations in SNP frequencies are seen between the major population groups (this is discussed in Chapter 1 in some detail). Association tests are statistical assays that detect differences in SNP or other genetic marker frequencies between cases and controls. Variation within and between populations may lead to false-positive associations. An example is a SNP in the **complement** factor H gene (*CFH*) which alters the susceptibility to age-related macular degeneration (AMD). The frequency of the AMD protective *A* allele in the Yoruba population of sub-Saharan Africa is four-fold higher than in Europeans. Similarly, another SNP in the complement system genes, the factor B gene (*CFB*), shows a higher frequency in Africans than Europeans. These differences may account for differences in susceptibility to AMD in these populations.

Recent associations reported an increased risk of Alzheimer's disease associated with genetic variation in two SNPs (rs11136000 and rs3818361) of the complement pathway genes for clusterin (*CLU*) and for complement receptor 1 (*CR1*). The *CR1* risk allele is frequent in the European population, but rare in African and Asian populations, while the *CLU* risk allele is much more frequent within Asian than African populations. It is not known whether this variation is a consequence of genetic drift or natural selection. A meta-analysis including white, African-American, Israeli–Arab, and Caribbean Hispanic individuals found an association between polymorphisms in *CR1* and *CLU* in populations with a European ancestry only. In addition, some mutations may be completely absent from some population groups, e.g. the *CARD15* mutations that are present in more than 30% of European Crohn's disease patients are more or less absent in Asian populations.

Meta-analysis

Meta-analysis is a means of compiling data from previously published studies to focus on a specific question. Recently, it has been applied to studies of the genetics of common disease and will be used more often as the number of GWAS grows. Meta-analysis has the advantage of being able to identify and confirm weak associations because with the vast numbers included the statistical power of these studies is considerably higher than any other individual center or single collaborative study. Meta-analysis is frequently used in clinical drug trials and in the analysis of psychological disorders.

Population structure or population stratification

Population stratification occurs when the study samples consist of several discrete subgroups of individuals who differ in their ancestry. In the presence of stratification, spurious associations can be reported. Thus, checking for population stratification in the study samples is important to avoid false-positive or false-negative associations. Statistical methods have been developed and implemented into software to control for population stratification. Several methods have been proposed for case control studies, including genomic control, **structured association**, **principal components analysis (PCA)**, and the **stratification score analysis**. Genomic control and the structured association methods both use SNPs that have not been associated with the disease to control for population stratification.

Genomic control

In the presence of population stratification, regular methods for testing association, such as Pearson's χ^2 method and the Cochran–Armitage test for a trend could produce excess false positives compared with the nominal significance levels. A popular method used in the presence of population stratification is genomic control. This method aims to control for population stratification by first estimating an inflation factor and then adjusting all of the test statistics. The logic behind genomic control is that the test statistic for association is inflated by a multiplicative factor λ, which is proportional to the level of population substructure present. λ is estimated using SNPs that are believed not to be in association with the phenotype tested (affection status) and are therefore expected to be random samples from the χ^2 distribution with 1 d.f.. One common solution to estimate genomic control is to align the median of Pearson's χ^2 or Cochran–Armitage test distributions with that of the χ^2 distribution with 1 d.f., which is about 0.456. A robust estimate of λ is therefore given by:

$$\lambda = \frac{\text{median}(\chi^2)}{0.456}$$

An inflation factor λ of 1 indicates that no population stratification is detected within the samples. We can use this value of λ to rescale all of the test statistics in a downward direction. In practice, the test statistics of each SNP are divided by the λ value then the corresponding adjusted *P* value can be re-estimated:

$$\chi^2_{\text{adjusted}} = \frac{\chi^2}{\lambda}$$

It is important to note that λ denoted above is not same as the λ value used in calculating disease incidence and prevalence (sibling relative risk).

Structured association

SNP genotypes that are not associated with disease can also be used in structured association analysis. This method has been implemented in STRUCTURE software (http://pritchardlab.stanford.edu/structure.html) and employs genotype data from the study to determine population structure in a similar method to that used in the genomic control analysis above. Predictive SNPs are those known to be associated with a specific ethnic population or other population subgroup. The software then performs tests for associations within each inferred subpopulation. STRUCTURE may also be used to identify individuals who do not cluster with the majority of the samples. The latter samples can then be eliminated from the analysis.

PCA

Another analytical methodology often incorporated into GWAS able to scan for population structures is called PCA. The basic idea behind PCA is to measure the data as principal components rather than on a normal *x*–*y* axis. Principal components are structures or directions where the data are most spread out and thus they indicate where the most variance is. If we are able to find those directions where most of the variance is, we can recognize structures in the data. Though we mainly use math to find the principal component throughout eigenvectors and eigenvalues, we can imagine plotting our data into a scatter plot and using PCA to find the direction where there is most variance. The PCA will first determine the straight line through the points that are able to catch the most variance among the data in a manner similar to linear regression. This inferred line is then used as the first principal axis or principal component to generate a second principal component. This process occurs in multidimensional space with one dimension for each of the variables (i.e. SNPs) included and the lines inferred through the points. At the end of the process PCA will synthesize the data from a mass of variables into a set of compound axes. The first axis will explain the most variation, then the second, and so on. Once the principal components are inferred, individuals can be plotted according to their principal components and the possible structures in the dataset will then be recognized.

PCA can be also be used in a principal components adjustment analysis to control for type I errors. Association tests performed on data affected by population stratification can be adjusted by performing PCA on different subsets of variants (or SNPs). We can first carry out a PCA using only common variants (set at less than 5%) with minor allele frequencies (set at 5%), or only low-frequency variants (minor allele frequencies between 1 and 5%) or only rare variants (minor allele frequencies of less than 1%) or a combination of the above. Then we can test if population stratification is present or not and eventually perform association inferences on the subset. PCA based on either common variants or low-frequency variants seems to provide effective control for population stratification, while rare variants performs less well. PCA based on low-frequency variants seems to adjust better than common variants, but the use of the former could result in over-adjustment producing a substantial loss of the power, especially in the absence of population stratification.

5.5 HOW TO INTERPRET A GWAS

The ultimate goal for all this effort is to identify genes associated with or linked to increased risk of disease. However, interpreting the results of these studies is a potential minefield.

There are several ways to interpret statistically significant genetic associations

Significant *P* values resulting from genetic association studies can indicate one of the following:

- A direct association, whereby there is a true causal relationship between the allele and the disease.
- An indirect association, whereby the identified allele is in linkage disequilibrium with the true causal allele on the same chromosome.
- A false positive result, whereby the association is simply a chance observation, perhaps resulting from some error in the study design, sample bias, or from population stratification.

Distinguishing between direct and indirect associations is challenging, and only sequencing the candidate gene region or dense fine-map genotyping of all available SNPs may resolve this issue. The detection of associations with a non-synonymous SNP (e.g. a SNP where variation results in an amino acid change) may indicate that it is the true causal variant in disease susceptibility. Functional studies, however, should be attempted in order to provide evidence before assigning a functional relationship to any polymorphism identified.

There are several diagnostic plots that can be used for the visualization of genome-wide association results

Quantile–quantile plots

In statistics, a **quantile–quantile (Q–Q) plot** is a plot for comparing probability distributions by plotting their quantile values. It is an effective graphical method to determine if the two distributions come from the same or two different statistical populations. This statistical tool is widely used in GWAS to assess the significance of observed associations (*y*-axis) compared with the expectations under no association (*x*-axis).

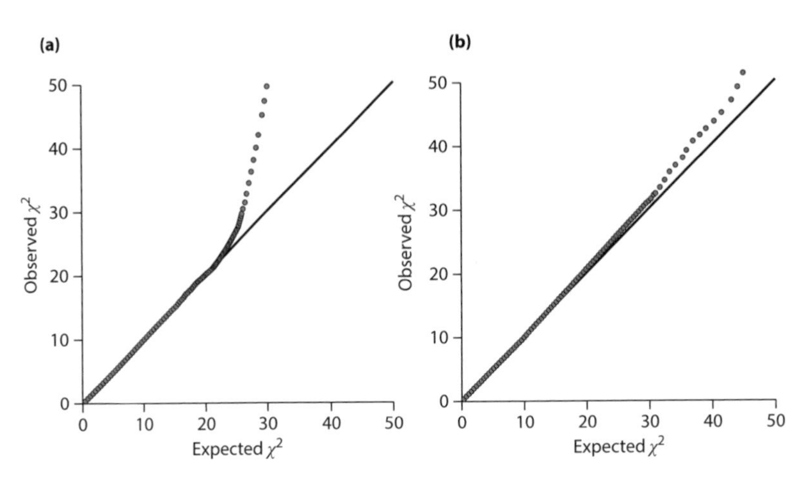

Figure 5.9: Hypothetical Q–Q plots in GWAS. The figure illustrates two different Q–Q plots. The plot compares probability distributions using their quantile values. This statistical tool is widely used in GWAS to assess the significance of observed associations comparing observed values (on the *y*-axis) and expected values (on the *x*-axis). Each dot represents a hypothetical SNP while the blank line is the expected null distribution. Plot (a) illustrates strong deviation from the expected values and therefore strong genetic association of the trait under study with SNPs in a heavily genotypes region or spurious associations. Plot (b) indicates very little deviation from the expected values and may suggest cryptic errors in the study population, in the genotyping, or either true association.

In **Figure 5.9**(a and b), the observed association statistics such as χ^2 or the calculated $-\log_{10}P$ values for each SNP are ranked from smallest to largest and plotted against the distribution that would be expected under the null hypothesis for no association. If the two compared distributions are similar, the points in the Q–Q plot will approximately lie on the null or identity line (denoted $y = x$), indicating that no association is detected for each SNP. These plots help to indicate whether the study has generated more significant results than expected by chance. Deviations from the identity line indicate either some bias in the observed (assumed) distribution or strong true associations.

As the underlying assumption in GWAS is that the vast majority of assayed SNPs are not associated with the trait, strong deviations from the null hypothesis suggest either a strongly associated locus, a heavily genotyped locus (Figure 5.9a) (i.e. an associated gene with many genotyped SNPs), significant undetected differences in the population structure (Figure 5.9b), cryptic relatedness in the study population, or errors in genotyping. A clean Q–Q plot (Figure 5.9a and b) should show a solid distribution matching the $y = x$ line until it sharply curves at the end, which represents the small number of true associations among the many thousands of genotyped SNPs.

Manhattan plots

Named after the Manhattan skyline, the plots are widely used to visualize data from GWAS studies. This type of scatter plot is particularly useful to display data with a large number of data points. A GWAS Manhattan plot outlines the $-\log_{10}P$ value of each SNP on the *y*-axis against the genomic coordinates on the *x*-axis (**Figure 5.10**). Each dot in the plot represents a different SNP, with the alternating bands of shading representing different chromosomes plotted on the *x*-axis. The *y*-axis indicates the strength of the association between a SNP and the trait under study. As the strongest association carries the smallest *P* value, it will be the highest ranked and will be visualized at the top of the plot. A good Manhattan plot visualizing a robust association study should show the highest ranked SNPs rising with a column of other associated SNPs symbolized as dots from the same

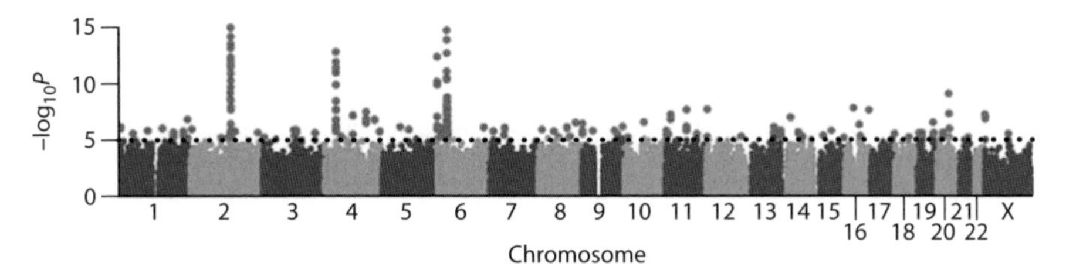

Figure 5.10: Manhattan plot displaying findings from a GWAS study. This Manhattan plot presents the findings of a GWAS. Each chromosome can be identified sequentially (1–22, then X and Y) in alternating light and dark gray coloring. The *y*-axis shows the $-\log_{10}P$. The significance threshold is set at 10^{-5} for moderate associations. Each tagged SNP is represented by a dot distributed according to the $-\log_{10}P$ along the 22 autosomes and the X. *P* is for 1 d.f. because with each SNP tested individually there are two possibilities. Points above the dotted line indicate SNPs with a $P < 10^{-5}$. $P < 10^{-5}$ or $P < 10^{-7}$ are significant; the greater the digit after the value 10, the more significant the observations are. (From The Wellcome Trust Case Control Consortium [2007] *Nature* 447:661–678. With permission from Macmillan Publishers Ltd.)

chromosomal location and haplotype. This is the result of linkage between nearby SNPs, all of which will have the same or similar association signals. A single dot visualizing as the highest ranked SNPs spread over the whole plot in isolation shows a pattern typical of genotyping errors and this may bring into question any reported associations.

Linkage disequilibrium is a useful tool in association studies provided you know how to handle it

Linkage disequilibrium and recombination are discussed in detail in Chapters 1 and 2, and therefore this subsection will be brief. Linkage refers to a situation where alleles or SNPs at specific loci are inherited together as a consequence of their physical proximity on a single chromosome. When two loci are close together, genetic recombination will occur less frequently than when they are further apart. Assuming recombination is a random event, all alleles would be independently assorted and all alleles would be in linkage equilibrium. However, recombination is not random. There are **recombination hot spots** and cold spots. Therefore, some alleles are found together more often than they would be by chance. These alleles are "out of equilibrium" and they are therefore said to be in "linkage disequilibrium". The level of linkage disequilibrium depends on the difference between observed (non-random) and expected (assuming random distributions) frequencies. By exploiting this feature of non-random association of alleles it is possible to detect genetic associations even if the causal polymorphism is not genotyped. Linkage disequilibrium is both a blessing and a curse. While it can help to identify associations with linked alleles, which may be for some reason difficult to genotype directly, it can make identification of the primary risk allele or locus very difficult, as we shall see in Chapter 6.

The ability to detect a significant association through linkage disequilibrium can increase the power of an association study

Several different measures of linkage disequilibrium have been proposed in the literature. The simplest one is the covariance *D*, but the two most commonly used are D' and r^2. In order to understand this we can consider two biallelic loci *A* and *B*, bearing alleles *A* and *a* at the *A* locus, and *B* and *b* at the *B* locus, which result in four possible haplotypes: *AB*, *Ab*, *aB*, and *ab*. The alleles have observed frequencies f_A, f_a, f_B, and f_b, as shown in the Punnett Square below. The Punnett Square also shows all possible haplotypes with the frequencies f_{AB}, f_{Ab}, f_{aB}, and f_{ab}.

Locus 1	Locus 2		Totals for rows
	B	b	
A	f_{AB}	f_{Ab}	f_A
a	f_{aB}	f_{ab}	f_a
Totals for columns	f_B	f_b	1

Calculating *D*

D is a measure of linkage disequilibrium, which is determined as the difference between the observed frequency of a two-locus haplotype and the frequency that we would expect if the alleles are randomly segregated. Suppose that f_{AB} is the observed frequency of the haplotype AB formed by the two alleles A and B. The expected haplotype frequency for AB is calculated by the product of the allele frequency of each of the two alleles:

$$f_{AB} = f_A \times f_B$$

If the alleles are in linkage equilibrium, the observed frequency of the AB in the test population will be the same as that which is expected. However, if they are in linkage disequilibrium, we would expect a departure from the expected frequency.

$$f_{AB} = f_A \times f_B + D$$

$$f_{aB} = f_a \times f_B - D$$

$$f_{Ab} = f_A \times f_b - D$$

$$f_{ab} = f_a \times f_b + D$$

The value of D can be obtained by anyone of the above equations as all will result in the same value of D. D can also be calculated by summarizing all of the above equations:

$$D = f_{AB} \times f_{ab} - f_{Ab} \times f_{aB}$$

When $D = 0$ there is no linkage disequilibrium. The method can be further illustrated by considering a population where the two loci above show the observed genotyping frequencies as:

$$f_{AB} = 0.85$$

$$f_{Ab} = 0.05$$

$$f_{aB} = 0.05$$

$$f_{ab} = 0.05$$

The frequencies of each allele are first calculated from the observed data as:

$$f_A = f_{AB} + f_{Ab} = 0.85 + 0.05 = 0.9$$

$$f_a = f_{aB} + f_{ab} = 0.05 + 0.05 = 0.1$$

$$f_B = f_{AB} + f_{aB} = 0.85 + 0.05 = 0.9$$

$$f_b = f_{Ab} + f_{ab} = 0.05 + 0.05 = 0.1$$

Then D can be calculated using the below equations:

$$D_{AB} = f_{AB} - f_A \times f_B = 0.85 - (0.9 \times 0.9) = 0.04$$

$$D_{aB} = f_a \times f_B - f_{aB} = (0.1 \times 0.9) - 0.05 = 0.04$$

$$D_{Ab} = f_a \times f_B - f_{Ab} = (0.1 \times 0.9) - 0.05 = 0.04$$

$$D_{ab} = f_{ab} - f_a \times f_b = 0.05 - (0.1 \times 0.1) = 0.04$$

$$D = f_{AB} \times f_{ab} - f_{Ab} \times f_{aB} = 0.85 \times 0.05 - (0.05 \times 0.05) = 0.04$$

Note that all the equations give the same value of D (0.04). Though D intuitively explains the concept of linkage disequilibrium, the numerical value is of little use as it does not really quantify the strength of linkage disequilibrium as D varies as allele frequencies change. Consequently, D is no longer used to estimate degrees of linkage disequilibrium, but it remains important as both D' and r^2 discussed below are derived from D.

Calculating *D′*

D' is a widely used measure of linkage disequilibrium and is the ratio of D compared with its maximum value. The absolute value of D' is determined by dividing D by its maximum value, given the allele frequencies at the two loci:

$$D' = \frac{D}{|D|_{\max}}$$

where $|D|_{\max} = \min(f_A \times f_b, f_a \times f_B)$. This value is calculated from D as:

If $D > 0$, $|D|_{\max}$ is equal to the smaller value between $f_A \times f_b$ and $f_a \times f_B$

If $D < 0$, $|D|_{\max}$ is equal to the smaller value between $f_A \times f_B$ and $f_a \times f_b$

A value of $D' = 1$ denotes complete linkage disequilibrium. Over time, repeated recombination results in the decay of D' towards 0. However, when $D' < 1$ there is no clear indication of the strength of linkage disequilibrium and it may be difficult to determine whether alleles are in linkage disequilibrium or not. As a rule of thumb, when $D' = 1$ there is linkage disequilibrium between the two loci, when $D' < 1$, the likelihood of linkage disequilibrium increases with the increasing proximity of D' to 1. Haploview software, for example, visualizes a strong linkage disequilibrium when $D > 0.95$. One disadvantage of using D' is that it is highly inflated in small samples and when alleles are rare.

Calculating *r²*

r^2 is the statistical coefficient of determination, a measurement of correlation between a pair of variables. The measure r^2 is in some ways complementary to D' as r^2 is equal to D^2 divided by the product of the allele frequencies at the two loci:

$$r^2 = \frac{D^2}{f_A f_a f_B f_b}$$

where f_A, f_a, f_B, and f_b are the frequencies of each allele, and D is the linkage disequilibrium measure introduced above. r^2 ranges between 0 (loci are in complete linkage equilibrium) and 1 (loci are in complete linkage disequilibrium).

If the value of r^2 between two loci equals 1 this implies the markers at the two loci are providing the same information. Consequently, the genotypes of alleles of one SNP are directly predictive of the genotypes of the other and genotyping either SNP or allele will result in the same P value. In the light of this it is sometimes reasonable to genotype only one of the two selected SNPs. This can be helpful if specific SNP sequences are difficult to genotype. A surrogate SNP used in this manner is often referred to as tagged SNP.

The relationship between D' and r^2

D' and r^2 are both measures that can be written in terms of allele frequencies. To illustrate this, consider D' such that $D \geq 0$ and $f_A \geq f_B$. In this case, $D_{\max} = f_a f_B$ (being $f_A \times f_b > f_a \times f_B$). In this case:

$$D' = \frac{D}{f_a f_B}$$

We can extrapolate D from the above equation as $D = D' \times f_a f_B$ and substitute r^2 in the equation, thus:

$$r^2 = (D')^2 \times \frac{f_a f_B}{f_A f_b}$$

The above equation describes the relationship between D', r^2, and allele frequencies when $D \geq 0$ and $f_A \geq f_B$. The upper boundary of r^2 is D' and is reached only when $f_A = f_B$. The implication of this is that D' provides the upper limits of r^2. D' is a commonly used measure of recombination and gives information on the physical extent of useful linkage disequilibrium. If we know the value of D' we can calculate r^2. For example, if a recombination point results in a $D' = 0.7$, the maximum possible r^2 for these alleles would be 0.49.

Most association analyses identify multiple SNPs, other genetic variants, and haplotypes

Analyses based on single SNPs have limited statistical power to detect true genetic associations. Most genes encode multiple SNPs (and other genetic variants) and it is more appropriate to consider haplotypes composed of multiple genetic variants (including SNPs). A haplotype is a set of genetic polymorphisms (mostly SNPs) found on the same chromosome. Some haplotypes exhibit extreme linkage disequilibrium (for example **HLA 7.2** and HLA 8.1 in Chapter 6). Haplotypes can be inferred throughout a process called **phasing**. This term is important in linkage analysis as described in the example of nail–patella syndrome at the beginning of this chapter. This method assigns each allele of the genotype to one or other chromosome and is based on the observation that certain haplotypes are common in certain regions. Several phasing algorithms have been developed, among which a popular algorithm is the **PHASE** program that was used in the HapMap project (http://hapmap.ncbi.nlm.nih.gov/). This program uses posterior probabilities implemented in a Bayesian approach in order to capture the tendency of haplotypes to cluster together over regions of the chromosome. A Bayesian approach incorporates the prior probability of association. As clustering can change as we move along the chromosome due to recombination, a flexible model for the decay of linkage disequilibrium is then used.

Once phased, the inferred haplotype can be assessed for associations using the same statistical methods as described previously. However, there are problems with haplotype-based analyses. For example, inclusion of rare haplotypes in studies increases the number of tests performed and reduces the statistical power of the testing. A common solution to this problem is to combine all of the rare haplotypes into a single category. By doing this we preserve the data and maintain the statistical power of the study. However, this could also introduce problems because individual rare haplotypes may be important in disease and these associations may be overlooked when clustered in this way.

CONCLUSIONS

In complex disease it is important to include statistics into study design and to do so at the earliest moment. Deciding what analysis packages and procedures to use after a study has been performed is not good practice and may prejudice the final conclusions. Data handling in studies of the genetics of complex disease has become complicated. This is the era of computerized science. Without computers it would not be possible to handle the data in the way we do. The Human Genome Project would have been impossible and the future would simply not exist for this type of science. Fortunately, it does exist and statistical analysis on various computer platforms helps us to achieve our goals. This is also the era of collaboration when competing centers work together for the greater good, by pooling samples so that statistical thresholds can be met and statistical power enhanced. Meta-analysis is the natural extension of this collaboration.

When looking at data analysis in complex disease we can start with calculating LOD scores in linkage studies and calculating ORs in association studies. Of course we need to consider the problems with false-positive and false-negative findings and the impact of sample size (large-scale studies) on statistical power. We need to be aware of sample relatedness, case call rate, and SNP call rate before we begin any study.

Whether we are testing allelic polymorphisms in a single gene or the whole genome we can use Pearson's χ^2 test, Fisher's exact probability test, and the Cochran–Armitage test. There are different models in the Cochran–Armitage test. The next step in complexity is logistic regression or linear regression analysis; however, these are complex tests and researchers are advised to use appropriate computer software and refer to genuine statistical genetics experts for guidance and help with their studies, but a little basic knowledge of the methods is also advised. It is also important to understand the limitations and potential pitfalls for different types of study, and consider how those pitfalls may be overcome by good study design and use of advanced analytical techniques. Other things to consider are how we present our data, e.g. should we use Q–Q plots or Manhattan plots.

FURTHER READING

Books

Strachan T & Read AP (2011) Human Molecular Genetics, 4th ed. Garland Science. Chapter 16 deals with identifying human genes and susceptibility factors, and provides excellent additional data for students wishing to delve deeper into medical genetics. There is also additional relevant information in the other chapters.

Articles

Armitage P (1955) Tests for linear trends in proportions and frequencies. *Biometrics* 11:375–386.

Balding DJ (2006) A tutorial on statistical methods for population association studies. *Nature Rev Genet* 7:781–791.

Barrett JC, Fry B, Maller J & Daly MJ (2005) Haploview: analysis and visualization of LD and haplotype maps. *Bioinformatics* 21:263–265.

Chapman JM, Cooper JD, Todd JA & Clayton DG (2003) Detecting disease associations due to linkage disequilibrium using haplotype tags: a class of tests and the determinants of statistical power. *Hum Hered* 56:18–31.

Clarke GM, Anderson CA, Pettersson FH et al. (2011) Basic statistical analysis in genetic case-control studies. *Nat Protoc* 6:121–133.

Clayton DG, Walker NM, Smyth DJ et al. (2005) Population structure, differential bias and genomic control in a large-scale, case-control association study. *Nat Genet* 37:1243–1246.

Collins FS & McKusick VA (2001) Implications of the human genome project for medical science. *JAMA* 285:540–544. This is a very important review that provides the reader with a guide to and an illustration of the potential for understanding complex disease in the immediate post-genome period.

de Bakker PI, Yelensky R, Pe'er I et al. (2005) Efficiency and power in genetic association studies. *Nat Genet* 37:1217–1223.

Dudbridge F & Gusnanto A (2008) Estimation of significance thresholds for genome-wide association scans. *Genet Epidemiol* 32:227–234.

Epstein MP, Allen AS & Satten GA (2007) A simple and improved correction for population stratification in case control studies. *Am J Hum Genet* 80:921–930.

Ermini L, Wilson IJ, Goodship TH & Sheerin NS (2012) Complement polymorphisms: geographical distribution and relevance to disease. *Immunobiology* 217:265–271.

Ewens WJ & Spielman RS (1995) The transmission/disequilibrium test: history, subdivision, and admixture. *Am J Hum Genet* 57:455–464.

Hill WG & Robertson A (1968) The effects of inbreeding at loci with heterozygote advantage. *Genetics* 60:615–628.

Hirschhorn JN, Lohmueller K, Byrne E & Hirschhorn K (2002) A comprehensive review of genetic association studies. *Genet Med* 4:45–61.

Ke X, Hunt S, Tapper W et al. (2004) The impact of SNP density on fine-scale patterns of linkage disequilibrium. *Hum Mol Genet* 13:577–588.

Lewis CM & Knight J (2012) Introduction to genetic association studies. *Cold Spring Harbor Protoc* 2012:297–306.

Lohmueller KE, Pearce CL, Pike M et al. (2003) Meta-analysis of genetic association studies supports a contribution of common variants to susceptibility to common disease. *Nat Genet* 33:177–182. This is a useful text on meta-analysis in common complex disease.

McIntosh I, Dunston JA, Liu L et al. (2005) Nail patella syndrome revisited: 50 years after linkage. *Ann Hum Genet* 69:349–363. This is a useful update on nail–patella syndrome genetics.

Morton NE (1956) The detection and estimation of linkage between the genes for elliptocytosis and the Rh blood type1. *Am J Hum Genet* 8:80–96.

Price AL, Patterson NJ, Plenge RM et al. (2006) Principal components analysis corrects for stratification in genome–wide association studies. *Nat Genet* 38:904–909.

Pritchard JK, Stephens M & Donnelly P (2000) Inference of population structure using multilocus genotype data. *Genetics* 155:945–959.

Purcell S, Neale B, Todd-Brown K et al. (2007) PLINK: a tool set for whole-genome association and population-based linkage analyses. *Am J Hum Genet* 81:559–575.

Sasieni PD (1997) From genotypes to genes: doubling the sample size. *Biometrics* 53:1253–1261.

Schaid DJ & Jacobsen SJ (1999) Biased tests of association: comparisons of allele frequencies when departing from Hardy–Weinberg proportions. *Am J Epidemiol* 149:706–711.

Stephens M & Scheet P (2005) Accounting for decay of linkage disequilibrium in haplotype inference and missing-data imputation. *Am J Hum Genet* 76:449–462.

Snyder LH (1932) Studies in human inheritance IX. The inheritance of taste deficiency in man. *Ohio J Sci* 32:436–468. This paper is considered to be the first direct pharmacogenetics study.

The 1000 Genomes Project Consortium (2010) A map of human genome variation from population-scale sequencing. *Nature* 467:1061–1073. This is a very exciting project with the promise of more yet to come.

The ENCODE Project Consortium (2012) An integrated encyclopedia of DNA elements in the human genome. *Nature* 489:57–74. This is an invaluable research paper for those starting out on the investigation of the genetics of human disease.

The Human Genome (2001) *Nature* 409:813–958. The complete issue of *Nature* mostly dedicated to the Human Genome Map with multiple papers, letters, and editorial commentary from multiple authors. It is full of useful insight and critical discussion. Any student of human genetics should read this issue.

The International HapMap Consortium (2005) A haplotype map of the human genome. *Nature* 437:1299–1320.

The International HapMap Consortium (2007) A second generation human haplotype map of over 3.1 million SNPs. *Nature* 449:851–861.

The International HapMap 3 Consortium (2010) Integrating common and rare genetic variation in diverse human populations. *Nature* 467:52–58. This is essential reading for those involved in complex disease genetics.

The Wellcome Trust Case Control Consortium (2007) Genome-wide association study of 14,000 cases of seven common diseases and 3,000 shared controls. *Nature* 447:661–678. This is another landmark paper highlighting the development and application of new technologies (GWAS) in complex disease research, and it provides a mass of useful information for those studying complex disease.

Wang WY, Barratt BJ, Clayton DG & Todd JA (2005). Genome-wide association studies: theoretical and practical concerns. *Nat Rev Genet* 6:109–118.

Wittke-Thompson JK, Pluzhnikov A & Cox NJ (2005) Rational inferences about departures from Hardy–Weinberg equilibrium. *Am J Hum Genet* 76:967–986.

Zhang K, Calabrese P, Nordborg M & Sun F (2002). Haplotype block structure and its applications to association studies: power and study designs. *Am J Hum Genet* 71:1386–1394. This is a good paper to read to understand how haplotypes can be used in GWAS.

Zhang Y, Guan W & Pan W (2013). Adjustment for population stratification via principal components in association analysis of rare variants. *Genet Epidemiol* 37:99–109.

Online sources

http://hapmap.ncbi.nlm.nih.gov
This is the site of the results of the International HapMap project.

http://pngu.mgh.harvard.edu/~purcell/plink
PLINK is a freely available, open-source, whole-genome association analysis toolset designed for quality and analysis of GWAS data.

http://pritchardlab.stanford.edu/structure.html
This site includes STRUCTURE software referred to in the text.
This is an essential site for updates and information on genetic disease of all types.

http://www.ncbi.nlm.nih.gov/projects/SNP/snp_summary.cgi
This site provides information about millions of validated reference SNP clusters in the genome.

http://www.ncbi.nlm.nih.gov/SNP
This an essential site for updates on SNPs.

www.r-project.org
This is another freely available and excellent package for statistical computing that can be used with or without existing commercial packages such as SPSS or SAS.

CHAPTER 6

The Major Histocompatibility Complex

The **extended major histocompatibility complex (xMHC)** located on the short arm of chromosome 6 (6p21.3) is perhaps the most complex genetic system in the human genome with 252 expressed genes in a region of approximately 7.6 Mb of DNA. The MHC is so named because it encodes genes with vital functions in determining the compatibility of tissues. In particular, it encodes the human leukocyte antigens (HLAs), matching of which is vital in some, but not all, forms of clinical transplantation. JJ van Rood, one of the pioneers of **immunogenetics**, suggested that all discoveries go through three phases: the discovery itself, the development of tools with which to investigate the discovery, and the final exposure of the discovery (the latter of which he likened to the charting of a new continent). The scientific process is an iterative cycle of discovery, hypothesis generation, and testing, continually looking back. This is particularly appropriate when we consider the human MHC.

This chapter will briefly examine the discovery of the MHC, the complex naming of HLA antigens and alleles, the structure and normal function of the products of the HLA alleles, and how genetic variation in HLA may determine susceptibility and resistance to disease using a number of key examples. We will also consider a selection of other non-HLA MHC-encoded immunoregulatory genes and discuss how they may also play a role in susceptibility to complex disease.

Selecting disease examples for this section is a very difficult task because there are many diseases with genetic associations with the MHC. Genetic associations with HLA were first reported in 1967 by Amiel, who reported an association between HLA and non-Hodgkin's lymphoma. Later, in the 1970s, Ceppellini et al. reported an association between endemic malaria and HLA in four Sardinian villages. Reports of other associations quickly followed: HLA-B8 and celiac disease by Stokes et al. in 1972, HLA-A3 and **multiple sclerosis** by Terasaki et al. in 1972, B13 and psoriasis by Russell et al. in 1972, and B27 and ankylosing spondylitis by Brewerton et al. in 1973. There are currently so many genetic associations with HLA that there are complete textbooks devoted to the subject. Therefore, we will focus here on a small number of illustrative examples, some of which are selected because they are rarely discussed, and some because they are frequently discussed and have become exemplary models. The selected examples in this chapter include hemochromatosis, psoriasis (type I and type II), severe cataplectic narcolepsy, three autoimmune liver diseases [primary sclerosing cholangitis (PSC), primary biliary cirrhosis (PBC) and autoimmune hepatitis (AIH)], multiple sclerosis, and systemic lupus erythematosus (SLE). The selection illustrates the impact of the MHC immunoregulatory genes in clinical disease, and informs the debate on disease diagnosis, treatment and care, and disease pathogenesis for a large variety of common and some less common diseases.

Type 1 and type 2 diabetes (T1D and T2D), rheumatoid arthritis, ankylosing spondylitis, cancer, and infectious disease are all discussed elsewhere in this book as part of other chapters or as specific chapters, and therefore the role of the MHC and MHC encoded genes in these diseases are not included here.

6.1 HISTOCOMPATIBILITY

Many geneticists avoid discussing the MHC because of its complexity and as a consequence it is seen by many as an area for specialists only. These specialists are mostly tissue-typers who provide data for transplant programs in particular. A summary of the history of the major developments in the MHC is shown in **Figure 6.1**.

The idea of histocompatibility first started with blood groups

ABO blood groups can technically be considered as the first detected histocompatibility antigens in man. Histocompatibility refers to the acceptance (compatibility) of tissue and is normally applied to the acceptance of tissue transferred between individuals via either blood transfusion or solid-organ transplantation. Histocompatibility can involve exchange between species (xeno-transplantation) or, more frequently, between individuals of the same species (allo-transplantation).

Interest in compatibility of tissues arises from the clinical problems associated with incompatibility, i.e. the failure of the host or donor tissue to tolerate this transfer. The ABO system is not encoded within the MHC, or even on the same chromosome, but it is important in clinical transplantation, especially in skin, kidney, heart, and liver transplantation. When blood transfusions are performed between A or B incompatible donors naturally occurring antibodies in the recipient attack the transferred blood from the donor and this causes agglutination and hemolysis (**Figure 6.2**). The problem has been known since James Blundell published his report of a blood transfusion in 1829, but the reasons for the transfusion reaction were not fully understood until Karl Landsteiner published his work on the ABO blood groups in 1900.

The MHC-encoded HLA antigens are the second major histocompatibility group

The first clues to the existence of the MHC came from mouse models of tumor immunology. It was clear from 1900 onwards that factors other than the blood groups influenced

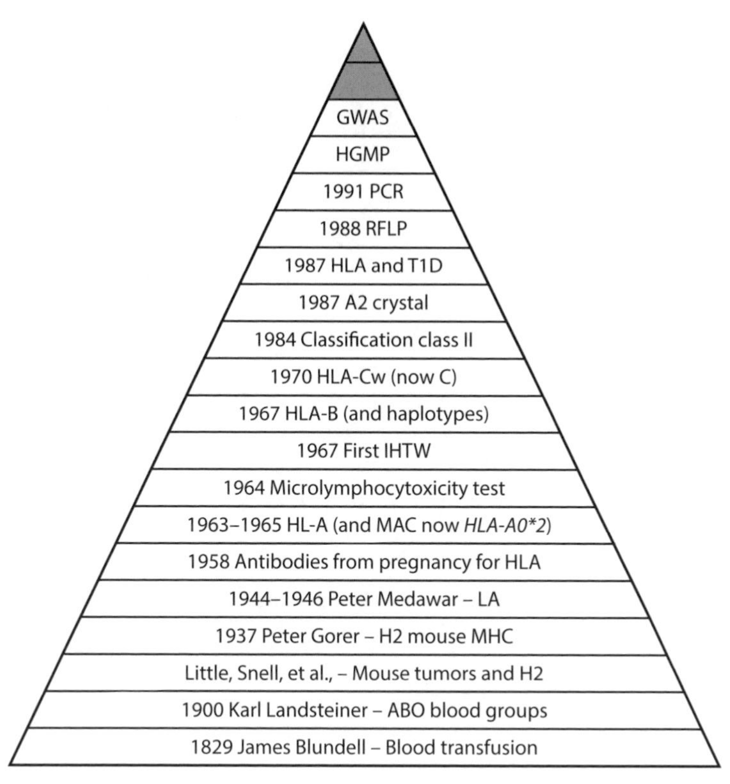

Figure 6.1: Time pyramid for major discoveries and developments in the world of HLA. The figure is a time pyramid illustrating the timing of some of the major discoveries and developments in the world of HLA. It shows selected developments over the last millennium from recent major technological advances to early discoveries that underpin our knowledge today. As students of science we should not forget the principle that today's work is based on the work of those who came before us. Many of the above achieved fame in their own lifetimes, but not all are as well remembered as they deserve. The list includes several Noble Prize winners. The top of the pyramid is left unfilled for the next great leap ahead.

the outcome of tissue grafts in both animals and in humans. These factors turned out to be encoded in the MHC. In the early days, at least, success or failure in clinical transplantation could depend on matching donor and recipient HLA types. The HLA antigens encoded in the MHC are important in bone marrow and kidney transplantation, and less important in heart and liver transplantation. The history of the discovery of the MHC is discussed in brief in **Box 6.1**.

Naming the HLA antigens and alleles up to and including the early molecular genotyping era

In addition to being one of the most complex systems in the human genome, some of the genes in the MHC are the most polymorphic. To understand the nomenclature applied to HLA we need to look back at the history; however, because this is neither an immunology book nor a history book, a prolonged discussion of early nomenclature is not appropriate here. **Table 6.1** indicates some of the chaos that was present in the early world of HLA typing when it came to naming new alleles and genes, and how this was adjusted to create current standard nomenclature. Therefore, this will only be dealt with in brief. The International Histocompatibility Testing Workshops (IHTWs) that were first set up in

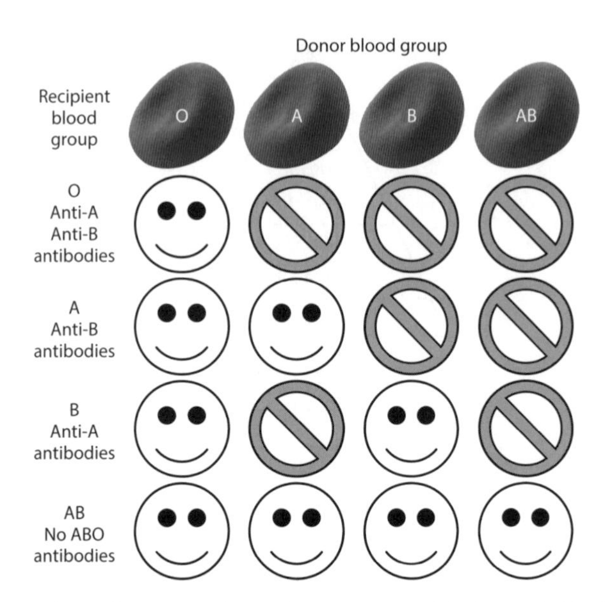

Figure 6.2: Donor and recipient interaction for ABO blood groups. The four main ABO blood groups are encoded on chromosome 9. Those with blood group O can donate to all other groups, but because they can produce antibodies to A and B antigens they can only receive blood from O donors. In contrast, those with AB can only donate to other AB carriers because O, A, and B carriers can all produce antibodies to either A and B (O blood group), or A (B blood group) or B (A blood group). However, AB-positive individuals do not produce antibodies against the other blood groups and can receive blood from any donor.

BOX 6.1 OF MICE AND MEN: HOW EARLY WORK WITH MICE REVEALED THE POSSIBILITY OF THE MHC AND THE HLA ANTIGENS

The first clues to the existence of the MHC came from mouse models of tumor immunology. As early as 1903, Jensen noted that different strains of mice were not equally susceptible to the growth of grafted tumor cells. Little and colleagues, who were also working on mice, proposed that acceptance of tumor grafts depended on the graft and host tissues possessing a large number of susceptibility factors in common. Bauer, working with human subjects in 1927, had also noted the acceptance of skin grafts in identical twins was consistent with the theory that compatibility is genetically determined. In 1937, Peter Gorer proposed a genetic theory for transplant rejection: "Normal and neoplastic tissues contain isoantigenic factors (which are) genetically determined. Isoantigenic factors present in the grafted tissue and absent (from the) host are capable of eliciting a response which results in the destruction of the graft."

These ideas may not seem revolutionary in 2015, but in 1937 they represented a great leap forward in both genetics and immunology, and Peter Gorer is considered one of the founding fathers of immunogenetics. Peter Medawar followed up Gorer's idea working on skin grafts in rabbits from which he was able to suggest that leukocytes and skin share important **transplantation antigens**. These were the first ideas pointing at HLA. Leukocytes are white blood cells—specifically, the mononuclear cells and these antigens were later called leukocyte antigens. After much work by many different groups, Jean Dausset published his hypothesis in 1952 suggesting that a similar antigenic system to that seen in mouse erythrocytes (red cells) by Gorer existed on the surface of human

BOX 6.1 OF MICE AND MEN: HOW EARLY WORK WITH MICE REVEALED THE POSSIBILITY OF THE MHC AND THE HLA ANTIGENS (*Continued*)

leukocytes. Subsequently the "antigens" were called "human leukocyte antigens" (HL-A or HL-antigen and finally HLA). One distinct feature of the HL-antigen is that, these antigens are not expressed on red blood cells, unlike the ABO blood groups. This system is the HLA system we know today and Dausset and several others were later awarded the Noble Prize for this work and related work.

Table 6.1 Early nomenclature for HLA.

Example early name(s) assigned	Locus or allele	Adjusted (current) name
Class I genes	Each **antigen** is constructed from a single α-chain supported by a β-2-microglobulin molecule encoded on chromosome 15	Each **gene** is highly polymorphic
LA1, LA2, LA3	HLA-A	*HLA-A*
HU-1	HLA-A	*HLA-A*
4a, 4b	HLA Bw4 and Bw6 (see Table 6.3)	Bw4 and Bw6
W5	HLA B35	*B*35:*
4c	HLA B5	Describes a series of families (superfamily)
Class II genes	Each **antigen** has two genes: A encoding the α-chain and B encoding the β-chain	Each **gene** is polymorphic; the level of polymorphism varies (see Figure 6.8)
D-related	HLA-DR	There are nine different DRB genes, four are expressed. DRB1 is the predominant DRB gene
DC1, MB1, (LB12), MT1	DQ	DQ1
DC3, MB2, LB-E17, Te24	DQ	DQ2
DC4, MB3, LB13, TA10, MT4, TB21	DQ	DQ3
LB13	DQ	DQ3 (split)
TA10	DQ	DQ3 (split)
MT2	DRw52	*DRB3*
MT3, BR3, BR4	DRw53	*DRB4*
SB	DP	DP

The table illustrates the complexity of the original naming systems for new antigens in the left-hand column. Through a series of committees and workshops, these differences have been clarified so that a single naming system works. A substantial restructuring of HLA naming has taken place in recent times.

1965 have ensured a consistent procedure for identification, reporting, and naming of all HLA alleles. The reason for this is so that different laboratories can perform HLA typing on the same sample and produce the same result. The history of how this came together is indicated in **Box 6.2**. Standardization of the naming system for HLA was essentially for clinical transplantation, where matching of donor and recipient is required.

BOX 6.2 EARLY HLA NOMENCLATURE: CREATING INTERNATIONAL STANDARDS

Early studies suggested a single locus for the histocompatibility antigen. Bodmer & Payne proposed a prefix of LA for leukocyte antigen. Dausset proposed the prefix HU for human system. Using the latter system, HU-1 meaning human system-1 was the first antigen. This was a time of considerable confusion. Each group began naming identified antigens differently: 4a and 4b (Van Rood and Leeuwen, 1963), LA1, LA2, and LA3 (Payne and Bodmer, 1964), Ao (Amos), Bt (Batchelor), To (Ceppellini, who is credited with coining the term "haplotype"), Da (Dausset) and Te (Terasaki, who was instrumental in developing the microcytotoxicity test). These early pioneers soon realized that it would be essential to standardize the systems for naming and testing these HLA antigens. In order to do this they assembled and decided on steps to unify and clarify the naming and identity of the HLA antigens. The first step was to set up regular International Histocompatibility Testing Workshops (IHTWs). The first of these was held in 1965. In 1968, the term HL-A standing for "Human Leukocyte A" was coined; the hyphen was dropped from the name following the recognition of other class I loci. The different antigen families (genes today) were called HLA A, HLA B, and HLA Cw. As part of standard procedure all newly identified antigens were given a "w" until confirmed at the next IHTW. HLA C antigens (and later alleles) retained the "w" in their name because complement components are also named using an alpha-numeric code with C as the letter and a single number. However, C alleles are no longer indicated by Cw, but by C alone.

Throughout the 1970s and early 1980s, a number of authors proposed the existence of novel class II loci and alleles; these included DC and MB (some of which are the same as LB), MT, SB, Te, and BR. This was very much like a repeat of the late 1960s and early 1970s. At the 1984 IHTW these novel class II antigens were reclassified and assorted, giving rise to the present DR, DQ, and DP picture for class II. Thus, DC, LB, and MB were all identified as DQ antigens; MT1, MB1, and DC1 were recognized as equivalent to DQ1; MT2 and MT3 were reassigned DRw52. BR3, BR4, Te24, and MB2 were reassigned DRw53. DC3 and LB-E17 were assigned as DQ2. MB3, MT4, DC4, and TB21 were reassigned DQ3. All SB antigens were identified as belonging to a single locus renamed DP to fit the sequence DR, DQ, DP. While some HLA DQ antigens were detectable by serology, HLA DP antigens (previously SB) were not.

Current nomenclature is focused on the genotype rather than the phenotype as before. There is an enormous amount of polymorphism in the HLA genes as seen in this chapter and consequently in 2010 the nomenclature was reformed. The group of genes is referred to in italics with the suffix *HLA* then each locus joined by a hyphen (e.g. *HLA-A*) next the allele or allele family is identified (e.g. *HLA-A*01*) with *A*01* referring to the first group or family of alleles of the *HLA-A* gene family. These first two digits refer to the antigen and are essentially the same as the serotype or phenotype. In early studies the example

BOX 6.2 EARLY HLA NOMENCLATURE: CREATING INTERNATIONAL STANDARDS (*Continued*)

above would be the A1 antigen. In the new era sub-typing this group means there has to be a greater number of possible alleles. This means more codes are needed to distinguish them all from each other. Therefore, the use of a colon and second set of digits is applied to identify variants of the *HLA-A*01* family thus *HLA-A*01:01*. These alleles must encode an amino acid change. If the variant does not encode an amino acid change (i.e. it is a synonymous polymorphism or noncoding polymorphism) it is listed using a third set of digits after the insertion of a second colon for example *HLA-A*01:01:01*. Alleles that encode changes in the intronic sequences are listed by use of a fourth set of digits after a third colon thus *HLA-A*01:01:01:01*. The new naming system is designed to cope with the most polymorphic genetic system in the human genome.

Antigens are proteins; in the case of HLA, these are usually expressed on the cell surface. Alleles are genetic variants, which in the case of HLA usually refers to the region of the gene that encodes these proteins. The two terms are often used interchangeably. Early work with serum and cells was based on detecting antigens. Later work based on **restriction fragment length polymorphism (RFLP)** or polymerase chain reaction (PCR) analysis detected genetic variants and therefore uses the term alleles. Occasionally, depending on the level of specificity applied, the term family may also be used, i.e. the *HLA* A2 family, where *HLA* A2 is tested rather than a range of different *HLA* A2 variants. The term family will be applied to both genes and antigens when appropriate.

HLA class I

Understanding nomenclature is difficult in any subject and no more so than with regard to *HLA* class I. Although there are many *HLA* class I genes, there are only three major *HLA* class I genes that have been studied in detail: *HLA-A*, *HLA-B*, and *HLA-C*.

HLA class II

The major HLA class II antigens are DR, DQ, and DP encoded by the gene pairs *DRA* and *DRB*, *DQA* and *DQB*, and *DPA* and *DPB*. The changes heralded by molecular genotyping, which allowed more accurate HLA typing (genotyping), also heralded a new era. They confirmed what serology had suggested, i.e. that there may be more than one expressed *DRB* gene on some, but not all, MHC haplotypes, and identified the DQ and DP families of genes with a single pair of expressed *DQA* and *DQB* genes encoding the DQ molecule and a single pair of expressed *DPA* and *DPB* genes encoding the DP molecule for all MHC haplotypes.

The second expressed *DRB* gene

There are nine *DRB* genes numbered *DRB1* to *DRB9*. The first of these, *DRB1*, is expressed on all haplotypes. Some haplotypes express only the *DRB1* gene (*DRB1*) and others express a second *DRB* gene. Which second *DRB* gene is encoded on each haplotype can be determined from the *DRB1* family encoded on each haplotype. The second expressed *DRB* gene on haplotypes encoding alleles of the *DRB1*15* and *DRB1*16* families is *DRB5*. *DRB3* is the second expressed *DRB* gene expressed on haplotypes encoding alleles of the *DRB1*03*, *DRB1*11*, *DRB1*12*, *DRB1*13*, and *DRB1*14* families. *DRB4* is the second expressed *DRB* gene on haplotypes encoding alleles of the *DRB1*04*, *DRB1*07*, and *DRB1*09* families. The *DRB1* families that express only a single *DRB1* gene are the *DRB1*01*, *DRB1*08*, and *DRB1*10* families.

The second expressed *DRB* genes are all thought to be only weakly expressed in each case. However, even though these genes are thought to be weakly expressed in comparison with *DRB1*, expression levels may alter under immune stress. The significance of having two, as opposed to one, expressed *DRB* gene, on some haplotypes, both of which may be polymorphic, is not properly understood in the context of complex disease, but the second *DRB* locus should not be overlooked.

The *DRB* pseudogenes

The final set of *DRB* genes to consider are *DRB2*, *DRB6*, *DRB7*, *DRB8*, and *DRB9*, all of which are pseudogenes and are not expressed. All haplotypes carry the *DRB9* pseudogene, but the number of other pseudogenes carried on each varies depending on the *DRB1* family. The different haplotypes (including pseudogenes) are illustrated in **Figure 6.3** and details of some *DRB* gene family nomenclature is given in **Table 6.2**.

Splits

These developments also helped to identify what were previously referred to as "splits" (more correctly "isoforms"). This is demonstrated in **Figure 6.4**. For example, the HLA B5 antigen was found to have three variants; these were first called Bw51, Bw52, and Bw53. The "w" in the name was dropped when their status was confirmed, leaving the labels B51, B52, and B53. These are now more correctly identified as *B*51*, *B*52*, and *B*53*. When they were discovered these three variants were named according to the next available unused number in the catalogue for HLA A and B antigens. Each of these variants has multiple subdivisions. However, not all new allele assignments were bone fide, some turned out to be incorrect and some alleles are null alleles. When an allele assignment is incorrect, the number is deleted from the official sequence record and the number is not reused. Consequently there are gaps in the allele sequences numbers, e.g. there is no *B*51:25*. In addition, the level of variation does not stop after four figures. Some alleles carry non-synonymous polymorphisms, which can lead to further splits and names like *B*51:01:01*. In fact, there are seven members in the *B*51:01* family (*B*51:01:01* to *B*51:01:07*). Though these are all members of the B5 family, not all of the alleles were detectable by serological typing and quite a few B5s identified by serology were incorrectly assigned as members of cross-reactive groups. DNA genotyping revolutionized HLA

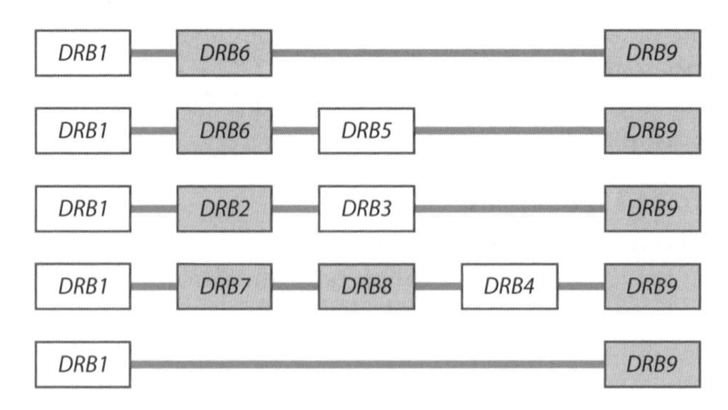

Figure 6.3: Organization of the HLA DRB genes. Each haplotype may carry two or more *DRB* genes. However, a significant number of the *DRB* genes are pseudogenes (*DRB2*, *DRB6*, *DRB7*, *DRB8*, and *DRB9*; shown in gray) and are not expressed.

Table 6.2 Commonly used DRB allele family names past and present.

Old name	Current equivalent allele families or genes	Gene or haplotype
DR2	*DRB1*15* or *DRB1*16*	*DRB1*
DR3	*DRB1*03*	*DRB1*
DR4	*DRB1*04*	*DRB1*
DR5	*DRB1*11* or *DRB1*12*	*DRB1*
DR6	*DRB1*13* or *DRB1*14*	*DRB1*
DRw52	*DRB3*	*DRB3*
DRw53	*DRB4*	*DRB4*
Ancestral haplotypes		
A1–Cw7–B8–DR3–DQ2		8.1
A3–Cw7–B7–DR2 (*DRB1*15*)–DQ1 (*DQB1*06:02*)		7.2

Names are used in this chapter as they have been historically and as they are currently; the evolution of HLA and the change in nomenclature are explained in some detail. To make reading easier, the simplest names have been used where possible, i.e. DR3 in place of *DRB1*03:01*; however, it is not always appropriate to make this exchange, especially when referring to molecular structures. For the ancestral haplotypes, a mixture of old and new nomenclature is used depending on published history and the nature of the studies performed. For example, B8, which is old nomenclature, is often used. This has been applied because although this example should be correctly labeled *HLA-B*08*, old studies did not use this naming system. In contrast most studies of DQ used the current nomenclature, e.g. *DQB1*06:02*.

typing to such an extent that the number of detectable alleles and variants (splits) catapulted from around 100 A and B antigens to several thousand alleles at each locus. The same process had the same effect on the HLA class II allele groups.

Taken altogether, i.e. identifying alleles on the new genes and the improved quality of typing (illustrated in **Tables 6.3** and **6.4**), which also enabled a greater number of variants to be accurately identified, these developments have been astounding. HLA typing is now well and truly in the molecular genetics era and nomenclature has been updated too.

The current naming system for HLA alleles and genes allows for a greater level of resolution to be reported

As HLA was at first thought to be a single locus system, antigens were first numbered sequentially A1, A2, etc., as they were discovered and approved. However, it soon became clear that there were at least two HLA loci: A and B. On close inspection when the antigen sets for the two loci (A and B) were pulled apart, it was discovered that there were quite a few B antigens that had A antigen labels. It was decided that for numbering purposes, A and B would share the same number set. Thus, no A antigen has the same number sequence as a B antigen and vice versa; the numbers assigned for A and B antigens are exclusive. Therefore, there are HLA A1, A2 and A3 antigens, but no B1, B2, and B3 antigens. This tradition has persisted. This way of naming new antigens and alleles has been subjected to a standard practice since the 1965 IHTW. It is also important to know that all B, and some A, antigens have a sequence variation that labels them as belonging to either the Bw4 or Bw6 family. This is illustrated in **Table 6.5**.

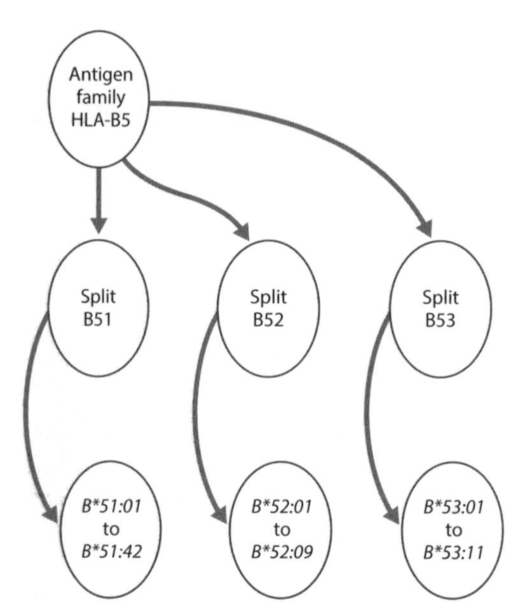

Figure 6.4: HLA nomenclature tree: splits. The figure illustrates the development of HLA nomenclature based on one member of the HLA-B antigen family. B5 was identified early on in studies of HLA (as indicated by the low number B5). Three variants of B5 were later identified. These variants were often called "splits" in tissue-typing laboratories and because at that time the numbering sequence for the combination of HLA A and B antigens had reached 50, the next numbers to be assigned were given to these three variants, i.e. B51, B52, and B53. As these antigens were first identified by serotyping, the original names reflect this naming, i.e. the phenotype (e.g. B51) not the genotype (e.g. *B*51*). In the molecular era further variation has been identified for each of these antigens with allele labels suggesting more than 40 B51 (*B*51:01* to *B*51:42*) variants and approximately 10 for both B52 (*B*52:01* to *B*52:09*) and B53 (*B*53*:01* to *B*53:11*).

In the molecular era we are dealing with alleles and genotypes. The alleles identified are named and numbered in a similar way to the antigens, but more information can be provided. For example, DR alleles of the *DRB1* family are referred to as *DRB1*01* and a further *:01* can be added to indicate a higher level of specificity (or resolution). The allele thus labeled is *DRB1*01:01*. The use of *DRB1* indicates that this is the product of the *DRB1* locus and not of the other expressed *DRB* loci (*DRB3*, *DRB4*, and *DRB5*). Considering the large number of HLA alleles so far identified it is important to be as precise as possible when naming alleles or allele families. **Table 6.6** illustrates the correct use of the current naming techniques. When looking at this we need to consider whether studies are being performed at low or high resolution, or somewhere in the middle. The term family can be useful when genotyping is being performed at low resolution. Current nomenclature is frequently updated and can be found at http://hla.alleles.org/antigens/index.html or http://www.ebi.ac.uk/imgt/hla/stats.html.

The MHC encodes a cornucopia of genetic diversity within the HLA genes

The importance of having a consistent method for assigning allele names is clearly illustrated when we consider the degree of genetic diversity at each of the HLA loci.

For example, in April 2015 there were 3107 *HLA-A* alleles, 3887 *HLA-B* alleles, 2623 *HLA-C* alleles, and 1726 *HLA-DRB1* alleles identified. Regular updates on all of these alleles (numbers and nomenclature, plus additional information) can be found at http://www.ebi.ac.uk/imgt/hla/stats.html.

Table 6.3 Analysis of the correlation between serological and RFLP DR typing in 1000 individuals.

Allotype	*N*	True positive	True negative	False positive	False negative	Equivocal	Concordance (%)
DR1	155	132	3	11	4	5	90
DR2	295	244	15	13	10	13	92
DR3	444	358	71	6	2	7	97
DR4	326	278	17	15	13	3	91
DR5	122	109	1	4	5	3	93
DR6	208	146	5	4	41	12	78
DR7	357	311	31	9	4	2	96
DR8	37	23	0	5	8	1	69
DR9	15	9	1	4	1	0	66
DR10	6	2	0	1	3	0	33
DRBr	35	–	35	–	–	–	–
Total count	2000	1612	179	72	91	41	91.8

This table shows the correlation between two different methods of HLA DR typing: serological phenotyping versus RFLP genotyping. In this model, the analysis shown in the table assumes the RFLP technique always correctly identifies the antigen/genotype. The earlier method is of lower quality due to the poor quality of antibodies for testing the less common DR antigens DR8, DR9, and DR10, and also the inability to accurately detect DR6, which was often unassigned especially in heterozygotes. True negative indicates true homozygotes.

The level of genetic polymorphism in the HLA genes is illustrated in **Figure 6.5**. However, the level of variation in the expressed gene is not always equal to the number of protein polymorphisms in the coding regions of the gene. For example, the DNA sequence may change, but the polypeptide composition may remain unchanged. This is because there is redundancy within the genetic code so that a single amino acid may be encoded by more than one DNA triplet (e.g. CCC, CCA, and CCG each encode the amino acid proline). Thus, the numbers of proteins encoded by the three HLA class I loci (*A*, *B*, and *C*) are 2185, 2870, and 1850, respectively. Each of these values is lower than the figures given for the number of alleles. Some of the alleles are null alleles with no expressed product; at present there are 147 null *HLA-A*s listed, 124 *HLA-B*s, and 86 *HLA-C*s.

Comparing the levels of genetic diversity at DR with those at DQ can make DQ look like a poor relation

Based on the level of variation at the *DRB* locus *DRB1* (1726 alleles) compared with *DQB1* (780 alleles), *DQB1* looks like a poor relation. This is misleading, however, because the *DRA* gene has only seven alleles and encodes only two protein variants. Generally, *DRA* is considered to be essentially invariant (though this is not actually correct). This lack of variability in the α-chain of the DR molecule means that almost all of the polymorphism is restricted to the DRβ-chain. This means that the 1726 alleles that are involved in producing 1262 protein variants in the *DRB1* gene encode almost all of the molecular variation in the complete expressed DR molecule. However, both *DQA* and *DQB* are widely polymorphic (54 and 780 alleles, respectively) and so there is a greater possibility

Table 6.4 Analysis of the correlation between RFLP and PCR oligonucleotide probing for HLA-DRB1 genotyping in 395 individuals.

Allotype	*N*	True positive	True negative	False positive	False negative	Equivocal	Concordance (%)
DR1	72	67	2	2	1	0	93
DR15	90	84	3	3	0	0	97
DR16	7	7	0	0	0	0	100
DR3-a	120	89	23	8	0	0	93
DR3-b	10	7	3	0	0	0	100
DR4	188	175	10	2	1	0	98
DR11	48	43	1	2	1	1	92
DR12	7	4	0	3	0	0	57
DR6-a	53	45	2	4	2	0	89
DR6-b	44	37	3	4	0	0	91
DR7	115	104	7	3	0	0	97
DR8	16	16	0	0	0	0	100
DR9	7	5	0	1	1	0	71
DR10	6	4	0	0	2	0	66
DRBr	10	10	0	0	0	0	100
Total count	790	694	51	29	11	2	95

This table shows the correlation between two different methods of *HLA-DRB1* genotyping: RFLP genotyping versus PCR-oligoprobing using 24 oligoprobes. PCR clearly identifies 18 *DRB1* allele families and this example includes the rarely discussed *DRBr* allotype (which is in fact one of the *DRB1*01* family members). In this model, the analysis shown in the table assumes the PCR technique always correctly identifies the antigen/genotype.

Table 6.5 Sequence differences between HLA Bw4 and Bw6.

Group	77	78	79	80	81	82	83
Bw6	S	L	R	N	L	R	G
Bw4	D	L	R	T	L	L	R
Bw4	S	L	R	T	L	L	R
Bw4	S	L	R	I	A	L	R
Bw4	N	L	R	I	A	L	R
Bw4	N	L	R	T	A	L	R

The sequences of the Bw4 and Bw6 epitopes vary at positions 77 to 83 of the second exon of the α1 domain on the HLA class I molecule. The critical amino acid difference between Bw4 and Bw6 appears to be a glycine (G) for arginine (R) exchange at position 83. Variations at the other positions in this amino acid sequence define the five Bw4 genotypes. Note that there is no variation at positions 78 and 79, which all carry leucine (L) and arginine (R), respectively.

Table 6.6 The current system for naming HLA genes and alleles.

Name	Explanation
HLA-DRB1	This indicates a specific HLA locus (*DRB1*)
*HLA-DRB1*01*	This indicates a group or family of alleles that encode the DR1 antigen
*HLA-DRB1*01:01*	This indicates a specific allele from the *DRB1*01* family (many scripts use *DRB1*0101* for this and omit the ":" sign)
*HLA-DRB1*01:01:02*	This indicates an allele that encodes a synonymous mutation of *DRB1*01:01*
*HLA-DRB1*01:01:01:02*	This indicates an allele that encodes a mutation outside the coding region of *DRB1*01:01*
*HLA-DRB1*07:10N*	This indicates a "null" allele (i.e. an allele that is not expressed)
*HLA-DRB1*01:01L*	This indicates an allele encoding a protein with significantly reduced cell surface expression

for formation of a wider range of variation in the expressed DQA molecules than initially indicated from the simple allele frequencies. In addition because every individual inherits two haplotypes—one from each parent—*DQA1* alleles of paternal origin can form functional DQ molecules with *DQB1* alleles of maternal origin (and vice versa), thus adding a further level of diversity for DQ molecules. The same is also true for HLA DP.

HLA class II molecules can be expressed in *trans* or in *cis*

For the most part HLA alleles are co-dominantly expressed, meaning that most individuals have the opportunity to express both paternal and maternal antigens. The possibility

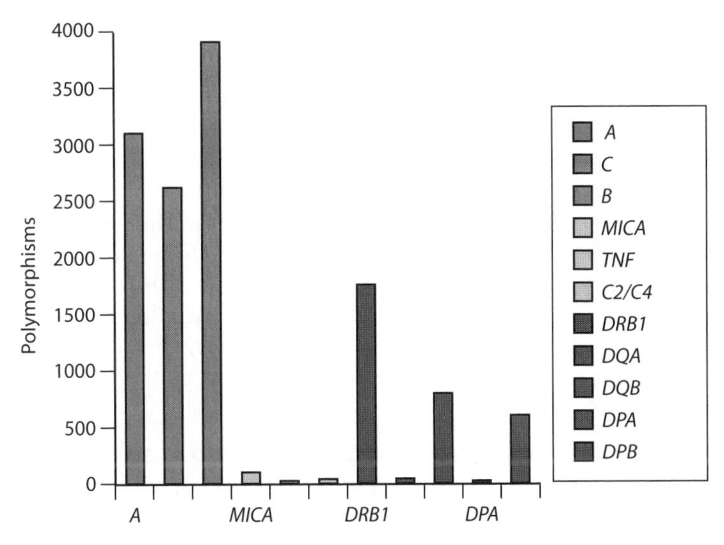

Figure 6.5: Frequency of HLA polymorphisms. The figure shows the extent of polymorphism in HLA as of April 2015. The very high frequency of *HLA-B* polymorphism gives a false impression regarding polymorphism for some of the other genes listed. There are 101 *MICA* alleles, for example. One hundred and one is a considerable level of polymorphism, it just looks small compared to the 3887 B alleles represented. The order of the genes is indicated on the right-hand descending column.

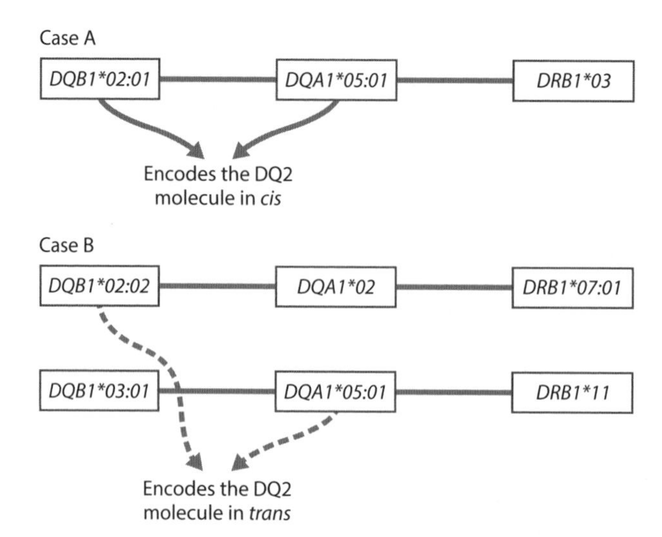

Figure 6.6: The HLA DQ2 molecule constructed in *cis* and *trans*. The figure shows the HLA molecule being constructed in *cis* and *trans*. The *cis* construction (case A) is illustrated by full gray arrows with each allele coming from a single haplotype from the same parent where both the α-chain and the β-chain of the final molecule are encoded on the *DRB1*03* haplotype by the alleles *DQA1*05:01* and *DQB1*02*. The *trans* construction (case B) is illustrated by the dotted arrows where more than one parental haplotype contributes a polypeptide chain to form the final molecule. In each case, the *DQA1* locus contributes the *DQA1*05:01* allele that encodes the necessary α-chain and the *DQB1* locus contributes the *DQB1*02* allele that encodes the necessary β-chain. The ability of the *DQ* genes to construct molecules in *trans* is important as it increases the number of potential genetic variations there are at the molecular level. HLA DR is less able to do this as the *DRA1* gene is essentially monomorphic with very little variation.

to increase the genetic variation through ***trans*** **expression** as opposed to the normal ***cis*** pattern can have important personal consequences (**Figure 6.6**). Sollid et al. found that patients with celiac disease have an increased frequency of HLA DR3 and DR7. The genetic association with DR3 was well known, but the association with DR7 was less well reported. Further analysis revealed that DR7 is only associated with celiac disease in the presence of either DR3 or DR5. DR3 is always found with DQ2 (*DQA1*05:01–DQB1*02:01*; sometimes called DQ2.5). DR5 is almost always found with DQ7 (*DQA1*05:–DQB1*03*) and DR7 occurs with either DQ9 (*DQA1*01:02–DQB1*03*) or DQ2 (*DQA1*01:02–DQB1*02:02*). The *DQA1* alleles of the DR3–DQ2 and DR5–DQ7 haplotypes are almost identical except for **codon** 135 in the membrane proximal domain of the DQ chain. DR5/DR7 heterozygotes with the DR7–DQ2 haplotype carry similar *DQA1* and *DQB1* genes encoding alleles with essentially the same sequences in the trans position as those found on the DR3– DQ2 haplotype in *cis*. The ability to form DQA/DQB in *trans* may explain the complex HLA association with celiac disease.

The final groups of genes that need to be considered are those called pseudogenes, gene fragments, and null alleles

Pseudogenes are genes that are not expressed. *DRB* pseudogenes have been discussed above, but there are also HLA class I pseudogenes (see **Table 6.7**). Gene fragments are exactly as described and are mostly fragments of once expressed genes, possibly the remains of our evolutionary past. Null alleles are generally considered to be alleles that are not expressed. In some cases there is doubt about expression: alleles may be found to be expressed in the

Table 6.7 A list of HLA other class I genes not discussed in the text.

Current names	Characteristic
HLA-E; HLA-F; HLA-G	Associated with class I *Hin*dIII fragments
HLA-J; HLA-K; HLA-L	Class I pseudogenes associated with *Hin*dIII fragments
HLA-N; HLA-P; HLA-S; HLA-T; HLA-U; HLA-V; HLA-W	Class gene fragments associated with *Hin*dIII
HLA-X	Class I gene fragment
HLA-Y	Class I pseudogene
HLA-Z	Class I gene fragment located in class II region
HLA-H	Class I pseudogene

Pseudogenes are genes that are not expressed. Others include the HLA class I pseudogenes (genes that do not encode functional products), gene fragments (fragments of once expressed genes) which are possibly part of our evolutionary past, and null alleles (alleles that are not expressed). There is some overlap in the use of these terms.

cell only and not at the cell surface or may be soluble, but not found attached to the cell surface. More information on this is given at http://hla.alleles.org/antigens/index.html.

6.2 THE EXTENDED HUMAN MHC MAP

The MHC was extended in 2004 to include a region telomeric of *HLA-A* as far as the hemochromatosis gene *HFE*. The details of the extended gene map (xMHC) can be found in Horton et al. (2004). Though the map includes 252 expressed genes, only a select few are appropriate for this book (**Figure 6.7**). It also includes 139 pseudogenes (i.e. genes whose products are not expressed) and a number of gene fragments. A list of HLA genes is given in **Tables 6.7** and **6.8**.

For historical reasons, the MHC can be divided into three subregions referred to as MHC class I, MHC class II, and MHC class III. These regions were originally defined on the

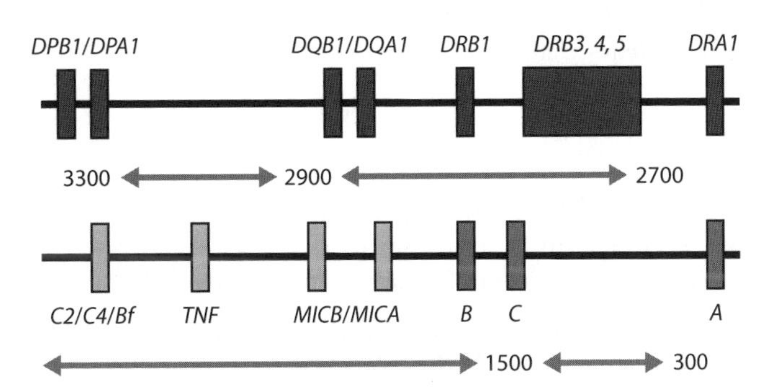

Figure 6.7: Gene map of HLA in MHC (showing selected expressed genes only). The figure is not to scale. The boxed figures and gray arrows mark the distances between loci. This figure covers the old 3.6-Mb pre-xMHC. Whether a haplotype carries another expressed *DRB* gene (*DRB3, 4,* or *5*) depends on the *DRB1* family encoded on the haplotype (see Figure 6.3).

Table 6.8 A list of HLA class II genes.

Current names	Characteristic
HLA-DRA	DRα-chain
HLA-DRB1	DRβ-chain encodes DR1–DR16 allele families
HLA-DRB2	Pseudogene
HLA-DRB3	Second expressed DRβ-chain found on haplotypes carrying DR3, DR11, DR12, DR13, and DR14 allelic families in Northern European populations
HLA-DRB4	Second expressed DRβ-chain found on haplotypes carrying DR4, DR7, and DR9 allelic families in Northern European populations
HLA-DRB5	Second expressed DRβ-chain found on haplotypes carrying DR15 and DR16 allelic families in Northern European populations
HLA-DRB6	DRB pseudogene found on DR1, DR2, and DR10 haplotypes
HLA-DRB7	DRB pseudogene found on DR4, DR7, and DR9 haplotypes
HLA-DRB8	DRB pseudogene found on DR4, DR7, and DR9 haplotypes
HLA-DRB9	DRB pseudogene isolated fragment
HLA-DQA1	DQα-chain
HLA-DQB1	DQβ-chain
HLA-DQA2	DQA-related chain not expressed
HLA-DQB2	DQB-related chain not expressed
HLA-DQB3	DQB-related chain not expressed
HLA-DOA	DOα-chain
HLA-DOB	DOβ-chain
HLA-DMA	DMα-chain
HLA-DMB	DMβ-chain
HLA-DPA1	DPα-chain
HLA-DPB1	DPβ-chain
HLA-DPA2	DQα-chain-related pseudogene
HLA-DPB2	DPβ-chain-related pseudogene
HLA-DPA3	DPα-chain-related pseudogene

basis that the genes within them encoded proteins with similar structures and functions. This terminology is no longer fit for purpose; all three regions encode a variety of proteins with variable structures and functions, including many none HLA antigens.

Of those genes listed in Figure 6.5, Tables 6.7 and 6.8, the most studied in this region have been the classical HLA genes *HLA-A*, *-B*, *-C*, *-DRB*, *-DQA*, *-DQB*, *-DPA*, and *-DPB*. In addition to these eight loci, some studies have also investigated a selection of other MHC genes. These include a group of genes encoding proteins of the classical and alternative complement pathways (*C2*, *C4A*, *C4B*, and *Bf*), tumor necrosis factor (TNF)-α (*TNFA*), and, more recently, the **MHC class I chain (MIC)**-like proteins (*MICA* and *MICB*).

6.3 MOLECULAR STRUCTURE OF HLA CLASS I AND CLASS II

A great deal can be learned about genetic susceptibility from looking at HLA antigen structure. As we will see in the section below, HLA plays a central role in the immune response. Structurally, HLA class I and class II antigens are different. The class I antigens have a single chain composed of three immunoglobulin domains supported by β-2-microglobulin (encoded on chromosome 15). All of the variance in the molecule is encoded on the HLA chain. The HLA class II molecules, however, are composed of two chains—an α-chain and a β-chain—both encoded on chromosome 6p21.3 and both with some (though limited in the case of *DRA*) polymorphism.

X-ray crystallography of HLA-A2 revealed the full structure and much about the function of HLA class I

HLA class I molecules on the cell surface are composed of three immunoglobulin domains (the HLA α-chain: α1, α2, and α3) and a fourth domain provided by β-2-microglobulin (encoded on chromosome 15). In addition, there is a short transmembrane section that anchors the expressed molecule to the cell surface. The **immunoglobulin domains** are numbered from the furthest out to the nearest in (α1 to α3). The α3 domain and the β-2-microglobulin form a supporting structure on which the two remaining immunoglobulin domains are mounted (**Figure 6.8**). These two domains (α1 and α2) form a complex structure with a series of β-pleated sheets and two α-helices (one each). Essentially, the β-pleated sheet forms the floor and the two opposing α-helices form the walls of a groove. X-ray crystallography shows that the groove is a closed structure of approximately 11 Å (1.1 nm) in length. Within the groove there are nine pockets for the binding of antigenic side chains (**Figure 6.9**). The publication in 1987 of the HLA-A2 crystal structure revealed the functional element of the HLA molecule to be the closed groove. This site is known to be the site at which **antigenic peptides** are bound and presented to the T cell receptor (TCR) of $CD8^+$ T cells—a crucial process in adaptive immunity. Furthermore, the antigens presented by HLA class I molecules to **T cells** are small peptides of no more than eight or nine amino acids. Genetic comparison between A2 and other HLA molecules revealed that the majority of allelic variation encodes amino acid changes in and around the antigen-binding groove. This provides a functional link to genetic diversity in HLA class I, and may provide an explanation for the very large number of genetic associations between MHC and disease.

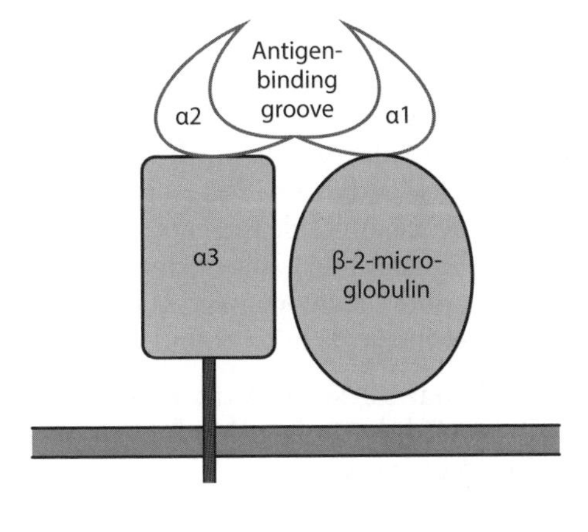

Figure 6.8: HLA class I molecule structure. The figure represents a simple diagram of the HLA class I molecule with three immunoglobulin domains (α1, α2 and α3) encoded by the HLA class I allele and a β-2-microglobulin molecule (encoded on chromosome 15). The α3 unit and β-2-microglobulin unit (in gray shading) form a support structure for the α1 and α2 globulin domains, which create a closed groove for the binding of antigenic peptides (the **antigen-binding groove**). In the case of class I molecules the antigenic peptides are usually eight to nine amino acids in length. The majority of HLA polymorphisms encode amino acid variations in and around the antigen-binding grove.

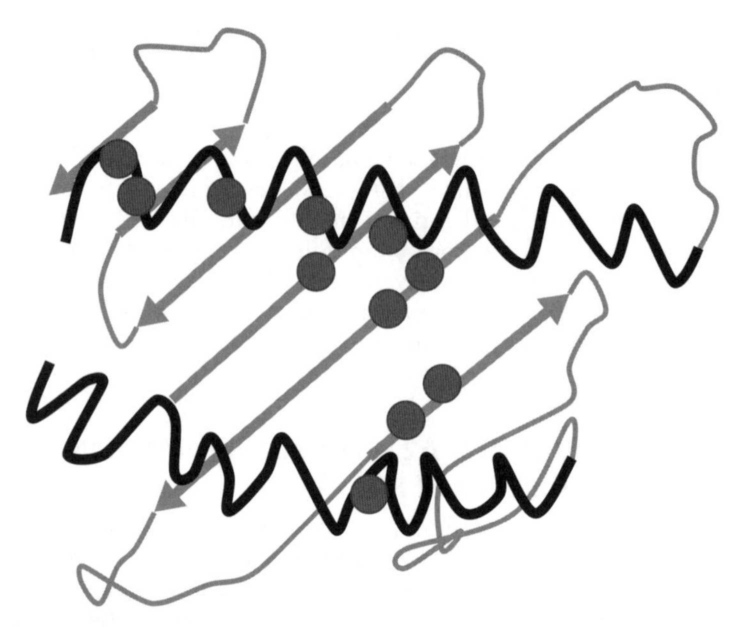

Figure 6.9: Top-down view of the HLA class I binding groove showing some of the sites at which different alleles encode amino acid variations. Some of the sites for amino acid variation are illustrated by dots. The figure shows the β-pleated sheets (in gray) which form the floor and two opposing α-helices (in black) which form the walls of the closed antigen-binding groove of the molecule. For HLA class I, short peptide antigens of around eight to nine amino acids in length are bound in this groove and presented to the T cell receptor in the formation of the immune synapse (Figure 6.11)–a key process in adaptive immunity. HLA class II molecules have a similar structure but the groove ends are open, allowing for longer peptides to be bound. (Adapted from Parham P [2009] The Immune System, 3rd ed. Garland Science.)

The X-ray crystallography structure of HLA class II structure revealed the critical difference between class I and class II

The first X-ray crystallography studies of HLA class II antigens were published in 1991. The critical difference between HLA class II and HLA class I is that the class I heterodimer is formed by the interaction of the class I α-chain with a β-2-microglobulin molecule. The result is a molecule with four domains as described above, but only the α-chain is polymorphic and encoded in the MHC. The class II heterodimer also has four domains all encoded on chromosome 6p21.3, and both α- and β-chains are polymorphic (though *DRA* polymorphism is limited). Two of the class II domains are provided by the A gene (α1 and α2 in each gene *DRA*, *DQA*, or *DPA*) and two are provided by the B gene (β1 and β2 in each gene *DRB*, *DQB*, or *DPB*). In each case, one of the two domains provides structural support for the other, and both the α-chain and β-chain have a short transmembrane tail that anchors them to the cell surface (**Figure 6.10**). The four-domain structure of the complete class II molecule is very like that of class I. Once again the lower domains (α2 and β2) support the higher domains (α1 and β1). The upper portion of the molecule presents as a series of β-pleated sheets and opposing α-helices which form the floor and walls of the antigen-binding groove. Unlike class I, the class II antigen-binding groove is an open-ended structure that enables larger polypeptides of approximately 13–22 amino acids to be bound.

In agreement with findings for HLA class I structural analysis revealed the majority of HLA class II polymorphisms are also encoded in and around the antigen-binding groove. Once again, this provided a basis for linking genotype to phenotype.

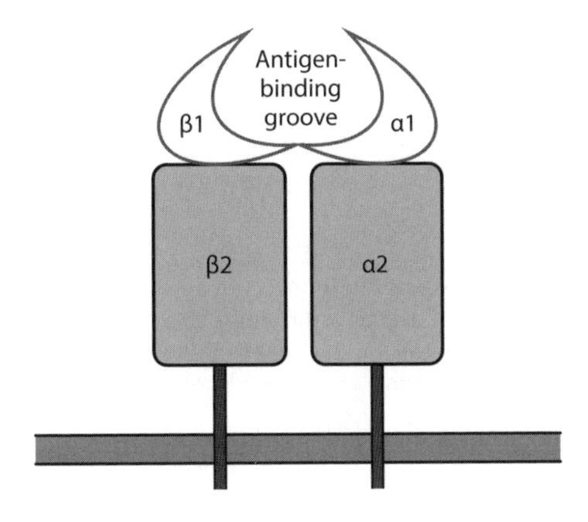

Figure 6.10: HLA class II molecule structure. The figure represents a simple diagram of the HLA class II molecule with four immunoglobulin domains. These are α1 and α2 encoded by the HLA class II A gene and β1 and β2 encoded by the class II B gene. As with class I molecules, two of the units form a support structure (α2 and β2 in gray), while the other two (α1 and β1) form the antigen-binding groove.

6.4 IMMUNE FUNCTION OF HLA CLASS I AND CLASS II

The X-ray crystallography studies on HLA class I and class II indicate significant differences between the molecules. These differences are likely to have significant functional importance and may be important in helping us to understand the different HLA associations in complex disease.

Class I molecules have distinct features

Class I molecules are present on all nucleated cells and encode a site for binding the CD8 protein on T cells. Class I molecules present short antigenic peptides to the TCR, and on presentation there is co-binding between the class I molecule and CD8. This HLA peptide–TCR interaction is often referred to as the formation of the **immune synapse** (**Figure 6.11**). This interaction activates **CD8$^+$ T cells**, which are crucial in adaptive immunity.

Class I antigens are not only involved in antigen presentation and adaptive immunity. They may also be involved in cell regulation in innate immunity. Natural killer (NK) cells use a variety of inhibitory receptors to detect alterations in the expression of HLA class I molecules. CD94 (or NKG2A) is an inhibitory receptor that monitors overall expression of HLA-A, HLA-B, and HLA-C proteins on human cells. This receptor uses the HLA-E system molecule, which expresses bound segments of HLA-A, -B, and -C, to consider the MHC status of the cell.

Unlike NKG2A, which is not MHC class I specific, the other form of inhibitory receptor expressed on NK cells, i.e. the **killer cell immunoglobulin-like receptor (KIR)** is highly selective and each NK cell must express at least one KIR that interacts with self HLA class I for normal functioning to occur. The normal function of the KIR HLA interaction is to monitor HLA class I function. Viral pathogens will often disrupt normal HLA expression as a method of evading the host's immune response, especially that of the CD8$^+$ T cells. Interestingly, it is thought that *HLA-C* is primarily involved in this process.

HLA class II is different to class I

HLA class II antigens are expressed on a selected variety of cells often referred to as **antigen-presenting cells (APCs)**, which include macrophages, dendritic cells, and B cells. Class II presents antigenic peptides to CD4$^+$ T cells and has a site for CD4 binding. The class II molecule has an open-ended binding groove that can accommodate longer peptide antigens.

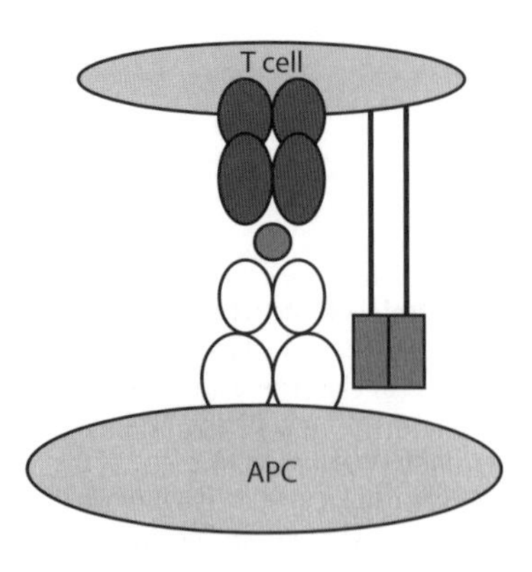

Figure 6.11: The immune synapse. The immune synapse is formed from the interaction of three key elements: the antigenic peptide (gray circle), the HLA molecule (in white), and the T cell receptor (TCR in dark gray). This figure could represent presentation by HLA class I or class II to $CD8^+$ or $CD4^+$ T cells, respectively.

Class II does not appear to be involved in NK cell activity. Furthermore, the mechanisms by which HLA class I and class II engage and present polypeptides are different. HLA class I presents antigenic polypeptides from within the cell (i.e. intracellular), whereas HLA class II presents antigenic peptides that are from outside the cell (i.e. intercellular). The mechanisms by which this occurs are complex and are illustrated in **Figure 6.12**.

HLA class I and class II have important similarities

There are several important similarities between HLA class I and class II. Both class I and class II antigens have binding grooves with pockets for the binding of antigenic side chains. In both cases the majority of HLA polymorphism encodes variation in and around the floor and walls of the binding groove, and in both cases this genetic variation can be functionally significant. The functional consequences of genetic variation in HLA alleles are attributable to changes in the amino acid sequence leading to variation in electrostatic properties of the binding groove, especially in and around the pockets for the binding of antigenic side chains.

HLA class I and antigen engagement in the cell is different from HLA class II

HLA molecules are rarely (if ever) presented on the cell surface without an occupied antigen-binding groove. Binding of intracellular peptides inside the cell is a competitive process. Peptides may be derived from normal cellular proteins broken down in the proteasome or from pathogenic proteins (antigenic peptides). These antigenic peptides are also generated by the **proteasome** (a large barrel-shaped complex found in the cytosol responsible for digesting, degrading, and recycling proteins). The peptides that result from this digestion are transported through the cytosol to the **endoplasmic reticulum (ER)** with the help of a specialized transporter molecule **TAP**. HLA class I antigens are constantly being produced in the cell and newly synthesized molecules are translocated to the ER where they are stabilized by binding to β-2-microglobulin. Finally, these newly synthesized HLA class I molecules bind one of the peptides delivered to the ER by TAP. This is a competitive process and is governed by the peptide-loading complex. If the loaded peptide is too long at the N-terminus, the ER amino-peptidase will remove excess amino acids from the amino-terminal end to ensure a better fit in the antigen-binding groove. Once the peptide has been bound, and if necessary trimmed, the completed molecule is exported from the ER in a membrane-enclosed vesicle through the **Golgi stacks** to the plasma membrane. The processes described above are continuous. Most of the peptides presented are self-antigens, in particular HLA class I that appears to be generated in large quantities in almost all tissues.

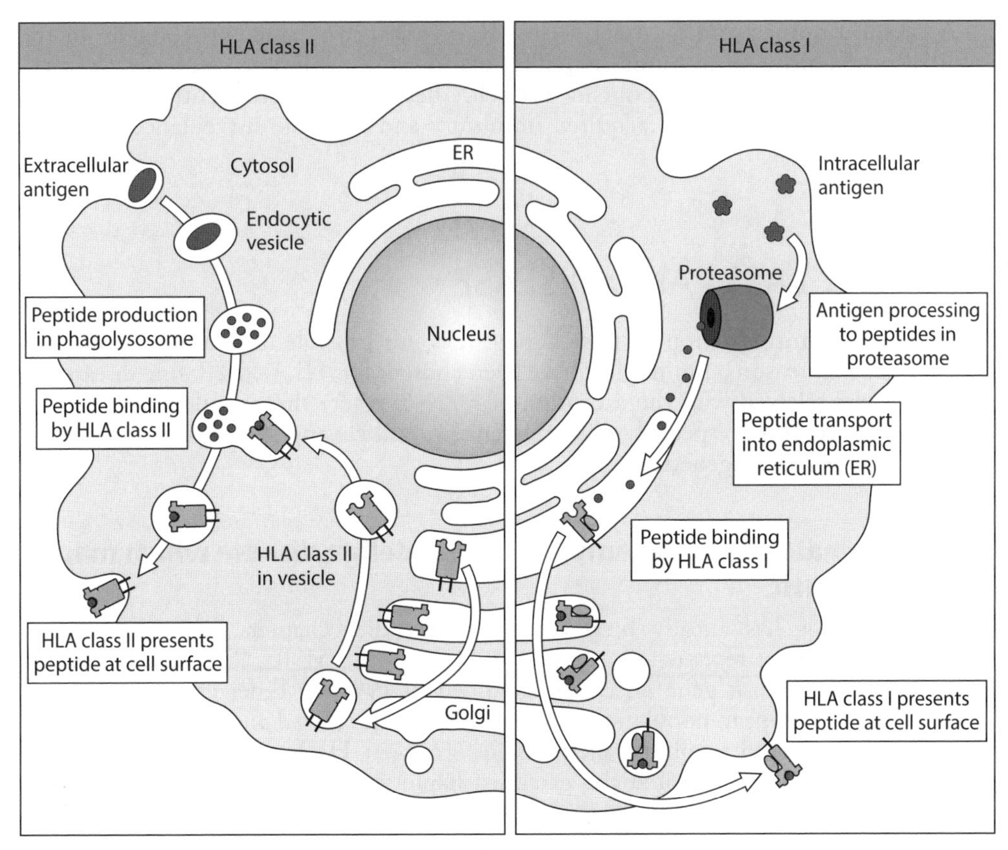

Figure 6.12: Antigen processing by HLA class I and II antigens. The right-hand side illustrates antigen processing by HLA class I. Intracellular antigens from the cytosol, including self-antigens and those of pathogens such as bacteria and viruses, are processed through the proteasome and transported through the endoplasmic reticulum (ER) to the Golgi apparatus. The cell is constantly producing new HLA molecules. Antigenic peptides contact these HLA molecules in the ER and the resulting HLA-peptide complex is transported through the Golgi apparatus to the cell surface. On the left-hand side peptides from outside the cell (extracellular peptides) are taken up by endocytosis and phagocytosis within endocytic vesicles into the cell. The figure illustrates a macrophage. In this situation proteases in the vesicles break down captured proteins and they are bound to newly synthesized class II molecules in the ER and transported through the Golgi apparatus to the cell surface. (From Parham P [2009] The Immune System, 3rd ed. Garland Science.)

HLA class II and antigen engagement in the cell is different

Unlike HLA class I, HLA class II presents peptides derived from proteins imported from outside the cell. The proteins are captured by **endocytosis** or phagocytosis and brought into the cytosol in a vesicle. Proteases break down the captured proteins within these vesicles and produce peptides. At the same time newly synthesized HLA class II molecules coupled with the **invariant chain** are exported into the cytosol in vesicles. The invariant chain acts as a molecular chaperone preventing premature binding of newly synthesized HLA class II to peptides in the ER. Only when the vesicle interacts with another vesicle bearing peptides is the invariant chain released through protease cleavage. This permits peptides to compete for binding with the HLA class II molecule. Once this process has occurred, the HLA class II–peptide molecule is transported to the cell surface for antigen presentation to the TCR.

HLA class I, on one hand, and HLA class II, on the other hand, provide two of the essential mechanisms of immune surveillance and induce immune responses to infectious agents within the cell and outside the cell. These processes frequently overlap, and are one of the main forces in acquired immunity and immune surveillance, which is essential for survival.

6.5 HLA CLASS I AND DISEASE

As stated in the Introduction to this chapter, it is not possible to consider all diseases here and for that reason a limited few have been chosen. For HLA class I, one disease per gene locus was selected with the exception of *HLA-B*, where there is adequate coverage elsewhere in the book, especially with reference to ankylosing spondylitis and also in Chapter 8 on pharmacogenetics.

Hemochromatosis is an example of a Mendelian disease which maps within the xMHC

Hemochromatosis has already been discussed in earlier chapters. This disease is a Mendelian autosomal recessive disease with linkage to the MHC. The precise location of the hemochromatosis gene *HFE* was identified in 1997 by Feder et al.. Early studies of Northern European populations found linkage with *HLA-A* and association studies identified the *HLA-A3* family of alleles as the strongest HLA association with hemochromatosis, but also implicated the extended haplotype HLA 7.2 that carries *A*03–C*07:02–B*07–DRB1*15:01–DQA1*01:02–DQB1*06:02* in disease susceptibility. Fine mapping around the MHC finally identified *HFE* at a distance of approximately 3Mb telomeric to the *HLA-A* locus indicating the extreme linkage disequilibrium found in some but not all MHC haplotypes. One explanation for this extreme linkage disequilibrium is the lower than average rate of recombination within the MHC, which has been calculated as 0.49 cM/Mb compared with 0.92 cM/Mb for the whole genome. This observation has given rise to the suggestion the MHC contains a greater number of recombination cold spots and a low number of hot spots. Overall the genetic association with *HFE* illustrates the idea that not all HLA-associated diseases are genetically complex diseases; some may be Mendelian.

However, this is not the whole story of hemochromatosis. In populations outside of Northern Europe, disease-causing mutations in other (non-MHC) genes have been proposed and penetrance for *HFE* homozygotes, even in populations with high levels of Northern European ancestry, has been shown to be very low, suggesting a more complex model than originally described.

Psoriasis proves the point that *HLA-C* is an important locus to consider in genetic studies of the MHC

HLA-C has often been overlooked by researchers. The reason for this is that in the early days HLA C antigens were difficult to detect and type. Even when molecular genotyping was introduced *HLA-C* remained difficult to genotype. As a result many studies have failed to include *HLA-C*, potentially giving the false impression that the C locus is not associated with disease. However, the history of HLA and psoriasis illustrates the importance of considering *HLA-C*.

Type I versus type II psoriasis

Clinically psoriasis occurs in two forms. Type I psoriasis is an early-onset disease (age 15–25 years) and type II psoriasis is an older-onset disease (approximately age 60 years). Type I psoriasis is often familial, whereas type II psoriasis is not. Early studies of type I psoriasis identified associations with *HLA-B*13* and *HLA-B*17* (later *HLA-B*57*). Studies in the class II region found associations with the *DRB1*07–DQA1*02–DQB1*03:03* haplotype. Altogether this implicates one of two common *DRB1*07* haplotypes. However, the association is not with *DRB1*07* itself; the strongest association is with *HLA-C*, specifically *C*06:02* that is carried on this haplotype. In contrast, studies on type II psoriasis have identified *C*02* and *B*27*, and have not reported associations with either *DRB1*07* or *C*06:02*. This genotype difference between early- and late-onset disease is important and frequently seen in complex disease.

Considering the pathogenesis of type I psoriasis from a functional perspective, it has been proposed that *HLA-C* may have a direct role in this disease. A molecular model based on alanine at position 73 has been proposed, but is not specific to *HLA-C*06:02* and remains controversial. It is, however, interesting that a member of the *HLA-C* family may be the major MHC-encoded risk factor for psoriasis, a dermatological disease, considering the role *HLA-C* plays in NK cell activation and innate immunity.

Before we leave HLA class I we need to consider Bw4 and Bw6

All HLA-B antigens and some HLA-A antigens carry either a Bw4 or Bw6 motif and historically it was found that most *HLA-B* alleles carry the same Bw4 or Bw6 motif as that carried by their family members, with a few exceptions. These exceptions were useful in helping to confirm certain specificities where there were two or more alternative serotypes (called splits). Which motif (Bw4 or Bw6) is expressed depends on the amino acid sequence at positions 77, 80, 81, 82, and 83 of the molecule. Work by Arnett et al. suggests that there may be more variation than previously recognized. For example, *HLA-B*08:01* and *HLA-B*08:02* encode different motifs even though they are both members of the same family (*HLA-B*08*). More importantly, further work also shows that the binding capacity of these two alleles (*HLA-B*08:01* and *HLA-B*08:02*) is quite different and that some of these differences are due to the Bw motif. This difference would not be predicted from the low-resolution genotyping of *B*08* or from the observation that the two alleles are from the same family but it does have potential functional consequence. This is important when considering early studies and planning current studies because it may be necessary to consider this when deciding on the level of resolution of genotyping to be applied in study.

6.6 HLA CLASS II AND DISEASE

A number of examples have been selected for looking at HLA class II. Once again, we cannot consider all diseases here. In this section we have decided to concentrate on two well-known examples: cataplectic narcolepsy and multiple sclerosis, and three less well known but interesting examples: AIH, PSC, and PBC. The latter three are all autoimmune liver diseases that have different and overlapping features.

Severe or cataplectic narcolepsy has one of strongest HLA associations ever reported

Among the neurological disorders, many have strong genetic associations with HLA, including multiple sclerosis and myasthenia gravis. However, the strongest reported association is that between severe **cataplectic narcolepsy** and the *DRB1*15–DQB1*06:02* haplotype. Narcolepsy is a relatively rare condition characterized by abnormal regulation of sleep. Symptoms may include hallucinations and sudden loss of muscle tone, which leads to a sleep-like or unconscious state (cataplexy). Patients with severe narcolepsy suffer from cataplexy following sudden and unexpected emotional stimulation such as fear or laughter.

In 1984, Langdon et al. reported that 100% of their cases were HLA-DR2 (*DRB1*15*)-positive. This astonishing finding was very soon confirmed in several different populations, including studies in France and Japan, while studies in Germany and the Czech Republic reported figures of 98% plus. Smaller studies based on African-Americans reported considerably fewer patients had HLA-DR2 (approximately 70%). Further analysis of African-Americans suggested that the true marker for HLA-encoded susceptibility to narcolepsy may be DQ2. The DQ2 molecule, in this case, is constructed from the products of the *DQA1*01:02* and *DQB1*06:02* alleles which encode the DQ α-chain and β-chain, respectively. *DQA1*01:02–DQB1*06:02* is carried on almost all *DRB1*15* haplotypes in Northern Europeans and was found in the majority (27 out of 28 cases) of the African-American *HLA-DR15*-negative cases.

There are different functional interpretations of the HLA association with narcolepsy

Narcolepsy in dogs is caused by a single autosomal recessive gene in the immunoglobulin heavy chain region. In canines, studies suggest that pathology is associated with the immunoglobulin switch-like sequence and enhanced microglial expression at age 1–3 months. This causes axonal pruning and cell necrosis. In humans, familial narcolepsy may occur in the absence of *HLA-DRB1*15*, suggesting there may be two forms of the disease: familial (not associated with HLA) and sporadic (associated with HLA). Whether narcolepsy is a single entity or not is unknown, but the *DQ* association may indicate that this is an immune-mediated disease and this may help us understand the pathology of the condition.

The *DQA1*02:01–DQB1*06:02* haplotype that predisposes to narcolepsy is protective of type 1 diabetes (T1D). T1D is in most patients, the consequence of autoimmune-mediated destruction of the insulin-secreting pancreatic β cells. Studies have revealed that susceptibility to and protection from T1D is determined by the amino acid at position 57 of the DQβ-chain. Aspartate 57 is protective whereas alleles encoding alanine, valine, or serine predispose to T1D. The reverse pattern appears to be true for narcolepsy. In T1D, a model based on position β57 has been proposed, whereby susceptibility and protection are based on the ability to form a salt bridge between β57 and α79 in the expressed DQ molecule. However, this model does not fit for narcolepsy; while *DQB1*06:02* confers susceptibility to narcolepsy, *DQB1*06:01* that carries the same *DQA1* allele confers protection. Comparing the polypeptide sequence encoded by the *DQB1*06:02* allele with that of *DQB1*06:01* reveals nine amino acid differences, one of which is outside the peptide-binding groove, while all the others are within the peptide-binding groove (of which five are listed in **Table 6.9**). The listed amino acids are all in close proximity to a putative T cell recognition surface or to a peptide-binding pocket within the groove.

There is also an important similarity here with rheumatoid arthritis as discussed in Chapter 3, in relation to *HLA-DRB1* and rheumatoid arthritis. The majority of genetic variation in HLA encodes amino acid differences in and around the antigen-binding groove of the expressed HLA molecule. These binding grooves each have up to nine pockets for the binding of side chains on the antigenic peptide. These pockets are labeled P1–P9 in

Table 6.9 Narcolepsy amino acid positions for *DQB1*06:01* versus *DQB1*06:02*.

HLA class II allele	Amino acid position				
	9	13	26	37	38
*DQB1*06:01*	L	G	L	D	V
*DQB1*06:02*	F	L	Y	Y	A

sequence. Depending on the associated allele, possession of a specific amino acid at a specific site can have profound effects on which antigenic peptides are bound and presented to the TCR. In narcolepsy, the P4 and P9 pockets appear to have the most pronounced effects. At P9 the critical change is with β37. At this position there is a tyrosine to aspartic acid exchange that introduces significantly more negative charge into the P9 pocket and this may subtly alter the anchor specificity. The P4 binding pocket is the largest pocket in the expressed *DQB1*06:02* structure. In contrast, the *DQB1*06:01* has a smaller P4 pocket. This is due to polymorphisms at positions β13 and β26, which result in exchange of glycine for alanine and leucine for tyrosine. The impact of these changes is considered to be substantial.

Considering the function of HLA class II molecules and these molecular models, it is plausible that narcolepsy is an immune-mediated disorder dependent on T cell responses. At a molecular level, changes in the chemical structure influence the functional dynamics of the antigen-binding groove in such a way that there may be expansion or contraction of the range of peptides that may be preferentially bound and presented to the TCR. In addition, subtle changes in the molecular structure of HLA molecules may influence the orientation of the bound peptide and the recognition of the HLA–peptide complex by the TCR. It is therefore interesting that three of the amino acid differences between *DQB1*06:01* and *DQB1*06:02* are in the putative TCR recognition sites, suggesting that both antigen binding (variation in the binding groove) and the TCR recognition site may be important in determining HLA-encoded susceptibility to narcolepsy.

Finally, more recent studies in narcolepsy aimed at identifying immune genetic markers in the Han Chinese population found several novel associations, including an association with *DQB1*03:01* and disease onset. The study also found that selected HLA-associated markers differed significantly in cases where onset occurred after the recent influenza pH1N1 pandemic. This study indicates three major points:

- The association in the Han Chinese population may be different to that in Northern Europeans.
- The association may be related to age of onset or disease severity.
- The association may be related to other environmental factors.

Multiple sclerosis is a disease with a strong genetic association with HLA class II

Multiple sclerosis is one of the most common neurological disorders and as many as one in every 1000 individuals may have multiple sclerosis in Northern European populations. It has been suggested that there are 400,000 cases in the USA alone. The disease presents in young adults and is more common in women than in men (approximately twice as common). There is chronic degeneration of the **myelin sheath** on the nerve cells that leads to the buildup of **sclerotic plaques**. Demyelination of the central nervous system leads to motor weakness, impaired vision, and lack of coordination. It is a highly variable disease and some sufferers experience slow progressive degeneration, while others have peaks and troughs with periods of recovery in between periods of degeneration.

It is generally accepted that multiple sclerosis is a genetically complex disease involving both genes and the environment with a considerable genetic load associated with specific MHC haplotypes, especially the *DRB1*15:01–DQA1*01:02–DB1*06:02* haplotype. This latter haplotype is normally found with *HLA-A*03:* and *HLA-B*07:*, and the extended haplotype is labeled with the abbreviation HLA 7.2 because of the linkage disequilibrium between *DRB1*15* (originally the major form of DR2) and *HLA-B7*. This haplotype is very common in European populations. Recent studies suggest the concordance rate for multiple sclerosis in monozygotic twins may be as high as 25%, while in dizygotic twins concordance is five times lower than in monozygotic twins and in siblings concordance is 10 times lower. However, concordance rates should be considered with caution as it appears that concordance rates mirror **prevalence rates**; therefore the lower the prevalence in the population, the lower the level of concordance. For example, there are significant differences in concordance rates between Northern and Southern European populations. This difference is reflected in the lower prevalence for multiple sclerosis in Southern Europe.

Early studies of multiple sclerosis

Early studies of the genetic basis of multiple sclerosis identified strong reproducible associations with the MHC and one haplotype in particular, the HLA 7.2 ancestral haplotype. The first studies to indicate an association with the MHC were published in 1972 and 1973 by the same group. Many others followed, but it would be wrong to give the impression that all early studies identified the same HLA associations or that all studies found any association. The reasons for these inter-study variations include differences in techniques applied and the availability and quality of anti-serum for the early HLA assays as well as sample size, diagnostic differences, and population differences (Chapters 3 and 5). For example, two studies based in the UK reported quite different results—one was based on the population of the remote Orkney Islands and the other based on the general UK population.

However, the association with the 7.2 haplotype has stood the test of time and remains a confirmed multiple sclerosis risk marker. The same group of HLA class II alleles, especially *DQB1*06:02*, are very strongly associated with narcolepsy (as we have discussed), but *DQB1*06:02* is protective against T1D and may have mild protective effects against other autoimmune conditions (though not all; see below).

The association with the MHC is suggestive of an autoimmune or at least immune-mediated etiology in multiple sclerosis. In fact, studies confirm this link—multiple sclerosis is an immune-mediated disease. Furthermore, it appears to be an autoimmune disease since there are myelin antibodies in cerebrospinal fluid. Other evidence to support this comes from studies of T cell function in multiple sclerosis. The activation of T helper 1 $CD4^+$ cells and production of interferon (IFN)-γ are implicated in the pathogenesis of multiple sclerosis. These cells are found to be enriched in both the blood and spinal fluid of patients, and activated macrophages that are found in sclerotic plaques can release cytokines and other enzymes that are able to cause demyelination. IFN-γ was trialed as a potential treatment for multiple sclerosis, but was found to worsen rather than improve symptoms.

With so much already known, some people question the value of further genetic studies in multiple sclerosis. There are two reasons why investigations should continue. The first is illustrated by the disappointing results of the IFN-γ trials, that is our knowledge to date is incomplete and not sufficient to aid in the formulation of effective therapies. The second is illustrated in the HLA association. Though we have this information, we have no idea what the auto-antigen or auto-antigens are, and how many immune-regulatory processes have failed or broken down to permit this disease to occur. Genetic studies offer one way to identify these.

Recent studies of multiple sclerosis including genome-wide association studies (GWAS)

Multiple sclerosis is an immune-mediated disease and so we would expect the MHC to be a dominant risk factor, but what other genetic risk factors may be hiding. Autoimmunity involves disordered immune regulation. There is a breakdown in T cell tolerance due to incomplete deletion of self-reactive T cells in the thymus and/or the periphery coupled with insufficient control of T cell co-stimulators. There is a whole family of additional immune regulatory genes that could be involved. Studies of other similar diseases (related syndromes) will provide clues to other potential risk alleles. Overlap is not uncommon, as illustrated in the example of rheumatoid arthritis (see Figure 4.10).

In 2007, the number of identified risk alleles for multiple sclerosis was 29. In 2011, Sawcer et al., as part of the International Multiple Sclerosis Genetics Consortium (IMSGC) together with the Wellcome Trust Case Control Consortium 2 (WTCCC2), published a follow-up GWAS on 9772 cases and 17,376 unaffected controls from 15 countries, and reported a total of 57 associations. This group included 23 previously suspected associations, 29 novel associations above the significance threshold and a further five probable associations at a lower significance threshold of $P < 5 \times 10^{-5}$ to $P > 5 \times 10^{-7}$. The evidence of these associations also supports the likelihood that the inflammatory processes seen in multiple sclerosis pre-empts the diagnostic symptoms seen and not the other way around as some have suggested. Candidates identified in these studies include various chemokines and cytokines, members of the tumor necrosis superfamily, various co-stimulatory and signaling genes, and one or two genes involved in neurological development.

This clustering of susceptibility loci from the same gene family suggests common pathways in the genesis of multiple sclerosis, perhaps indicating that the polymorphisms and mutations all with low individual risk work in an additive model to create a risk portfolio for multiple sclerosis, i.e. the sort of interactive network that is illustrated in **Figure 6.13**. It is relatively easy to create mind-maps linking potential candidates from association studies for multiple sclerosis and these genetic discoveries can be used to inform the debate on disease pathogenesis.

HLA class II and autoimmune liver disease

Liver disease is often thought of as a self-induced disease resulting from excess alcohol consumption. However, for the majority of cases nothing could be further from the truth. Liver disease affects the old and the very young, and comes in all shapes and sizes. There are Mendelian autosomal liver diseases such as hemochromatosis and Wilson's disease, there are drug induced liver diseases such as those that occur following an adverse drug reaction, there are autoimmune liver diseases, and there are viral liver diseases to consider. Altogether, liver disease accounts for a very significant proportion of mortality and morbidity within the population, and most forms of liver disease have a genetic association with HLA (**Figures 6.14** and **6.15**).

Essentially there are three different autoimmune liver diseases: autoimmune hepatitis (AIH), primary sclerosing cholangitis (PSC), and primary biliary cirrhosis (PBC). The disease names are to some extent descriptive of the signs of the disease.

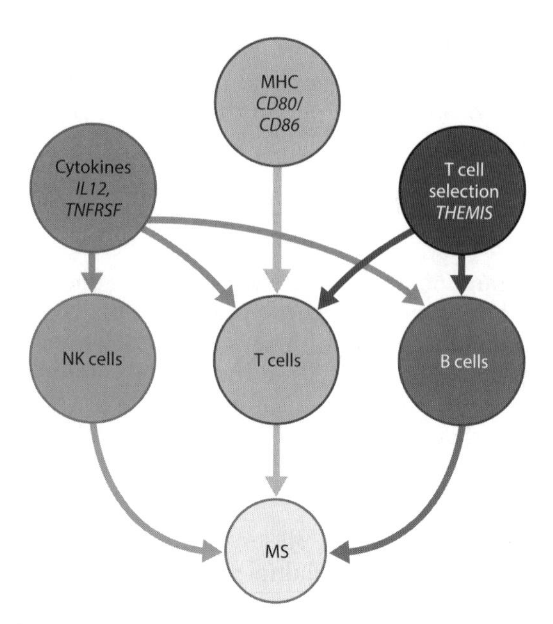

Figure 6.13: Illustrating risk genes and cell targets for multiple sclerosis. This figure illustrates the complexity of interaction between a few selected immune response genes associated with an increased risk of multiple sclerosis (MS). There are several different explanations for this. Risk alleles may work independently or in an additive model. The additive model may include a number of different groups. In an additive model, it may be necessary to inherit several common polymorphisms before there is significant risk of multiple sclerosis. Alternatively, under the correct set of environmental conditions a solitary polymorphism may be sufficient to significantly promote the disease or protect from it.

AIH is a relatively rare classical autoimmune disease of the liver

AIH presents in children and in young and mature adults. The disease has all the classical features of an autoimmune disease. AIH has a strong female preponderance. It can be controlled in the majority of cases by use of immunosuppressive drugs; however, the disease is progressive and not all cases respond to immune suppression. Antibodies against a variety of antigens have been identified, though there is some controversy over which antibody is most helpful in diagnosis and which is most informative in determining the pathogenesis of the disease. The disease can be subclassified into different types, but we are concerned only with type 1 AIH and will use the term AIH throughout this section.

A genetic association between AIH and HLA was first reported in 1972 (*HLA-A1* and *HLA-B8*). Subsequent studies confirmed this association, and added Dw3 (*HLA-DR3*) and *HLA-DR4* (in late onset cases) to the list of susceptibility alleles. A series of studies published by the same group from 1991 onwards, confirmed the association with *HLA-DR4* in late-onset cases, and extended the susceptibility haplotypes to include *DQA1-DQB1* as well as *HLA-C* and the *C4A** null sequence (*C4A*Q0*). A summary of this work is presented in **Table 6.10**.

A shared epitope hypothesis in AIH

The studies on populations from the UK and Northern USA identified *HLA DRB1* as the primary susceptibility locus and *DRB1*03:01* and *DRB1*04:01* as the key alleles. As these studies had used molecular genotyping, it was possible to compare the amino acid sequences encoded by the susceptibility alleles with those encoded by the protective alleles. After a thorough statistical analysis, one particular amino acid sequence stood

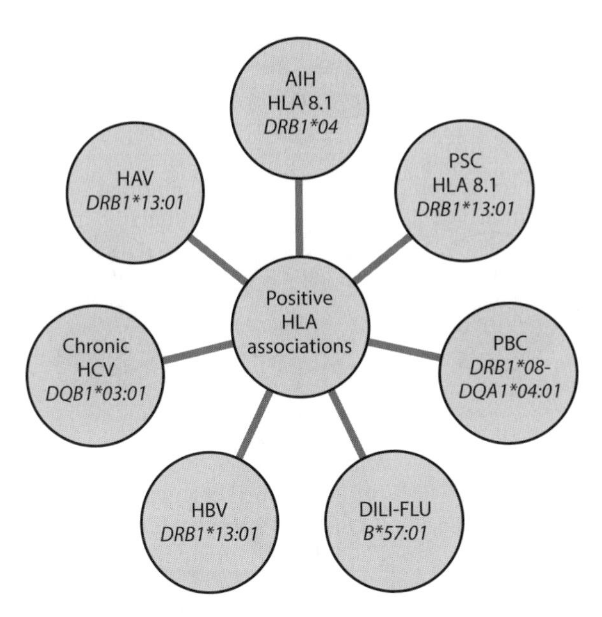

Figure 6.14: HLA and liver disease. The figure illustrates the major positive HLA associations with seven different forms of liver disease. Clockwise these are: autoimmune hepatitis (AIH), primary sclerosing cholangitis (PSC), primary biliary cirrhosis (PBC), drug-induced liver injury specifically flucloxacillin (DILI-FLU), hepatitis B virus (HBV), hepatitis C virus (HCV), and hepatitis A virus (HAV). The first three are all autoimmune diseases; the last three are viral liver diseases. For AIH, *DRB1*04* represents *DRB1*04:01* in European and North Americans of European origin and *DRB1*04:05* in Japanese, Korean, and South American adult cases. The association between PSC and the HLA 7.2 haplotype is weak and controversial, and is not reported in the figure. The association with PBC in Japan and China may involve a different *DQA1-DQB1* haplotype: *DQA1*01:03-DQB1*06:01*. DILI-FLU refers to the extraordinarily strong association with *HLA-B*57:01* seen in a minority of DILI patients treated with the commonly prescribed antibiotic flucloxacillin.

out: leucine–leucine–glutamic acid–glutamine–lysine–arginine (LLEQKR) at positions 67–72 of the DRβ-chain. This sequence, which is referred to as a **shared epitope** for AIH, is not unique to *DRB1*03:01* and *DRB1*04:01*, but is present on a number of different *DRB1* allele sequences. Alleles associated with resistance to AIH encode isoleucine–leucine–glutamic acid–glutamine–alanine–arginine (ILEQAR) at positions 67–72, respectively. There are only two major differences between these two sequences: the first at position 67 and the second at position 71. The change of leucine for isoleucine at position 67 is unlikely to be of major significance, but the change of lysine (a highly charged polar amino acid) for alanine (a non-polar uncharged amino acid) is a major change. Therefore, it is reasonable to hypothesize that this change in the amino acid sequence in the DRβ-chain at position 71 may affect antigen presentation. The residue at position DRβ71 interacts with the P4 binding pocket of the HLA DR molecule. This difference may be critical in determining which antigens are preferentially bound and the orientation of the antigen for presentation to the TCR. The same epitope is also found to be associated with an increased risk of rheumatoid arthritis. Genetic overlap is common in complex diseases. However, it is likely that there are both disease specific and non-specific risk alleles that determine which diseases occur in different individuals.

Studies of other populations in AIH

Looking at other populations can be very helpful in deciding where to place weight on these findings. Studies in Japan and Korea identified *DRB1*04:05* as the primary susceptibility

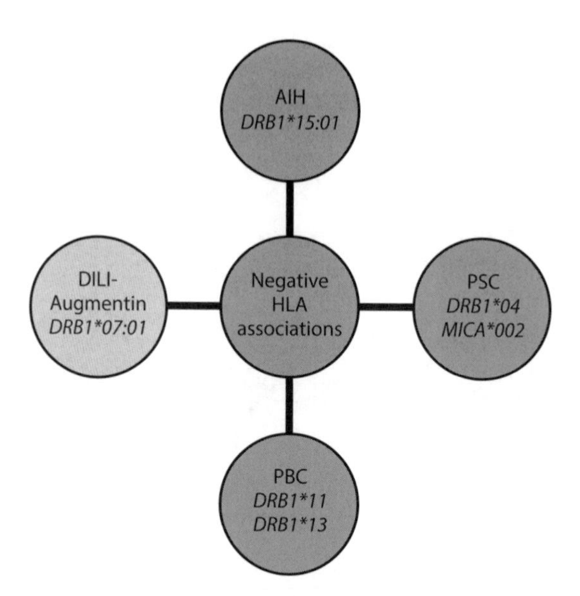

Figure 6.15: Protective HLA associations with three major autoimmune liver diseases and one example of drug-induced liver injury. The figure illustrates four major negative (protective) HLA associations with three autoimmune liver diseases and a selected example of drug-induced liver injury (DILI). Clockwise these are autoimmune hepatitis (AIH), primary sclerosing cholangitis (PSC), primary biliary cirrhosis (PBC), and drug-induced liver injury, specifically augmentin (DILI-Augmentin). Comparing Figure 6.14 with figure 6.15, *DRB1*04* which is associated with an increased risk in AIH protects from PSC and *DRB1*13* which is associated with an increased risk of PSC, chronic HAV infection in South America, and AIH in children from Argentina is associated with a reduced risk of PBC. The *DRB1*07:01* haplotype that protects from augmentin DILI is the one which carries *HLA-B*57:01* - the major risk allele for flucloxacillin DILI.

allele in AIH. The Korean and Japanese populations have a shared ancestry and therefore this was not surprising. However, compared with the European and North American findings, initial analysis suggested the results were discordant and do not agree with the shared epitope model for AIH. *DRB1*04:05* does not encode LLEQKR at positions 67–72, it encodes LLEQRR. However, under further scrutiny it was found that the only difference is at position 71 where *DRB1*04:05* does not encode lysine but it encodes arginine and this amino acid is very similar to lysine, being a highly charged polar amino acid of similar structure. Therefore, at a functional level these two alleles may encode molecules with very similar preferences for antigen binding.

Extending the shared epitope model for HLA-encoded susceptibility to AIH

At present the LLEQKR model with emphasis on DRβ-71 is the best model to explain HLA-encoded susceptibility to AIH. However, there may be scope to extend the model to include amino acids upstream of position 71 as far as position 74. This idea was first promoted in a paper from Korea by Lim et al. in 2008. In the extended shared epitope model, LLEQKRGR is encoded by the *DRB1*03:01* allele and LLEQK-(or R)-RRA is encoded by the *DRB1*04:01* and *DRB1*04:05* alleles. In the extended model there is a critical difference between *DRB1*03:01* (on one side) and *DRB1*04:01* and *DRB1*04:05* (on the other). *DRB1*03:01* encodes arginine at position 74, while both *DRB1*04:01* and *DRB1*04:05* encode alanine. As we have already seen, these are two very different amino acids with different polarity. Essentially, Lim et al. argue that *DRB1*04:01* and *DRB1*04:05*, which both carry arginine 71 and are positively charged, can form a **salt bridge** with the negatively charged P4 binding pocket. *DRB1*03:01* is also able to form

Table 6.10 Autoimmune Hepatitis and HLA alleles and haplotypes.

Haplotype	HLA locus						
	A	*C*	*B*	*DRB3/4/5*	*DRB1*	*DQA1*	*DQB1*
Haplotypes that encode susceptibility to AIH							
8.1	**01*	**07*	**08*	*3*01:01*	**03:01*	**05:01*	**02:01*
DR4–DQ7	–	–	–	*4**	**04:01*	**03*	**03:01*
DR4–DQ4	–	–	**54*	*4**	**04:05*	–	**04:01*
A11–DR4	**11*	–	–	*4**	**04* **04:05*	–	–
DR13	–	–	–	–	**13:01*	**01:03*	**06:03*
Haplotypes that encode resistance (protection) to AIH							
7.2	–	**07*	–	*5*01:01*	**15:01*	**01:02*	**06:02*
DR4–DQ7	–	–	–	*4**	**04:06*	–	**03:01*
DR13	–	–	–	–	**13:02*	–	–

The first two haplotypes are associated with AIH in UK, Northern Europe and USA. The third haplotype is associated with AIH in Japan and Korea. The fourth and fifth haplotypes are found in pediatric AIH in Argentina. In Argentina, *DRB1*13:01* is the major susceptibility determinant in pediatric AIH, but not in adult cases. In adult AIH, from South America *DRB1*04:05* is the major allele. Not all studies are listed, only enough to give a cross-sectional view of the data on AIH and those which contribute most to the scientific discussion of this subject. In cases where individual groups have produced multiple studies, the study with the largest number of patient cases is shown. Alleles are not listed at all loci for all haplotypes, in each case a dash is entered to indicate this.

this salt bridge as it has lysine at position 71. This confirms the idea that the interaction of the DRβ-71 residue with P4 is of primary importance in AIH. This implies that the amino acid at position 74 is of secondary importance. Thus, the alanine carried by both *DRB1*04:01* and *DRB1*04:05* at position DRβ-74 does not appear to influence the effect of the amino acid at position DRβ-74 on the P4 binding pocket due to its small size and charge, whereas *DRB1*03:01* has arginine at position 74 and this can also interact with the P4 binding pocket to form a second salt bridge. In this extended model those carrying *DRB1*03:01* receive a double-dose effect through structural variations encoded by a single allele. This may also go some way to explaining the differences in HLA distribution reported in early- and late-onset disease in AIH (**Figure 6.16**).

A slightly more difficult situation arose when studies from South America reported associations with *DRB1*13:01* in children with AIH. However, other studies in South America have suggested that chronic hepatitis A virus is associated with *DRB1*13:01* and this virus has been identified as a potential trigger for AIH. Interestingly, adult patients from Argentina have an excess of *DRB1*04:05*, that is the same allele that is associated with susceptibility to AIH in Japan and Korea. Thus, at least the data for South American adults fits into the shared epitope hypothesis. The different data in children from South America remains unexplained.

Haplotypes and clinical severity in AIH

The two AIH major susceptibility haplotypes *DRB1*03:01* and *DRB1*04:01* or *DRB1*04:05* (depending on nationality) each carry a second expressed *DRB* gene: *DRB3* for *DRB1*03:01*

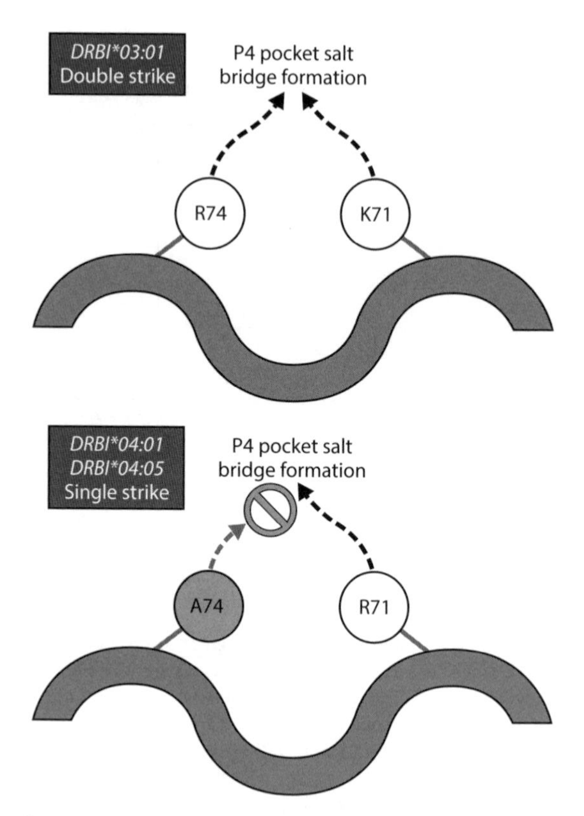

Figure 6.16: Salt bridge formation over the P4 binding pocket in AIH may explain genetic associations with different HLA alleles. The figure is a simplified illustration of the peptide-binding groove of an HLA molecule focusing on the interaction between peptides at positions 71 and 74 on the *DRB1*-encoded β-chain. The figure compares the formation of the salt bridge at the P4 binding pocket (fourth binding pocket for antigen peptide side chains) in two groups; those with *DRB1*03:01* and those with either *DRB1*04:01* or *DRB1*04:05*. These alleles are the major susceptibility alleles for autoimmune hepatitis in the European, European American (the first two), and Japanese populations (the last). All alleles are able to form a salt bridge; however, those carrying *DRB1*03:01* have a double strike because they have positively charged amino acid at positions 71 (lysine K) and 74 (arginine R), while those with *DRB1*04:01* and *DRB1*04:05* have only a single strike because though they have arginine at position 71, they have neutral alanine (A) at position 74. This may account for the observation that AIH patients with *DRB1*03:01* tend to present earlier and may have more severe disease. The model presented is similar to that seen for other autoimmune diseases. In those with *DRB1*04:06*, there is glutamate at position 74 and even though they have arginine at position 71 they are unable to form a salt bridge because the negatively charged glutamate repels the negatively charged P4 peptide.

and *DRB4* for *DRB1*04:01/04:05*. While *DRB3* encodes LLEQKRGR, *DRB4* does not. Therefore patients with the *DRB1*03:01* haplotype have at least two *DRB* alleles, each with the same critical sequence, whereas those with *DRB1*04:01* or *DRB1*04:05* do not. It is therefore possible that the increased number of LLEQKRGR-expressing alleles restricts the selection of antigenic peptides available for binding to the expressed DR molecule. This **multihit hypothesis** may explain why patients with *DRB1*03:01* present earlier in life and those with *DRB1*04:05* present later. In fact, early-onset AIH is rarely seen in Japan or Korea and the association with *DRB1*04:01* is with late-onset cases in European populations. Based on all of the information above, the possible gene dose effect of having a single *DRB1*03* gene will increase to two hits, i.e. lysine 71 (first hit) and arginine 74 (second hit) for the *DRB1*03* allele alone, and this will increase from two to four hits in most cases, as they will also all have the *DRB3* gene that encodes the LLEQKRGR sequence. Both genes on the same haplotype have a potential two-hit locus (two times two equals four). Homozygotes for *DRB1*03-DRB3* haplotype may have eight hits.

This does not rule out the possibility that other genes within the same MHC haplotype are having an impact on disease risk. It is worth noting that recent findings on rheumatoid arthritis indicate the possibility of another three MHC genes encoding five amino acids as risk factors. With respect to AIH, it is also important to understand when comparing populations that the *DRB1*03:01* allele and the **ancestral HLA 8.1 haplotype** to which most *DRB1*03:01* alleles are connected are rare in Japan and Korea as is the *DRB1*04:01* allele. It is therefore not surprising when the associations reported in Europe and the USA cannot be replicated in these quite distinct populations, and it is a major advance when unifying hypotheses (such as that above) can be developed from what may appear to be conflicting data.

PSC is not a classical autoimmune disease

PSC is a chronic cholestatic syndrome of unknown etiology characterized by diffuse inflammation, biliary obstruction, and cholestasis, leading to fibrosis of the biliary tree. Even though PSC is quite rare (around 1:10,000 adults, which allowing for adult onset, equates to approximately 4000 cases living in the UK in 2014), because there is no effective therapy it remains a leading cause for liver transplantation. The intrahepatic and extrahepatic bile ducts are destroyed in the disease process, and biliary cirrhosis, portal hypertension, and liver failure follow. First described in 1924 by Delbet, it was until recently considered a rare condition, but it is now being identified more frequently due to advances in diagnostic medicine. It is for the most part a severe disease that progresses quite rapidly to a clinical endpoint. The disease is frequently found to be associated with mild ulcerative colitis. In fact, some centers report that 100% of cases have ulcerative colitis. The relationship between ulcerative colitis and PSC is unclear; however, even though a large proportion of patients with PSC have ulcerative colitis, only a small proportion of ulcerative colitis patients develop PSC.

PSC and genes

PSC is not a classical autoimmune disease. It is found predominantly but not exclusively in adult males and does not respond to normal immune suppressive therapy. However, genetically PSC looks like an autoimmune disease.

In terms of genetics, PSC has often been referred to as an autoimmune disease and therefore early studies investigated possible relationships with the MHC. Studies of HLA based on very small numbers of patients in 1982 and 1983, reported associations with *HLA-A1* and *HLA-B8*. These studies were followed up by a series of studies based on a relatively large collection of cases from a single center. The studies first used serological phenotyping, then applied RFLP-based genotyping, and finally used a variety of PCR-based genotyping methods to investigate *HLA-A*, *-B*, *-C*, *-DRB1*, *-DRB3*, *-DRB4*, *-DRB5*, *-DQA1*, and *-DQB1*. These studies and studies of other populations reported three HLA haplotypes associated with an increased risk and two associated with a decreased risk of PSC. The data from these studies is summarized in **Table 6.11**.

Comparing studies, the associations with haplotypes 2 (7.2) and 5 (MICA*002) remain controversial. However, even these large-series studies are quite small by modern standards and much larger numbers may be needed to confirm or refute these weaker associations with a realistic degree of confidence.

Table 6.11 Primary sclerosing cholangitis and HLA alleles and haplotypes.

Haplotype	HLA locus						
	A	*C*	*B*	*DRB3/4/5*	*DRB1*	*DQA1*	*DQB1*
Haplotypes that encode susceptibility to PSC							
8.1	**01*	**07*	**08*	*3*01:01*	**03:01*	**05:01*	**02:01*
7.2	**03*	**07*	**07*	*5*01:01*	**15:01*	**01:02*	**06:02*
*DRB1*13*	–	–	–	*3*01:01*	**13:01*	**01:03*	**06:03*
Haplotypes that encode resistance (protection) to PSC							
*DRB1*04*	–	–	–	*4*01:03*	**04:01*	**03*	**03:02*
*DRB1*07*	–	–	–	*4*01:03*	**07:01*	**02:01*	**03:03*
*MICA*002*	–	–	–	–	–	–	–

Note that the haplotype 7.2 is only weakly associated with PSC as is the *DRB1*07* haplotype. The *MICA*002* association is discussed later in the chapter. Note that both the 8.1 and 7.2 haplotypes carry the *MICA*008* allele. Alleles are not listed at all loci for all haplotypes, in each case a dash is entered to indicate this.

Leucine DRβ-38 versus asparagine DRβ-37

Analysis of these associations through consideration of the amino acid sequences encoded by the different alleles initially suggested that the key amino acid was leucine at position 38 of the DRβ-chain. Leucine DRβ-38 is encoded on both of the second expressed *DRB* genes for the HLA 8.1 and the **HLA 7.2 ancestral haplotypes** (i.e. *DRB3* and *DRB5*), but is not encoded by the *DRB1* alleles. The protective haplotypes carry the *DRB4* alleles that encode alanine at this position and all other alleles encode valine. Further exploration of the amino acid sequences in the DQ region revealed a potential role for DQβ-55 and DQβ-87, explaining the genetic susceptibility with haplotypes 2 and 3. In 2011, however, Hov et al. reinvestigated the electrostatic modifications of the HLA DR molecules associated with inheritance of different HLA *DRB* alleles in 356 PSC patients and 366 matched controls, and identified DRβ-37 in the *DRB1* gene and not DRβ-38 on the *DRB3* and *DRB5* genes as the primary susceptibility determinant. This study reconsidered previous studies in detail, pointing out that the second expressed *DRB* gene is expressed at a low frequency and that the *DQB1* alleles were different for each haplotype, though there are shared sequences at critical sites. As the association with the DRβ-38 model is based on the *DRB3* and *DRB5* genes, and the DQβ model makes no allowance for the association with the *DRB1*03:01–DQB1*02:01* haplotype (which is the strongest of all associated HLA haplotypes in PSC), the overall weight of evidence is in favor of the findings of Hov et al. The key amino acids are asparagine (which increases the risk of PSC) at position 37 that induces a positive charge in the P9 binding pocket, and tyrosine (which protects against PSC) at position DRβ-37 that induces a negative charge in the P9 binding pocket. Variation at position DRβ-86 was also found to influence the electrostatic properties of the P1 binding pocket, and appears to have a secondary role in susceptibility and resistance to PSC (**Figure 6.17**).

Just like the previous examples (narcolepsy and T1D), amino acid changes that alter the electrostatic charge of the antigen-binding pockets (in this case P9) appear to be critical in determining HLA-encoded susceptibility to the disease. Though in narcolepsy and T1D it is the *DQB* locus and not *DRB* that is the primary susceptibility locus, the basic concepts

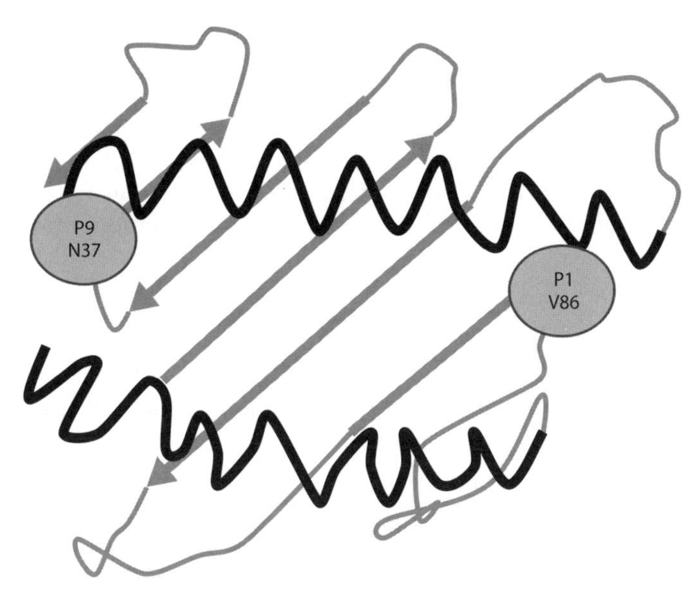

Figure 6.17: HLA DR binding groove and HLA-encoded susceptibility to PSC. The figure shows a simplified version of the HLA peptide-antigen binding groove from a top-down perspective. The amino acids asparagine (N) at position 37 and valine (V) at position 86 are highlighted. These two amino acids interact at the P9 and P1 binding pockets, respectively. Studies indicate that HLA *DRB1* alleles carrying the amino acids asparagine 37 and valine 86 are most strongly associated with PSC in European populations and account for the majority of HLA-encoded susceptibility to PSC. (Adapted from Hov JR, Kosmoliaptsis V, Traherne JA et al. [2011] *Hepatology* 53:1967-1976. With permission from John Wiley and Sons.)

remain the same, and it is interesting that both AIH and PSC favor associations with *DRB* rather than *DQ* (as does rheumatoid arthritis).

MICA in PSC

The ***MICA*** genes are encoded in the so-called MHC class III region, and are stress inducible and expressed on the gastric epithelium. *MICA* can activate NK cells and may also activate γδ T cells. All of these factors make them ideal candidates to consider when looking at the MHC in PSC. There is considerable polymorphism in *MICA*. Based on their putative function, their location, and the level of polymorphism in the *MICA* and *MICB* genes, two studies in 2001 investigated *MICA* in PSC: one from the UK and the other from Norway. The UK study described a strong association with *MICA*008* and a strong dominant protective effect of *MICA*002*. In the UK study there were two patient groups totaling 112 altogether and both patient groups reported statistically significant associations with these alleles. Overall, the *MICA*008* association appeared to be due to a very high frequency of homozygous patients.

The Norwegian study described associations with the *MICA5.1* and *MICB24* markers. The method employed for MIC genotyping in this study was different from that above. *MICA5.1* is a marker for the ancestral 8.1 haplotype. This study did not agree with that from UK. One reason for this is the different techniques; the other lies within a subtle difference between the two populations. These two populations though similar are not the same. At the time of this study the Norwegian PSC patients were all diagnosed to be positive for ulcerative colitis, whereas in the UK only 70% of cases of PSC are diagnosed to be positive for ulcerative colitis.

The UK set have a weak but confirmed association with the ancestral 7.2 haplotype. This association is not seen in either Sweden or Norway. The *MICA*008* allele is carried on both the HLA 8.1 and 7.2 ancestral haplotypes (haplotypes 1 and 2 in Table 6.11). The absence of the weak association with the 7.2 haplotype in the Scandinavian population may account for this difference, and it may also explain the difference reported at the molecular level regarding electrostatic charge and position DRβ-37 versus position DRβ-38. This difference may also reflect genuine population differences and case ascertainment criteria.

In the case of DRβ-38 versus DRβ-37, clearly the latest data in favor of DRβ-37 are very convincing and clearly correct, but the potential role of MICA has yet to be determined in a positive or negative manner. There are four reasons for this:

- The Scandinavian team did not identify *MICA*008*, but a marker for the HLA 8.1 haplotype *MICA*5.1*.
- The sample sizes in the two studies were comparable (112) versus (130).
- The location of *MICA* expression in relation to the disease.
- The function of *MICA* and the fact that the liver has very high numbers of NK and γδ T cells.

These observations are likely to be only a piece of the final picture and other MHC genes may have a role to play in PSC. Until very large-series studies of the whole MHC region have been performed in PSC we will have to wait to find out what else is hiding within this region. Finally, we do not yet know the antigenic peptides that are preferentially bound by these HLA molecules. Identifying these is the essential theme for this research as this will provide insight into the disease pathogenesis and will hopefully lead to novel therapies.

PBC is an autoimmune liver disease with a genetic component

PBC is inflammatory non-suppurative disease of the intra-hepatic bile ducts. PBC leads to biliary fibrosis, cirrhosis, and liver failure. This disease is more common than PSC, occurring in approximately 1:3000 people. It is more common in women than in men, with between 90–95% patients being female and it usually presents over the age of 40. The disease is considered to be an autoimmune disease because of the female preponderance and the presence of high titers of auto-antibodies and the auto-reactive $CD4^+$ and $CD8^+$ T cells, specific for epitopes derived from the pyruvate dehydrogenase complex E2 antigen (PDC-E2). However, the disease does not respond to classical therapy with corticosteroids or other immunosuppressive therapy.

Evidence of a genetic component to this disease is based on:

- Gender bias (up to 95% patients are female).
- Geographic and familial clustering (it has been estimated that 6.4% of cases have at least one other affected family member).
- The sibling relative risk (λ) is 10.5 (with this figure rising to 58.7 for daughters of affected mothers).
- Concordance in monozygotic twins is 64% (though this is based on a small self-reporting sample).
- Overlap with other autoimmune diseases including autoimmune thyroid disease and SLE.

Early studies of PBC found no reproducible HLA associations

Contrary to the findings in AIH and PSC, early studies of PBC failed to find an association with any of the markers of the HLA 8.1 or 7.2 ancestral haplotypes. Even though two small phenotyping studies reported associations with DR2, DR3, and DR4, these were not confirmed. During the 1980s there were multiple studies of *HLA-A* and *-B* in PBC, but none proved fruitful. Studies of DR using serological phenotyping were of variable quality and most were negative, with one exception. In 1987, Gores et al., working at the Mayo Clinic in Rochester Minnesota, USA, struck gold. In their study, they used high-quality serotyping and detected a previously unreported association with DR8. Prior to this study, this association had been missed because most commonly used serotyping trays were poorly equipped to detect the DR8 antigen. This missed association illustrates the importance of considering the limitations of methodology, before drawing conclusions from genotyping and phenotyping studies from the past. The differences in methods are illustrated in Table 6.3, which shows a 69% concordance rate for serologically typed *HLA-DRB1*08*-positive samples versus RFLP, which was found to be 100% concordant with PCR-based methods (Table 6.4) on the same samples. These data are from the mid-1990s. Prior studies were for the most part unable to detect DR8 with even this degree of accuracy. Following the study of Gores et al., multiple studies were able to confirm the association with DR8 using RFLP analysis, which had only recently been introduced and is an excellent method for detection of *HLA-DR8* as indicated in Table 6.4, where all 16 *DR8* RFLP-positive samples were confirmed by PCR. The association between PBC and *DR8* is unusual in two ways.

- Few diseases are associated with *HLA-DR8*.
- The association accounts for less than 25% of the total patient group.

Throughout the 1980s and 1990s studies continued to find *DR8* associated with PBC. In Northern and Southern European populations, and in their descendants in North America, the key allele appears to be *DRB1*08:01*, but in Japan the key allele is *DRB1*08:03*. These two alleles differ by a single amino acid at position 67, where *DRB1*08:01* encodes phenylalanine and *DRB1*08:03* encodes isoleucine. Extending the studies to the other class II genes identified *DQA1*04:01* and *DQB1*04:02* as the primary DQ elements of the *DRB1*08:01* haplotype in Europeans and *DQ3* as the primary DQ element in Japanese patients. The finding of different DQ alleles on the two *DR8* haplotypes and the absence of a second expressed *DRB* gene on the *DR8* haplotypes suggested that *DRB1* is the primary susceptibility locus.

Recent studies in UK and Italy

More recent studies (2006 onwards) on larger series in the UK and Italy identified previously overlooked protective associations with *DRB1*11* and *DRB1*13* family members. This made it possible to review the data from a molecular point of view. Comparing the alleles associated with increased susceptibility to those associated with reduced risk of PBC shows there are four significant amino acid residues. Presented in the order *DRB1*08* versus *DRB1*11* or *DRB1*13* below, these are:

- Glycine for serine at position 13.
- Tyrosine for histidine at position 16.
- Tyrosine for phenylalanine at position 47.
- Leucine for alanine at position 74.

As the associations with *DRB1*11* and *DRB1*13* are weak in some populations this model is speculative, but it provides a basis for considering the molecular biology of this association and the binding groove, and moves the discussion away from the simple process of genotype collecting. Three of the four positions above are of potential functional importance. The amino acid at position 13 affects the binding of antigenic side chains associated with both the P4 and P7 binding pockets, while the amino acids at positions 47 and 74 affect P4 and P6, respectively.

*HLA-DQA1*04:01* may be the primary PBC risk allele in the MHC

Prior to 2012, the situation for PBC looked very like that for AIH and PSC with susceptibility mapping to *DRB1*. However, work on PBC based on imputed HLA genotypes from GWAS studies in 2012 revealed that the likely primary MHC encoded susceptibility alleles are on the *HLA-DQA1* locus, not *DRB1*. The allele in question is *DQA1*04:01*, which is almost always found with *DRB1*08:01*, though a very small proportion of *DRB1*08:01* haplotypes do not carry *DQA1*04:01*. The reason for the difference between the 2012 work and earlier studies is simple. Earlier studies were based on smaller numbers (compared with this GWAS-based study) and given the extreme linkage disequilibrium between *DRB1*08:01* and *DQA1*04:01*, early studies will have been unable to distinguish between the influence of *DRB1* and *DQA1* because there were so few *DRB1*08:01* cases without *DQA1*04:01*. In fact, few studies included *DQA* in their analysis, though most recent studies have included *DQB*. Separating *DQA1*04:01* from *DRB1*08:01* has required a very large cohort of cases. In addition, the *DQA1*04:01* association currently reported is based on imputed data, not HLA genotyping of samples, and this too may have had an impact.

It is important to compare genetic associations in different populations before coming to a final conclusion in PBC and other diseases, such as AIH and PSC above. If the association with *DQA1*04:01* is the primary association with PBC, then the role of *DRB1* needs to be reconsidered. However, we should not cast aside the work of earlier studies and indeed there is some discord over this latter finding. In the Japanese and Han Chinese populations, the most common *DRB1*08* family member is *DRB1*08:03*. In these populations the *DRB1*08:03* allele most frequently occurs with *DQA1*01:03-DQB1*06:01* and it rarely occurs with *DQA1*04:01*.

It is important to consider these findings and differences between studies with care. Current studies are the fruits of earlier work and even though the focus of attention has now moved from DR to DQ, there is no doubt about the initial studies being correct in that there is a strong association with *DRB1*08* in PBC. Whatever the final interpretation of the data on HLA in PBC, future studies need to consider other populations, and also leave room for recently discovered protective associations with *DRB1*11* and *DRB1*13*. In addition when considering HLA we must not forget that the MHC is complex and there is always room within a haplotype for a second bite of the susceptibility cherry, as we have seen in AIH. Identifying the key to this strong association with HLA will ultimately help us to unpack the pathogenic mechanisms that lead to PBC.

Non-MHC-encoded regulatory genes in PSC and PBC and in PSC and UC

A number of markers have been associated with both diseases. Currently, there are 12 non-MHC risk loci for PSC (nine of which were identified in 2013) and 25 in PBC (three of which were new in 2012). In the 2012 PBC study by Liu et al., only 16 of the 22 previously identified associations were significant at $P < 5 \times 10^{-7}$. In both PSC and PBC there is overlap of genetic risk with other diseases, e.g. *TYK2* is found in both

PBC and ankylosing spondylitis, but there does not appear to be any overlap between PSC and PBC, though some members of both risk groups have roles in cytokine production and regulation (*IL12A* in PBC; *IL2* and *IL2RA* in PSC). Ulcerative colitis (usually mild) occurs in 70% of all PSC cases. These two diseases share a number of risk alleles and have some that are not shared. Two examples of PSC/ulcerative colitis risk alleles are *MST1* and *IL21*; others are found in PSC only, e.g. *IL2RA* and *SIK*.

6.7 COMPARING THE HLA ASSOCIATIONS OF THE THREE LIVER DISEASES

There is a great deal of overlap between diseases. This can be clinical overlap where there are two or more diseases that fall into the same group or syndrome, or it can be genetic overlap. Diseases often share the same risk alleles but also have differences. Looking at HLA in these three diseases we see a lot of genetic differences and some overlap (Figures 6.14 and 6.15). The primary susceptibility haplotype for AIH and PSC is the ancestral 8.1 haplotype. However, that is where the similarity between AIH and PSC stops. In PSC, the 7.2 haplotype is associated (albeit weakly) with increased risk, whereas in AIH, the 7.2 haplotype is associated with reduced risk. In complete contrast to both AIH and PSC, neither of these two ancestral haplotypes has been associated with PBC. In PBC, *DRB1*13* is associated with a reduced risk of disease, whereas in PSC, one of the *DRB1*13* haplotypes is strongly associated with increased risk of disease. These differences are useful in evaluating the different molecular models proposed for these diseases and provide further insight into the different mechanisms of disease pathogenesis.

However, even more interesting may be consideration of other diseases, such as the viral liver diseases hepatitis A, hepatitis B, and hepatitis C. Chronic infection with hepatitis B virus is associated with an increased frequency of some *DR13* haplotypes and *DR13* is also associated with sustained hepatitis A virus infection in South American children. These associations, especially when they are shared, may give us a clue to the potential of viral or other infectious triggers for autoimmunity.

6.8 NON-HLA MHC GENES AND DISEASE

While the MHC encodes 252 expressed genes, the majority are not HLA genes. Overall, only a few genes within this region have been investigated in detail. One consequence of this is that GWAS studies which often use tagged single nucleotide polymorphisms (SNPs) across the MHC, seeking to identify novel associations in this area and to confirm existing associations, are beginning to identify some non-HLA genes as strong contestants for the title of primary susceptibility locus. These tagged SNPs often have very high odds ratio and *P* values associated with them and in some cases these are higher than those associated with specific HLA alleles. However, we should not over-interpret this. In the majority of cases the HLA associations reported do stand the test of high-resolution analysis with tagged SNPs. One problem with the MHC is the high level of linkage disequilibrium seen across the region that makes dissecting the 6p21.3 region especially difficult. Very large studies will be needed to do this, but the rewards could be great. The MHC does indeed contain a cornucopia of interesting genes such as those in the MHC class III region discussed below.

The MHC class III region complement, *MICA*, and *TNFA* genes in complex disease

The gene products of the MHC class III region are an assortment of proteins, many with important roles in immunity and immune regulation. We will consider six genes but only one disease example. The genes are:

- Complement *C2*, *C4A*, *C4B*, and *Bf*.
- *MICA* (MHC class I chain-related A).
- *TNFA* (TNF-α).

MHC-encoded complement genes and disease

Complement is best described as a system of plasma proteins that mark pathogens for destruction. It is a critical element in innate and adaptive immunity. Innate (or in-born) immunity and adaptive (or acquired) immunity are the two central pillars of immunity. The innate system is active at birth, whereas the adaptive system develops as we grow. Though there are more than 30 proteins that make up the complement system, C3 is one of the most important. There are three pathways for complement activation: the classical pathway, the alternative pathway, and the lectin pathway. For C3 to be activated it needs to be cleaved into its subcomponents C3a and C3b. Of these two subcomponents, C3b tags bacteria for destruction, while C3a is involved in recruitment of phagocytes. Though the 30 complement proteins are encoded by genes located throughout the genome, the genes that encode the components that cleave C3 in both the alternative and classical pathways are located in the MHC class III region (see **Figure 6.18**). The MHC-encoded

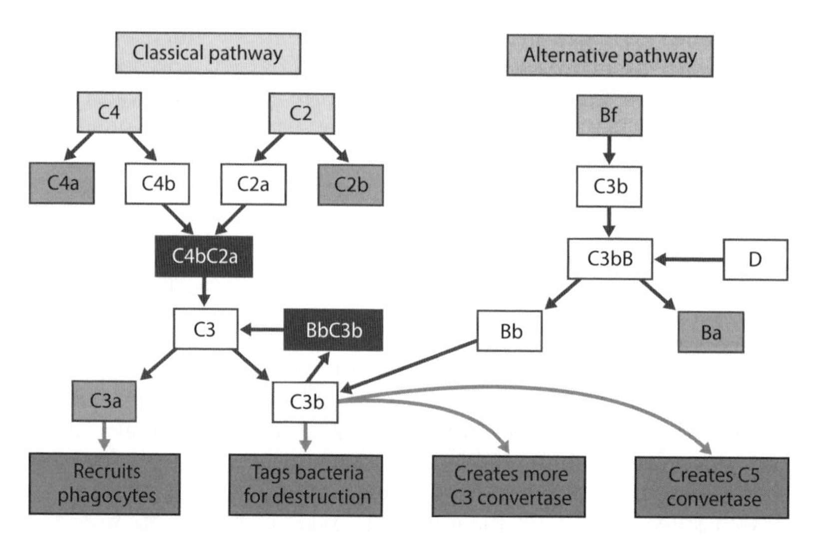

Figure 6.18: Complement system and C3 conversion. The two major complement pathways—the classical pathway and the alternative pathway—converge at the point at which C3 is converted into subunits C3a and C3b. C2a and C4b subunits are primarily responsible for this process in the classical pathway. Thus, converted C3 has many different functions. C3a is active in macrophage recruitment. C3b interacts with the factor B of the alternative pathway and along with factor D cleaves the protein into two subunits: Ba and Bb. The Bb subunit interacts with C3b and forms a C3 convertase. Through this interaction C3 once converted creates a positive feedback loop for further conversion of C3. C3 conversion producing C3b also leads to C5 conversion in the classical pathway. In many ways C3 conversion is the critical stage in the complement pathways.

complement genes are *C2*, *C4A*, and *C4B* for the classical pathway and *Bf* (factor B) in the alternative pathway.

Polymorphisms in the *C2* and *C4* genes have been associated with a number of autoimmune diseases. One reason for this is the very strong linkage disequilibrium in the HLA 8.1 ancestral haplotype. Studies of complement genes in diseases where there are associations with this haplotype will also report associations with *C4A* null alleles. These nulls are large-scale deletions of the *C4A–CYP21A* region of 6p21.3 and they do not have a functional product. Though these are called null alleles by some, they are in fact null genes as the whole gene sequence is deleted. However, we each inherit two copies of our genome and in complement there are two functional forms of *C4*, designated *C4A* and *C4B*. As these have similar functions, the partial absence of an expressed *C4A* gene (heterozygote) or even complete absence (homozygote) frequently has no phenotypic consequences. *C4B* simply makes up for *C4A*. Interestingly, there are also two forms of *C4B*, designated long and short, which differ by as much as 6 kb in size and there are also null alleles for *C2*. Copy number variation (CNV) also occurs with these genes whereby some haplotypes carry multiple copies of the genes. As *C4A* null is frequently carried on the HLA 8.1 haplotype it is difficult to select whether or not this null allele has any effect in diseases with the HLA 8.1 association. However, there is one outstanding exception: systemic lupus erythematosus (SLE).

Complement and SLE

SLE is an autoimmune disease where the immune response appears to target a whole system rather than a particular cell type or organ. SLE is characterized by deposition of immune complexes in the small blood vessels, kidneys, and joints, leading to intense outbreaks of inflammation. It is thought to occur as a result of complement deficiency; indeed, 75% of cases have familial disease due to C1q and C2 deficiencies. The remaining cases have sporadic disease and this has been associated with inheritance of the HLA 8.1 ancestral haplotype. However, the association with HLA does not explain all of the sporadic cases of SLE. Fielder et al., by retyping patients focusing on C4, found a higher percentage of patients had *C4A* null alleles than for any of the previously associated HLA alleles, suggesting the true association was with C4A null rather than with the HLA class I or class II genes. Further studies in the Japanese population confirmed this finding. In Northern Europeans the 8.1 haplotype carries the following complement alleles *C4A*Q0–C4B*1–C2*C–Bf*S*.

Of course, the association above makes sense at the functional level and it is a good example of both an MHC class III-associated disease and the importance of considering the functional relationship between the identified susceptibility allele and disease. Almost all individuals who carry the ancestral 8.1 haplotype carry a *C4A* null (*C4A*Q0*) but few of these develop SLE. This is because most people are heterozygotes and also because most people have a functional *C4B* allele on the haplotype (*C4B*1*). C4A and C4B are two different isoforms of C4 with similar overlapping functions. As most HLA 8.1 carriers are heterozygotes and because there is the potential for *C4B* to make up for the absence of *C4A* to some degree, most *C4A*Q0* carriers do not develop SLE. It is possible that in SLE disease may only arise when the immune system is stressed and the level of production of C4 cannot meet demand. This hypothesis may explain the low level of penetrance for *C4A*Q0*.

MHC class I chain-related A

There are two expressed **MHC class I chain (*MIC*)-related** genes (*MICA* and *MICB*) located on the centromeric side of *HLA-B* and three *MIC* pseudogenes (*MICC*, *MICD*, and *MICE*). Polymorphisms in the *A* and *B* genes have been related to a number of autoimmune diseases including Behçet's disease and Addison's disease. *MICA* encodes an MHC class I-like molecule with three extracellular immunoglobulin domains, a transmembrane segment, and a cytoplasmic tail to anchor the molecule to the cell. *MICA* is stress inducible and expressed on gastric epithelium. *MICA* can activate NK cells and may also activate γδ T cells. There is considerable polymorphism in *MICA*. Currently, there are 101 *MICA* and 41 *MICB* alleles (encoding 80 and 27 proteins, respectively). A potential association between *MICA* polymorphism and PSC is reported above.

TNF-α

TNF-α is encoded by the *TNFA* gene in the MHC class III region. It is a pro-inflammatory cytokine with powerful effects that can be localized to infected tissue or produced systemically through the body. TNF-α is released by macrophages following **Toll-like receptor** stimulation. It can be beneficial or harmful depending on whether it is local or systemic.

The gene encodes a number of well-characterized SNPs that have been investigated in a number of diseases. The most commonly investigated SNP is the A/G –308 SNP. The *TNF-308A* allele has been identified as being associated with increased TNF-α production. Early studies found it was increased in most diseases studied. However, interest in *TNFA* soon faded and then stopped when it was realized that the *TNFA-308A* allele was in linkage on the HLA 8.1 haplotype. Interest in *TNFA* did, however, spark a general interest in **cytokine** gene polymorphisms as potential potent **immune-regulatory genes**.

6.9 A SINGLE GENE OR A RISK PORTFOLIO

Though the MHC encodes 252 expressed genes, most studies have considered less than 14 of these: *HLA-A*, *-B*, *-C*, *-DRB1*, *DRB3*, *-DRB4*, *-DRB5*, *DQA1*, *-DQB1*, *-DPA1*, *-DPB1*, *MICA*, *C4*A*, and *TNFA*. The clustering together of so many important immune-regulatory genes into some haplotypes may be of great significance in terms of human survival against infectious disease.

One key question therefore is whether MHC-associated genetic susceptibility to disease is determined by single genes, or by the collective activity of several or all the members carried on extended haplotypes as part of our individual risk portfolios. We can consider two possibilities.

A single gene may explain MHC-encoded genetic susceptibility to disease

The MHC almost always seems to have a large risk impact. In all studies, the size of effect depends on many factors, including the size of the population studied, the specificity of the disease studied, and a number of other factors related to sample collection and testing. Even so, the MHC seems set aside or different. The question is why.

Part of the answer to this question lies with considering the central role that is played by the HLA molecules in T cell activation and in innate immunity. Interestingly, many of the HLA associations described above were initially described as associations with the same common ancestral haplotype. However, it is difficult to believe that one or two haplotypes are the major determinants of genetic risk in the majority of human diseases. It is possible that these links to common haplotypes reflect the fact that each carries more than one potential risk allele. However, this does not suggest that the disease results from the effect of several genes; on the contrary, it suggests perhaps a range of diseases are associated with the same haplotype because the haplotype encodes several risk loci. For example, an allele that is associated with an increased risk of SLE may be associated with AIH through carriage on the common 8.1 haplotype. In SLE, it is the *C4A* gene that is thought to be important in disease pathogenesis, whereas in AIH it is thought to be DRβ-71 and DRβ-74. In this hypothesis the association between AIH and *C4A*Q0* simply reflects linkage disequilibrium in the 8.1 haplotype, and the real disease risk is explained by the association with *DRB1* alleles, which carry lysine (and arginine) at DRβ-71. This suggests that the association with *C4A*Q0* may be false and due simply to linkage disequilibrium. In SLE, the situation is reversed—the association with HLA genes on the 8.1 haplotype may be simply due to linkage disequilibrium with the *C4AQ0* allele.

Alternatively there is always room for a second bite of the cherry: a multihit hypothesis

In contrast to the hypothesis above, the magnitude of the disease risk encoded within the MHC may be due to the clustering of so many highly important immune response genes within one area (**Figure 6.19**). Here, a single allele may not be enough to explain the MHC-associated risk. Instead, we may consider the possibility that within these extended haplotypes there is a cumulative effect on disease risk. Thus, patients with AIH or PSC who have the ancestral 8.1 haplotype have inherited a particular risk portfolio that includes several independent risk alleles. This multihit hypothesis offers an explanation

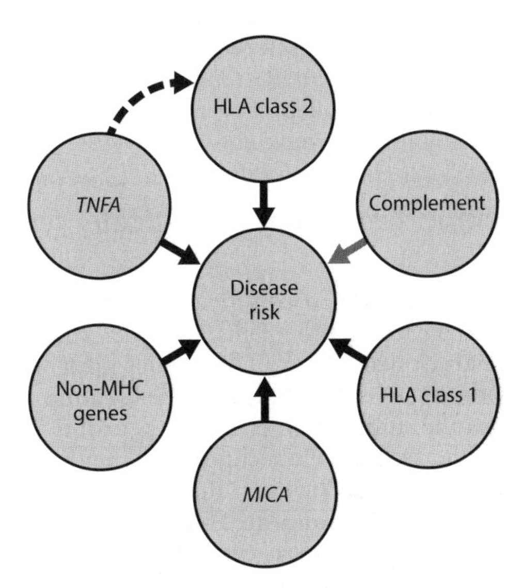

Figure 6.19: Multihit hypothesis. The figure shows the potential for multiple genes to increase disease susceptibility. Here the figure shows five different MHC gene families or genes. However, these are not the only risk genes for the traits discussed in this chapter. There is an indication that genes may interact, e.g. *TNFA* and MHC class II. Thus, it is possible that polymorphisms only increase disease risk when other alleles are present or overexpressed. Finally, the figure also allows for redundancy. In this case, the complement system is used as an example. The gray arrow indicates that even in the presence of complement gene-encoded risk alleles, possession of such an allele may not have an effect unless the biological system that the gene products serve is stressed. Under such conditions having a null allele for *C4A*, for example, may lead to failure to produce adequate complement levels and this may have downstream consequences in terms of immune complex clearance, etc..

of why certain haplotypes are associated with several diseases. For example, the extended 8.1 haplotype includes *HLA-A*01–C*07:01–B*08–MICA*008–TNFA-308**–C4A*Q0–C4B*1–C2*C–Bf*S–DRB3*01:01–DRB1*03:01–DAQ1*05:01–DQB1*02:01*, one or any combination of these individual alleles may increase or reduce a particular disease risk. A combination effect is partly illustrated by the example of AIH above. This possibility may explain the predominance of this haplotype in autoimmune disease. Of course, this is a common haplotype and most carriers do not develop autoimmune diseases, but possession of this group of risk alleles under **immune stress** may be sufficient to tip the balance in the wrong direction. The converse is also possible under certain circumstances, for example during a major epidemic it may be beneficial to have this haplotype because possession of the haplotype may result in a mild response to the infectious agent. This may explain the high frequency of some haplotypes in the population.

According to this explanation there are multiple risk alleles within the MHC. Each haplotype will carry multiple hits, but the same haplotype does not necessarily operate in the same way in each disease. The same DR, DQ, and C4 polymorphisms may impact on different diseases differently, each making a large, small, or no contribution to the risk portfolio. The multihit model can be exclusive.

6.10 HOW TO COMPARE AND CRITICALLY EVALUATE CONTRASTING STUDIES

History tells us about our past and informs our future. Only when we know our past can we consider our future. This is as true in science as it is in global politics. We learn from each publication. Most scientists are focused on narrow areas, developing hypotheses and research plans based on the findings of previous studies. Thus, the scientific past also informs the scientific present and the future.

Knowing history is important when we critically review and design studies

PSC provides an interesting example to consider. PSC is relatively rare. Consequently there are few centers that have studied the genetic basis of PSC, and even less that have considered HLA or the MHC. Most of the UK studies have taken place at one center, often in collaboration with other centers, but nevertheless almost always as a single unit. This means that with time, the studies that have been performed, have benefited from an ever-increasing DNA bank as well as new technologies that have become available. The same is true for the Norwegian series, albeit that there is strong collaboration with a major Swedish center, this group can still be considered as a single unit. In many ways there is great benefit in this because this evolution within the research units ensures that technical advances are applied to the study series and also it ensures a greater degree of consistency in the studies.

However, it is also a potential trap unless one keeps a critical eye on one's own work as well as that of others. This is because when reviewing and comparing studies of genetic associations in disease it is all too easy to consider earlier studies as incorrect or flawed compared with more recent studies. However, this is not always appropriate. We are inclined to be apologetic about early studies, but there is no need for apologies—early analyses were most often based on the best available methods for the time. As methods have improved it has become possible to increase the level of resolution and to focus on areas previously ignored, allowing hitherto undetected, often undetectable relationships to be identified. Therefore in studies of the MHC we often see the focus of research move from one location to another. We must remember that had there been no initial interest in HLA, these associations would not have been identified and perhaps in many cases the idea that these complex diseases would be worthy of genetic investigation would have gone unmarked. Studies of HLA in disease go back at least as far as 1967, and flowered in the 1970s and 1980s. These were periods during which even with poor technology, great advances were made in a range of diseases and these were the stepping stones on which present-day advances in the MHC and elsewhere began.

CONCLUSIONS

This chapter contains many key messages for the student of the genetics of complex diseases. The early history of the MHC illustrates the importance of having standardized nomenclature for complex systems. This will be of increasing importance in the post-genome era with high-resolution genotyping being applied to more and more genes, creating a potential for personalized naming systems to arise.

The MHC is very complex with extreme levels of linkage disequilibrium and polymorphism. The high levels of linkage disequilibrium can be both helpful and problematic. In particular, problems arise when data based around a single gene are over-interpreted. The MHC illustrates the importance of studying whole haplotypes and not focusing on single genes until there is sufficient evidence to be confident that the candidate has been identified. However, systems interact and genetic interaction on extended haplotypes can turn the most carefully formulated hypotheses into nonsense.

In a more positive frame of mind, the examples discussed enable us to see the link between genotype and phenotype, and the way genetics informs the debate on disease pathogenesis at the molecular level. The key among these is the role of HLA in the formation of the immune synapse—a process that is essential in adaptive immunity.

Though a key example, it would be wrong to consider the formation of the T cell synapse as the only immune process influenced by MHC genes. HLA alleles have important roles in innate immunity through NK cell activation and regulation.

Genetic associations with the MHC are not all the result of HLA polymorphism. This large region encodes many immune-regulatory genes with diverse functions. Polymorphism in any of these non-HLA immune-regulatory genes has the potential to impact on disease susceptibility. For example, complement is a key element in immune complex clearance and bacterial destruction, *MICA* genes impact on innate immunity, and the list goes on.

Finally, the examples discussed in this chapter also allow us to consider the difficulties and complexity of assessing genetic studies in this area, and how we can approach these subjects with a critical mind.

In summary, the MHC is a large densely packed genetic region within the human genome. The genes within this region are particularly important in immune regulation, and there are very strong links between polymorphisms in these genes and many complex diseases. The region has been the focus for a considerable amount of research; however, much remains to be done and, despite some very promising findings and hypotheses, the link between genotype and phenotype is yet to be fully established. Interestingly, with current technologies we are beginning to unpick the MHC to a greater extent than before and identify multiple associations within the region. For example, the WTCCC2 study in 2011 suggested that there may be an independent association with an *HLA-A* allele that confers some protection from multiple sclerosis. These technologies will be discussed in more detail later in the book.

FURTHER READING

Books/Theses

Donaldson PT. The influence of donor recipient histocompatibility and immunogenetics on the outcome of clinical liver transplants. PhD Thesis, Faculty of Clinical Medical Sciences, King's College London. 1995. This thesis gives a very thorough history of the development of transplantation from 1829, with Blundell's first use of blood transfusion, up to 1995.

Hillert J & Fogdell-Hahn A (2000) HLA and neurological diseases. In HLA in Health and Disease (Lechler R and Warrens A eds), pp 219–230. Academic Press. This chapter is full of data on HLA, and both multiple sclerosis and narcolepsy.

Parham P (2009) The Immune System, 3rd ed. Garland Science. This is an excellent basic immunology book for the student of human immunology including links to immunogenetics.

Van Rood JJ (2000) The history of the discovery of HLA. In HLA in Health and Disease (Lechler R & Warrens A eds), pp 3–22. Academic Press. This is an excellent overview of the early days of HLA, and the flowering of HLA nomenclature and typing methods, by one of its founding fathers.

Articles

Arnett KL, Haung W, Valiante NM et al. (1998) The Bw4/Bw6 difference between HLA-B*08:02 and HLA-B*08:01 changes the peptides endogenously bound and the stimulation of alloreactive T cells. *Immunogenetics* 48:56–61.

Bauer KH (2007) Homiotransplantation von epidermis bei eineiigen zwillingen. *Bruns Beit Klin Chir* 141:442–447. This is an example of the early work on histocompatibility and raises awareness that not all papers were published in English.

Blundell J (1829) Observations on transfusion of blood. *Lancet* 12:321–324. A fascinating insight into very early clinical investigation.

Bjorkman PJ, Saper MA, Samraoui B et al. (1987) Structure of the human class I histocompatibility antigen, HLA-A2. *Nature* 329:506–512.

Bodmer WF, Albert E, Bodmer JG et al. (1985). Nomenclature for factors of the HLA system, 1984. *WHO Bull* 63:399–405. This is the original report of 1984 IHTW identifying HLA DQ and DP for the first time.

Brown JH, Jardetzky TS, Gorga JC et al. (1993) Three dimensional structure of the human class II histocompatibility antigen HLA-DR1. *Nature* 364:33–39.

Donaldson PT (2011) Electrostatic modifications of the human leucocyte antigen DR P9 peptide-binding pocket primary sclerosing cholangitis: Back to the future with human leucocyte antigen DRβ. *Hepatology* 53:1798–1800. This is the Editorial for the paper by Hov et al. (2011) written by one of the authors of this book. The Editorial explains the history of HLA and PSC.

Donaldson PT & Norris S (2002) Evaluation of the role of MHC class II alleles, haplotypes and selected amino acid sequences in primary sclerosing cholangitis. *Autoimmunity* 35:555–564.

Han F, Faraco J, Dong XS et al. (2013) Genome wide analysis of narcolepsy in China implicates novel immune loci and reveals changes in association prior to versus after the 2009 H1N1 influenza pandemic. *PLoS Genet* 9 (10):e1003880.

Horton R, Wilming L, Rand V et al. (2004) Gene map of the extended human MHC *Nat Rev Genet* 5:889–899.

Hov JR, Kosmoliaptsis V, Traherne JA et al. (2011) Electrostatic modifications of the human leukocyte antigen-DR P9 peptide-binding pocket and susceptibility to primary sclerosing cholangitis. *Hepatology* 53:1967–1976. This is an excellent article that discusses the past and present interpretations of HLA associations with PSC, introducing some very up-to-date methods for data analysis and providing a new model to account for these genetic associations.

Landsteiner K (1900) Zur kenntniss der antifermentativen lytischen und agglutinierenden. Wirkungen blutserums und der lymphe. *Zbl Bakt* 27:357–362. Once again this paper reminds the student that not all science was and is published in English. We should endeavor to be aware of this especially when we consider that Landsteiner was awarded a Nobel Prize for his work.

Liu JZ, Almarri MA, Gaffney DJ et al. (2012) Dense fine-mapping study identifies new susceptibility loci for primary biliary cirrhosis. *Nat Genet* 44:1137–1141.

Liu JZ, Hov JR, Folseaas T et al. (2013) Dense genotyping of immune-related disease regions identifies nine new risk loci for primary sclerosing cholangitis. *Nat Genet* 45:670–675.

Norris S, Kondeatis E, Collins R et al. (2001) Mapping MHC-encoded susceptibility and resistance in primary sclerosing cholangitis: The role of MICA polymorphism. *Gastroenterology* 120:1475–1482.

Siebold C, Hansen BE, Wyer JR et al. (2004) Crystal structure of HLA-DQ0602 that protects against type 1 diabetes and confers strong susceptibility to narcolepsy. *Proc Natl Acad Sci USA* 101:1999–2004. This original paper compares two diseases and discusses complex molecular models to explain the HLA associations with them. Not for the faint-hearted, but a great paper that requires some consideration.

Terasaki PI & McClelland JD (1964) Microdroplet assay of human serum cytotoxins. *Nature* 204:998–1000. This original report opened the door to the use of micro assays for HLA typing—a method almost universally adopted for HLA class I typing in the 1970s and 1980s until molecular genetics techniques became available.

The International Multiple Sclerosis Genetics Consortium & The Wellcome Trust Case Control Consortium 2 (2011) Genetic risk and a primary role for cell-mediated immune mechanisms in multiple sclerosis. *Nature* 476:214–219.

Todd JA, Bell JI & McDevitt HO (1987) HLA-DQβ gene contributes to susceptibility and resistance to insulin-dependent diabetes mellitus. *Nature* 329:599–604.

Wiencke K, Spurkland A, Schrumpf E & Boberg KM (2001) Primary sclerosing cholangitis is associated to an extended B8-DR3 haplotype including particular MICA and MICB alleles. *Hepatology* 34:625–630. The study by Norris et al., above and this study by Wiencke et al., show slightly different results. This is common in association studies between competing centers. Confounding results may be based on many factors. In this example the studies are both based on similar-sized patient pools, but use different methods and different populations. This is an example that illustrates many of the concepts of studies in complex diseases, and together these two papers provide an excellent basis for critical appraisal.

Online sources

http://hla.alleles.org/antigens/index.html
This is an excellent source for up-to-date information on HLA nomenclature.

http://www.ebi.ac.uk/imgt/hla/stats.html
This is an excellent source for up-to-date information on HLA allele numbers and nomenclature. The website includes text on the history of HLA nomenclature.

CHAPTER

7

Genetics of Infectious Disease

Infectious diseases account for the majority of morbidity and mortality in human populations. In a classical sense, we would not normally consider infectious diseases as genetic diseases. However, variation in the human genome can lead to variation in resistance and susceptibility to infectious disease, and in this sense it is correct to say that there is a genetic component in infectious disease. In this chapter, we will consider how the evolutionary pressure of infectious disease has led to particular genetic variants being maintained in a human population and will examine molecular mechanisms that link particular polymorphisms with susceptibility to infection. Two key examples, malaria and human immunodeficiency virus type 1 (HIV-1), will be used to illustrate how different clinical aspects of infectious disease are influenced by host genetic variation. In addition, we will discuss the use of **hypothesis-driven** and genome-wide studies to identify specific human genes that influence the severity of infectious disease.

7.1 THE INFECTION PROCESS AND DISEASE

Before embarking on a discussion of the genetics of infectious disease it is worth briefly reviewing the infection process. Infectious diseases occur when a susceptible host is exposed to and infected by a pathogenic microorganism. We can divide these microorganisms broadly into five types: viruses, bacteria, fungi, and **unicellular** or **multicellular eukaryotic parasites**. Viruses, as well as some bacteria and unicellular parasites, are intracellular pathogens, meaning that at least a part of their life cycle takes place inside the host cells. The remainder are extracellular pathogens and colonize the extracellular environment either inside the host organism or on its surface.

Mechanisms of infection vary widely but common steps in the process can be identified

Though the mechanisms through which infectious pathogens cause disease vary, there are several key steps in this process (**Figure 7.1**). First, an individual must be exposed to the pathogen, and the likelihood of this depends on a number of factors including the health status of the individual, the behavior of the individual, route of transmission of the pathogen, population size and density, sanitation, and healthcare. Once exposed, the pathogen must be able to propagate and colonize the host. This may take place only at the site of infection, leading to a localized infection, or alternatively the pathogen may spread from the initial site of infection via the blood and **lymphatic system** to infect other organs and generate a systemic infection.

The immune response combats infectious disease

The efficiency with which a pathogen is able to establish an infection is strongly influenced by the counter-effect of the host's immune response. The innate and adaptive arms of the immune response are both deployed in an attempt to resist infection and to eliminate the pathogen from the body. The majority of infections are asymptomatic, usually because the host immune response is successful in quickly containing and clearing the infection before symptoms appear. However, not all infections are dealt with so efficiently and those that are not eradicated may result in disease symptoms of varying severity. In some cases, a pathogen may establish a persistent infection, continuing to live in the host for many years with only mild or even no symptoms. It is worth noting that many common symptoms of infection, e.g. fever and tissue damage, may not be a result of damage induced by the pathogen, but be the direct effects of an active host immune response.

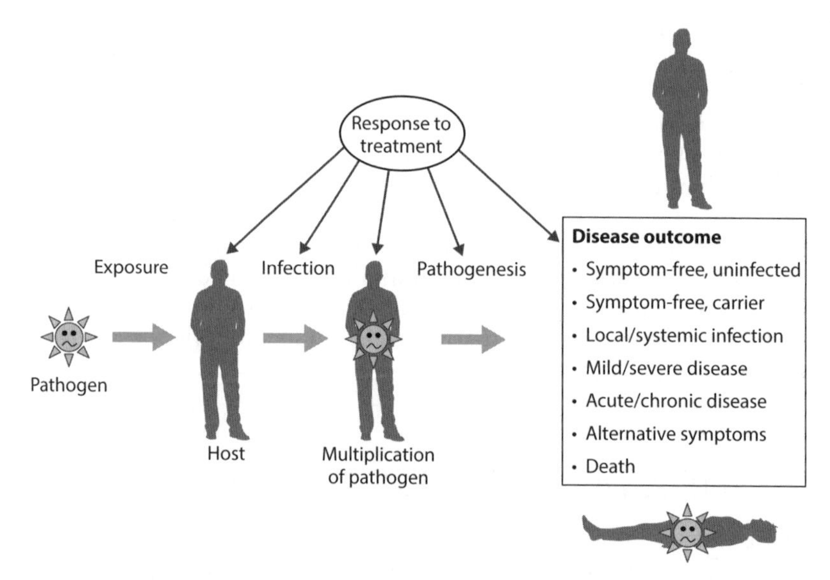

Figure 7.1: The infection process. All infectious diseases follow a similar process. First, the host must be exposed to the pathogen that may then infect the host, replicating either on host cell surfaces or inside cells. The infection process often requires the pathogen to bind to a host cell receptor protein. The outcome of infection varies enormously, depending upon the specific interaction between host and pathogen, and ranges from asymptomatic infection to mild disease, severe disease, and death. Treatment of infection (either before or after exposure) may inhibit the infection process at one or more steps in the pathway, and there is significant variation in the response of the patient and the pathogen to treatment.

Individuals infected by the same pathogen may experience different outcomes

The symptoms that result from an infection depend to a large extent on the phenotype of the pathogen. More interestingly, in the context of this book, the outcome of infection can also vary among different individuals infected by the same pathogen. We will briefly consider two important key examples that illustrate this point:

- HIV-1 is a retrovirus that causes acquired immunodeficiency syndrome (AIDS). HIV-1 is a major global health threat and was responsible for up to 1.6 million deaths in 2012 according to the World Health Organization (WHO). Most people who are repeatedly exposed to HIV-1 through, for example, unprotected sexual intercourse will become infected and go on to develop AIDS within approximately 10 years. In December 2013, there were an estimated 35.3 million cases of HIV-1 infection worldwide. Some individuals, however, are repeatedly exposed to HIV-1 in the same way, but never become infected.
- Malaria is another global killer responsible for an estimated 627,000 (range 473,000–789,000) deaths in 2012, mostly among African children. Malaria is caused by eukaryotic parasites (protozoa) of the *Plasmodium* genus. The total number of exposed individuals exceeded 3.4 billion with 207 million infected cases in 2012. Some individuals infected with the *Plasmodium* parasite experience no symptoms, whereas some experience severe malarial anemia and some develop cerebral malaria.

It is clear from these two examples that pathogens can have very different effects in different individuals. Non-genetic factors, including age, living conditions, health status, and presence of other infections, may contribute to the range of outcomes. Some of this diversity also arises due to novel variations in the genome of the pathogen. Most microorganisms multiply quickly and in doing so accumulate mutations. Viruses with an RNA genome accumulate mutations rapidly because RNA polymerases lack proofreading ability. Variation in the pathogen genome may increase or decrease the pathogenicity of the microorganism and any mutations that confer a survival advantage will be quickly selected. Non-genetic factors and genetic variation in the pathogen genome are, however, by no means the whole story and cannot account for all observed variation in the human response to infectious disease. We must also consider the effect of variation in the genome of the human host, which is the basis of many differences in individual susceptibility and resistance to infectious disease.

7.2 HERITABILITY OF RESISTANCE AND SUSCEPTIBILITY TO INFECTIOUS DISEASE

What evidence is there that resistance and susceptibility to infectious disease is **heritable**? Even before the development of modern techniques for molecular genetic analysis, historical and **epidemiological** studies provided clues that genetic variation in human populations affected their response to infection.

Different populations infected by the same pathogen may experience different outcomes

As well as variation in individual responses to infection by a particular pathogen, variation is evident at a population level. This is best illustrated by looking back in history, to a time when populations were more geographically isolated than they are today. In 1519, Spanish Conquistadors led by Hernando Cortes landed in Mexico. Unknown to the Conquistadors, they were taking with them a number of infectious agents, including measles, smallpox, and the influenza viruses. They also carried bacteria, including *Rickettsia*, which can cause typhus. The invading populations were relatively resistant to these pathogens, whereas the indigenous South American populations, including the Mexican Aztecs, were highly susceptible to these infectious agents and died in vast numbers as a result. For example, smallpox became epidemic in Mexico between 1520 and 1521, and may have killed an estimated 10–50% of the local population. Several factors are likely to have contributed to the difference in susceptibility between the invading and indigenous populations. The invaders had already been exposed to these pathogens and would therefore have acquired immune memory. In contrast, the indigenous populations who had not previously been exposed to these pathogens had no acquired immune memory, making them more vulnerable to infection. However, host genetic differences between the two populations may also have played a significant role. Pathogens such as smallpox had been common among European populations for many years, and individuals with allelic variants conferring resistance to smallpox and other common pathogens would have had a survival advantage over those without such alleles. As a consequence of this selective pressure, resistance alleles may have been present at a higher frequency in the invading populations than in South American populations, where no such selection pressure had occurred. Thus, a difference in the relative frequency of resistance-associated alleles in the two populations may have contributed to the greater susceptibility of the South American people to the imported pathogens.

There are many other examples in history of naïve populations being devastated by infectious diseases brought to them by immigrant populations. Haldane, in 1949, suggested that "Europeans have used their genetic resistance to such viruses as that of measles (rubeola) as a weapon against primitive peoples as effective as fire-arms". Yet the reverse rule also applies: Europeans, migrating to colonial regions, have found themselves more susceptible to local infectious agents and in some cases early colonies were wiped out by exposure to infectious disease.

Leprosy and tuberculosis were once believed to be inherited diseases

Leprosy and tuberculosis are diseases that we now know to be caused by infection with the intracellular bacteria *Mycobacterium leprae* and *Mycobacterium tuberculosis*, respectively. The distinct clustering of leprosy and tuberculosis in families, however, resulted in an early assumption that both diseases were inherited. Gerhard Hansen challenged this view of leprosy in 1873, when he identified *M. leprae* as the causative agent. At the time this was the first bacterial agent of human disease to be identified. Hansen stated "your opinions about leprosy are completely wrong. You believe that the disease is hereditary but not infectious. The truth is that it is infectious but not hereditary." This was followed by the identification of *M. tuberculosis* by Robert Koch in 1882.

More recently, twin studies indicated that concordance rates of leprosy and tuberculosis were higher in monozygotic twins than dizygotic twins, though the accuracy of the tuberculosis study has been questioned. Although leprosy and tuberculosis are infectious diseases, the familial clustering and concordance rates still provide a strong indication that susceptibility to both diseases has a heritable component that is likely to include host genetic variation. This suggests that Hansen was not entirely correct and this may be why some thought these were heritable diseases. Altogether, this reminds us not to look for simple answers to complex questions. Twin studies of susceptibility to polio, febrile malaria, and hepatitis B virus carrier status, among others, add further support to the hypothesis that susceptibility to infectious disease is at least partially determined by the genetic makeup of the host.

Adoption studies indicate that susceptibility to infectious disease has a heritable component

A study of Danish adoptees revealed that their relative risk of death due to infectious disease was six times greater if their biological parent died of infectious disease at an early age (less than 50 years old) compared with those whose biological parent was alive at 50 years old. No increased risk was observed for the adoptee if the adopted parent succumbed to infectious disease. A follow-up study revealed that the odds ratio (OR) for death of a full sibling due to infection was greater than 9, compared with half siblings or adopted siblings, further strengthening the case for a genetically determined susceptibility to infection.

Rare monogenic defects in immunity can cause primary immune deficiencies

The ability to resist infection is dramatically impaired in some individuals, due to monogenic defects in genes involved in the immune response. More than 200 monogenic disorders of this type, known as **primary immune deficiencies**, have been identified. Though these are not the subject of this book, these Mendelian disorders do illustrate an important concept. Depending on the specific mutation, a primary immunodeficiency may result in an increased susceptibility to infectious disease in general, to a particular subset of infections, or to a single pathogen (**Table 7.1**).

Most primary immune deficiencies are rare **Mendelian recessive traits** occurring in less than 1:50,000 births and, as such, these disorders are beyond the scope of this book, which is about common genetically complex disease. It is, however, important to appreciate the existence of these primary immune deficiencies, firstly in order to have a full picture of how the host genome influences the outcome of infection, and secondly because more common polymorphisms in the same genes, with lower penetrance, may have a more subtle and complex effect on the host response to infection.

Table 7.1 Selected examples of rare monogenic disorders that cause defects in the immune response leading to increased susceptibility to infectious disease.

Name of syndrome	Immune defect	Susceptibility
Severe combined immunodeficiency	No T cells (and in some cases B cells)	General
(Bruton's) X-linked agammaglobulinemia (XLA)	No mature B cells, therefore no immunoglobulin	Extracellular bacteria and viruses, especially pyogenic bacteria and enteric viruses
Mendelian susceptibility to mycobacterial diseases (MSMD)	Defects in IL-12–IFN-γ signaling circuit (macrophage–T cell–macrophage)	Mycobacterial infection
Epidermodysplasia verruciformis (disseminated cutaneous warts)	Mutations in EVER1 and EVER2	Human papilloma virus

7.3 IDENTIFYING ALLELES THAT AFFECT RISK OF SUSCEPTIBILITY AND RESISTANCE TO INFECTIOUS DISEASE

In complex disease we are dealing with elevations and reductions in risk rather than absolute values. Having a risk allele is neither sufficient nor necessary to develop the disease. Nowhere is this more obvious than in infectious disease where the primary determinant of disease is the infectious agent itself.

Risk alleles can be identified using a hypothesis-driven or genome-wide approach

As for complex diseases described previously in this book, there are broadly two strategies for identifying alleles that modify the risk of susceptibility or resistance to an infectious disease. The first is to adopt a hypothesis-driven approach (sometimes referred to as a candidate gene approach) in which existing knowledge of the infectious process is used to select candidate genes that are predicted to influence susceptibility to infectious disease. Genes involved in the immune response to a pathogen are obvious candidate genes. Hypothesis-driven studies have successfully identified a number of genetic variants associated with resistance and susceptibility to infectious disease, some of which are discussed below. These candidate gene studies are, however, constrained by the limits of current knowledge of disease pathogenesis, and have often proved difficult to replicate for a variety of reasons and have tended to involve small study sizes with consequently low statistical power.

The second strategy is to take a genome-wide approach. This usually involves either genome-wide linkage studies (GWLS) or genome-wide association studies (GWAS), techniques which are described in Chapter 3. The advantage of genome-wide approaches is that they are not **hypothesis constrained** and do not rely on prior knowledge. As a consequence, they have the potential to identify novel genetic variants that were not previously known to play a role in the infectious process. Genome-wide approaches can thus open up new avenues for research into infectious disease, but do require large numbers to reach adequate levels of statistical power.

Studies on infectious disease resistance/susceptibility alleles have been commonly based on a hypothesis-driven approach. The reasons for this are two-fold. (1) There are some clear candidate genes to target based on a general knowledge of the infectious process, including genes that encode receptor proteins and genes that are involved in the immune response to infection. (2) Genome-wide studies require large numbers of well-phenotyped infected individuals, which can be difficult to collect in developing countries where infectious diseases are often most prevalent. Genome-wide studies in Africa, where infectious disease is a significant burden, are particularly challenging. African populations are often more genetically diverse than other populations and linkage disequilibrium can be less prominent, leading to a number of practical difficulties. For example, false-positive associations may arise as a result of population stratification and multicenter studies may be compromised by differences within the haplotypes in different participating communities.

Improvements in high-throughput genotyping technology have led to increased use of genome-wide association studies, in which thousands of tagged single nucleotide polymorphism (SNPs) and other genetic markers across the entire genome are tested for association with a specific infectious disease. This has allowed researchers to cast a much wider net in the search for relevant alleles.

The outcome of infectious disease being tested must be clearly defined

As in all studies of complex disease genetics it is important when searching for risk alleles that influence resistance and susceptibility to infectious disease, that the phenotype (outcome) of the disease being tested for association is clearly defined. For example, is the study testing for association with infection per se or for a specific clinical outcome in an infected population? Lack of clear definition of the parameters of the groups being tested may lead to important effects being masked, and will also present significant difficulties when combining results in collaborative studies and meta-analyses. This is illustrated by studies of malaria resistance, which may focus on either association with infection per se or on disease severity following infection. Disease severity may be defined simply by hospitalization or by more stringent clinical parameters. It is also important that the method of testing the phenotype is clear. For example, in studies of HIV-1 infection, infection per se may be defined by polymerase chain reaction (PCR) detection of the viral genome or seroconversion and studies on the speed of progression to AIDS may define progression based on **CD4$^+$ T cell** count or viral load (measured as viral genome copies per milliliter). Using different parameters in different studies to measure these traits makes direct comparison of different studies impossible.

7.4 MALARIA

Malaria is caused by protozoa of the *Plasmodium* genus, comprised of four species: *Plasmodium falciparum*, *Plasmodium vivax*, *Plasmodium ovale*, and *Plasmodium malariae*. Of these *P. falciparum* and *P. vivax* are the most common, with *P. falciparum* causing more severe disease than *P. vivax*. The *Plasmodium* parasite is carried by the female *Anopheles* mosquito, which thrives in wet, humid areas.

The life cycle of the *Plasmodium* protozoa is complex

When bitten by a malaria-infected mosquito, sporozoites in the salivary gland of the mosquito are transmitted into the bloodstream of human hosts. On entering the bloodstream these sporozoites travel to the liver. In the liver they mature and are released back into the bloodstream as merozoites, which infect and destroy the host red blood cells (erythrocytes). Inside erythrocytes the parasites differentiate into male and female **gametocytes** that are taken up once more by a mosquito. The sexual stage of the parasite reproduction takes place within the mosquito, culminating in the production of new sporozoites and completing the life cycle (see **Figure 7.2**).

Symptoms of malaria typically include fever, chills, and anemia. These coincide with the destruction of red blood cells by the parasite, which leads to loss of hemoglobin and production of a surge of cytokines, including tumor necrosis factor (TNF)-α and interleukin (IL)-1, which trigger the fever and chills. Severe malaria in children is often characterized by severe anemia, respiratory distress (due to lactic acidosis), and neurological symptoms, which may culminate in coma (cerebral malaria). In adults, severe symptoms include organ failure, particularly renal failure.

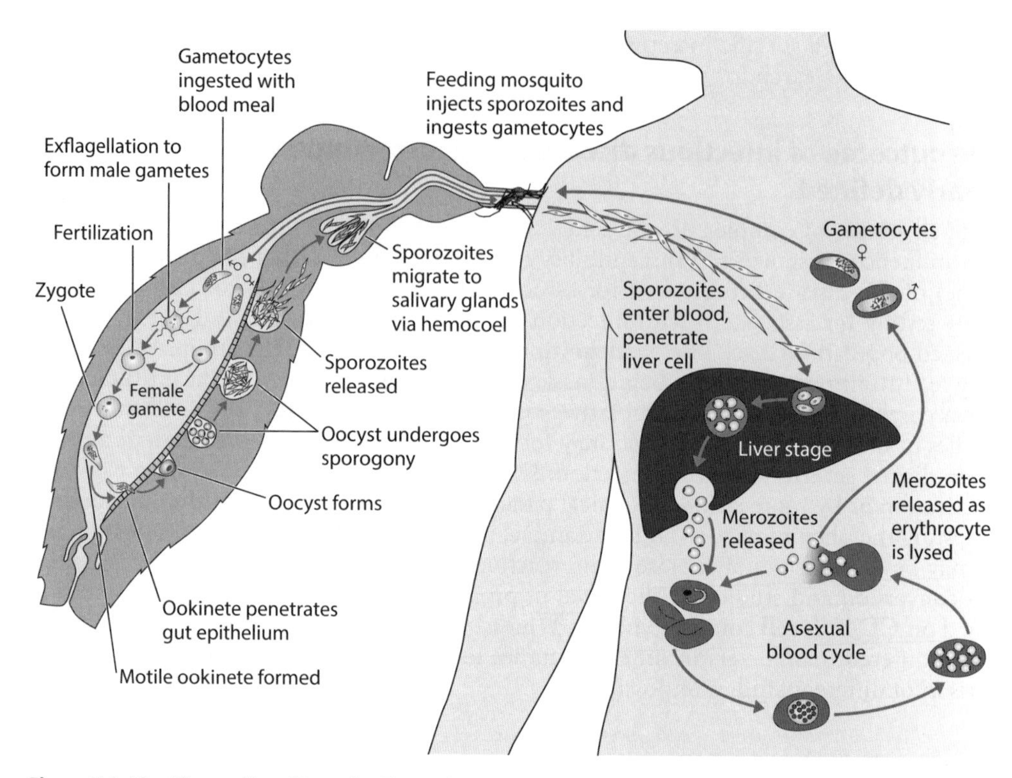

Figure 7.2: The *Plasmodium* life cycle. *Plasmodium* parasites that cause malaria have a complex life cycle involving a mosquito insect vector. When an infected mosquito bites a human host, *Plasmodium* sporozoites may be introduced into the blood and travel to the liver. Here, the *Plasmodium* enters the next phase in its life cycle, producing merozoites that infect red blood cells. Gametocytes are released from the red blood cells and can be ingested when the host is bitten by another mosquito. The sexual phase of the *Plasmodium* life cycle takes place in the gut of the mosquito and ultimately new sporozoites are produced. These travel to the salivary glands of the mosquito from where they can be transmitted to the next host. (From Loker ES & Hofkin BV [2015] Parasitology. Garland Science.)

Hemoglobinopathies confer resistance to malaria

Mutations in the genes for hemoglobin were the first specific polymorphisms recognized to be associated with resistance to an infectious disease. Hemoglobin is the protein that packs erythrocytes and carries oxygen around the body (**Figure 7.3**). The most common form of hemoglobin is hemoglobin A (HbA) which is a tetramer of four polypeptides, consisting of two α-globin chains and two β-globin chains, each with a heme prosthetic group that is able to bind oxygen. Hemoglobinopathies, which include sickle cell anemia and **thalassemias**, result when mutations in the globin genes lead to changes in hemoglobin structure or expression levels.

Sickle cell anemia

Sickle cell anemia is caused by a SNP in the β-globin gene, leading to a substitution of valine instead of glutamate at amino acid 6 of the β-globin molecule. This causes polymerization of the hemoglobin at low oxygen concentrations, which in turn leads to the affected erythrocytes developing a characteristic sickle shape. Individuals who are homozygous for the sickle cell allele (*HbS*) suffer from debilitating symptoms associated with

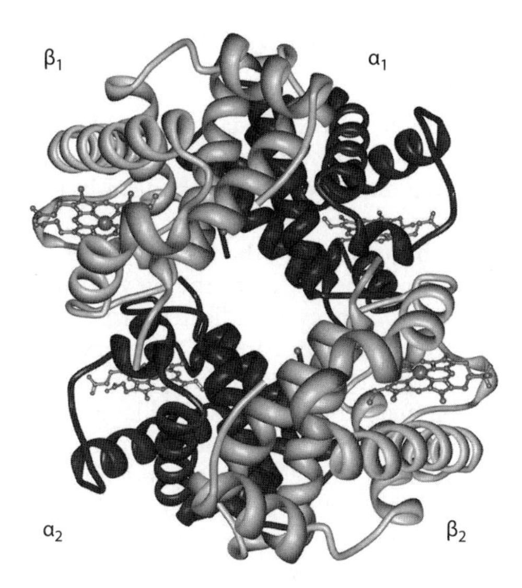

Figure 7.3: Structure of hemoglobin. Hemoglobin is a tetrameric protein composed of four polypeptide units that is found in red blood cells and is responsible for the transport of oxygen in the blood. The most common form of hemoglobin in adults is HbA, which consists of two α polypeptides and two β polypeptides. Each of these polypeptides forms a complex structure with an iron-containing heme prosthetic group, which is able to bind one oxygen molecule (shown above). Diseases caused by mutations that affect the structure or expression levels of hemoglobin polypeptides are called hemoglobinopathies.

sickle cell anemia. Heterozygotes (*HbAS*), however, have sickle cell trait and suffer few if any symptoms.

Thalassemia

Thalassemia is caused by mutations in globin genes that reduce the expression of α-globin (α-thalassemia) or β-globin (β-thalassemia). There are two α-globin genes, closely linked on chromosome 16, and therefore two types of α-thalassemia: α^+-thalassemia occurs when one α-globin gene is defective and α^0-thalassemia occurs when both are defective. There are five possible allele combinations for α-thalassemia, as illustrated in **Table 7.2**.

β-thalassemia is caused by defects in the β-globin gene. β-thalassemia heterozygotes (β/–) suffer mild anemia and morphological changes to red blood cells, whereas β-thalassemia homozygotes suffer severe anemia that can be fatal without treatment.

Table 7.2 Thalassemia: genotypes and associated phenotypes.

Genotype	Phenotype
Normal homozygote (αα/αα)	Normal
α^+-thalassemia heterozygote (–α/αα)	No symptoms
α^+-thalassemia homozygote (–α/–α)	Mild anemia and small red blood cells with reduced hemoglobin
α^0-thalassaemia heterozygote (– –/αα)	Mild anemia
α^0-thalassasemia homozygote (– –/– –)	Lethal

Haldane's malaria hypothesis proposed that thalassemia confers protection against malaria

In 1949, the biologist and geneticist JBS Haldane proposed that resistance to infectious disease is an important driving force in the evolution by natural selection of the human genome. Haldane proposed that an important mechanism in the process is what we now call heterozygote advantage, whereby the heterozygote is fitter than either homozygote. He went on to suggest that the high frequency of thalassemia observed in Mediterranean countries such as Greece and Italy could be linked to the prevalence of malaria in those countries. Knowledge of the *Plasmodium* life cycle indicated that erythrocytes could be the link: part of the *Plasmodium* life cycle takes place inside erythrocytes and thalassemia is caused by defective production of the protein hemoglobin leading to erythrocyte abnormalities. In what became known as the malaria hypothesis, Haldane proposed that erythrocytes of individuals who were heterozygous for the thalassemia gene may be resistant to infection by the *Plasmodium* parasite. In areas where malaria is endemic this would confer a survival advantage, leading to the thalassemia allele being maintained at high frequencies in those populations. The thalassemia heterozygote is therefore fitter than both the wild-type homozygotes, i.e. those without the thalassemia mutation, who will lack the increased protection against malaria, and the thalassemia homozygote, who suffers from severe or fatal anemia.

Allison demonstrated that sickle cell trait confers resistance to *P. falciparum*

Haldane's malaria hypothesis was at the time largely speculative, with no supporting experimental evidence. Indeed, evidence for a link between thalassemia and resistance to malaria only began to appear in the early 1970s, as discussed in the following section. A link between malaria resistance and another hemoglobinopathy, sickle cell anemia, was more quickly established.

In 1954, Allison recognized that the geographical distribution of sickle cell trait coincided with that of malaria (**Figure 7.4**) and went on to demonstrate experimentally that

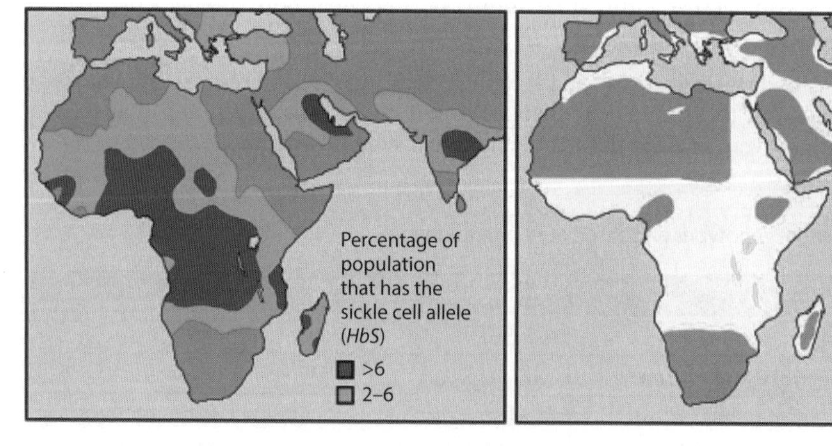

Figure 7.4: Comparison of global distribution of the HbS allele and malaria. The global distribution of the *HbS* sickle cell allele corresponds strikingly with the distribution of *P. falciparum*. This correlation was first recorded in 1954 by Allison, who proposed that the *HbS* allele conferred resistance to *P. falciparum* and went on to demonstrate that red blood cells from *HbS* heterozygotes were resistant to infection by *P. falciparum*. (From Berg JM, Tymokczo JL & Stryer L [2007] Biochemistry, 6th ed. With permission from W. H. Freeman.)

the erythrocytes of individuals who are heterozygous for the sickle cell allele (*HbAS*) are more resistant to infection by *P. falciparum* than wild-type (*HbA*) homozygotes. Allison concluded that "those who are heterozygous for the sickle-cell gene will have a selective advantage in regions where malaria is hyper-endemic." His work provided strong evidence to support the malaria hypothesis, although at the time Allison was unaware of Haldane's earlier work and had reached the same conclusions independently.

More recent studies on the distribution of sickle cell trait indicate that 9% of the population carry the *HbS* allele in parts of Africa stretching from southern Ghana to northern Zambia and a frequency of 18.25% was recorded in northern Angola. A case control study in West Africa in 1991 indicated that children with the *HbS* allele had a 92% reduction in the relative risk for severe malaria compared with those without the *HbS* allele.

Studies on Pacific Island populations provided experimental evidence that thalassemia confers protection from malaria

Although Haldane proposed, as early as 1946, that thalassemias offered protection against malaria, experimental evidence to support his hypothesis was slow to emerge. Studies undertaken in the 1950s and 1960s to investigate the link between the two diseases were inconclusive, partly because it was difficult, with the techniques available at the time, to accurately diagnose the milder forms of thalassemia. Misdiagnosis of thalassemia (either false positives or false negatives) is likely to have skewed the results of these early studies. The advent of molecular diagnostic techniques in the 1970s allowed more accurate diagnosis and more convincing evidence began to emerge. For example, in 1997 the results of a case control study of children in Papua New Guinea showed that the relative risk of developing severe malaria was 0.4 for α^+-thalassemia homozygotes when compared with their non-thalassemic peers. The risk in α^+-thalassemia heterozygotes was 0.7 compared with those without the thalassemia mutation.

The mechanism of resistance to malaria conferred by hemoglobinopathies is still not fully understood

Since the 1950s, numerous studies have investigated the mechanism for malaria resistance associated with thalassemia and sickle cell trait, but no single clear mechanism has emerged for either allele (see **Table 7.3**).

Both the innate and adaptive immune responses are important in destroying free *Plasmodium* parasites and infected cells. *In vitro* studies indicate that *Plasmodium*-infected erythrocytes from α-thalassemia and sickle cell patients may be more efficiently eliminated by the immune response. A range of mechanisms have been proposed, including increased phagocytosis, **complement-mediated lysis**, and **antibody-dependent cytotoxicity**.

A possible mechanism for *Plasmodium* resistance in sickle cell patients involves the *P. falciparum* erythrocyte membrane protein-1 (PfEMP-1), a protein encoded by the *Plasmodium* genome and expressed on the surface of infected erythrocytes. PfEMP-1 is an important antigen that is recognized as part of the antibody-mediated immune response to *Plasmodium*. PfEMP-1 also mediates the binding of infected erythrocytes to nearby uninfected erythrocytes, forming rosettes, and to endothelial cells that line the post-capillary venules. This cell-to-cell binding, known as **cyto-adherence**, causes the infected cells to accumulate in the small blood vessels, thereby evading immune clearance in the spleen and is often linked to development of severe malaria. *Plasmodium*-infected HbAS erythrocytes express lower levels of PfEMP-1 on their surface compared with infected erythrocytes from normal individuals; less PfEMP-1

Table 7.3 Selected examples of hemoglobinopathies that confer susceptibility or resistance to malaria and some of the proposed mechanisms of action.

Variant	Protein affected	Effect	Proposed mechanisms
Hemoglobin S (*HbS* variant of *HBB* gene)	β-globin (Glu6Val substitution)	Reduced risk of severe malaria	Reduced invasion of erythrocytes by *Plasmodium*; reduced growth of *Plasmodium* inside erythrocytes; enhanced phagocytosis and clearance of infected erythrocytes; reduced cyto-adherence of infected erythrocytes
Hemoglobin C (*HbC* variant of *HBB* gene)	β-globin (Glu6Lys substitution)	Reduced risk of severe malaria	Reduced cyto-adherence of infected erythrocytes
Hemoglobin E (*HbE* variant of *HBB* gene)	β-globin (Glu26Lys substitution)	Reduced risk of severe malaria	Reduced invasion of erythrocytes by *Plasmodium*; reduced growth of *Plasmodium* inside erythrocytes; enhanced phagocytosis and clearance of infected erythrocytes
α-thalassemia (variants of *HBA1* and *HBA2* genes)	α-globin (reduced expression of α-globin)	Reduced risk of severe malaria	Reduced growth of *Plasmodium* inside erythrocytes; increased antibody binding to infected cells; reduced rosetting of infected erythrocytes; increased susceptibility to *P. vivax* in infancy, enhancing later immunity to *P. falciparum*; increased number of small erythrocytes means that less hemoglobin is lost per infected erythrocyte
β-thalassemia (variants of *HBB* gene)	β-globin (reduced expression of β-globin)	Reduced risk of severe malaria	Increased antibody binding to infected cells; enhanced phagocytosis

means less cyto-adherence, offering a clear mechanism for protection against severe malaria. It has also been suggested that the polymerization of hemoglobin in HbAS erythrocytes may limit the growth of the *Plasmodium* parasite within the cells.

Rather surprisingly, in 1996 Williams et al. found that α^+-thalassemia children on the Pacific island of Espiritu Santo were more likely to suffer from milder forms of malaria compared with non-thalassemic children. The effect was particularly significant in younger children and for malaria caused by *P. vivax*. A possible explanation for this apparent paradox is that α^+-thalassemia makes these children more susceptible to infection by the less dangerous *P. vivax* at a very young age, when they are still protected by maternal antibodies, so that they develop immune memory that may protect them from more severe malaria later in life.

Another rather counterintuitive suggestion is that the microcytic anemia experienced by α^+-thalassemic individuals may itself offer protection from more severe malarial anemia. Malarial anemia is caused by destruction of *Plasmodium*-infected erythrocytes. α^+-thalassemic patients have a high number of small circulating erythrocytes, each with a relatively low concentration of hemoglobin. The rationale for the proposed mechanism is that less hemoglobin is lost when a microcytic erythrocyte is destroyed, compared with destruction of a normal erythrocyte. The resulting malarial anemia is therefore less severe.

Resistance to malaria conferred by HbS and thalassemia is a complex genetic trait

It is worth noting that sickle cell anemia and the thalassemias are diseases caused by recessive alleles that follow a classical Mendelian pattern of inheritance. Resistance to malaria conferred by HbS and thalassemia alleles is, however, a complex genetic trait—the alleles increase the probability of resistance to malaria, but do not guarantee it. The polymorphisms that cause HbS and thalassemia therefore conform to the extended definition of complex disease, based on that of Haines and Pericak-Vance (1998), as those where alterations in more than one gene (allele) alone or in concert either increases or decreases the risk of developing a trait (see Chapter 2)—the trait in this case being susceptibility to malaria.

Other malaria resistance alleles have been identified via epidemiological or hypothesis-driven studies

In addition to the thalassemia and sickle cell alleles, a number of other polymorphisms have been identified on the basis of epidemiological studies and existing knowledge of malaria pathogenesis, examples of which are summarized in **Table 7.4**.

GWAS suggest that polymorphisms in immunity-related genes may affect outcome of *Plasmodium* infection

GWAS have revealed significant linkage between the intensity of the outcome following malaria infection and chromosomes 10p15.3–p14 and 13q. Suggested linkage has also been identified for clinical malaria disease with chromosome 5p15 and chromosome 13q13, and for parasite density in asymptomatic infection with chromosome 5q31 and chromosome 12q21. The specific genes involved have yet to be identified, but there is some overlap between these regions and regions linked to asthma, suggesting that immune related genes may be involved. The 5q31–q33 region on chromosome 5 includes many genes involved in the immune response to infection, including those encoding granulocyte colony stimulating factor (CSF2), colony stimulating factor 1 receptor (CSF-1R) and the cytokines IL-3, -4, -5, and -9.

GWAS searching for malaria resistance alleles highlight the challenges of GWAS in African populations

As discussed above, there are practical difficulties in carrying out GWAS in developing countries where malaria is prevalent and particularly in African populations, which have a relatively higher level of genetic diversity and lower level of linkage disequilibrium than seen in other populations. In 2009, a GWAS study carried out by members of the Malaria Genomic Epidemiology Network (MalariaGEN) addressed some of these difficulties. They were able to test their methodology using the known association between HbS and malaria resistance. If the methods were robust there should be a strong association between malarial resistance and a SNP labeled rs334, which causes the critical substitution of valine instead of glutamate in the β-globin chain. The initial study, on children from the Gambia, investigated 500,000 SNPs in 1060 children with severe malaria compared with 1500 controls. Being aware that genetic variation in the African population could lead to false-positive results in the study, the investigators first used principle components analysis (PCA) to determine the population structure of the study subjects and found that the population

Table 7.4 Selected examples of non-hemoglobin polymorphisms that confer susceptibility or resistance to malaria and some of the proposed mechanisms of action.

Variant	Protein affected	Effect	Proposed mechanisms
*FY*O*	Duffy antigen (*P. vivax* receptor protein)	Reduced risk of severe malaria	FY*O Duffy-negative individuals express no Duffy antigen on the erythrocyte surface, therefore *P. vivax* is unable to enter cell; confers 100% protection against *P. vivax*
G6PDH	Glucose 6-phosphate dehydrogenase (enzyme that protects against oxidative stress), reduced expression	Reduced risk of severe malaria	Reduced growth of *Plasmodium* inside erythrocytes; increased cell damage due to oxidative stress; enhanced phagocytosis
CR1	Complement receptor 1 (binds PfEMP-1; mediates rosetting), reduced expression	Reduced risk of severe malaria	Reduced CR1 on surface of uninfected erythrocytes leads to reduced rosetting
TNF-308A and *TNF-376A* (promoter polymorphisms)	TNF-α (pro-inflammatory cytokine)	Increased risk of cerebral malaria	TNF-α induces production of adhesion molecules and pro-inflammatory cytokines; increased levels of TNF-α implicated in development of cerebral malaria
*HLA-B*53*	HLA-B (involved in antigen presentation to cytotoxic CD8⁺ T cells)	Reduced risk of severe malaria	Enhanced presentation of *Plasmodium* antigens

could be divided into four distinct genetic subgroups. This is a clear example of population stratification discussed in the opening chapters of this book. Once aware of the subpopulations, they could be taken into account in the analysis. A total of 19 regions were identified that showed genetic association with malarial resistance, one of which was the *HbS* locus. The association with marker SNPs in the *HbS* region was, however, relatively weak ($P < 4 \times 10^{-7}$) which was puzzling, given the high frequency of the *HbS* allele in the Gambian population and the strong protection that it confers against *P. falciparum*.

Could weak linkage disequilibrium between the *HbS* tagged SNPs and the *HbS* causal SNP in the Gambian population be the reason for the unexpectedly weak association signal? To address this question the investigators needed more detailed genotype information and so they carried out high-resolution association mapping at the *HbS* locus, using **multipoint mapping/imputation**. Imputation analysis is a statistical method used to infer a genotype by combining information on sequence variation and haplotype structure—essentially it is a way of filling in missing genotype information. Once again the investigators recognized that there was considerable local variation in haplotype structure in Africa and therefore they sequenced the *HbS* region in 62 of the subjects and used this as reference data for the imputation analysis, instead of using the HapMap samples from the Yoruba people, which are more commonly used in African GWAS. The result was a far more convincing association signal for the SNP rs334 ($P = 4.5 \times 10^{-14}$).

The MalariaGEN GWAS provides an excellent example of the difficulties of carrying out GWAS in African populations. The Wellcome Trust Case Control Consortium GWAS (2007 study), which investigated seven diseases in the British population, used $P < 5 \times 10^{-7}$ as a significance threshold. Without the multipoint imputation-based fine-resolution mapping described above, the association of malarial resistance with the *HbS* locus would only just meet this level of significance, based on the earlier study, and this is despite the *HbS* allele being present at a high frequency and known to be a strong resistance determinant. This type of approach may now be applied to future GWAS, but it is clear that denser, population-specific genotyping arrays for GWAS or direct genome sequencing are needed for successful studies in Africa and it is hoped that sequencing initiatives such as the 1000 Genomes Project will help to achieve this.

The aim of GWAS in malaria, of course, is not simply to confirm known risk alleles, but, more importantly, to identify novel associations with genes not previously implicated in malaria pathogenesis. The MalariaGEN study identified a number of novel associations, the most significant being with rs1451375, close to *SCO1* that encodes a protein involved in cytochrome oxidase function, and rs7803788, which lies in an intron of *DDC*, the gene for DOPA decarboxylase that is involved in synthesis of dopamine and serotonin. More work is needed, however, to establish whether these and other newly identified loci play a role in the progression of malaria.

7.5 HIV-1

HIV-1 is a retrovirus that infects macrophages and helper T cells ($CD4^+$ T cells). The virus is transmitted by close contact with infected blood and other body fluids, most commonly during sexual intercourse, or via use of contaminated needles either in a healthcare setting, intravenous drug abuse, piercing, or tattooing. It can also be passed from mother to child across the placenta, during birth, and in breast milk. In December 2013, there were 35.3 million people infected with HIV-1 worldwide.

On entering the host, the gp120 glycoprotein on the surface of the virus binds to its primary receptor, the CD4 protein, on the surface of the $CD4^+$ T cells and macrophages. This initial binding event results in a conformational change in the gp120 protein, allowing it to interact with a second receptor, or co-receptor. Further conformational changes in the receptor proteins and viral glycoproteins facilitate fusion of the viral envelope with the cell membrane and the viral nucleocapsid is released into the cytoplasm of the cell. Once inside the cell, the viral genome is uncoated and the RNA genome is used as a template in the synthesis of complementary DNA (cDNA) by the reverse transcriptase enzyme carried in the virus particle. Immediately after this process, a second viral enzyme, integrase, catalyzes the integration of the viral cDNA into the host cell chromosome. Once integrated, the viral genes can be transcribed and translated in order to generate new virus particles, which leave the T cell by budding out from the cell membrane (see **Figure 7.5**).

C-C chemokine receptor 5 (CCR5) acts as a co-receptor for HIV-1 in the early stages of infection

CCR5 is a chemokine receptor, located predominantly on the surface of macrophage and memory T cells. Though not its natural function, CCR5 acts as a co-receptor for HIV-1 in the early stages of infection. At this time the virus is largely limited to these $CCR5^+$ cell types and is referred to as macrophage or M tropic. M-tropic strains of HIV-1 predominate

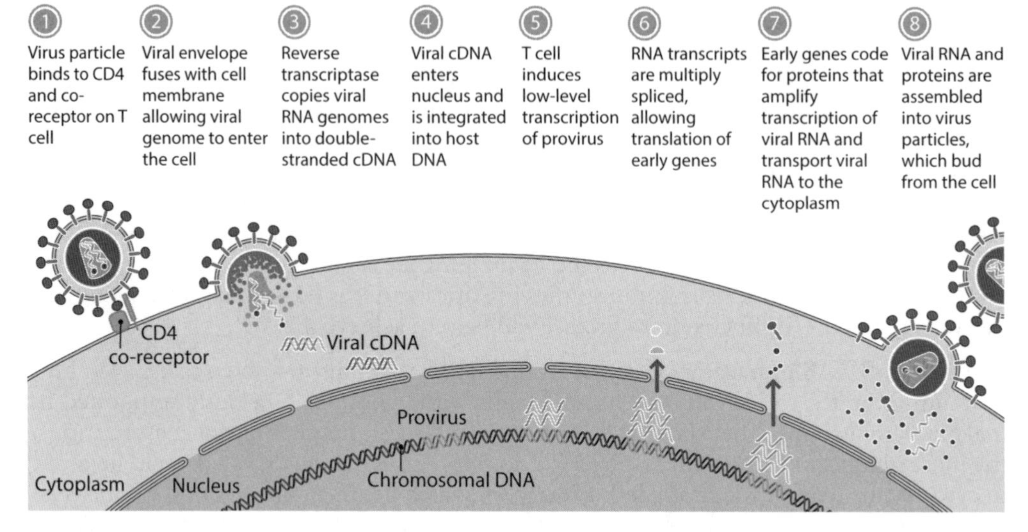

Figure 7.5: The life cycle of HIV. HIV infects predominantly macrophages and T cells expressing CD4 on the cell surface. The virus initially binds to CD4 (primary receptor) and to a co-receptor (CCR5 or CXCR4). The viral envelope then fuses with the cell plasma membrane and the viral nucleocaspid enters the cytoplasm. After uncoating, a complementary DNA (cDNA) copy of the viral RNA genome is synthesized by reverse transcriptase and this cDNA is able to integrate into the host cell chromosome. The integrated genome is transcribed and translated to produce viral proteins, which package newly synthesized viral RNA genomes to form new virus particles that bud from the host cell. (From Strelkauskas AJ, Strelkauskas JE & Moszyk-Strelkauskas D [2009] Microbiology: A Clinical Approach. Garland Science.)

during the asymptomatic phase of infection and are believed to be the strain commonly transmitted by sexual contact. Over time, however, the viral RNA genome accumulates mutations and changes in the gp120 protein alter its binding affinity such that it recognizes the CXCR4 chemokine receptor instead of the CCR5 receptor (**Figure 7.6**). The CXCR4 receptor is found on a wider range of $CD4^+$ T cells, including naïve T cells, and is associated with more rapid depletion of the $CD4^+$ T cell population. The CXCR4-adapted HIV-1 is known as the T-tropic strain and emerges in around 50% of infected individuals. $CD4^+$ T cells are essential in both antibody-mediated and cell-mediated arms of the immune response, so as more T cells become infected and their numbers fall, the host develops deficiency in both antibody and cell-mediated immunity. This acquired immune deficiency syndrome is called AIDS. In most cases, without treatment, infection with HIV-1 is fatal within approximately 10 years. Fatality is not caused by the HIV-1 virus directly, it occurs as a result of overwhelming opportunistic infections and tumors which occur due to the failure of the host immune response, especially $CD4^+$ T cell activity.

Some individuals are naturally resistant to HIV infection

Despite the usually high level of HIV-1 infectivity, some individuals who are repeatedly exposed to HIV-1 are remarkably resistant to infection. Conversely, there are others who are highly susceptible and progress to AIDS within as little as 1 year. Individuals who show either unusual resistance or unusual susceptibility can be divided into four phenotypic subsets:

- Exposed uninfected (EU), also known as exposed seronegative: these people show no sign of HIV infection even after many years of repeated, high-risk exposure to the virus.
- Long-term non-progressors: are infected with HIV-1, but show a delayed progression to AIDS of several decades.

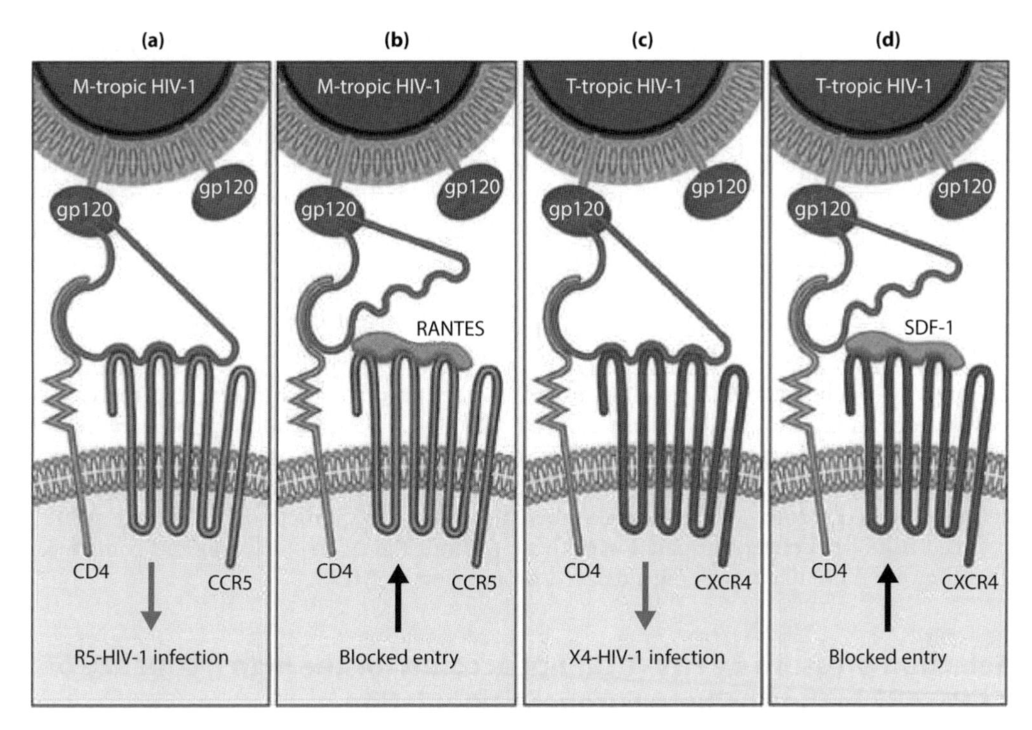

Figure 7.6: Attachment and entry of HIV-1. (a) CCR5 is a chemokine receptor expressed on the surface of macrophages and T memory cells. The gp120 surface protein binds to CD4, inducing a conformational change that allows gp120 to bind to the co-receptor CCR5. Binding to the co-receptor initiates fusion of the viral envelope with the cell membrane allowing the virus nucleocapsid of the M-tropic HIV-1 to penetrate the cell. (b) RANTES chemokine is a natural ligand for CCR5 and has the potential to block attachment of M-tropic strains of HIV-1, thereby inhibiting infection. (c) As infection progresses, accumulated mutations in the gp120 gene lead to a change in binding affinity. The resulting T-tropic strains of HIV-1 bind CXCR4 protein, expressed on the surface of a wider range of $CD4^+$ T cells and macrophages, as a co-receptor instead of CCR5. (d) SDF-1 chemokine is a natural ligand for CXCR4 and has the potential to block attachment of T-tropic strains of HIV-1, thereby inhibiting infection. (From O'Brien SJ & Nelson GW [2004] *Nat Genet* 6:565–574. With permission from Macmillan Publishers Ltd.)

- Fast progressors: are unable to control the virus and develop AIDS in less than 2 years.
- Elite controllers (EC): are infected but able to control the viral replication to an extremely low level (less than 50 genome copies/ml).

We have already seen how variation in the host genome can influence outcome in malaria. Could this spectrum of extreme outcomes following HIV-1 infection be attributed to genetic variation in the human hosts?

A 32-bp deletion in the *CCR5* gene confers resistance to HIV-1 infection

In the 1990s, three research groups, working independently, identified a key genetic variant in EU individuals. Based on knowledge of the HIV-1 life cycle, they proposed that variation in one or more of the HIV-1 receptor proteins could reduce the efficiency with which the virus binds to and penetrates the target cell. 1996 saw a flurry of publications testing this hypothesis: firstly, in early 1996, the newly identified CCR5 chemokine receptor was shown to be the major co-receptor for M-tropic strains of HIV-1. Just a few months later, it was reported that T cells from two EU individuals were resistant to infection by M-tropic HIV-1, but were susceptible to T-tropic strains. This result hinted that CCR5 may be the key to the remarkable resistance of these two subjects and in August 1996 it was reported that the same two EU subjects were homozygous for a 32-bp deletion (Δ32) in the *CCR5* gene (*CCR5*-Δ32). The deletion causes a **frameshift** that leads to a premature truncation of the CCR5 protein such that it is no longer trafficked to the cell membrane. Within a matter of weeks two further publications independently reported that the *CCR5*-Δ32 homozygotes among EU subjects were entirely absent from the HIV-infected populations tested. Further analysis of the HIV-infected subjects compared the genotype of those who showed a delayed progression to AIDS (long-term non-progressors) with those who progressed to AIDS at the usual rate and found *CCR5*-Δ32 heterozygotes were more frequent among the long-term non-progressors.

The overall conclusions at the end of this remarkably productive year were that the *CCR5* polymorphism is a key player in infection by HIV-1. *CCR5*-Δ32 homozygotes are highly resistant to M-tropic strains of HIV-1 because macrophage and T cells in these individuals do not express CCR5 on the cell surface, and so the virus is unable to bind to and enter the cell via the CCR5 route. Further work showed that *CCR5*-Δ32 heterozygotes have a 70% reduced risk of infection compared with those without the Δ32bp deletion and confirmed that they are more likely to show a delayed progression to AIDS.

Selection pressure by HIV-1 cannot account for the high frequency of *CCR5*-Δ32 in the northern European population

The *CCR5*-Δ32 allele is present in approximately 10% of Europeans, but is very rare in African and Asian populations. The deletion has been detected in DNA recovered from skeletal remains up to 2900 years old from various parts of Europe. It seems likely that, at some time in the past, *CCR5*-Δ32 provided a survival advantage to the population, possibly by offering resistance to an infectious pathogen or multiple pathogens. HIV-1 is very unlikely to be responsible, as HIV-1 infection has only been recorded in human populations since the early 1980s—not long enough to exert selection pressure. Numerous epidemics have struck European populations over the last 3000 years and could be implicated, particularly hemorrhagic fever viruses and smallpox. There is circumstantial evidence for a connection between smallpox and *CCR5* polymorphism, e.g. myxoma pox viruses are able to infect cells expressing CCR5 on the cell surface and there is a five-fold reduction in the replication of M-tropic HIV-1 in lymphocytes from subjects vaccinated against smallpox compared with non-vaccinated controls. There are, however, arguments against smallpox being the driving force in selection of *CCR5*-Δ32, including the fact that smallpox has in the past been endemic in parts of India and yet *CCR5*-Δ32 is rare in Indian populations. It cannot be ruled out that selection advantage has not, in fact, been conferred by *CCR5*-Δ32 itself, but by an alternative closely linked allele.

CCR5-Δ32 affects the outcome of infection by West Nile virus

As discussed in the previous section, CCR5 polymorphism plays a key role in infection by HIV-1. The loss of CCR5 expression on the surface of macrophage and T cells, as seen in *CCR5*-Δ32 homozygotes, provides a remarkable level of resistance to HIV-1. CCR5 therefore presents a potential target for therapeutic intervention: could CCR5 be blocked with, for example, anti-CCR5 monoclonal antibodies in order to prevent HIV-1 infection? It is important to remember that, in addition to acting as a co-receptor for HIV, the primary function of CCR5 is to act as a chemokine receptor and help to orchestrate the immune response to infection. Blocking CCR5 action may therefore compromise the immune response. Intriguingly, *CCR5*-Δ32 homozygotes appear be entirely healthy and generally show no signs of immune suppression. This is presumably due to redundancy in the chemokine system, whereby if CCR5 is unavailable then its role is taken on by other chemokine receptors, making CCR5 an attractive therapeutic target. Redundancy in biological systems is very important when considering the genetics of complex disease.

Although not essential to the immune response in general, CCR5 does have an important role in the immune response to West Nile virus (WNV). WNV is a flavivirus transmitted by mosquitoes that pass the virus on to horses, birds, and humans. It was first identified in Africa in 1937 and cases have been reported across the world since then. 80% of WNV infections are asymptomatic and the remaining 20% result in West Nile fever, with patients suffering flu-like symptoms. As many as 1:150 infections result in more severe disease. In these cases the virus infects the brain causing encephalitis, meningitis, or West Nile poliomyelitis. In 1999, WNV cases began to appear in the USA, the virus having been carried by flamingos imported into New York's Bronx Zoo. This developed into a serious epidemic with approximately 20,000 human symptomatic cases and 782 deaths, prompting an increased interest in the biology of WNV.

In 2006, Glass et al. reported a higher frequency of *CCR5*-Δ32 in patients admitted to hospital with severe WNV infection than in uninfected controls. It was apparent that the *CCR5*-Δ32 allele, rather than conferring resistance as in the case of HIV-1, was actually making the patients more susceptible to WNV-related disease. Further work established that the *CCR5*-Δ32 variant had no effect on WNV infection per se, i.e. *CCR5*-Δ32 was found with the same frequency in WNV-infected patients as in non-infected controls. The *CCR5*-Δ32 allele did, however, confer susceptibility to the more severe, neuro-invasive form of WNV-related disease. CCR5 appears to play an important role in the immune response to WNV in the brain. Infected neurons secrete cytokines that bind to CCR5 on the surface of lymphocytes in the peripheral blood, recruiting them to the site of infection. In *CCR5*-Δ32 individuals, who have reduced levels of CCR5 expressed on the cell surface, this signaling process is compromised and insufficient lymphocytes are recruited to the brain, allowing the WNV infection to take hold and cause neurological symptoms (see **Figure 7.7**).

Although WNV is a relatively rare infection, this study does highlight the fact that CCR5 is not an entirely redundant chemokine receptor and that it may be important in the immune response to other, as yet unidentified pathogens. For example, a subsequent study in mice suggested that CCR5 may also be important in defense against Japanese encephalitis virus, another mosquito borne flavivirus. Therapeutic agents that block CCR5 function must therefore be designed and used with caution. In addition, even though there is often redundancy in biological systems that may be sufficient to cope with everyday circumstances, these systems may break down when under stress. The compensation offered by other elements of a pathway, which may normally make up for deficiency in one component of the system, may then be insufficient to prevent morbidity.

CCR5-Δ32 cannot account for all HIV-1 resistance

It is clear that the *CCR5*-Δ32 variant cannot explain all observed HIV-1 resistance. *CCR5*-Δ32 is present at very low frequencies in non-European populations and yet the full spectrum of HIV-1 infection phenotypes, from exposed non-infected to fast progressors, can be observed in all infected populations. Other genetic variants must contribute to the overall heterogeneity in susceptibility to HIV-1 infection and some of these are discussed in the following sections (see **Table 7.5**).

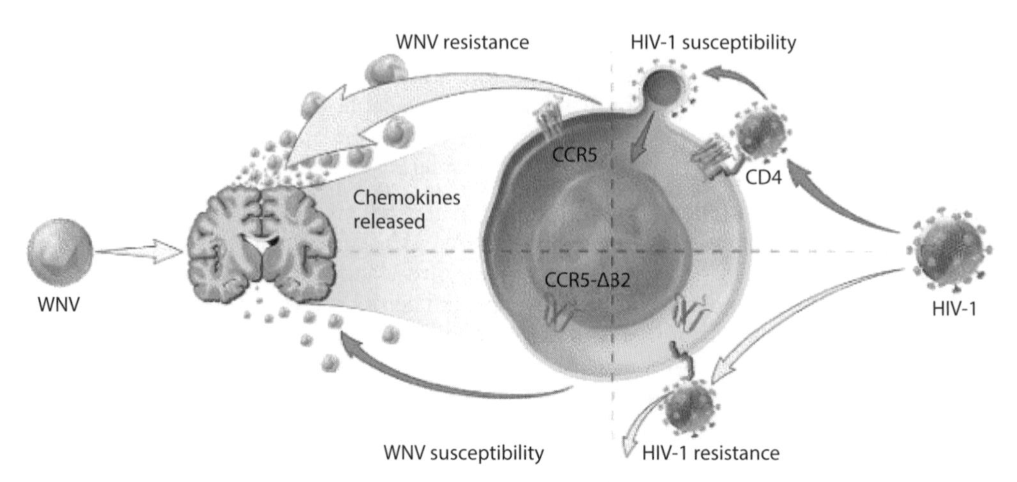

Figure 7.7: The contrasting effect of *CCR5*-Δ32 on infection by WNV compared with HIV-1. CCR5 is a chemokine receptor expressed on the surface of macrophages and some CD4+ T cells. It acts as a co-receptor for HIV-1, facilitating entry of HIV-1 into the host cell. A 32-bp deletion in the *CCR5* gene (*CCR5*-Δ32) inhibits trafficking of CCR5 to the cell surface and this reduces susceptibility to infection by HIV-1. By contrast, *CCR5*-Δ32 increases susceptibility to infection of the brain by WNV. CCR5 appears to play an important role in the immune response in WNV infection, detecting cytokines secreted in response to infection of the brain by WNV and mediating recruitment of lymphocytes to the site of infection. Reduced expression of CCR5 in cells with the *CCR5*-Δ32 allele leads to suppression of the immune response to WNV. (From Lim JK, Glass WG, McDermott DH & Murphy PM [2006] *Trends Immunol* 27:308–312. With permission from Elsevier.)

CCR5 promoter polymorphisms affect HIV-1 control

A number of SNPs in the CCR5 promoter are associated with increased or decreased risk of HIV-1 progression to AIDS. One example, identified as part of the Multicenter AIDS Cohort Study (MACS), is the *CCR5 59029A/G* variant (also referred to in the literature as *-2459 A/G* and *-303 A/G*). In a cohort of homosexual and bisexual men in the USA, those with a *59029G/G* genotype progressed to AIDS on average 3.8 years later than those with the *59029A/A* genotype. When the promoter activity of the *59029G* variant was tested *in vitro* it showed lower activity and lower levels of CCR5 expression on the surface of lymphocytes than the *59029A* variant. These findings suggested that the slower disease progression observed in the *59029G/G* individuals may be the result of reduced transcription of the *CCR5* gene leading to less CCR5 on the surface of the target cells. The same SNP was later reported to influence resistance to HIV-1 infection in *CCR5*-Δ32 heterozygotes. Individuals with just one copy of the *CCR5*-Δ32 allele are able to resist infection and display an exposed-uninfected phenotype if the promoter of the wild-type *CCR5* on the other chromosome has the *59029G* variant, resulting in reduced expression of the one functional *CCR5* allele.

CCR5/CCR2 haplotypes have a complex effect on HIV-1 control

There is strong linkage between *CCR5*-Δ32 SNPs in the CCR5 promoter region and a polymorphism in the *CCR2* gene that encodes an isoleucine for valine exchange at amino acid 64 (*CCR2-V64I*). *CCR2* is another chemokine receptor and the *CCR2* gene is located less than 20 kb upstream of the *CCR5* gene. The CCR2 protein is sometimes, though rarely, used as a co-receptor in HIV infection and there is some evidence that CCR2 may form dimers with CXCR4, thereby reducing the amount of CXCR4 available to bind HIV-1 in later stages of infection. Early studies indicated that *CCR2-V64I* was associated with delayed progression to AIDS in African-Americans, but not in Europeans,

Table 7.5 Some examples of genetic variations in the genes encoding the chemokine receptor and chemokine ligands that influence the outcome following infection with HIV-1.

Variant	Protein affected and role	Effect	Proposed resistance mechanisms
CCR5-Δ32	C-C chemokine receptor 5 (co-receptor for HIV-1 M-tropic strains)	Δ32/Δ32: prevents infection by HIV-1 M-tropic strains; Δ32/+: delayed progression to AIDS	Reduced CCR5 on surface of target cells leads to reduced HIV-1 entry
CCR2 V64I	C-C chemokine receptor 2	Delayed progression to AIDS	Dimerization with CXCR4, reducing CXCR4 available for HIV-1 binding; linkage disequilibrium with CCR5
CCL5 In1.1C	C-C chemokine ligand 5 (RANTES; natural ligand of CCR5)	Accelerated progression to AIDS	Reduced RANTES expression leads to reduced competition with HIV-1 for CCR5 binding sites
CXCL12 3'A (3′ UTR)	C-X-C chemokine ligand 12 (SDF-1; natural ligand of CXCR4)	Delayed progression to AIDS	Increased SDF-1 expression leads to increased competition with HIV-1 for CXCR4 binding sites

but subsequent studies in various populations and ethnic groups have given very mixed results—some indicating protection, some no effect, and some increased risk. It is therefore unclear whether *CCR2-V64I* has a causal effect on AIDS progression or whether the association is due to linkage disequilibrium with the *CCR5* gene.

A key paper published in 1999 by Gonzalez et al. shed some light on the complexity of associations between *CCR5* and *CCR2* genotypes and the outcomes of HIV-1 infection. The paper identified nine **haplogroups**, based on polymorphisms in *CCR2* and *CCR5*, and showed that the frequency of the haplogroups varied considerably between ethnic and racial groups (**Figure 7.8**). Moreover, the effect of the various haplotype combinations on the outcome of HIV-1 infection also varied depending upon ethnicity. For example, HHA and HHA/HHF*2 haplotypes were associated with delayed progression to AIDS in African-Americans but had no effect in European-Americans. Conversely, HHC haplotypes were associated with delayed progression to AIDS in Europeans and European-Americans, but with accelerated progression in African-Americans. To add to the complexity, haplotype pairing is also important. For example, accelerated progression to AIDS in African-Americans with the HHC haplotype could be mitigated if combined with the protective effects of HHA or HHF*2, thus HHC/HHA and HHC/HHF*2 African-Americans showed delayed progression to AIDS. Subsequent studies on the effect of these haplogroups on different populations have added to the complexity of the story but the important message was eloquently summarized by Gonzalez et al., who observed that "analysis of a single mutation or haplotype in isolation may obscure the complexity underlying *CCR5* genotype–phenotype relationships".

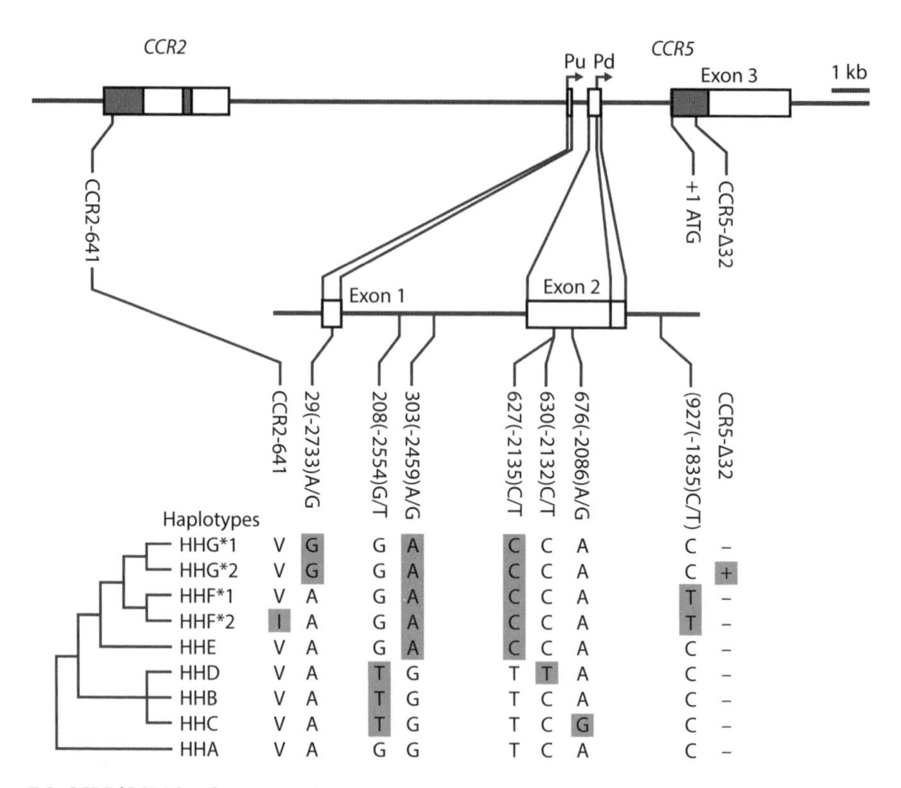

Figure 7.8: CCR5/CCR2 haplogroups. The gene for *CCR5*, a chemokine receptor that also acts as an important co-receptor for HIV-1, is located on the short arm of chromosome 3. The gene for *CCR2* (another chemokine receptor) lies less than 20 kb upstream of the *CCR5* gene. Nine *CCR2/CCR5* haplogroups were identified by Gonzalez et al., based on polymorphisms in the *CCR5* gene, in the promoter region, and in the *CCR2* gene. The effect of each haplogroup on the outcome of HIV-1 infection was shown to vary depending upon the ethnic or racial background of the population studied. (Adapted from Arenzana-Seisdedos F & Parmentier M [2006] *Semin Immunol* 18:387–403. With permission from Elsevier.)

Polymorphisms in chemokine receptor ligand genes influence HIV-1 control

The chemokine receptors CCR5 and CXCR4 have a number of natural chemokine ligands (CCLs). These ligands are chemo-attractant molecules that bind to CCR5 and CXCR4, recruiting CD4$^+$ T cells to the site of an infection and initiating signaling pathways involved in the immune response. The ligands include RANTES (CCL5), MIP-1α (CCL3), MIP-1αP (CCL3L1), and MIP-1β (CCL4), all of which are ligands for CCR5, and SDF-1 (CXCL12), which is a ligand for CXCR4. Increased levels of these natural ligands have been known for some time to compete with HIV-1 for the receptor targets and thus to reduce HIV-1 infectivity both *in vitro* and *in vivo*. Based on this knowledge, the genes that encode the chemokine ligands were investigated as candidate genes in HIV-1 resistance and susceptibility.

Several SNPs in chemokine ligand genes have been identified that associate with HIV-1 resistance or susceptibility. For example, the *In1.1C* SNP within intron 1 of the *RANTES* gene was found to be associated with accelerated progression to AIDS. *In vitro* studies suggest that this SNP confers an altered affinity for nuclear-binding proteins and a four-fold down-regulation of transcription. Therefore, a possible mechanism for the increased rate of progression to AIDS in individuals carrying the *In1.1C* allele is that the production of RANTES is downregulated. Competition for the CCR5 receptor-binding site is consequently reduced and HIV-1 is able to bind to CCR5 and enter the target cell more frequently.

As alluded to in the previous section, however, it is important to consider the effects of these chemokine ligand SNPs not only in isolation, but also in combination with CCR5 receptor polymorphisms. The effect of *CCR5/CCR5*-ligand haplotypes on HIV-1 infection is therefore an important ongoing area of study.

Polymorphisms in HLA genes affect outcome of HIV infection

As discussed in Chapter 6, the key function of the HLA is to present antigenic peptides to $CD4^+$ and $CD8^+$ T cells, leading to the generation of pathogen-specific acquired immunity. HLA works well in this process due to a number of factors: the relatively large number of expressed HLA genes on each haplotype, the extensive polymorphism found within populations, and the fact that each HLA molecule is capable of presenting a range of different antigenic peptides and is not restricted to a single peptide antigen. Thus, it stands to reason that, from the earliest days of HIV-1 research, the HLA genes were likely to be important candidates for resistance and susceptibility to HIV. Polymorphisms in HLA class I alleles were the first to be identified as influencing the outcome of HIV-1 infection (see **Table 7.6**).

HLA class I homozygosity is associated with accelerated progression to AIDS; the effect is observed for all three HLA class I loci (*HLA-A*, *HLA-B*, and *HLA-C*) and is more

Table 7.6 Some HLA class I gene polymorphisms that have been shown to influence the outcome following infection with HIV-1.

Variant	Protein affected and role	Effect	Proposed resistance mechanisms
HLA class I homozygosity	HLA class I (HLA-A, -B, and -C); antigen-presenting protein	Accelerated progression to AIDS	Reduced variety of epitopes presented to cytotoxic $CD8^+$ T cells
HLA class I concordance	HLA class I (HLA-A, -B, and -C); antigen-presenting protein	Increased risk of mother to child and sexual transmission	Escape epitopes in mother (or infected partner) are able to escape cytotoxic T cell response in child (or newly infected partner)
*HLA-B*35-Px*	HLA-B; antigen-presenting protein	Accelerated progression to AIDS	Reduced epitope binding decreases cytotoxic T cell response to HIV-1
*HLA-B*57*	HLA-B; antigen-presenting protein	Delayed progression to AIDS; increased risk of hypersensitivity to abacavir	Enhanced antigen presentation
KIR3DS1 + *HLA-Bw4*	Killer immunoglobulin-like receptor 3DS1 + HLA-B; NK cell activation	Delayed progression to AIDS	Enhanced activation of NK cells leading to improved killing of HIV-1-infected cells

pronounced if two or three loci are homozygous. HLA class I proteins present viral antigens to cytotoxic CD8$^+$ T cells, leading to destruction of the infected cell. Homozygosity in the HLA class I allele genotype means that the individual is able to present a reduced number of viral epitopes compared with a heterozygous individual. Mutation of the HIV-1 genome allows the virus to escape immune surveillance over time and this process may happen more quickly in HLA class I homozygotes, which may explain the faster progression to AIDS. When there is concordance in the HLA genotype of the mother and child, there is also an increased risk of transmission from mother to child. Concordance in this case means that both mother and child share the same HLA genotype, so that **HIV-1 escape epitopes** arising in the mother are also able to escape the cytotoxic CD8$^+$ T cell response in the child. Similarly, a high risk of transmission is reported to be associated with concordance between sexual partners.

HLA class I homozygosity is not always bad news

All HLA-B (and some HLA-A) proteins display one of two motifs called Bw4 or Bw6, and *HLA-Bw4* homozygosity has been associated with delayed progression to AIDS. This apparent contradiction to the accelerated progression described above may be explained by the missing-self hypothesis which describes an important interaction between HLA class I proteins and natural killer (NK) cells during the innate immune response to viral infection. This missing-self hypothesis is not related to the genetic missingness discussed elsewhere in this book. HIV-1 and many other viruses are able to suppress the cytotoxic T cell response by downregulating the expression of HLA class I proteins on the surface of the infected cell. To counter this NK cells are able to detect and destroy infected cells displaying reduced HLA class I expression. This very specific killing relies on interaction between killer immunoglobulin-like receptor (KIR) proteins, which are expressed on the surface of NK cells, and HLA class I proteins on the infected cell. KIRs can be classified as activating or inhibitory. The destruction of an HIV-1-infected cell by an NK cell may be initiated by stimulation of an activating KIR or suppressed by an inhibitory KIR (**Figure 7.9**). Thus, the lack of self-major histocompatibility complex (MHC) class I molecules on the cell surface leads to NK destruction of infected cells – the missing-self

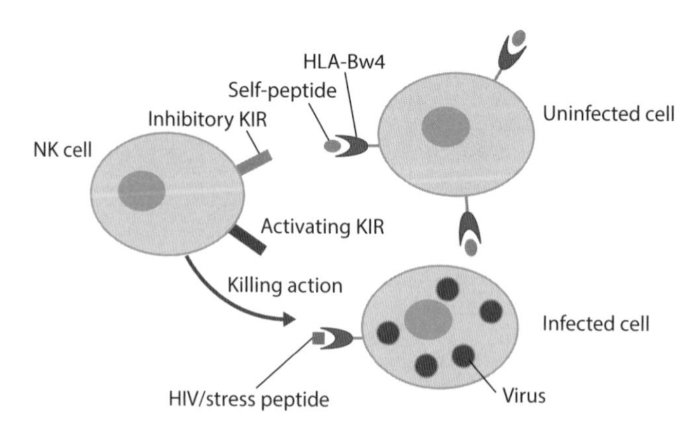

Figure 7.9 Interaction between KIRs on NK cells and HLA-Bw4 on uninfected and HIV-1-infected cells. NK cells express KIRs on the cell surface, which may activate or inhibit NK cell activity in response to ligand binding. This interaction may be modified by the HLA molecule. HLA-Bw4 epitopes are found on some HLA-B proteins, as well as some HLA-A proteins, and act as ligands for both inhibitory and activating KIRs. This process mediates the destruction of HIV-1-infected cells by NK cells and may prevent destruction of non-infected cells. (Adapted from Pelak K, Need AC & Fellay J [2011] *PLoS Biol* 9(11):e1001208. With permission from the Public Library of Science.)

hypothesis. The HLA-Bw4 motif plays a key role in this process, acting as an important KIR-ligand. *HLA-Bw4* homozygosity may therefore offer protection from HIV by enhancing NK control of HIV-1 infection, although the mechanism for this is complex and not yet fully understood.

Other HLA class I polymorphisms consistently associated with delayed progression to AIDS are *HLA-B*57* and *HLA-B*27*. The proposed mechanisms underpinning these associations are thought to be linked to the peptide-presenting role of the HLA class I proteins. *HLA-B*57* is believed to have a broad peptide-binding specificity, which makes it harder for HIV-1 escape mutants to evolve. *HLA-B*27* is known to recognize and present an epitope from the HIV-1 gag protein and escape mutations with this epitope appear to reduce the fitness of the virus. Again, it is therefore difficult for the virus to escape the $CD8^+$ T cell response. It is worth noting that, as discussed in Chapters 6 and 8, *HLA-B*57* is also associated with hypersensitivity to the anti-retroviral drug Abacavir and therefore may also influence the outcome of HIV-1 infection by modifying the patient's response to treatment.

Peptide presentation may not, however, be the whole story. Interestingly, both, *HLA-B*27* and *HLA-B*57* carry the Bw4 motif discussed above, and there is growing evidence that interaction with KIRs may offer an alternative, or complimentary, protective mechanism. The KIR genes themselves are highly polymorphic and a marked protective effect has been reported, e.g. for individuals with high-expressing alleles of *KIR3DL1* in combination with *HLA-B*57*.

*HLA-B*35* is associated with rapid progression to AIDS; the effect appears to vary depending on the ethnicity of the subjects, with significantly faster progression in European and European-Americans, but not in African-Americans. *HLA-B*35* individuals can be divided into two subtypes based on peptide-binding preference: *HLA-B*35*-PY preferentially binds peptides with a proline at position 2 and a tyrosine (Y) at position 9; *HLA-B*35*-Px also binds peptides with a proline at position 2 and has no specific preference (x) at position 9, but will not accept tyrosine. This difference in peptide binding can be attributed to a single amino acid change at position 116 of the *HLA-B*35* allele. Rapid progression to AIDS is associated with *HLA-B*35*-Px, but not *HLA-B*35*-PY. This may explain the ethnic variation in this association, as *HLA-B*35*-PY is the most common *HLA-B*35* variant in African-Americans. It should be noted, however, that there are more than 60 listed *HLA-B*35* alleles and this relationship may be much more complex than first reported.

GWAS confirms the protective role of *HLA-B* in HIV-1 infection

As for malaria, the advent of improved technology has allowed GWAS to be employed in the search for HIV-1 resistance and susceptibility alleles. The first GWAS of HIV-1 was carried out by Fellay et al. in 2007 and reported association between resistance and the HLA complex P5 (*HCP5*) gene on chromosome 6p21.3, within the MHC very close to the *HLA-B* locus where alleles predicted to protect against HIV-1 are encoded. Later work, however, showed that the apparent protective effect of *HCP5* could be attributed to linkage disequilibrium with *HLA-B*57*. The *HCP5* gene (now used as a proxy for *HLA-B*57*) was identified in at least two other GWAS and *HLA-C* was also found to be associated with protection against HIV-1 in at least four GWAS. There was therefore striking reproducibility between several of the GWAS, although this is perhaps less surprising when taking into account the fact that all the studies mentioned above were carried out on Europeans and those with European ancestry. Two GWAS of African and African-American cohorts carried out in 2010 failed to identify any alleles reaching the significance threshold. This once again highlights the challenges of conducting GWAS in the genetically diverse African ethnic populations. As discussed for malaria above, more extensive SNP arrays, the application of imputation analysis, and more detailed sequence data such as that provided by the 1000 Genomes Project, will help improve the success of GWAS in these very important cohorts. As Africa bears by far the largest burden of HIV-1 infection, such developments will be critical in the fight against AIDS.

Amino acids in the HLA-B binding groove are associated with HIV-1 control

A GWAS published in 2010 by the International HIV Controllers Study identified four SNPs associated with control of HIV-1, including marker SNPs for *HLA-B*57:01* and *HLA-C*. The study went on to define the *HLA-B* amino acids most significantly associated with protection to be 67, 70, and 97, with amino acid 97 showing the strongest impact. Amino acids 62 and 63 were also found to be significant. These amino acids are all located in the peptide-binding groove of the HLA-B protein (**Figure 7.10**), which again provides convincing evidence that variation in the specificity and efficiency of peptide presentation may play a central role in HIV-1 control.

GWAS revealed, for the first time, association of HLA-C with HIV-1 control

As discussed above, *HLA-B* and *HLA-C* showed the most consistent association with HIV-1 control in GWAS. However, whereas the *HLA-B* association confirmed previous results from hypothesis-driven studies, the findings regarding *HLA-C* were new. *HLA-C*, like the other HLA class I proteins, presents peptides to CD8$^+$ T cells, but is not down-regulated during HIV-1 infection, and interacts with KIR on NK cells to a greater extent than the HLA-A and HLA-B antigens. The newly associated tagged SNP is located 35bp upstream of the *HLA-C* gene, but is not proposed to be the causal polymorphism. Instead

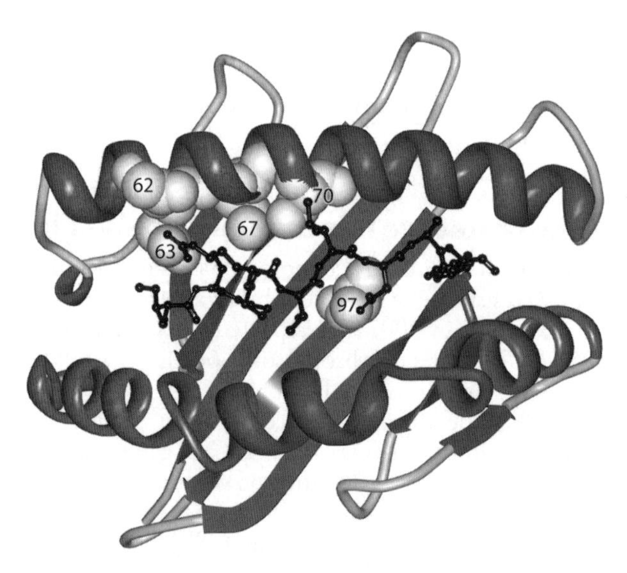

Figure 7.10: A three-dimensional ribbon diagram of the peptide-binding groove of the HLA-B protein. Variations at amino acid positions 62, 63, 67, 70, and 97, all located in the peptide-binding groove, have been shown to affect HIV-1 resistance. Variation in amino acid 97, located in the floor of the peptide-binding groove, has the greatest impact on HIV-1 resistance.

the mechanism of control appears to involve a second SNP, 263D/I, located in a binding site for micro-RNA in the 3′ untranslated region (UTR) of the *HLA-C* gene, which is in linkage disequilibrium with the –35 SNP in Europeans. Binding of micro-RNA destabilizes mRNA transcripts and is an important control mechanism for expression of many genes. The 263D/I polymorphism affects the amount of *HLA-C* expressed on the cell surface and presumably thereby moderates the immune response to HIV-1.

Some SNPs previously implicated in HIV-1 control have not yet been confirmed by GWAS

As polymorphisms in the *CCR5* gene and its **chemokine** ligands are clearly associated with HIV resistance we would predict that association with these alleles would be identified in GWAS. Interestingly, only a weak association with *CCR5/CCR2* was detected and no significant association was seen with any of the other SNPs identified in hypothesis-driven studies (except those in the MHC region). MHC variants have been proposed to account for around 19% of the variability seen in viral load in HIV-1-positive subjects, which rises to 23% when combined with the *CCR5/CCR2* variants. There are clearly a large number of as yet undetermined factors, probably both genetic and environmental, which affect infection by HIV-1 and progression to AIDS.

CONCLUSIONS

Infectious disease, unlike most of the other complex diseases discussed in this book, is not generally considered to be genetic in the common use of the term. However, it does have a heritable component. It is important to remember that infectious disease has played a major role in human evolution and genetic diversity. Indeed, one of the important scientific issues that developed from present studies has been a better understanding of how the pressure from infectious disease on human populations has played a significant role in the evolution of the human genome and also the maintenance of particular alleles that confer resistance to pathogens.

Heritability in infectious disease can be illustrated by considering human history. For example, consider the impact of pathogens carried by the conquistadors on the indigenous populations of South America, and later scientific observations on the apparent heritability of leprosy and tuberculosis. Both of these examples illustrate how the outcome of exposure to a pathogen may be influenced by the host genome. There is an enormous variety of pathogens that can challenge *Homo sapiens*, and new viruses and bacterial strains arise frequently, e.g. the hepatitis viruses (A, B, and C) and HIV-1 are all quite recently emerged human pathogens. In a genetically diverse population when new infectious agents develop, diversity ensures that someone is likely to survive a pandemic, whereas in a less diverse population the likelihood of survival will be reduced.

The genetics of infectious diseases can be more difficult to study than the genetics of non-infectious chronic disease. The causative agents of most infectious diseases are known. In some cases exposure almost always leads to infection, so investigating the role of genes in infection per se is not possible or appropriate. The real points for genetic research often lie with understanding response to infection, progression, outcome, and response to treatment, and not initial susceptibility or resistance to infection, which may be governed by other factors. However, this is not true for all infectious diseases, e.g. HIV-1 and the *CRR5*-Δ32 deletion discussed above.

The same processes that have been applied to non-infectious disease have been applied to infectious disease and with great success in some cases, especially when the right question is set. Substantial existing knowledge of the infectious process has allowed scientists to use a hypothesis-driven approach to identify genes that confer susceptibility or resistance to infectious disease. More recently, non-hypothesis-constrained approaches, particularly GWAS, have been used to identify susceptibility and resistance alleles in infectious disease, and this is shedding new light on the mechanisms by which some pathogens cause disease. There are, however, significant technical challenges to be overcome in carrying out GWAS in non-European and mixed populations, especially in Africa, where the burden of infectious disease is greatest. Despite considerable progress in our understanding of the interplay between the human genome and infectious disease, there are as yet few examples of this knowledge leading to effective preventions or treatments.

By concentrating on two selected examples of infectious disease, this chapter shows how variation in the host genome plays a significant role in determining the disease phenotype. The first example, malaria, illustrates how genetic variation associated with Mendelian recessive traits can influence complex traits. In the second example, HIV-1, we considered how inherited variation in genes can directly influence infection with a virus. In this example we also considered how genetic variation can influence the host response to a virus and briefly how our genes may determine our response to treatment. The last point is covered in more detail in chapter 8. All of this holds a clue to how we may one day conquer many different infections through a better understanding of disease pathogenesis and the development of more efficient treatment programs, including better vaccines where appropriate.

Future use of whole-genome sequencing, epigenetics, and expansion of existing databases using imputation may produce new insights into the genetics of infectious disease. We are making progress towards fulfilling some of the promises of the Human Genome Project that is to improve patient management and care and improve our understanding of disease pathology. However, as we have seen, and shall see again in the next chapter there is still plenty of work to be done.

FURTHER READING

Books

Engleberg NC, DiRita V & Dermody T (2012) Schaechter's Mechanisms of Microbial Disease, 5th ed. Lippincott Williams & Wilkins. An accessible entry-level text book on microbial disease, including chapters on the immune response, retroviruses (including HIV-1), and protozoa (including *P. falciparum*).

Haines JL & Pericak-Vance MA (1998) Overview of mapping common and genetically complex human disease traits. In Approaches to Gene Mapping in Complex Human Disease (Haines JL & Pericak-Vance MA eds), pp 1–16. Wiley. This book provides the best definition of complex disease and is a good starter text on major issues in complex disease.

Articles

Allison AC (2004) Two lessons from the interface of genetics and medicine. *Genetics* 166:1591–1599. This is an interesting autobiographical account of the discovery of the protective role of sickle cell anemia against malaria.

An P & Winkler CA (2010) Host genes associated with HIV/AIDS: advances in gene discovery. *Trends Genet* 26:119–131. This is an excellent overview of the field that includes a discussion of genome-wide RNA interference screens, not included in this chapter.

Arenzana-Seisdedos F & Parmentier M (2006) Genetics of resistance to HIV infection: role of co-receptors and co-receptor ligands. *Semin Immunol* 18:387–403.

Carrington M & Walker B (2012) Immunogenetics of spontaneous control of HIV. *Annu Rev Med* 63:131–145. This is a clear and readable review on the effect of HLA genotype on response to HIV.

Chapman SJ & Hill AVS (2012) Human susceptibility to infectious disease. *Nat Rev Genet* 13:175–188. A comprehensive overview of recent developments in the field; includes a discussion of single-gene variants associated with susceptibility and resistance to infectious disease, and a very useful glossary of terms.

Driss A, Hibbert JM, Wilson NO et al. (2011) Genetic polymorphisms linked to susceptibility to malaria. *Malaria J* 10:271–281. An overview of polymorphisms associated with susceptibility and resistance to malaria.

Fellay J, Shianna KV, Ge D et al. (2007) A whole-genome association study of major determinants for host control of HIV-1. *Science* 317:944–947.

Glass WG, McDermott DH, Lim JK et al. (2006) CCR5 deficiency increases risk of symptomatic West Nile virus infection. *J Exp Med* 203:35–40.

Gonzalez E, Barnshad M, Sato N et al. (1999) Race-specific HIV-1 disease-modifying effects associated with *CCR5* haplotypes. *Proc Natl Acad Sci USA* 96:12004–12009. This is a key study that first defined and investigated *CCR2–CCR5* haplogroups and their influence on HIV progression to AIDS.

Hill AVS, Allsop CEM, Kwiatkowski D et al. (1991) Commonest African HLA antigens are associated with protection from severe malaria. *Nature* 352:595–600.

International HIV Controllers Study (2010) The major genetic determinants of HIV-1 control affect HLA class I peptide presentation. *Science* 330:1551–1557. This paper reports a major collaborative GWAS that implicates *HLA-C* and the binding groove of *HLA-B* in the response to HIV-1 infection.

Jallow M, Teo YY, Small KS et al. (2009) Genome-wide and fine-resolution analysis of malaria in West Africa. *Nat Genet* 41:657–665. This paper describes a GWAS study carried out by members of the MalariaGEN consortium that provides an excellent example of the challenges of GWAS in African populations.

Lim JK, Glass WG, McDermott DH & Murphy PM (2006) CCR5: no longer a "good for nothing" gene – chemokine control of West Nile virus infection. *Trends Immunol* 27:308–312. This is an interesting review of the role of CCR5 in the control of WNV infection.

Lopez C, Saravia C, Gomez A et al. (2010) Mechanisms of genetically based resistance to malaria. *Gene* 467:1–12. A review of potential mechanisms by which polymorphisms associated with susceptibility and resistance to malaria exert their effects.

Vannberg FO, Chapman, SJ & Hill AVS (2011) Human susceptibility to intracellular pathogens. *Immunol Rev* 240:105–116. A review focused on intracellular pathogen genetics that provides an interesting range of pathogens beyond the examples given in this chapter.

Williams TN, Maitland K, Bennett S et al. (1996) High incidence of malaria in alpha-thalassaemic children. *Nature* 383:522–525.

Online sources

http://www.who.int
The WHO provides up-to-date information about the infectious diseases discussed in this chapter.

http://www.malariagen.net
MalariaGEN Genomic Epidemiology Network. A data-sharing community focusing on genes involved in resistance to malaria.

http://www.1000genomes.org
An ambitious project run by the European Bioinformatics Institute that aims to find most genetic variants that have frequencies of at least 1% in 27 populations worldwide.

http://www.hivcontrollers.org
A collaborative study that aims to investigate why some people infected with HIV are able to control the infection without medication.

http://statepi.jhsph.edu/macs/macs.html
Multicenter AIDS Cohort Study.

CHAPTER 8

Pharmacogenetics

Individuals vary considerably in their response to prescribed drugs and other foreign compounds. Genetic factors have been shown to make an important contribution to this variability. The subject of pharmacogenetics is concerned with this relationship. Though considered to be a specialized discipline, pharmacogenetics can also be considered to be a sub-branch of complex disease genetics, because even though many pharmacogenetic traits behave as Mendelian traits, some are complex traits with low levels of penetrance and more than one risk allele. In this chapter, we will concentrate on genetic factors that affect either drug metabolism or the interaction of the drug with its target within the body and also with adverse drug reactions, which are unwanted and unexpected effects of a prescribed drug that can sometimes be life-threatening to the patient. A number of different drug classes will be included in the examples we consider, but we will not consider pharmacogenetic aspects of individual physiological systems or deal with particular medical specialties. Pharmacogenetics can also be extended to other areas not concerned with response to prescribed medicines such as genetic susceptibility to diseases where chemical exposure is a risk factor, including cancer. This area will be considered briefly in the Chapter 9. Similarly, the cancer genome as opposed to the patient genome may help predict the outcome of drug treatment in cancer. This specialized aspect of pharmacogenetics will also be considered separately in Chapter 9.

8.1 DEFINITION AND A BRIEF HISTORY OF PHARMACOGENETICS

The term pharmacogenetics has been in use since the 1950s and can be defined broadly as the study of genetic factors affecting the response to drugs. **Box 8.1** summarizes the development of the subject from the early twentieth century to the present day. There is considerable overlap between pharmacogenetics and the more recent discipline of **pharmacogenomics** and both terms are used interchangeably. Pharmacogenetics has traditionally been concerned with single-gene effects, whereas pharmacogenomics is described as the whole-genome application of pharmacogenetics. In this chapter, the term pharmacogenetics will be used to cover both genetic and genomic examples.

BOX 8.1 MILESTONES IN PHARMACOGENETICS

1902	Garrod discusses chemical individuality in relation to the disease alkaptonuria.
1932	Synder shows the ability to taste the bitter-tasting compound phenylthiocarbamide is genetically inherited.
1952–1956	The antimalarial drug primaquine shown to give rise to acute hemolysis in some individuals who lack the enzyme glucose 6-phosphate dehydrogenase.
1953–1960	Metabolism of the antituberculosis drug isoniazid shown to vary between individuals. Those who convert the drug slowly to acetylisoniazid are more likely to develop peripheral neuritis, with this trait being inherited recessively.
1956–1957	The adverse drug reaction succinylcholine apnea, where rare individuals given this muscle-relaxing drug develop breathing difficulties, shown to be due to an inherited defect in the enzyme pseudocholinesterase, which metabolizes the drug.
1959	Term "pharmacogenetics" first used by Friedrich Vogel.
1975–1977	Working separately, researchers in the UK and Germany show that some individuals are unable to oxidize the drugs sparteine and debrisoquine.
1980	Genetic polymorphism in metabolism of 6-mercaptopurine by thiopurine methyltransferase described.
1984	New pharmacogenetic polymorphism affecting mephenytoin metabolism described.
1988	*CYP2D6* cloned and genetic polymorphisms responsible for poor metabolism of sparteine and debrisoquine identified.
1993	Gene duplication and other amplifications of the *CYP2D6* gene shown to occur in an early example of CNV. This results in increased rates of drug metabolism (ultrarapid metabolism).

BOX 8.1 MILESTONES IN PHARMACOGENETICS (*Continued*)

1994	*CYP2C19* gene shown to code for the enzyme that metabolizes mephenytoin and gene defects identified.
1995	Thiopurine methyltransferase gene cloned and polymorphisms responsible for absence of activity identified.
1997	Term "pharmacogenomics" first used.
1998	Trastuzumab (Herceptin®) licensed as a targeted treatment for breast cancer.
2001	First draft of the Human Genome Map.
2004	*VKOR* gene cloned. Polymorphism in this gene is a key determinant of warfarin dose requirement.
2007	First GWAS on complex diseases are followed by GWAS on pharmacogenetic traits.
2008	Clinical trial shows *HLA-B*57:01* to be a highly sensitive predictor of abacavir hypersensitivity. Genotyping for this allele prior to treatment becomes common.
2009	*IL28B* genotype shown to be key predictor of interferon-α response in hepatitis C infection.

8.2 CYTOCHROME P450

There are two stages in drug metabolism: phase I and phase II. In phase I, reactions are carried out by monooxygenases that work by adding an oxygen atom. The effect is to make it easier for the second phase to be carried out. Phase II reactions involve transferase enzymes that conjugate (add) small polar molecules such as glucuronic acid, sulfate, or glycine to the drug. Phase I and phase II reactions are illustrated in **Figure 8.1**.

Early studies on drug metabolism showed that many phase I metabolic reactions occur in the endoplasmic reticulum (ER) of cells in the liver. In 1962, Omura and Sato isolated a protein they named **cytochrome P450** from rat liver ER. This protein showed an absorbance peak at 450 nm and contained a molecule of heme. Further studies confirmed that this protein was an enzyme with a role in the metabolism of a range of drugs. It was initially assumed that cytochrome P450 was a single enzyme, but evidence soon emerged to suggest multiple cytochrome P450s, with at least 50 different human cytochrome P450s.

There is a clear relationship between genotype and phenotype for several forms of cytochrome P450

In parallel with continuing biochemical studies on cytochromes P450 in the 1970s, studies on the drugs debrisoquine and sparteine, showed that approximately 10% of Europeans could not metabolize these drugs. The term **poor metabolizer** was used to describe this phenotype. The deficiency in metabolism of both drugs was found to co-segregate in families and the trait was found to be inherited recessively. Subsequently, the enzyme responsible for metabolism of these two drugs was shown to be a member of the cytochrome P450 family of enzymes, now referred to as CYP2D6. This is one of the most important

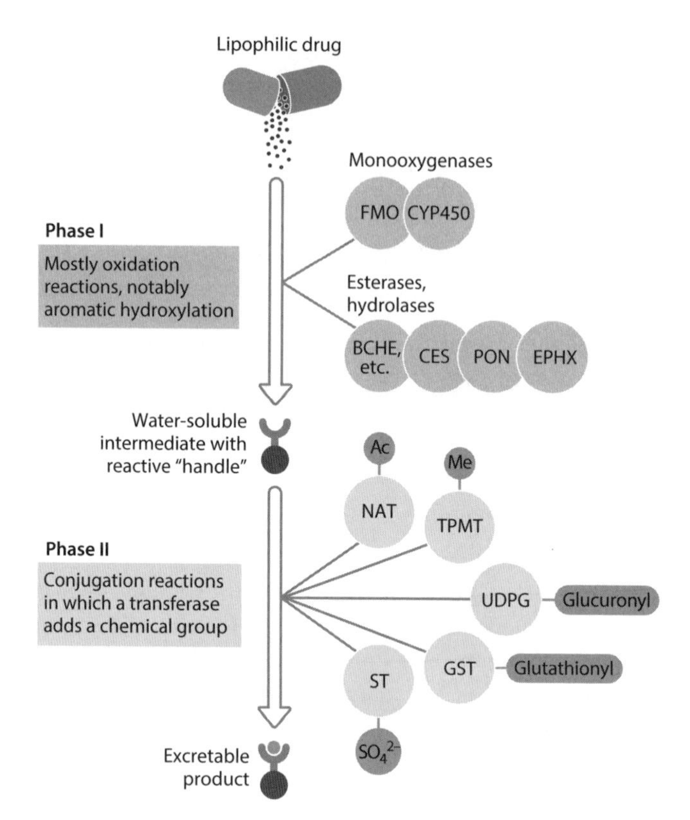

Figure 8.1: Phase I and phase II drug metabolism. Phase I drug reactions create more polar drugs with a reactive group that prepares the drug for phase II. They involve a variety of enzymes, including CYP450. The phase II reactions involve addition of chemicals to the molecule to facilitate excretion of the final product. Note that phase II reactions do not always occur after phase I reactions; they can occur without a phase I reaction. BCHE, butyrylcholinesterases; CES, carboxyesterase; EPHX, epoxide hydrolases; FMO, flavin-containing monooxygenases; GST, glutathione *S*-transferases; NAT, *N*-acetyltransferases; PON, paraoxonases; TPMT, thiopurine methyltransferases; UGT, UDP-glucuronosyltransferases; ST, sulfotransferases. (From Strachan T, Goodship J & Chinnery PF [2014] Genetics and Genomics in Medicine. Garland Science.)

cytochrome P450s involved in drug metabolism, contributing to metabolism of around 20% of all of the drugs metabolized by the cytochrome P450 family. A large number of different polymorphisms in *CYP2D6* associated with the poor metabolizer phenotype have been described, but the most common variant involves a base substitution at an intron/exon boundary that affects RNA splicing, leading to production of a truncated protein with no enzyme activity. In addition to the poor metabolizer phenotype, there is also an **ultrarapid metabolizer** phenotype associated with higher than normal CYP2D6 activity due to the presence of one or more additional copies of the *CYP2D6* gene adjacent to the wild-type gene in germ-line DNA. The finding of extra copies of *CYP2D6* in germ-line DNA was one of the first reports of copy number variation (CNV) in the human genome. It is particularly significant that copy number is related to function in this example as this is not always the case when CNV is observed in other genes. Additional polymorphisms in the *CYP2D6* gene that give rise to amino acid changes that decrease, but do not abolish, enzyme activity have been identified. Individuals homozygous for these alleles or heterozygous for both an absence of activity allele and a decreased activity allele fall into the category of **intermediate metabolizers**. As shown in **Figure 8.2**, there is a clear relationship between drug-metabolizing activity for the antidepressant nortriptyline and the number

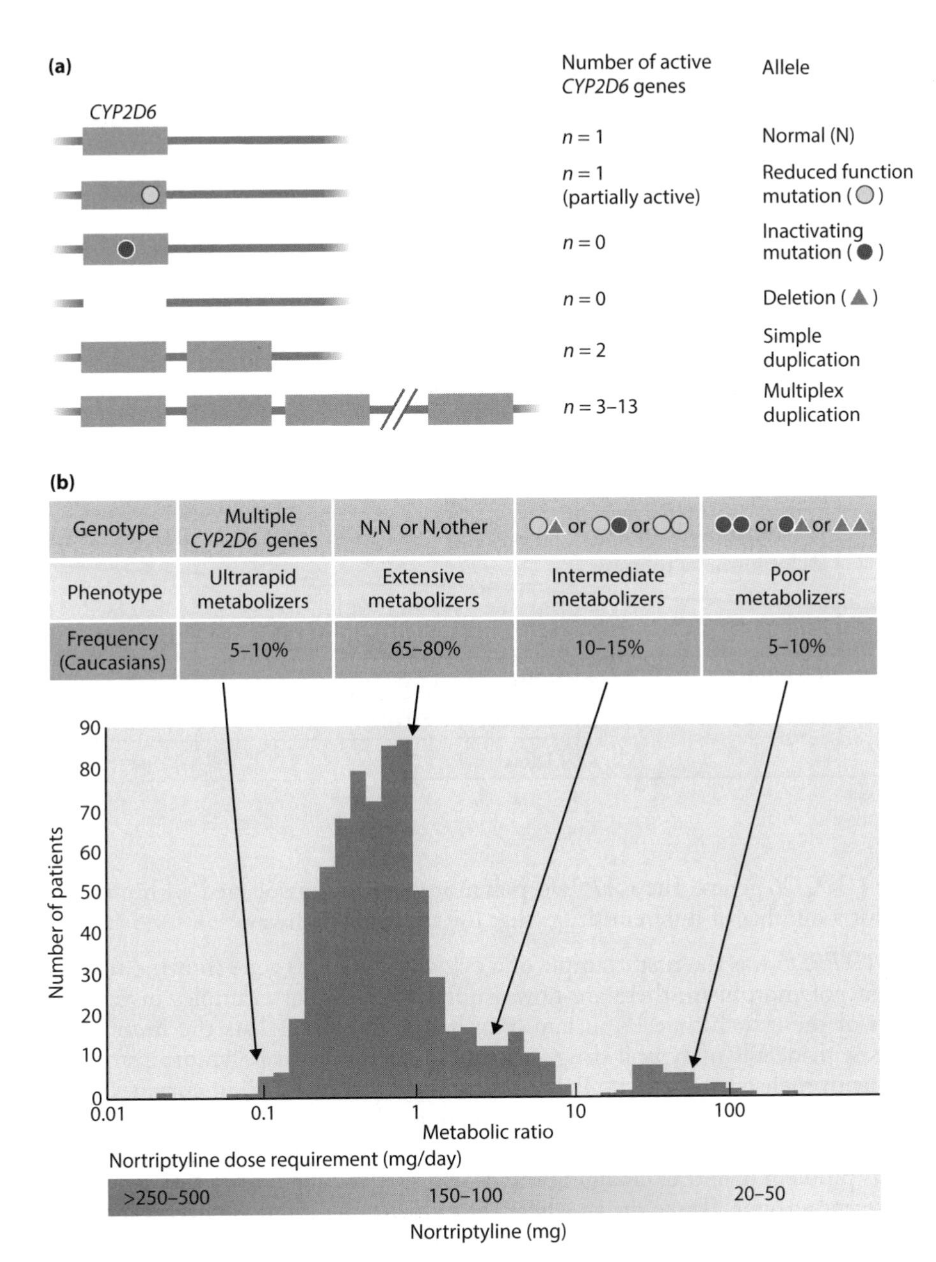

Figure 8.2: CYP2D6 alleles and relationship between genotype and metabolism of the antidepressant drug nortriptyline. (a) CNV in *CYP2D6* genes is common in some populations and creates different phenotypes according to the copy number. There are also deletions, inactivation, and reduced function mutations. (b) CNV results in four phenotypes: ultrarapid metabolizers, extensive metabolizers, intermediate metabolizers, and poor metabolizers. The figure shows urinary concentration of the drug substrate, in this case the antidepressant nortriptyline, with high metabolic ratios representing poor metabolizers and low metabolic ratios representing ultrarapid metabolizers. Extensive and intermediate metabolizers are shown in between these two extremes. (Adapted from Meyer UA [2004] *Nat Rev Genet* 5:669–676. With permission from Macmillan Publishers Ltd.)

Table 8.1 Properties of selected cytochromes P450 important in human drug metabolism.

Isoform	Substrates	Examples of polymorphisms (including rs number where appropriate)	Effect of polymorphism
CYP2D6	Debrisoquine, tricyclic antidepressants, codeine	Splice site (rs3892097)	No activity
		Deletion	No activity
		Missense (rs165852)	Decreased activity
		Duplication	Increased activity
CYP2C19	Clopidogrel, diazepam, omeprazole	Splice site (rs4244285)	No activity
		Promoter region base substitution (rs12248560)	Increased activity
CYP2C9	Warfarin, diclofenac, ibuprofen	Missense (rs1057910)	Decreased activity
CYP3A4	Midazolam, nifedipine, cyclosporin, tacrolimus	Missense (rs55785340)	Decreased activity
		Intronic base substitution (rs35599367)	Decreased activity
CYP3A5	Tacrolimus	Splice site (rs776746)	No activity
CYP2A6	Nicotine	Missense (rs1801272)	No activity
		Deletion	No activity

of active *CYP2D6* genes. The *CYP2D6* polymorphisms are associated with metabolizing activity for a number of different drugs that use the same pathway.

Though *CYP2D6* was the first example of a cytochrome P450 gene showing functionally significant polymorphism, there are now a number of similar examples involving other members of the cytochrome P450 family of genes. **Table 8.1** lists the most important examples with details of typical drug substrates and the main polymorphisms involved together with their overall effects. Polymorphisms in genes encoding enzymes that affect drug metabolism such as cytochrome P450 may result in circulating drug levels that are either too low or too high to achieve the normal therapeutic response. It is possible to overcome this problem by either changing the dose or prescribing a drug that is metabolized by a different enzyme. Three drugs where the cytochrome P450 genotype appears to be clinically relevant are considered in more detail below: codeine, warfarin, and clopidogrel.

The conversion of the analgesic drug codeine, which is administered as a pro-drug and is activated to morphine by CYP2D6, is of clinical importance

Codeine is used widely as an analgesic, but it is a **pro-drug** that needs to be converted to morphine by CYP2D6 for the analgesic effects to occur. This means that it is generally not an effective analgesic in those who lack CYP2D6 activity. This is illustrated in Figure 8.2

in relation to a different drug (nortriptyline) that is metabolized by this same enzyme. Studies of this phenomenon for codeine have been relatively limited, partly because it is difficult to accurately define the severity of pain. However, there are a number of recent reports that suggest that in addition to lack of or low levels of CYP2D6 activity being a problem when codeine is administered individuals who have additional *CYP2D6* copies (ultrarapid metabolizers) may have high levels of CYP2D6 and hence high levels of morphine. Under certain conditions high levels of morphine generated in those with multiple copies of the *CYP2D6* gene may have significant consequences, with serious, even potentially fatal, adverse outcomes when codeine is administered. This seems to be a particular problem with infants and children, though there are also some reports of adverse reactions in adults. One relatively recent study from Koren et al. (2006) in Canada described the case of a breast-fed baby who died 13 days after birth. Further investigations found that the mother's breast milk, given to the baby, contained a high level of morphine. This high level of morphine was generated by elevated CYP2D6 activity in the mother. Genetic investigations found that the mother carried more than one copy of the *CYP2D6* gene and was therefore an ultrarapid metabolizer for codeine. The baby had a normal *CYP2D6* genotype. Subsequently there were a number of similar reports of serious problems following treatment with codeine where either children or adults were ultrarapid metabolizers.

The cytochrome P450 CYP2C9 metabolizes warfarin - a very widely used drug

Warfarin, which was developed originally as a rat poison, is an anticoagulant drug that acts by inhibiting the enzyme vitamin K epoxide reductase (VKOR) (see Section 8.4). Vitamin K is essential in the activation of several factors in the blood coagulation cascade. If VKOR is unable to regenerate reduced vitamin K from vitamin K epoxide produced during the clotting process (see **Figure 8.3**), the normal coagulation process is disrupted. In patients who are at risk of developing blood clots due to some types of cardiovascular disease, disrupting coagulation can be help prevent the formation of clots that might lead to strokes and heart attacks. However, if there is too much inhibition of VKOR, the patient is at risk of developing uncontrolled and potentially fatal bleeding due to the inability of blood to clot. This means that the dose of warfarin given to the patient needs to be titrated by monitoring the drug response on a regular basis and especially during initial treatment. The dose needed by different patients to achieve the required level of anticoagulation varies.

As shown in Figure 8.3, warfarin is metabolized to the inactive 7-hydroxywarfarin by the cytochrome P450 CYP2C9. Two common polymorphisms which each result in amino acid substitutions have been detected in the *CYP2C9* gene. Individuals who have one or two of the variant alleles show slower than normal warfarin metabolism. Some years ago, it was demonstrated that patients being treated with warfarin who were positive for one or more *CYP2C9* polymorphisms generally needed lower doses of warfarin than patients who were homozygous wild-type. This finding has been confirmed in a large number of subsequent studies and there is general agreement that *CYP2C9* genotype is an important determinant of warfarin dose requirement, especially in those of European or African ethnicity.

CYP2C19 activates clopidogrel - a drug widely used to prevent strokes and heart attacks

Clopidogrel is a drug used in the prevention of strokes and heart attacks in high-risk patients, especially those who have already suffered such events and have undergone

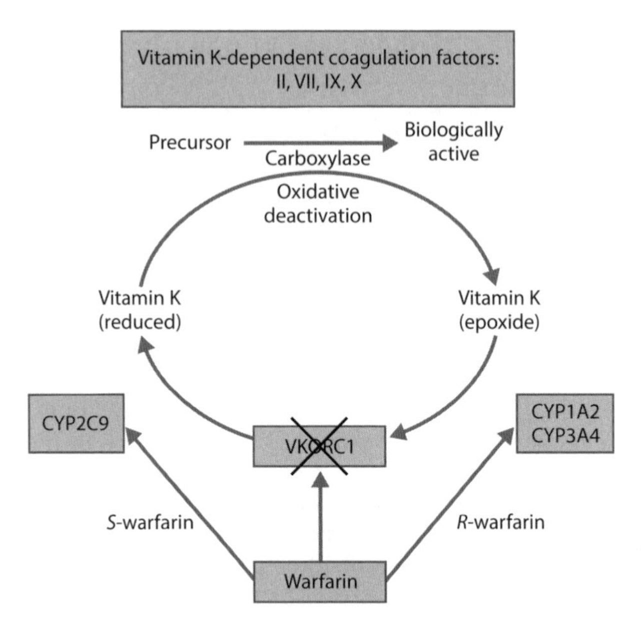

Figure 8.3: The mechanism of action of warfarin and role of VKORC1 in the pathway. The role of VKORC1 in the conversion of vitamin K epoxide to oxidized vitamin K is shown. The carboxylase enzyme transfers carboxyl groups to glutamic acid residues of selected clotting factors with a requirement for oxidized vitamin K. Warfarin inhibits the VKORC1 reaction. There are two chemically distinct enantiomers of warfarin: the *R* and *S* forms. The *S* enantiomer is more biologically active than the *R* form and is also metabolized only by CYP2C9, whereas other P450 isomers metabolize the *R* form.

arterial stenting. Cytochrome P450 genotype may be important in predicting individual responses to clopidogrel, which is a pro-drug that needs to be metabolized to have biological activity. One of the enzymes involved in this metabolism is the cytochrome P450 CYP2C19. CYP2C19 activity is completely absent in around 3% of white Europeans and 20% of people from East Asia, who are often referred to as poor metabolizers. CYP2C19 deficiency is an autosomal recessive trait that is most commonly due to a splice site mutation resulting in a truncated protein with no enzyme activity (Table 8.1). A number of studies suggest that patients treated with clopidogrel who are homozygous for *CYP2C19* variant alleles associated with absence of activity have an increased risk of death and myocardial infarction, though not all studies agree on this. There is also more limited data suggesting that those carrying one or more copies of the *CYP2C19* allele *CYP2C19*17* (ultrarapid metabolizers) are at increased risk of suffering bleeding when treated with clopidogrel due to higher than normal circulating levels of the activated drug.

The practical consequences of these findings are such that the US Food and Drug Administration (FDA) now refers to the *CYP2C19* polymorphism on their approved label for clopidogrel. Those homozygous for *CYP2C19* poor metabolizer alleles could be given an alternative drug as there are several drugs with similar properties to clopidogrel that do not require activation by CYP2C19.

These examples provide a direct link to the promises of the Human Genome Project (HGP), where it was suggested that we would in future use genetics to develop individualized therapies for patients and in patient management and care. Several of the findings described above, however, predate the publication of the first complete human genome sequence.

8.3 OTHER DRUG-METABOLIZING ENZYMES AND TRANSPORTERS

Though the cytochrome P450 enzyme family is perhaps the most important enzyme family in drug metabolism, other enzyme families also contribute. Several genetic polymorphisms affecting drug metabolism by conjugation (phase II metabolism) have been well studied.

In addition to phase I and phase II, several proteins known as drug transporters affect entry and efflux of drugs into a variety of cell types, including enterocytes, hepatocytes, and renal epithelial cells. These contribute to drug absorption, distribution and excretion within the body, and are also subject to genetic polymorphisms relevant to drug treatment. The overall effects of these polymorphisms are on drug levels, broadly similar to those described for the cytochromes P450 in Section 8.2.

For phase II conjugation reactions, the UDP glucuronosyltransferase family makes the largest contribution

The UDP glucuronosyltransferases (UGTs) are one of the major players in phase II conjugation reactions. As with the cytochrome P450 family, a number of polymorphisms with functional effects have been detected in the UGT gene superfamily, but one of the most studied is the *UGT1A1* gene. In addition to a role in drug metabolism, *UGT1A1* contributes to metabolism of endogenous compounds by conjugating bilirubin, produced during heme degradation with glucuronic acid, which increases its solubility and facilitates biliary excretion. Bilirubin is a toxic compound whereby higher than normal plasma levels result in jaundice, which can lead to toxic effects in a number of organs including the brain. Approximately 5% of Europeans show mildly elevated plasma bilirubin levels because of slower than normal conjugation with glucuronic acid in a condition called Gilbert's syndrome, which is generally considered benign. This syndrome is due to lower than normal levels of UGT1A1 activity. This decreased activity is associated with homozygosity for a T > A insertion in the promoter region of the *UGT1A1* gene close to the TATA box resulting in lower than normal transcriptional activity. UGT1A1 has an important role in the metabolism of irinotecan—an anticancer drug used in the treatment of certain forms of advanced cancer, especially colon cancer and lung cancer. Individuals homozygous for the *UGT1A1* promoter region polymorphism have been demonstrated to be more likely to develop neutropenia, a serious adverse reaction where **neutrophil** counts are very low, if treated with irinotecan. In the USA, it is recommended by the FDA that patients undergoing chemotherapy with irinotecan should be genotyped for this promoter region polymorphism and a lower dose of the drug used in those homozygous for this variant.

Methyltransferases are also important in phase II drug metabolism

Conjugation with a methyl group is another type of phase II drug metabolism reaction. It does not occur as commonly as conjugation with glucuronic acid and there are a number of different methyltransferase enzymes that carry out these reactions. One of the most frequently studied pharmacogenetic examples from this group is thiopurine methyltransferase (TPMT), which has a rather specialized role in the conjugation of 6-mercaptopurine. 6-mercaptopurine is commonly used in the treatment of childhood leukemia. Azathioprine is a pro-drug precursor of 6-mercaptopurine and is quite widely prescribed for the treatment of common diseases with an immune etiology, such as inflammatory bowel disease,

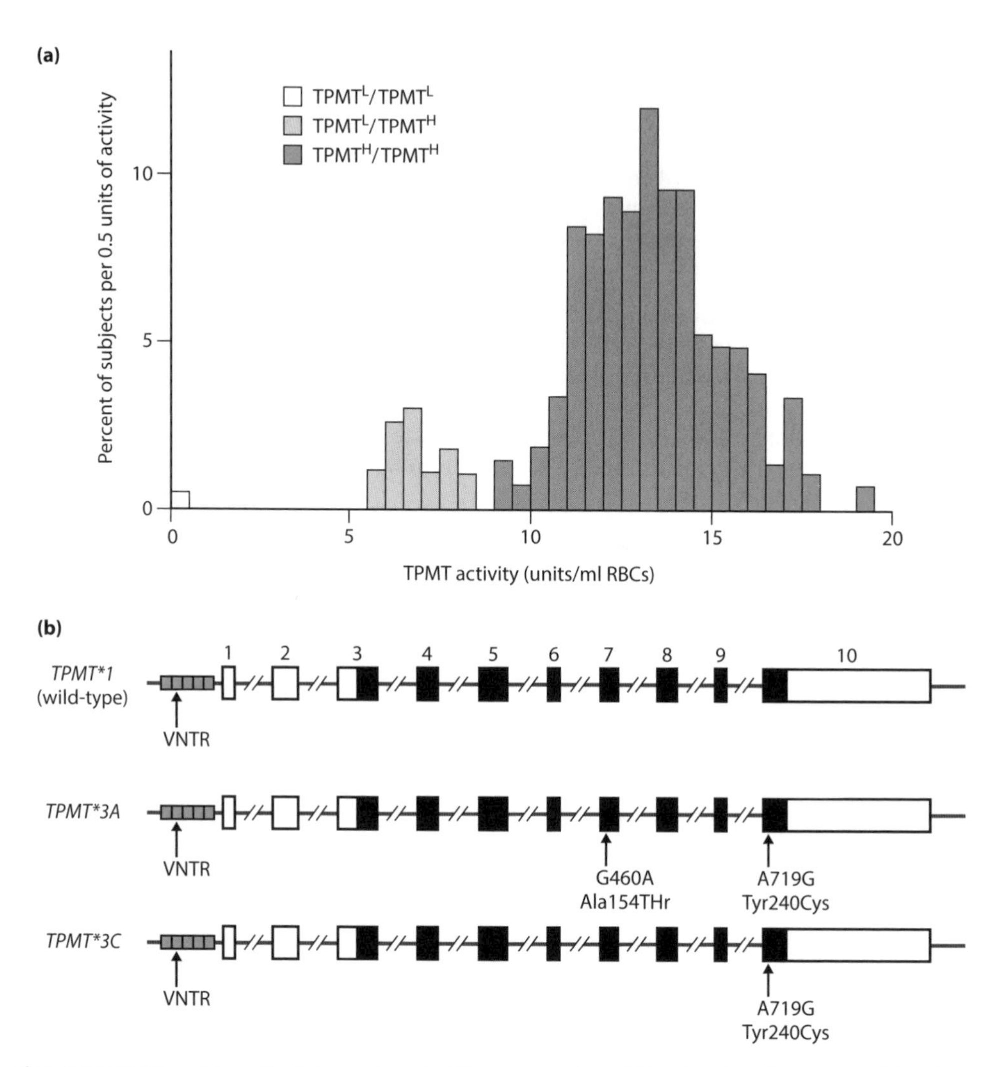

Figure 8.4: TPMT: relationship between levels of the enzyme in red blood cells and genotype. (a) Activity of TPMT in red blood cells (RBCs) from 298 European blood donors. Presumed genotypes for the *TPMT* gene polymorphisms are also indicated. *TPMTL* and *TPMTH* are designations for alleles resulting in "low" and "high" activity, respectively. These allele designations were used before the molecular basis for the polymorphism was understood. (Adapted from Weinshilboum RM & Sladek SL [1980] *Am J Hum Genet* 32:651–662. With permission from Elsevier.) (b) Common *TPMT* alleles. *TPMT*1* is the most common allele (wild-type). *TPMT*3A*, with two non-synonymous coding SNPs, is the most common variant allele in Caucasian European subjects. *TPMT*3C* is the most common variant allele in East Asian subjects. Rectangles represent exons with the gray scale representing the open reading frame.

eczema, and rheumatoid arthritis. Both 6-mercaptopurine and azathioprine are thiopurine drugs that were originally developed by Gertrude Elion—a pioneering biochemist who received the Nobel Prize for Medicine in 1988. TPMT is polymorphic (**Figure 8.4**). Approximately 1:300 individuals lack TPMT activity. This appears to be a **recessive trait** usually due to the presence of one or more non-synonymous mutations. These individuals will develop severe bone marrow toxicity if prescribed the normal recommended dose of either 6-mercaptopurine or azathioprine. It is therefore recommended that patients should be genotyped for *TPMT* before they are treated with 6-mercaptopurine or azathioprine and those predicted to lack activity should either be given a lower drug dose or treated with another drug that is not metabolized by TPMT.

This is probably the best current example of a pharmacogenetic polymorphism where genotyping is routinely performed prior to drug prescription and the result used to guide treatment. The example once again illustrates the potential role for pharmacogenetics in clinical practice.

Polymorphisms in drug transporters also play a role in pharmacogenetics

Hepatocytes are the main cell type in the liver and an important site for drug metabolism. Some drug targets are also located in these cells. Though some drugs can enter the hepatocyte by passive diffusion, most require active transport that is achieved using transporter proteins located on the cell membrane of the hepatocytes facing the sinusoids which form the main blood vessels within the liver. Several different transporter families exist. For anionic drugs, the SLCO family is the main group of transporters. SLCO1B1 has a role in transport of a range of different drug classes, including statins and antimicrobial agents. Several common polymorphisms result in decreased transport by SLCO1B1. One of these appears to increase the risk of statin-induced myopathy, a serious adverse drug reaction (see Section 8.5), and may also be relevant in the response to methotrexate, a widely used immunosuppressant and anticancer drug.

8.4 DRUG TARGETS

In addition to the genetic factors affecting the entry and elimination of drugs within the body discussed in Sections 8.2 and 8.3, genes encoding the targets for drugs may also encode some common polymorphisms that may affect response to treatment. Depending on the individual drug, the target gene product may be an enzyme, a receptor, or a membrane transporter. Polymorphisms affecting drug targets may affect the response of the individual to a particular drug, so the overall effect will be different to that seen for polymorphisms in the genes encoding metabolizing enzymes discussed in Sections 8.2 and 8.3. A higher or lower than normal drug dose may be needed to achieve the desired response. In some patients, the presence of a particular polymorphism in the gene encoding the drug target may make the drug unsuitable for clinical use.

Among drug targets, one of the most consistent pharmacogenetic associations reported has been that for the gene encoding the VKOR with coumarin anticoagulants. Evidence for other pharmacogenetic associations with drug targets is more limited. However, there is some positive data regarding an association between the genotype for *ADRB2*, which encodes the β_2-adrenergic receptor, a target for β-adrenergic receptor agonists that are used in the treatment of asthma.

The relationship between VKOR and coumarin anticoagulants is one of the most consistently reported genetic associations involving drug targets unrelated to cancer

As described in Section 8.2 and in Figure 8.3, VKOR regenerates reduced vitamin K from vitamin K epoxide produced during γ-carboxylation of coagulation factors in the blood clotting cascade. The human enzyme is an integral membrane protein within the endoplasmic reticulum. *VKORC1*, the gene encoding the enzyme VKOR in humans, was only identified quite recently. Coumarin anticoagulants including warfarin (see Section 8.2) inhibit VKOR enzyme activity directly. A very small proportion of patients treated with coumarin anticoagulants fail to respond at doses within the normal range and are termed coumarin resistant. This lack of response appears to be due to rare non-synonymous mutations in *VKORC1* that appear to affect binding of the anticoagulant to the enzyme. Some coumarin-resistant individuals cannot be given coumarin anticoagulants successfully, but others can be stabilized on a very high dose. Polymorphisms in the *VKORC1* coding sequences are rare, but several non-coding polymorphisms, including *-1639G>A* (rs9923231) affect levels of *VKORC1* expression. The A variant, for example, is associated with lower transcription than the G variant. VKOR protein levels appear to directly determine the required dose of an anticoagulant. A clear association between *-1639G>A* genotype and warfarin dose requirement has been reported in a large number of independent studies, and similar associations for the other coumarin anticoagulants such as acenocoumarol and phenprocoumon have also been detected. There is ethnic variation in the frequency of the *-1693A* variant. Though the *-1639A* allele is less common than *-1693G* in Europeans, it occurs at a high frequency in East Asians and, as a result, the average dose of coumarin anticoagulant required in East Asian patients is significantly lower than that required in Europeans.

Approximately 20–30% of variability in coumarin anticoagulant dose requirement can be explained by *VKORC1* genotype. Anticoagulant dosage is routinely individualized by measurement of coagulation rate, but it is possible that problems with excessive bleeding or lack of response during the initial treatment period may be decreased by genotyping for *VKORC1* and for *CYP2C9* (see Figure 8.3). Genotype and other patient-specific factors can be used to calculate an individualized dose. The FDA suggests that genotyping for *VKORC1* and *CYP2C9* should be considered in patients receiving warfarin treatment, and clinical trials to assess the value of genotype-based dosing are in progress.

The efficacy of β-adrenergic receptor agonists widely used in the treatment of allergies may also be genetically determined

β_2-Adrenergic receptors are G-protein-linked receptors that are expressed in smooth muscle. Activation involves elevation of intracellular cAMP. *ADRB2*, a member of the family of β-adrenergic receptors, appears to be unusually polymorphic compared with other adrenergic receptors, especially in the coding sequence. As a result of this, and also because β-adrenergic agonists are widely used to treat asthma and other common lung diseases, there has been considerable interest in assessing the relevance of these polymorphisms to drug response.

There are at least five polymorphisms in the *ADRB2* coding region with four resulting in amino acid substitutions. In addition, there is a single non-synonymous polymorphism in the region encoding a leader peptide and several upstream polymorphisms close to the transcription start site (**Figure 8.5**). Four common haplotypes with frequencies less than 5% have been described but the frequency of these varies between ethnic groups. Among Europeans, three haplotypes are common. Two of these differ mainly at codon 16 with one having a sequence encoding glycine, while the other encodes arginine at this position. The glycine-encoding form can be further subdivided into two haplotypes: one encoding glutamine at position 27 and the second, which is less common, encoding glutamic acid. These three haplotypes are also common in other ethnic groups, though overall frequencies differ to those seen in Europeans. A fourth haplotype is also common in African-Americans. Most clinical studies to date have focused on the codon 16 polymorphisms.

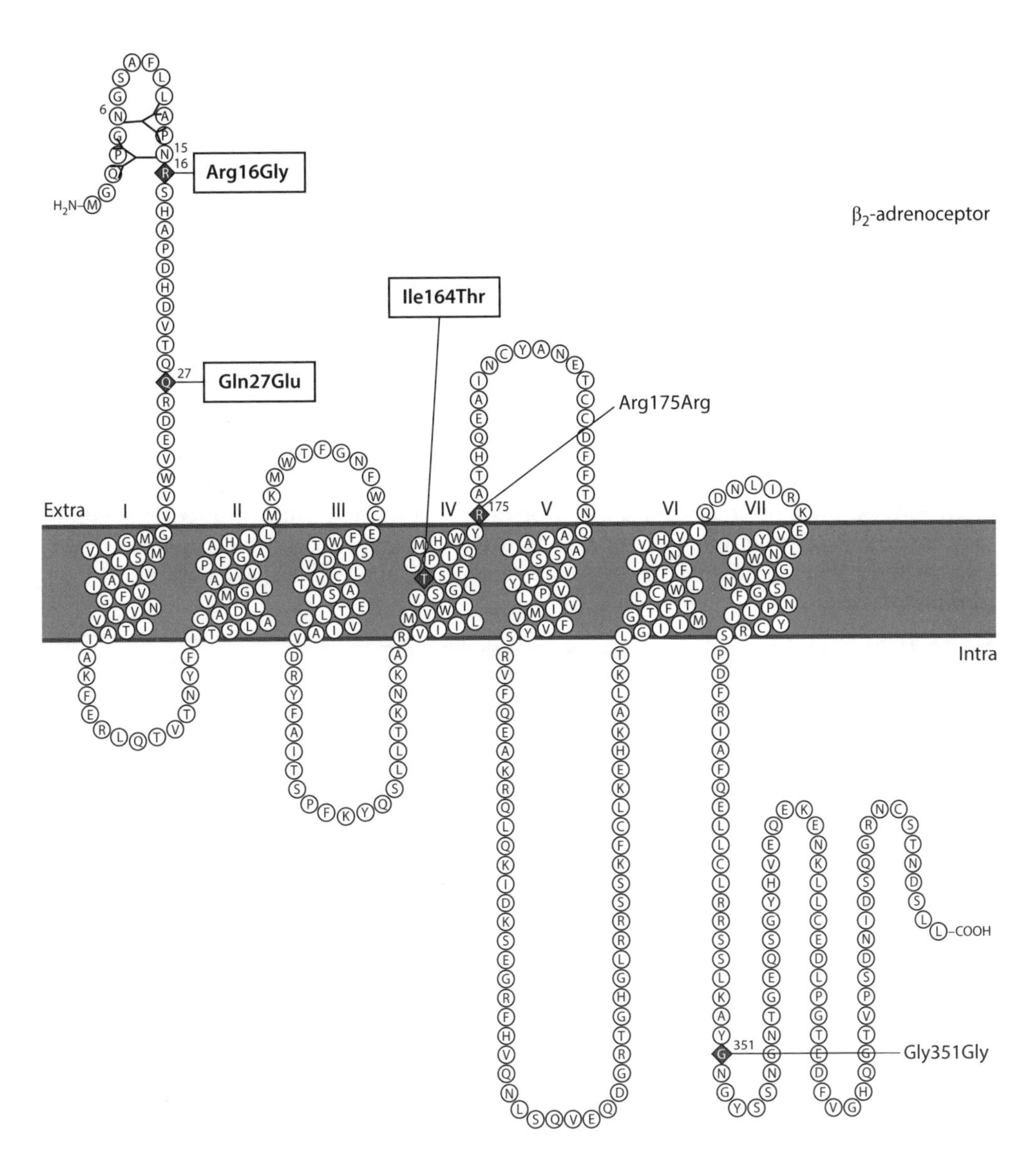

Figure 8.5: *ADBR2* gene polymorphisms. The figure illustrates main polymorphism in the *ADBR2* gene. The β_2-adrenoreceptor (*ADBR2*) gene is located on chromosome 5q31-q32 and consists of only one exon. There are five common variants to the coding sequence indicated in the figure. The R16G polymorphism (rs1042713), with the Arg (R) allele occurs in 25–40% of all individuals and the Q27E variant (rs1042714), with the Glu (Q) allele occurs in 50–60% of all white individuals. The I164T variant (rs1800888) is associated with a strong and stable phenotype. Additional common variants of unknown functional relevance are R175R (rs1042718) and G351G (rs1042719). (From Rosskopf D & Michel MC [2008] *Pharmacol Rev* 60:513–535. With permission from The American Society for Pharmacology and Experimental Therapeutics.)

Data on the functional significance of the codon 16 polymorphism is limited but the arginine-16-containing form of *ADRB2* appears to be associated with increased agonist-induced downregulation and with a poorer response to inhaled short-acting β-agonists. This response to β_2-adrenergic receptor agonists appears to be specific to short-acting agonists as a large study involving patients treated with longer-acting agonists failed to show any difference in outcome based on the codon 16 genotype. In general, it is now accepted that the codon 16 genotype does appear to affect response to short-acting agonists. However, this may not be important in clinical practice as current treatments mainly involve use of longer-acting agonists.

8.5 ADVERSE DRUG REACTIONS

Polymorphisms affecting drug metabolism and the normal function of drug targets can lead to adverse drug reactions. This may occur if, for example, as discussed above in earlier sections, the plasma drug concentration is unusually high due to slow metabolism or the effect on the target is greater than expected due to a low concentration of the target protein. In these cases toxic levels of the drug may build up. However, in addition to these predictable and generally concentration-dependent effects, other forms of less predictable adverse drug reactions may occur. These reactions, which are often referred to as idiosyncratic adverse drug reactions, have a less obvious relationship with drug concentration, may involve more than one genetic risk factor, and often involve the immune system. Idiosyncratic adverse drug reactions are generally rare, but are quite problematic for a number of reasons:

- Rarity makes it less likely they will be seen during clinical trials.
- There are generally no appropriate cell- or animal-based models that can be used to screen new drugs for propensity to cause these reactions early in drug development.
- As they are rare, they can be difficult to recognize in the initial stages. Stopping the drug will often improve symptoms, but idiosyncratic adverse drug reactions can be fatal or require major clinical interventions, such as an organ transplant.

There have been a number of high-profile cases during the last 20 years where new drugs have been withdrawn from the market relatively soon after licensing because of reports of serious adverse drug reactions.

Idiosyncratic adverse drug reactions can affect a number of different organs, including the liver, skin, kidney, heart, and muscle, and with some drugs, more generalized hypersensitivity reactions can occur. HLA genotype is an important predictor of some though not all idiosyncratic adverse drug reactions. The development of genome-wide association studies (GWAS) has been useful in increasing our understanding of genetic susceptibility to idiosyncratic adverse reactions because of the limitations of exploring biologically plausible candidates. The basis for many idiosyncratic adverse drug reactions appears to involve covalent binding of a chemically reactive metabolite or possibly the actual drug to cellular proteins. This may cause an inappropriate immune response, direct cellular damage, or affect normal cellular function. The reason why particular drugs are associated with one type of organ toxicity remains unclear, though the location at which reactive metabolites are created may explain organ-specific toxicity.

HLA genotype is a potent determinant of susceptibility to several different types of adverse drug reactions

HLA genotype has been associated with susceptibility to many forms of hypersensitivity including hypersensitivity to drugs. Hypersensitivity is an inappropriate immune reaction to an otherwise non-toxic agent. Type 1 hypersensitivity is normally seen as a skin rash and can be mild or severe. In pharmacologically induced hypersensitivity, the mildest form of reaction involves a skin rash without additional symptoms and drug withdrawal normally results in rapid improvement. Blistering skin reactions are considered the most severe manifestation of drug-induced hypersensitivity reactions affecting the skin and are potentially fatal. A number of other symptoms are also manifest in severe cases, including fever and effects on organs such as the liver, kidney, lungs, and bone marrow. However, the extent of involvement of other organs varies considerably between patients even when treated with the same drug.

Liver injury is a common symptom of drug-induced hypersensitivity and can in extreme cases be fatal. The genetic associations with HLA are currently among the best-established inherited risk factors for serious idiosyncratic adverse drug reactions, though other genetic risk factors are also likely to contribute. It is also important to realize that HLA genotype is unlikely to contribute to all types of adverse reaction. HLA genes and alleles, and their relationship with other forms of complex diseases and traits, are discussed in depth in Chapter 6.

The anti-human immunodeficiency virus (HIV-1) drug Abacavir gives rise to hypersensitivity in some patients

In terms of hypersensitivity, a particularly well-studied HLA association is that with the anti-HIV-1 drug abacavir. Hypersensitivity to abacavir affects approximately 5% of all patients treated. This reaction can be fatal, particularly if a patient is re-exposed to the drug after initial withdrawal and resolution of the original symptoms. An association between abacavir hypersensitivity and the *HLA-B*57:01–HLA-DRB1*07–HLA-DQB1*03* haplotype was initially demonstrated in two small independent studies. The association was then confirmed in a large randomized controlled trial by Mallal et al. in 2008. The clinical trial demonstrated that over 50% of patients positive for *HLA-B*57:01* will develop symptoms of hypersensitivity if exposed to abacavir, which means that genotyping for this allele to decide whether to treat with abacavir has a high sensitivity and specificity. Testing for *HLA-B*57:01* is now routine prior to treatment in many countries and patients found to have the risk allele are offered alternative therapies. In 2008, the FDA added information to the drug label recommending that individuals be tested for abacavir sensitivity prior to prescription. This recent development is another example of the translation of pharmacogenetic findings into clinical practice.

This is not the only strong association between HLA and adverse drug reactions. A large number of HLA alleles and haplotypes have now been shown to be associated with adverse drug reactions (**Table 8.2**), but there are some limitations. The associations reported for some of these other adverse drug reactions are not as strong as that reported for abacavir and *HLA-B*57:01*, which means that in practice testing is not as sensitive or specific when predicting toxicity. However, there are some exceptions. The *HLA-B*15:02* allele is an excellent predictor of susceptibility to carbamazepine-induced skin injury in certain ethnic groups, including the Han Chinese and Thai populations. HLA typing for this allele is frequently performed in individuals from these ethnic groups prior to carbamazepine prescription. Another HLA allele, *HLA-A*31:01*, is a risk factor for skin injury due to carbamazepine in both Japanese and Europeans, though the specificity is not as high as that for *HLA-B*15:02*.

Drug-induced liver injury is a rare, but clinically important problem

Many different drugs can cause drug-induced liver injury (DILI), with the precise pattern of injury varying between drugs. DILI reactions are usually classified as

Table 8.2 HLA associations with serious adverse drug reactions.

Type of reaction	Drug involved	HLA allele associated
Generalized hypersensitivity/ skin rash	Abacavir	*B*57:01*
	Allopurinol	*B*58:01*
Liver injury	Co-amoxiclav	*A*02:01*, *DRB1*15:01–DQB1*06:02*
	Flucloxacillin	*B*57:01*
	Lapatinib, ximelagatran	*DRB1*07:01–DQB1*02:02*
	Lumiracoxib	*DRB1*15:01–DQB1*06:02*
	Ticlopidine	*A*33:03*
Skin injury	Carbamazepine, phenytoin	*B*15:02*[a]
	Carbamazepine	*A*31:01*[b]

[a]Han Chinese, Thai, Korean, and other East Asian populations.
[b]Japanese and European populations.

hepatocellular when the injury is focused on the hepatocyte and cholestatic when the damage occurs at the hepatocyte canalicular membrane or further downstream in the biliary tree (**Figure 8.6**). The underlying mechanism by which DILI develops may involve both direct toxic effects by the drug, e.g. involving oxidative stress or cellular damage, and formation of reactive intermediates resulting in an inappropriate immune

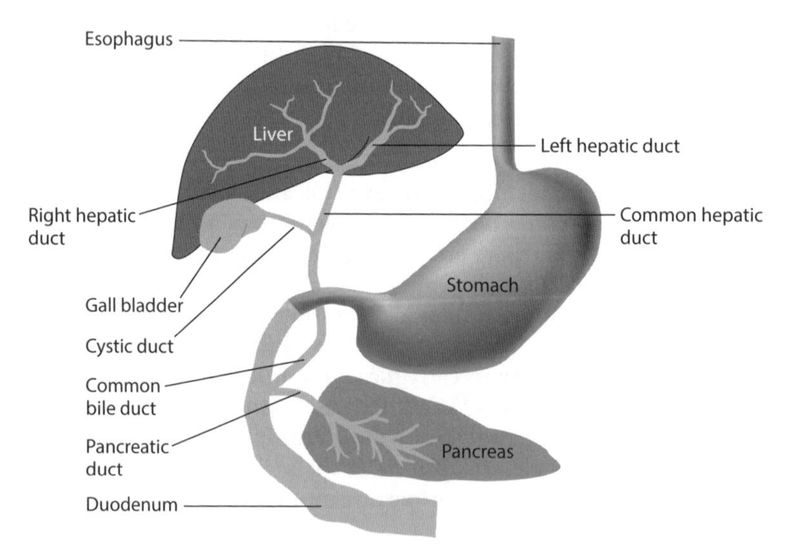

Figure 8.6: The biliary tree. When the liver cells secrete bile, it is collected by a system of ducts that flow from the liver through the right and left hepatic ducts to the common hepatic duct, which joins with the cystic duct to form the common bile duct. Approximately half of the bile formed flows into the duodenum or small intestine and half into the gallbladder. The stored bile is used to break down fats following the ingestion of food. (From Ahmed N, Dawson M, Smith C & Wood E [2006] Biology of Disease. Garland Science.)

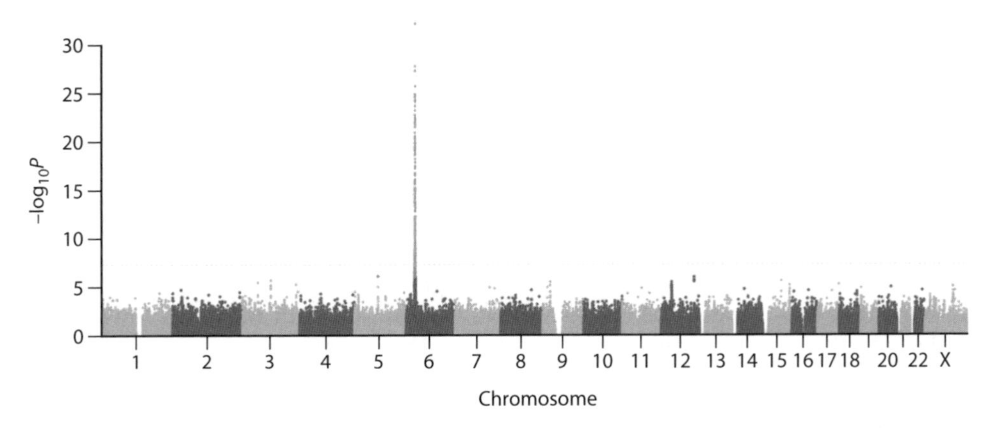

Figure 8.7: A Manhattan plot for a GWAS of flucloxacillin-induced liver injury. This Manhattan plot indicates a number of statistically significant associations with flucloxacillin-induced liver injury. In particular, the $-\log_{10}P$ value shows that a number of polymorphisms in the MHC region have genome-wide significance. The study involved 51 cases of DILI and 282 population controls. The polymorphism (rs2395029) showing the most significant difference in frequency between cases and controls is in complete linkage disequilibrium with the HLA class I *HLA-B*57:01* allele. (From Daly AK, Donaldson PT, Bhatnagar P et al. [2009] *Nat Genet* 41:816–819. With permission from Macmillan Publishers Ltd.)

response. Up to 10% of patients with DILI develop liver failure, which may be fatal unless a liver transplant is performed, though the majority of patients recover after the causative drug is discontinued. There is sometimes overlap between hypersensitivity reactions as discussed above and DILI.

Several different HLA alleles are risk factors for DILI, but the strongest effect for any HLA allele reported up to the present is that with *HLA-B*57:01*, which has been associated with flucloxacillin-induced liver injury (see **Figure 8.7**). Though this is the same risk allele as that identified for abacavir hypersensitivity and the overall specificity is high, sensitivity is low because most individuals who are prescribed flucloxacillin (a commonly used antibiotic, especially in the UK) and are *B*57:01*-positive do not appear to develop liver injury. The underlying mechanisms for toxicity with abacavir and flucloxacillin may also differ. Another example where the same HLA alleles and haplotypes appear to be risk factors for liver injury induced by different drugs (Table 8.2) is lapatinib and ximelagatran, two drugs with very different chemical structures that share the same association with the *HLA-DRB1*07:01–DQB1*02:02* haplotype and the development of liver toxicity. A common feature of many of these associations discussed here is that compounds with very different chemical structures show the same HLA association.

There are many other susceptibility factors for serious adverse drug reactions

Based on a number of studies, it appears that liver injury associated with some widely used drugs such as statins and isoniazid is not HLA-associated. Risk factors for these forms of DILI may include:

- Genotypes for enzymes affecting drug metabolism, e.g. *N*-acetyltransferase 2 (*NAT2*), which metabolizes isoniazid.

- Genes relevant to oxidative stress, e.g. superoxide dismutase (*SOD2*), and genes relevant to the innate immune response, e.g. signal transducer and activator of transcription 4 (*STAT4*), which encodes a transcription factor that regulates T cell responses.

Adverse reactions to commonly used statins provide a key example of non-HLA-related adverse drug reactions

Statins are widely prescribed drugs that are highly effective in lowering cholesterol. However, they can cause a toxic reaction involving muscle pain in some individuals. A number of different drugs can give rise to **myopathy**—a general term used to describe adverse reactions involving the muscle. These reactions can range from muscle weakness or pain, through **myositis** where the enzyme creatine phosphokinase is elevated, to rhabdomyolysis where creatine phosphokinase is greatly elevated and breakdown of muscle fibers can lead to renal damage. Most cases are not serious and are reversible through drug withdrawal. Statins are a relatively common cause of muscle toxicity. The ability of different statins to cause myopathy varies, but an overall estimate of an incidence of 0.1% has been made, though some say as many as 2% of all patients encounter muscle pain.

Individual statins have been shown to vary in their ability to cause myopathy, with cerivastatin being the most toxic followed by simvastatin, lovastatin, pravastatin, atorvastatin, and fluvastatin. Drug interactions where patients are taking more than one drug seem to be an important contributor to statin-induced myopathy. A GWAS on a subgroup of patients suffering from elevated creatine phosphokinase in a clinical trial involving use of simvastatin found a signal for a tagged single nucleotide polymorphism (SNP) (rs419056) close to the *SLCO1B1* gene (see Section 8.3) and in complete linkage disequilibrium with the *SLCO1B1*15* valine-174-alanine allele. This valine for alanine exchange has been shown previously to affect transport activity with several statins. This association has been confirmed for other statins, but possession of the variant allele only explains around 18% of the risk. The *SLCO1B1* risk (wild-type) allele has been associated with a four-fold risk in heterozygous individuals and a 16-fold risk in homozygotes according to some authors. Additional genetic risk factors possibly relating directly to the muscle rather than drug levels determined by *SLCO1B1* may also contribute.

Cardiotoxicity reactions to drugs do not appear to involve an immune or inflammatory response

Drug-induced **cardiotoxicity** usually involves prolongation of the QT interval (**Figure 8.8**), which is the time taken for one complete cycle of cardiac ventricular depolarization and repolarization, and can be measured using an electrocardiogram. A number of different drug classes can cause this type of toxicity, which can result in sudden death in susceptible individuals and it is a frequent reason for new drugs in development being withdrawn before they are licensed. Many of the causative drugs do not act primarily on the cardiovascular system and range from antimicrobial agents to drugs used in psychiatry, though drugs such as **amiodarone** that are used to treat cardiac arrhythmias can also cause QT prolongation. At present, genes found to predict drug-induced QT prolongation include those encoding proteins, such as ion channels expressed in heart tissue and the nitric oxide synthase 1 adaptor protein (*NOS1AP*), which interacts with nitric oxide synthase 1 to accelerate cardiac repolarization by inhibition of calcium channels. These risk factors overlap with those reported for syndromes associated with long QT interval independent of drug exposure.

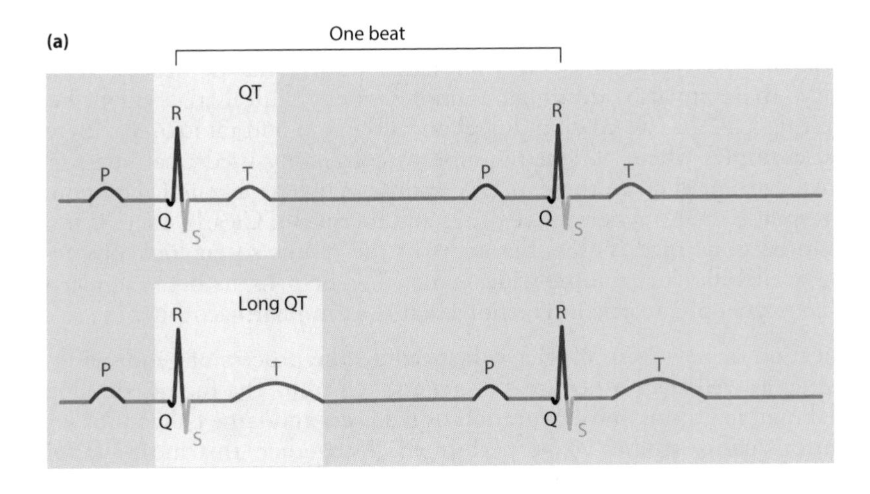

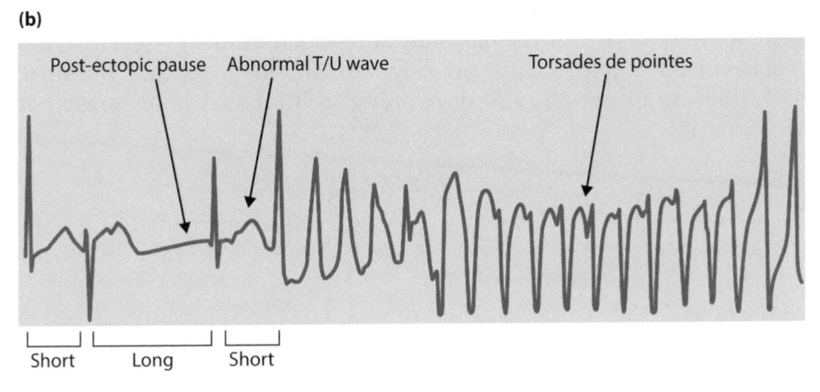

Figure 8.8: Drug-induced prolongation of the cardiac QT interval and torsades de pointes. (a) The cardiac depolarization and repolarization cycle. The QT interval is shaded; it starts with the onset of QRS and terminates at the end of the T wave. This is the time taken for one complete cycle. (b) Some drugs can prolong the QT interval and induce rapid heart beating, which is seen in the torsades de pointes (TdP; polymorphic ventricular tachycardia). The abnormal TU wave preceding the TdP indicates cardiac arrhythmia, which can potentially be fatal. (From Strachan T, Goodship J & Chinnery PF [2014] Genetics and Genomics in Medicine. Garland Science.)

CONCLUSIONS

There are now a large number of well-described and replicated pharmacogenetic associations. Recent developments in genomics such as the widespread application of GWAS have helped increase knowledge in this area. Despite the large body of knowledge, there has been less than hoped for progress in the clinical implementation of these findings, with a few important exceptions. Currently, genotyping for *HLA-B*57:01* prior to abacavir prescription and *TPMT* genotyping or phenotyping prior to azathioprine or 6-mercaptopurine prescription are the main examples of tests that are performed routinely in many countries. For many of the other examples described above, the sensitivity of the test may not be sufficient to make genetic testing useful in diagnosis or treatment choice. As there are no data based on some of these randomized clinical trials, assessing the clinical utility of genetic associations is not yet possible. In the case of warfarin and other coumarin anticoagulants, several clinical trials to assess the value of genotyping for *CYP2C9* and *VKORC1* prior to prescription have now been completed, but unfortunately the findings are inconsistent with the two largest studies disagreeing on whether genotyping is useful. Even larger studies may be needed to resolve the discrepancy. There are also additional examples where clinical trials might be helpful, such as testing *CYP2C19* and clopidogrel and *CYP2D6* and tamoxifen. There are also additional examples where patient response to commonly used treatments is variable and impossible to predict. In these cases, response may have a genetic component. This includes response to antihypertensive drugs and to aspirin. Clearly, there is much more work yet to be done in this area, but some of the results so far look promising. The increasing availability of genome-wide sequencing help us to better understand the complex genotype–phenotype interactions that exist in pharmacogenetics.

The application of new technologies will speed up the process of genotyping in pharmacogenetics as well as in other areas of genetic research. The further development of large collaborative groups and integration of data sets and large collections will enable larger, better quality studies to be performed. Altogether, this means faster, better, and more detailed studies. The potential to identify major genetic links in pharmacogenetics and develop better pharmacological agents using that information, or to identify genetic links to an individual's response to commonly used drugs, and then produce personalized therapy plans, are only two of the goals that the new genetics offers. The future use of present and developing technologies holds great possibilities in pharmacogenetics.

FURTHER READING

Books

Coleman MD (2010) Human Drug Metabolism. An Introduction. Wiley-Blackwell.

Maitland-van der Zee A-H & Daly AK (eds) (2012) Pharmacogenetics and Individualized Therapy. Wiley.

Strachan T, Goodship J & Chinnery PF (2014). Genetics and Genomics in Medicine. Garland Science.

Articles

Brodde OE (2008) Beta-1 and beta-2 adrenoceptor polymorphisms: functional importance, impact on cardiovascular diseases and drug responses. *Pharmacol Ther* 117:1–29.

Daly AK (2004) Pharmacogenetics of the cytochromes P450. *Curr Topics Med Chem* 4:1733–1744.

Daly AK & Day CP (2012) Genetic association studies in drug-induced liver injury. *Drug Metab Rev* 44:116–126.

DeGorter MK, Xia CQ, Yang JJ & Kim RB (2012) Drug transporters in drug efficacy and toxicity. *Ann Rev Pharmacol Toxicol* 52:249–273.

Guillemette C, Levesque E, Harvey M et al. (2010) UGT genomic diversity: beyond gene duplication. *Drug Metab Rev* 42:24–44.

Jonas DE & McLeod HL (2009) Genetic and clinical factors relating to warfarin dosing. *Trends Pharmacol Sci* 30:375–386.

Killeen MJ (2009) Drug-induced arrhythmias and sudden cardiac death: implications for the pharmaceutical industry. *Drug Discov Today* 14:589–597.

Koren G, Cairns J, Chitayat D et al. (2006) Pharmacogenetics of morphine poisoning in a breastfed neonate of a codeine-prescribed mother. *Lancet* 368:704.

Mallal S, Phillips E, Carosi G et al. (2008) HLA-B*5701 screening for hypersensitivity to abacavir. *N Engl J Med* 358:568–579.

Marsh S & Van Booven DJ (2009) The increasing complexity of mercaptopurine pharmacogenomics. *Clin Pharmacol Ther* 85:139–141.

Meyer UA (2004) Pharmacogenetics – five decades of therapeutic lessons from genetic diversity. *Nat Rev Genet* 5:669–676.

Niemi M (2010) Transporter pharmacogenetics and statin toxicity. *Clin Pharmacol Ther* 87:130–133.

Relling MV, Gardner EE, Sandborn WJ et al. (2011) Clinical Pharmacogenetics Implementation Consortium guidelines for thiopurine methyltransferase genotype and thiopurine dosing. *Clin Pharmacol Ther* 89:387–391.

Rubboli A, Becattini C & Verheugt FWA (2011) Incidence, clinical impact and risk of bleeding during oral anticoagulation therapy. *World J Cardiol* 26:351–358.

Scott SA, Sangkuhl K, Gardner EE et al. (2011) Clinical Pharmacogenetics Implementation Consortium guidelines for cytochrome P450-2C19 (*CYP2C19*) genotype and clopidogrel therapy. *Clin Pharmacol Ther* 90:328–332.

Uetrecht J (2007) Idiosyncratic drug reactions: current understanding. *Annu Rev Pharmacol Toxicol* 47:513–539.

van Schie RM, Wadelius MI, Kamali F et al. (2009) Genotype-guided dosing of coumarin derivatives: the European pharmacogenetics of anticoagulant therapy (EU-PACT) trial design. *Pharmacogenomics* 10:1687–1695.

Online sources

http://www.cypalleles.ki.se

This is a useful website which provides information of CYP450 gene nomenclature and alleles.

CHAPTER 9

Cancer as a Complex Disease: Genetic Factors Affecting Cancer Susceptibility and Cancer Treatment

Cancer is a major cause of morbidity and mortality worldwide. Statistics from the USA and UK reported in 2013/14 showed that 1,690,000 and 331,487 new cancer cases were diagnosed in the period 2012 and 2011 in the USA and UK, respectively. More than 500,000 American and 159,178 UK citizens die from cancer each year. In the USA and UK, lung cancer is the leading cause of cancer-related deaths in both men and women, accounting for 28 and 22% of all cancer mortalities, respectively. In the USA, breast cancer is the second-leading cause of cancer-related deaths for women (approximately 23 deaths per 100,000 women annually) and prostate cancer is the second-leading cause for men (approximately 20 deaths per 100,000 men annually), followed by colorectal cancer for both sexes (15 deaths per 100,000 population annually). In the UK, figures for breast cancer and prostate cancer are similar to those reported in the USA; however, in the population overall bowel cancer accounted for 10% of all cancer-related deaths, whereas breast and prostate cancer accounted for 7% in 2011.

The examples above are all examples of solid tumors, but it is important to remember that not all cancers are solid tumors (e.g. leukemias). Some cancers are caused by viruses, e.g. human papilloma virus is linked to cervical cancer and human herpes virus 8 is associated with Kaposi's sarcoma, which is also common in human immune deficiency virus (HIV-1)/acquired immune deficiency syndrome (AIDS). The viruses that can cause tumors are often referred to as oncoviruses.

This chapter will address selected examples of recent advances in the field of cancer genetics, but will not attempt to cover all cancers and genetic risk factors. However, before we start looking at the selected few it is worth considering the basic statistics on other forms of cancer. **Table 9.1** presents selected currently available data for cancer deaths per 100,000 individuals for a range of different countries for men and women published by the World Health Organization (WHO) in 2014 (based on data from 2008; http://apps.who.int/gho/data/node.main.A864). This shows some interesting contrasts, both between countries and between males and females. In most cases, the incidence of death from cancer is lower in women compared with men. The size of this difference can vary from country to country, with figures being twice as high in men as in women. In some countries, the figures are very similar for men and women, particularly where the overall incidence is low, for example in Namibia and Kuwait. Looking in more detail at rates for different cancers, **Figure 9.1** illustrates the 10 most common cancers in men and women in the UK in 2011 as published by Cancer Research UK in 2014 (http://www.cancerresearchuk.org/home/). These clearly show the impact of the selected examples on the UK population only.

However, we cannot discuss all forms of cancer here. In the selected examples, we will look at genetic risk factors in three categories: novel genetic risk factors i.e. those that have

Table 9.1 Mortalities from cancer per 100,000 head of population in 2014.

Country	Male	Female
Australia	141	93
Austria	154	95
China	182	105
France	183	94
Germany	156	99
Italy	158	91
Japan	150	77
Kuwait	62	70
Namibia	64	50
Panama	111	98
Qatar	101	84
Thailand	115	96
UK	155	114
USA	141	104

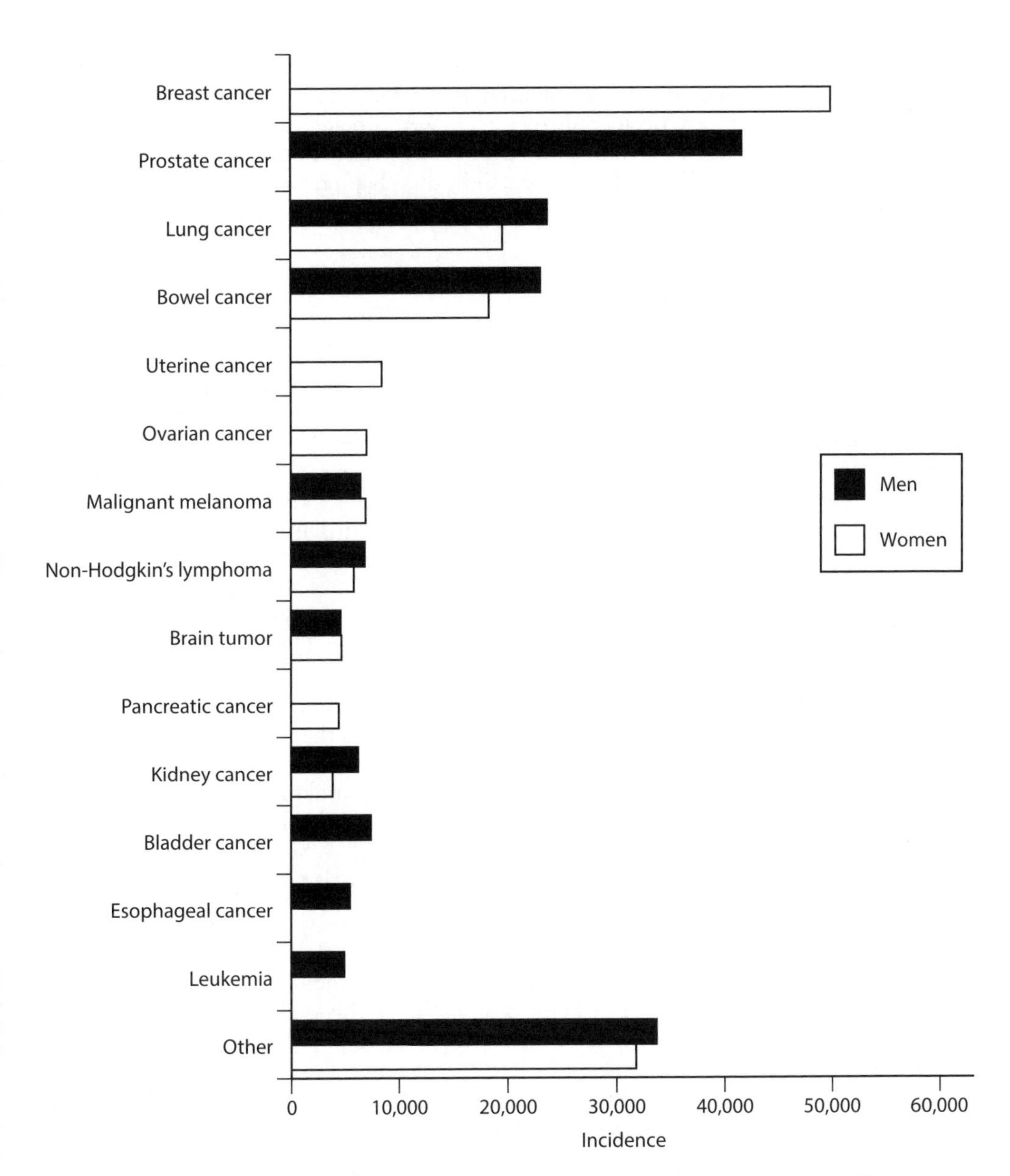

Figure 9.1: Incidence of the 10 most common cancers in women and men in UK in 2011. This figure shows the reported incidence of each form of cancer in 2011 in the UK. Note that some of the cancers do not occur in both men and women (for example ovarian and prostrate cancer), or are very rare in one sex compared to the other (for example, breast cancer is rare in men). In addition, some though common, are not found to be in the 10 most common for both groups. For example pancreatic cancer is listed in women but does not make the top 10 in men for this period. Whereas in women where ovarian, uterine, and pancreatic cancer are in the top 10, bladder, esophageal cancer, and leukemia are not listed. The less common cancers are all included in the group labeled "other." Breast cancer accounts for 30% of cancers in women in this figure, while prostate cancer accounts for 25% of cancers in men, with bowel and lung cancer making up approximately 25% of cancers in both men (28%) and women (23%). The frequencies for the other listed cancers range from around 5 to 2% in descending order. (Data from Cancer Research UK.)

been detected only by genome-wide association studies (GWAS), risk factors detected by GWAS affecting more than one type of cancer, and risk factors previously detected by candidate gene studies that have been confirmed by GWAS. In addition, the role of genetics in understanding response to therapy and advances in treatment of cancers as a result of genetic research will also be considered in detail in this chapter.

9.1 DEFINING CANCER

A cancer or a tumor is by definition an uncontrolled growth of abnormal cells in any part of the body, that is tissues developing outside the normal regulatory boundaries or at unrequired levels within those boundaries. Rapid growth is common. Some tumors follow simple Mendelian patterns of inheritance, but the majority are sporadic. Estimates of the percentage of total cancers that show Mendelian inheritance vary from 5 to 15%, though the genes responsible in some Mendelian cases may still be unknown. One well-studied example is breast cancer. Here, it is well established that inherited mutations in genes such as *BRCA1*, *BRCA2*, *TP53*, *ATM*, and *CHEK2* account for approximately 25% of the genetic risk. Current data suggests that these genes are mainly relevant in familial disease. Examples of other common types of cancer where heritability is relatively high include colorectal cancer and prostate cancer, with estimates of 35% and 42% heritability based on twin studies from Scandinavia. Despite these findings, the majority of cancers, like most other diseases, can be considered to be genetically complex. In the last 20 years, our understanding of cancer as a complex genetic disease and the role that common polymorphisms play in cancer susceptibility has increased. As with other complex diseases, both genes and the environment play a role. For example, smoking is the main environmental risk factor for lung cancer, but a number of other environmental factors may be important contributors in cancer development (**Figure 9.2**). When considering the term "heritability" it is also important to remember that heritability is not constricted to genetic inheritance—other factors are heritable.

Cancer predominantly affects the aging population. For example, as shown in **Figure 9.3**, the greatest number of new breast cancer cases in women in the UK is in the 60–65 years age group and the disease is very rare below age 40 years. This is also true in other developed countries. This means that with an increasingly aging population, the incidence of many cancers is increasing. The relatively late age of onset also means that performing genetic studies on families in cancer is challenging as parents and grandparents of affected individuals are less likely to be available for testing.

Before considering individual genetic risk factors for cancer in more detail, it is important to consider that in cancer, as with infectious disease and unlike most other complex diseases, we are dealing with two different genomes—that of the normal host and that of the cancer cells. There will be important differences between the germ-line DNA and **tumor DNA**. The differences between germ-line and tumor DNA are not inherited differences, but they are differences that develop during **carcinogenesis**. During this process by which a somatic cell is changed to a cancer cell, a variety of changes can occur in the normal host DNA, including point mutations, loss or gain of DNA, and epigenetic changes involving altered DNA methylation or other DNA or histone modifications (see Chapter 12 for more detail on epigenetics). Such changes lead to altered gene expression with the result that the patterns of both mRNA and protein expression will be different in a tumor cell compared with the equivalent somatic cells. Powerful new methods such as **whole-genome sequencing** have increasingly enabled the detection and study

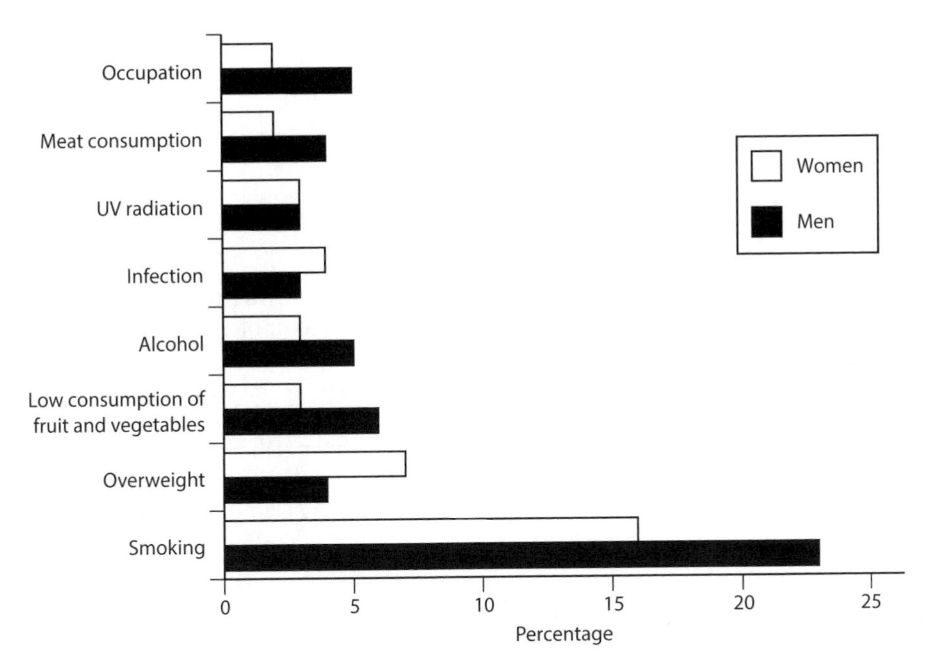

Figure 9.2: Fraction of cancer attributable to selected lifestyle and environmental factors. The percentage of all cancers where specific risk factors contribute. Men and women are considered separately. The data is for the UK and is based on information collected for 2010. (Data from Parkin DM, Boyd L & Walker LC [2011] *Br J Cancer* 105 (Suppl 2):S77–S81.)

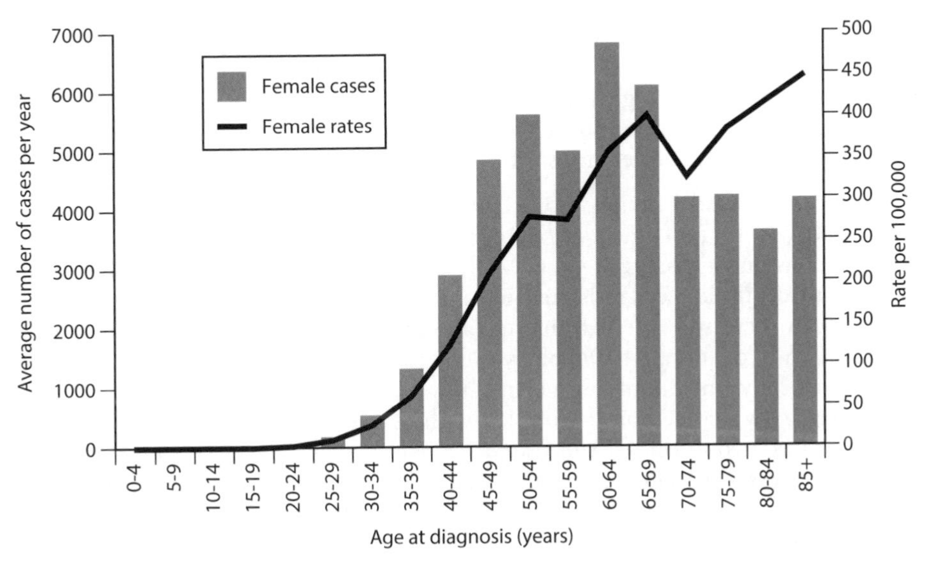

Figure 9.3: Average age at diagnosis of breast cancer. The histogram shows the average number of cases per year in the UK in 5-year age ranges. The solid line shows the rate per 100,000 women plotted against age. (Data from Cancer Research UK.)

of these changes in tumor material. However, the emphasis in this chapter is on cancer susceptibility and individualized treatment of cancer, not the process by which normal cells become tumor cells.

9.2 CANCER AS A COMPLEX DISEASE

In general, sporadic cancers, usually defined as cancers occurring in individuals without a family history of cancer, can be considered as typical complex diseases with both a strong contribution from the environment and a genetic component. Data from families indicate a two- to three-fold increased risk in first-degree relatives for a number of common cancers and there is higher concordance between monozygotic twins compared with dizygotic twins. Despite this, attempts to identify specific genetic risk factors for most common (usually sporadic) cancers have been relatively unsuccessful. However, the recent application of large-scale, high-resolution GWAS has led to considerable progress in this area.

Early studies of cancer found evidence of genetic associations with risk

The risk for a number of common cancers is strongly associated with environmental factors including smoking, occupational exposure to chemicals, and diet. Candidate gene-based case control studies performed between 1985 and early 2000 led to the discovery of a large range of apparently significant genetic associations with polymorphisms in genes that contribute either to the activation or to the detoxification of proven or possible carcinogens. However, many of the initial findings were not reproduced in subsequent studies. Most of these studies shared the same limitations, including use of relatively small numbers of cases and controls, and the selection of single genes and single polymorphisms for study. There are particular problems in studies based on small numbers, which are related to statistical power. In these early studies, false-positive results were common and false negatives equally so. By restricting the analysis to selected candidates, no thought was given to other polymorphisms. Though the polymorphisms studied were frequently biologically plausible candidates in relation to the cancer under investigation and in some cases bone fide associations were identified, most of these early association studies lacked consistency.

Improvements in study design from 2000 onwards, particularly the use of better characterized case groups, larger numbers of cases and controls, and the introduction of non-hypothesis-constrained analysis (e.g. GWAS), has helped to identify gene polymorphisms involved in susceptibility and resistance to cancer. One of the important features of recent work has been increased international collaboration to achieve very large case cohorts, e.g. the COGS (Collaborative Oncology Gene-environment Study) consortium that has focused specifically on identifying genetic risk factors for breast, ovarian, and prostate cancers (http://www.cogseu.org). Coupled with major technological advances, this type of approach has revolutionized the search for alleles associated with cancer as it has in many other diseases. Overall, it appears that the role of the polymorphisms identified in early studies may be less important than originally suggested, with a few notable exceptions. These include, for example, an association between aerodigestive cancers and alcohol dehydrogenase genes and an association between bladder cancer susceptibility, and polymorphisms in the *GSTM1* and *NAT2* genes that code for enzymes that contribute to the detoxification of chemicals (see Section 9.5).

GWAS has revolutionized the search for cancer-promoting alleles in non-familial cancers

The application of GWAS to cancer has resulted in the identification of more novel alleles that are not obvious candidate risk alleles for cancer, but in many cases are still biologically plausible. As with other complex diseases where GWAS has identified new risk factors, the impact tends to be small with typical odds ratios (ORs) being lower than 2. In addition, GWAS has identified large collections of different risk alleles in different forms of cancer as well as other diseases. It is generally not feasible, based on current knowledge, to predict an individual's risk of developing a cancer by genotyping for these newly identified risk alleles and mutations, but the detection of novel risk variants is important in understanding the biological basis of these diseases and may also enable new treatments to be developed in the future.

In cancer, as with most complex diseases, good quality phenotypic data is very important in studies aimed at identifying genetic risk factors. Tumors can be categorized according to their phenotype. Tumors in particular organs show considerable biological heterogeneity and the cells affected may also differ from one individual to another. Detailed information on tumor pathology is particularly useful in studies of cancer. Other types of phenotypic data that are collected in studies typically include age of onset, clinical stage, and response to treatment.

9.3 GENETIC RISK FACTORS FOR PARTICULAR CANCERS DETECTED BY GWAS

Various cancers have featured strongly in the list of diseases studied by GWAS approaches. Breast cancer, lung cancer, and prostate cancer have been particularly well studied, and a few specific examples of individual genetic risk factors for these cancers will be considered here to illustrate the value of the GWAS approach.

GWAS has identified a number of biologically plausible genetic risk factors for breast cancer

Though there are clear Mendelian cases of breast cancer, most cases (75–80%) are sporadic with no family history. It has been observed that the incidence of breast cancer is higher in women with affected siblings and the onset in this group is often earlier, pointing to genetic susceptibility. However, because it is a common condition, affecting up to 1:12 women in the developed world, there is often concordance for breast cancer in families by simple coincidence rather than by inheritance. Therefore, it is important to be aware of this before embarking on any studies of breast cancer. It is also important to note that breast cancer is not restricted to women and male breast cancer (not discussed here) does occur, though it is relatively rare.

Breast cancer was one of the first diseases to be studied by GWAS and since 2007 a series of independent GWAS have identified more than 70 different genetic signals [mostly tagged single nucleotide polymorphisms (SNPs)] that are significantly associated with risk of breast cancer, though the precise genes relating to these tagged SNP signals have not yet all been identified. Of those that have been identified, the majority that contribute to the increased risk of sporadic breast cancer code for proteins that have functions in cell growth and cell signaling (see **Table 9.2** for further examples). Of course it is not the genes themselves that increase the risk, as these are present throughout the population, but it is possession of specific sequence variations in the form of mutations and polymorphisms that confer risk. The *FGFR2* gene on chromosome 10q26 codes for the fibroblast

Table 9.2 Selected examples of identified potential susceptibility loci for breast cancer and the gene function.

Gene	Biological function
Fibroblast growth factor receptor 2 (*FGFR2*) (OR = 1.26)	Receptor that has tyrosine kinase activity and is located on the plasma membrane; binds fibroblast growth factors that have a diverse range of biological effects including roles in angiogenesis, mitogenesis, cellular differentiation, cell migration, and tissue injury repair
TOX high mobility group (HMG) box family member 3 (*TOX3*) (also referred to as *TNRC9* and *CAGF9*) (OR = 1.11)	Codes for a protein containing an HMG box that may modify chromatin structure by bending and unwinding DNA
Estrogen receptor 1 (*ESR1*)	Codes for nuclear receptor for estrogen that acts as a transcriptional activator of a range of target genes following estrogen binding
Mitogen-activated protein kinase kinase kinase 1, E3 ubiquitin protein ligase (*MAP3K1*) (OR = 1.13)	Codes for a serine/threonine protein kinase enzyme that is part of some signal transduction cascades, including the ERK and JNK kinase pathways, and the NF-κB pathway
Caspase 8 (*CASP8*)	Caspase 8 is a member of the cysteine-aspartic acid protease (caspase) family, and is involved in the programmed cell death induced by Fas and various apoptotic stimuli
Cytochrome *c* oxidase assembly homolog 11 (*COX11*)	Cytochrome *c* oxidase is a mitochondrial protein complex that transfers electrons from reduced cytochrome *c* to molecular oxygen; *COX11* codes for a protein which is not a structural subunit, but may be a heme A biosynthetic enzyme involved in cytochrome *c* oxidase formation
Solute carrier family 4, sodium bicarbonate co-transporter, member 7 (*SLC4A7*)	This plasma membrane protein transports sodium and bicarbonate ions; these processes may be important in regulating intracellular pH
RAD51 homolog B (*RAD51B*) (also known as *RAD51L1*; *REC2*; *R51H2*)	*RAD51* family members are proteins with roles in DNA repair by homologous recombination

growth factor receptor 2 and in European populations has been shown to have one of the strongest genetic associations of all the risk variants reported up to now. However, the per allele odds ratio of approximately 1.26 even for the SNP showing the strongest association (rs2981578) is relatively modest, though statistically significant. As we may expect, there are several other SNPs within the gene showing roughly equivalent effects and the molecular basis for this association is still not completely clear. However, the SNP rs2981578 is in a site for transcription factor binding that is conserved among species and may affect gene expression. It is important to note that the SNPs in *FGFR2* that were found to be associated with breast cancer risk were all more strongly related to estrogen receptor-positive cancer. Differences in associations depending on the estrogen receptor status of the tumor have been found to extend to other genetic risk factors for breast cancer, though one of those listed in Table 9.2, *TOX3*, which encodes a transcription factor, shows a similar risk for both estrogen receptor-positive and -negative tumors.

Novel insights into lung cancer involving the target for nicotine were detected by GWAS

Prior to the advent of GWAS, genetic susceptibility to lung cancer had been studied quite widely using the candidate gene case control approach, though there was considerable inconsistency between reports. The impact of this cancer worldwide and the availability of existing DNA collections enabled GWAS studies to be performed and data collected from large numbers of cases. However, the number of loci found to be significantly associated with the disease has been smaller than for some other cancers, including breast cancer (summarized in **Table 9.3**). The most significant association reported for lung cancer is in a region of chromosome 15 where several genes encoding nicotinic acetylcholine receptors (*CHRNA3* and *CHRNA5*), the targets for nicotine, are located. It has been proposed that these genes contribute to an increased risk of becoming addicted to nicotine through the expression of these receptors in the brain. In addition, expression of these receptors in the epithelial cells in the lung has been implicated in the development of lung cancer. More recently, studies of lung cancer in individuals who have never smoked have shown that this region is not associated with disease development. This observation provides further evidence for a direct effect of the presence of nicotine in the lungs, which is unlikely to happen in non-smokers, playing a key role in lung cancer development.

A large number of genetic risk factors for prostate cancer have been revealed by GWAS

Prostate cancer is another cancer that has been well studied by GWAS. Up to now, GWAS studies on this cancer have led to the identification of the largest number of genetic risk factors for any common cancer, accounting for approximately 30% of total familial risk of the disease. Some of the most established associations with specific gene loci are listed in **Table 9.4**. One of the stronger associations described is with *MSMB*, which codes for the prostate-specific protein β-microseminoprotein (PSP94). The gene product is secreted into seminal plasma after synthesis in the prostate epithelial cells. *MSMB* appears to function

Table 9.3 Selected potential susceptibility loci for lung cancer and their function.

Gene	Biological function
CHRNA3/5	These genes encode nicotinic acetylcholine receptor subunits; these receptors are members of a superfamily of ligand-gated ion channels that mediate fast signal transmission at synapses
BCL2-associated athanogene 6 (*BAG6*) [also called *BAT3*/mutS homolog 5 (*MSH5*)]	*BAG6* codes for a nuclear protein that is cleaved by caspase 3 and involved in the control of apoptosis; *MSH5* codes for a member of the mutS family of proteins that are involved in DNA mismatch repair and meiotic recombination
TERT–CLPTM1L	More general risk factor for several cancers (see Table 9.4)
Tumor protein p63 (*TP63*)	This gene codes for a member of the p53 family of transcription factors

Table 9.4 Selected potential susceptibility loci for prostate cancer and their function.

Gene	Biological function
β-Microseminoprotein (*MSMB*) (*PSP94*)	Prostate-specific protein that is secreted into seminal plasma
Kallikrein 3 (*KLK3*)	Codes for prostate-specific antigen (PSA), a protein produced exclusively by the prostate
NKX3.1 (human homolog of the *Drosophila NK3* gene)	Androgen-regulated transcription factor that is specific to the prostate
Integrin α6 (*ITGA6*)	First integrin gene to be associated with human cancer susceptibility
Hepatic nuclear factor-1β (*HNF1B*)	Transcription factor shown previously to be associated with maturity onset diabetes of the young (MODY); haplotypes associated with diabetes appear to be protective against prostate cancer

as a **tumor suppressor gene**, it is believed that the SNP is associated with prostate cancer risk and it affects transcription, possibly leading to decreased expression. Several other genes that are expressed in the prostate only, are also risk factors. Though the number of different loci associated with prostate cancer is relatively high, the biological basis of these associations is not as well understood as for some of the breast cancer risk factors discussed previously. Several separate loci on chromosome 8q24 have been identified that may also contribute to prostate cancer risk (see **Figure 9.4**). It is possible that the genetic risk is outside the **exome** as a large amount of genetic variation is encoded within intronic sequences. It is also possible that epigenetics plays a significant role in this and some other

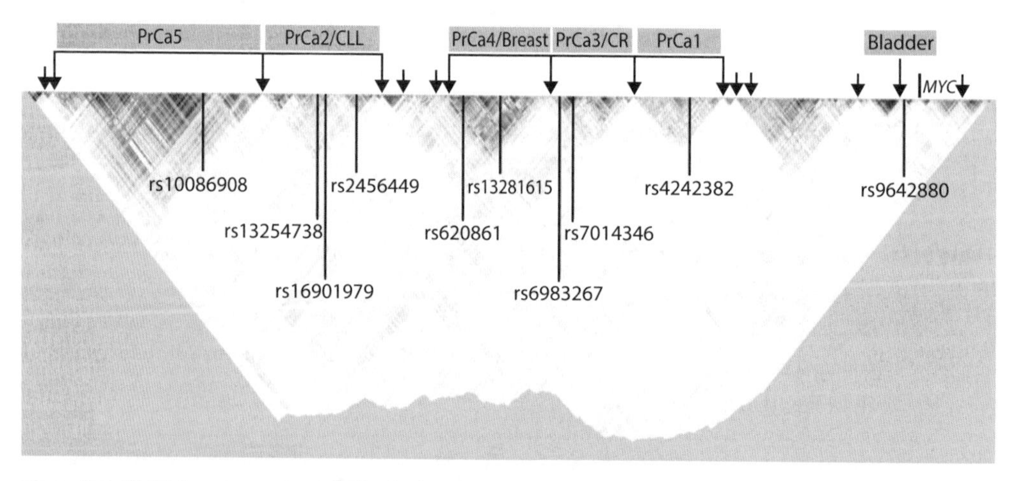

Figure 9.4: Multiple cancer susceptibility loci on 8q24 defined by cancer GWAS. Approximate locations of GWAS-reported cancer susceptibility loci are indicated by vertical arrows on the linkage disequilibrium pattern of the 1000 Genomes Project "CEU" (Northern Europeans from Utah) data (November 2010 release, chr8:127,878–128,880 kb genomic region, reference build 37.1). The arrowheads indicate probable recombination hotspots as per HapMap 1 and 2. Five distinct regions have been associated with prostate cancer risk (regions 1–5). Region 3 is also conclusively associated with colorectal cancer. Region 4 also harbors a breast cancer susceptibility locus rs13281615, and a bladder cancer susceptibility locus rs9642880 is telomeric to region 1 and approximately 30 kb centromeric to the *MYC* oncogene. (From Chung CC & Chanock SJ [2011] *Hum Genet* 130:59–78. With permission from Springer.)

cancers. Finally, some but not all of the associations appear to be with polymorphisms that act as risk factors in several cancers and they are not organ-specific risk alleles confined to one organ or one system (see Section 9.4).

9.4 GENERAL CANCER RISK LOCI DETECTED BY GWAS

A number of different genome regions have been found to associate with susceptibility to more than one form of cancer (**Table 9.5**). The major histocompatibility complex (MHC) on chromosome 6p21.3 has long been established as a region that includes risk alleles for several cancers. More recent GWAS have identified other chromosomal regions that are associated with a general increase in risk of cancer. Two examples that will be considered in more detail are the 8q24 region, discussed in Section 9.3 in relation to prostate cancer, and the region 5p15.33, which includes the *TERT* and *CLPTMIL* genes.

The 8q24 region is relatively close to the *MYC* oncogene, but the SNPs associated with cancer are over 300 kb away from this gene, are not in linkage disequilibrium with SNPs within *MYC*, and there are no other known genes nearby. There are five independent loci in this region associated with prostate cancer, and some of these loci are also associated with chronic lymphocytic leukemia, colorectal cancer, and breast cancer. There are also additional loci associated with bladder cancer and ovarian cancer in this area. It is possible that the disease associations relate to long-range interactions through DNA looping with the *MYC* oncogene. Some evidence that this might occur has been obtained.

Two genes, *TERT* and *CLPTMIL*, are located in the 5p15.33 region. *TERT* encodes a subunit of the telomerase gene and has been associated with some rare genetic diseases in addition to cancer. **Telomerase** and the control of telomere length are intimately linked to the process of tumor genesis in humans. Telomerase appears to contribute to tumor progression and **metastasis** by activation of the glycolytic pathway and suppression of cancer

Table 9.5 Selected potential general susceptibility loci for cancer.

Chromosomal region	Biological function and types of association
6p21 (HLA region)	Associations with pancreatic cancer, non-Hodgkin's lymphoma, nasopharyngeal cancer, and hepatocellular carcinoma; some, but not all, relate to cancers where viral infection is known to contribute to disease etiology
8p24 (in general area of *MYC* oncogene) (OR in breast cancer = 1.08)	Associations with prostate, breast, colorectal, bladder, and ovarian cancer together with CML reported; may involve long-range interactions with *MYC* oncogene
5p15.33 (*TERT–CLPTM1L*)	Associations with lung adenocarcinoma, pancreatic, testicular, bladder, brain, and testicular cancer, as well as basal cell carcinoma, melanoma, cervical cancer, prostate cancer, breast cancer, and several leukemias; associations may involve increased or decreased susceptibility for the same polymorphism depending on the individual cancer; effects may relate to the telomerase gene, but not proven
11q13	Associations with prostate, renal, and breast cancer reported; biological basis unclear

cell differentiation. Abnormal telomere length has also been demonstrated in many cancers. The biology of *CLPTM1L* is less well understood, but there are limited data suggesting a role for involvement in apoptosis following **genotoxic stress**. The cancers that show associations with the *TERT–CLPTM1L* region include lung, brain tumors involving **glial cells (glioma)**, bladder, testicular, and pancreatic cancer together with basal cell carcinoma and melanoma, which affects the skin. The relationship between polymorphisms in this region and susceptibility to particular cancers is quite complex. A single SNP, rs401681, has been identified as a risk factor for lung cancer and **basal cell carcinoma**. This SNP has also been suggested to be protective against another skin cancer, melanoma, and pancreatic cancer. For lung cancer, several different SNPs across the entire 100 kb region show associations. One of these, rs2736100, lies within *TERT* and appears to be a risk factor for lung cancer in both smokers and non-smokers, with the effect strongest in adenocarcinoma a tumor type that is particularly common in non-smokers. It is also a risk factor for glioma.

Discovery of these common genetic risk factors for various cancers is an interesting development and should add to our understanding of cancer biology. It is useful to consider the situation in research into the genetics of immune-mediated diseases where many of the risk loci are shared between diseases. This may indicate common pathways in disease and illustrates how understanding genetics of disease, even when the risk ratios are small, can help us to understand disease pathogenesis.

9.5 PREVIOUSLY ESTABLISHED CANCER RISK FACTORS CONFIRMED BY GWAS

Candidate gene case control studies on cancer predated GWAS by approximately 20 years and were generally unsuccessful in detecting robust genetic associations for the reasons cited above and elsewhere in this book. Early studies also expected (mostly incorrectly) to identify alleles with large effects. The cost of genetic testing and lack of information about the human genome and gene polymorphisms were major restrictions for geneticists seeking associations in complex disease. Consequently, the literature is generously peppered with papers reporting apparently plausible associations that were not confirmed. Despite these limitations, occasionally studies struck gold and some associations originally reported in candidate gene association studies have been subsequently confirmed in GWAS. The cancer-related genes detected as risk factors in these earlier studies are generally those concerned with carcinogen metabolism and so are different biologically to, for example, those identified recently as risk factors for breast and prostate cancer. In this section, polymorphisms relevant to carcinogen metabolism and to risk of upper aerodigestive cancer and bladder cancer will be considered (**Table 9.6**).

Alcohol, smoking, and chemical exposure increase the risk of cancer

Cancer of the oral cavity, larynx, and esophagus

Upper aerodigestive cancers comprising squamous cell carcinomas of the oral cavity, pharynx, larynx, and esophagus are known to be associated with both alcohol consumption and tobacco smoking. The possibility that polymorphisms in the genes encoding alcohol dehydrogenases, which convert alcohol to acetaldehyde, and aldehyde dehydrogenases, which convert acetaldehyde to acetate, could affect susceptibility to a large number of diseases associated with high intake of alcohol has received considerable attention. Up to the present, the main association reported by several independent research groups is that

Table 9.6 Genetic associations relevant to carcinogen metabolism detected by GWAS and candidate gene approaches.

Gene	Associations
Alcohol dehydrogenases *ADH1B* and *ADH7*	Both contribute to ethanol metabolism; SNPs within these genes have been shown to contribute to risk of upper aerodigestive cancer by both GWAS and candidate gene studies
N-acetyltransferase 2 (*NAT2*)	Polymorphisms in this gene associated with absence of enzyme activity are risk factors for bladder cancer development in both GWAS and candidate gene studies
Glutathione *S*-transferase M1 (*GSTM1*)	Null allele is associated with increased risk of bladder cancer

individuals positive for variant alleles in two alcohol dehydrogenase isoforms, *ADH1B* and *ADH7*, show a decreased risk of upper aerodigestive cancer. Original findings based on candidate gene case control studies have now been confirmed by GWAS.

NAT2 and bladder cancer

Tobacco smoking is an important risk factor for development of bladder cancer. *N*-acetyltransferase 2 (NAT2) is an enzyme that conjugates certain chemicals (including some prescribed drugs) with acetyl groups. Typical substrates include the aromatic amines that are found in tobacco smoke. It has been known since the 1950s that NAT2 activity is absent in approximately 50% of individuals, who are often referred to as **slow acetylators**. The genetic basis of this recessive defect is now well understood with a number of polymorphisms in the coding gene (*NAT2*), including several **non-synonymous SNPs**, shown to be associated with absence of activity. From the late 1970s onwards, there were isolated reports that slow acetylators were over-represented among cases of bladder cancer. A recent large well-controlled study in the USA found that slow acetylators who smoked heavily were over-represented among bladder cancer cases compared with those with one or two wild-type alleles, though the overall increase in risk was small. The association between tagged SNPs predicting the *NAT2* slow acetylator phenotype and increased risk of bladder cancer in smokers has been confirmed by GWAS.

In 1982, a small study in the UK based on cases of bladder cancer among former workers in the dye industry, who were likely to have been exposed to the carcinogen benzidine, showed that the slow acetylator phenotype was significantly over-represented. However, slow acetylators without occupational exposure to benzidine also appeared to have a slightly increased risk of developing bladder cancer. The occupational exposure reported in the 1982 study dated back to the 1950s. It is likely that subsequent improved regulations on occupational exposure to chemicals mean that the currently lower incidence of bladder cancer, which is down by 14% in women and 18% in men in the UK, may be partly due to these regulatory changes. Nevertheless, the slightly increased risk of this tumor among slow acetylators continues to be reported.

GSTM1 and bladder cancer

There is also report of a relationship between the *GSTM1* genotype and increased risk of bladder cancer. The *GSTM1* gene encodes glutathione *S*-transferase M1—an enzyme that conjugates and detoxifies chemicals. Approximately 50% of many populations lack this enzyme because they are homozygous for a large deletion (null allele) in the *GSTM1* gene.

There are a relatively large number of published studies showing that those homozygous for the null allele are at increased risk of bladder cancer development. This observation was confirmed in the same GWAS study from the USA as that mentioned above for *NAT2*, though the overall effect for *GSTM1* was smaller, with a 1.28-fold increased risk for those predicted to have two null *GSTM1* alleles on the basis of genotyping using tagged SNPs specific for the null variant compared with other genotypes. However, a separate GWAS has shown that the *GSTM1* null genotype is a risk factor for bladder cancer in non-smokers only. This is a slightly unexpected finding, but there is at least partial agreement between the early candidate gene studies and more recent GWAS.

In summary, the importance of polymorphisms in genes relevant to carcinogen metabolism as risk factors for cancer has been overestimated in the past with a small number of exceptions, including those discussed above. There are a number of possible explanations for this smaller than anticipated role, one of which is redundancy in biological systems, whereby the existence of a number of different proteins with an ability to carry the same function makes up for the deficiency created by the non-expressed or absent gene. This compensation acts as a safeguard against extinction. It is possible that the biological effect (the trait or disease) is only seen when the system is stressed to a point where compensation alone is not enough to make up for the deficiency.

9.6 INDIVIDUALIZING DRUG TREATMENT BASED ON TUMOR GENOTYPE

As discussed in detail in Chapter 8, genetic factors are important in determining response to drug treatment. In particular, some patients treated for cancer with drugs such as 6-mercaptopurine and irinotecan will develop toxicity, but this can be predicted by genotyping prior to treatment. In treating cancer, however, there is an added dimension. For a number of drug treatments, especially those developed recently, the tumor genotype or phenotype needs to be considered in addition to the germ-line genotype of the patient. In the case of the newer targeted therapies, knowledge of the tumor genotype is essential to determine whether the patient will benefit from the treatment. Targeted anticancer drugs can be subdivided into small molecules similar to traditional drugs and antibodies. A few representative examples of each class where tumor phenotype or genotype is determined before initiation of treatment will be considered in more detail below.

Newly developed drugs inhibit the function of mutated proteins in cancer cells

A well-established example of targeted therapy relates to the use of endocrine hormone therapy for breast cancer. To benefit from this treatment, the tumor needs to express the estrogen receptor-α (estrogen receptor-positive phenotype). The original drug used to treat estrogen receptor-positive breast tumors was tamoxifen, though specific aromatase inhibitors are now used in preference to tamoxifen. In contrast to estrogen receptor-positive breast cancer, it is generally accepted that there is little benefit in treating estrogen receptor-negative breast cancer with hormone therapy. This is an early example of targeting of therapy to a particular cancer phenotype (phenotype-guided cancer treatment).

Approximately 95% of cases of **chronic myelogenous leukemia (CML)** are positive for leukemic cells that have undergone a chromosomal rearrangement whereby there is reciprocal translocation between chromosomes 9 and 22, which is known as the **Philadelphia**

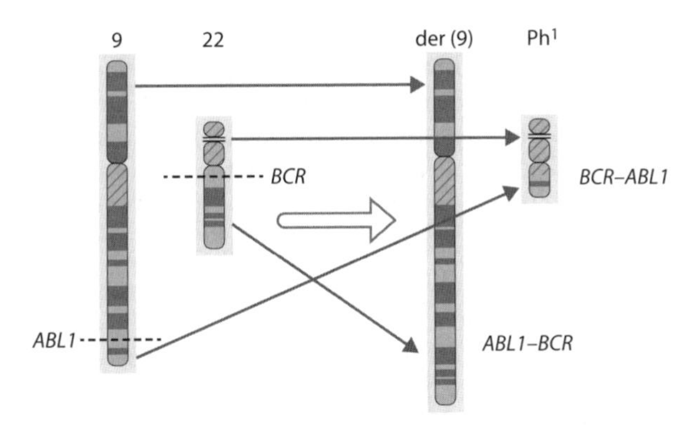

Figure 9.5 The Philadelphia chromosome. The Philadelphia chromosome is created by a translocation between chromosome 9 and chromosome 22. The transformed chromosomes are labeled as "der (9)" for chromosome 9 and "Ph1" for chromosome 22. Ph1 is the Philadelphia chromosome. The translocation joins exons of the breakpoint cluster region (*BCR*) gene and the *ABL1* oncogene. (From Strachan T, Goodship J & Chinnery P [2014] Genetics and Genomics in Medicine. Garland Science.)

chromosome. Chromosome 9 becomes longer while chromosome 22 is shortened in this process. This rearrangement results in a hybrid gene that encodes a fusion protein of *ABL,* which codes for a protein kinase, with the breakpoint cluster region or *BCR*, termed *BCR–ABL* (**Figure 9.5**). This results in a constitutively active tyrosine kinase, with much higher enzyme activity than that observed for the normal *ABL* gene product, and this in turn results in increased and poorly controlled cell proliferation. The drug imatinib was developed to specifically inhibit the *BCR–ABL* tyrosine kinase and has proved to be extremely successful as a treatment for CML. The Philadelphia chromosome targeted by imatinib is present in the vast majority of CML cases, so no additional test would normally be required to assess suitability of imatinib treatment. The main limitation of imatinib is that resistance can develop, but in such cases new tyrosine kinase inhibitors are now available and can provide suitable alternatives to imatinib. Imatinib is also an inhibitor of the *KIT* gene product, another receptor with tyrosine kinase activity. The c-Kit protein encoded by *KIT* is overexpressed on the cell surface of gastrointestinal stromal tumor cells and imatinib is a useful treatment for many, though not all, cases.

Another tyrosine kinase inhibitor, gefitinib, targets the epidermal growth factor receptor (EGFR). This drug was originally developed as a general treatment for non-small cell lung cancer (NSCLC), but was subsequently found to be effective only when the EGFR gene was mutated at certain positions. In practice, few NSCLC tumors are found to be positive for these mutations and the prescription of gefitinib is only recommended if tumors are found to be positive by DNA sequencing for these activating mutations. Similarly melanomas may have mutations at certain positions in the *BRAF* gene. The drug vemurafenib interrupts the B-Raf/MEK step on the B-Raf/MEK/ERK cell signaling pathway leading to apoptosis or programmed cell death, but only if the *BRAF* gene in the tumor has the common V600E mutation. As this drug may stimulate growth of melanoma cells with the wild-type or other *BRAF* sequence, determining the *BRAF* V600E genotype of the tumor is essential prior to use of vemurafenib.

Specific antibodies can target tumor-specific proteins and inhibit tumor growth

Tumor cells often express different proteins on their cell surfaces compared with nonmalignant cells and specific targeting of these proteins with antibodies has been used as an experimental technique ever since methods to prepare **monoclonal antibodies** in mice became available. Initial attempts to treat human cancers with these antibodies were generally unsuccessful, but the development of **chimeric** and **humanized antibodies** has increased success rates. To predict the likelihood of a response to antibody treatment, testing of tumor material by either phenotypic or genotypic methods is often performed before use. Treatment with targeted antibodies is often combined with conventional cancer treatments.

Trastuzumab is the most widely used targeted antibody treatment in cancer and was the first drug to be approved by regulatory authorities for use in the treatment of advanced cases of breast cancer in association with a diagnostic test that determines whether the tumor expresses the target protein HER2 or ERBB2, encoded by the *ERBB2* gene, and is thus responsive to this monoclonal antibody. ERBB2 is a cell surface growth factor receptor that is overexpressed in 20–25% of invasive breast cancers. When combined with conventional chemotherapy, trastuzumab treatment of *ERBB2*-positive tumors results in an objective response rate of approximately 20% for an average of 1 year. Though originally licensed for treatment of advanced disease, recently there have been reports of good responses in *ERBB2*-positive early-stage tumors. Currently, the licensed tests for *ERBB2* expression involve analysis of tumor material by either **immunohistochemistry** or fluorescence *in situ* hybridization (FISH). A major limitation of trastuzumab treatment is that patients generally become resistant after or during treatment at about one year. The underlying mechanism for this resistance is still not completely understood.

Cetuximab is a monoclonal antibody that targets EGFR. EGFR is a member of the same group of growth factor receptors as HER/ERBB2. It was developed as a treatment for advanced colorectal cancer where other treatments had failed, and is only suitable for use as a treatment for tumors that express EGFR at a high level. It is also now used as a treatment for advanced head and neck cancer. Receptor expression is assessed by immunohistochemistry. However, in addition to assessing expression of EGFR in the tumor prior to treatment, the tumor *KRAS* gene sequence is also determined. Certain mutations in the *KRAS* gene are common in colorectal cancer and result in constitutive KRAS activity. In these cases, blocking EGFR is ineffective at preventing cell proliferation and cetuximab is therefore not an effective treatment.

Epigenetic changes in the tumor involving methylation may affect response to conventional drug treatments

In studies on the clinical response of gliomas to treatment with the drugs carmustine (BiCNU®) and temozolomide, which modify DNA by alkylation, it was found that the best outcome was seen in patients with tumors where the promoter region of the DNA repair gene O^6-methylguanine-DNA methyltransferase (MGMT) was methylated, leading to absence of expression or gene silencing. It may be possible to inhibit *MGMT* where methylation has not occurred as a means of improving treatment outcome. Analysis of methylation patterns in tumors may therefore be useful in determining the most appropriate treatment. In breast and ovarian tumors, the *BRCA1* gene, which codes for another DNA repair protein, may also be subject to methylation and silencing. There is some evidence that a better response to treatment with **poly(ADP) ribose polymerase inhibitors** is seen in patients with methylated *BRCA1*.

Gene expression profiling may enable personalized cancer treatment

The examples discussed in this section up to now have focused on single genes. However, increasingly there are possibilities for individualizing patient treatment for cancer on the basis of the transcriptional profile of the tumor for expression of a number of genes. Clinical trials of several different gene expression profiling protocols for breast tumors have been performed. The number of different genes whose expression is profiled is typically between 20 and 70, and the main use of such a profile is currently to assist with decisions on whether patients should receive chemotherapy in addition to other treatments, such as endocrine therapy and radiotherapy. Use of this profiling is currently under investigation in the treatment of several cancers, and in the future, designing individual care plans based on questions, such as which chemotherapeutic agents to apply, whether or not to use radiotherapy, and which other drugs to prescribe, could become part of standard clinical practice.

Before we close the chapter on cancer it is important to recognize that there are many forms of this disease

As stated at the beginning of the chapter, it is not possible to discuss all the different types of cancer in one chapter. Even with this restriction, the chapter gives only a snippet of current data based on three selected common examples. However, there are other examples and other data are available. Cancer patterns are changing. Some forms are becoming more common, while others are becoming less common. This is illustrated for 20 common cancers in men and women in the **Figures 9.6** and **9.7**. Though the data presented

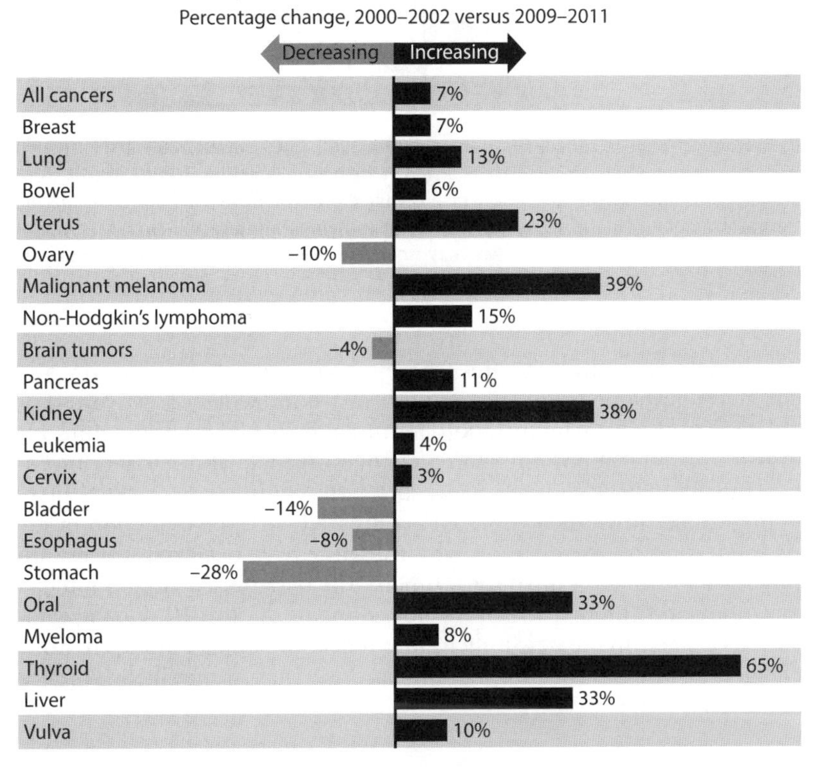

Figure 9.6 Percentage change in cancer frequency in women 2000-2002 versus 2009-2011. The figure shows how frequencies of some common cancers have changed in the new millennium. There are some differences compared with men (see Figure 9.7). These are mostly based on anatomy, but some are surprising. Thus, while bladder and stomach cancer have fallen in both groups, lung cancer has risen in women and fallen in men. These clearly reflect lifestyle changes, whereas others are more difficult to pinpoint. By comparing sexes we may be able to identify sex-specific risk factors in common cancers. (Data from Cancer Research UK.)

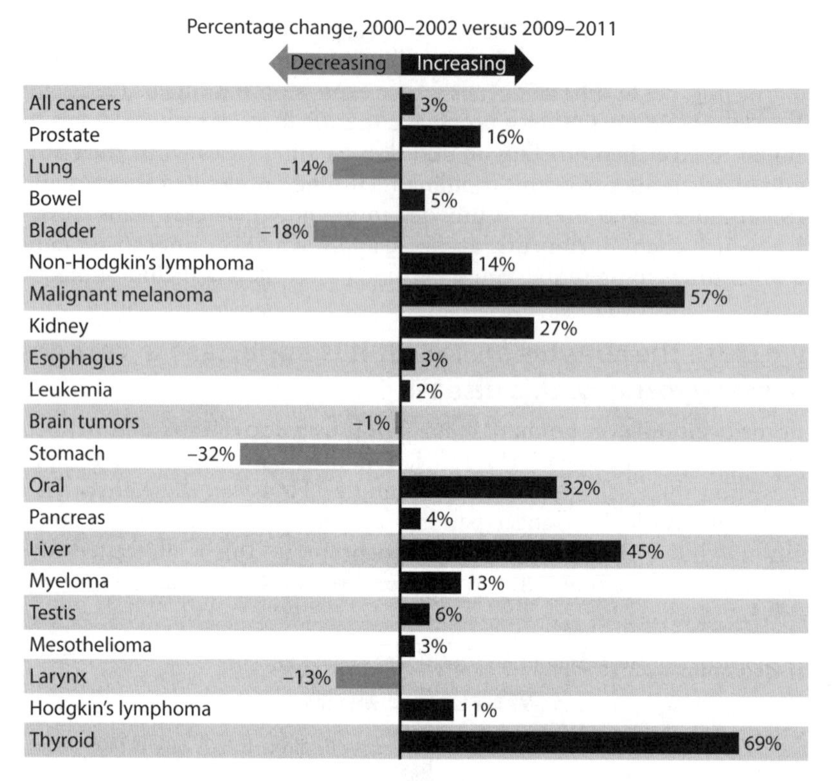

Figure 9.7 Percentage change in cancer frequency in men 2000-2002 versus 2009-2011. The figure shows how frequencies of some common cancers have changed in the new millennium. Bladder cancer, stomach cancer, lung cancer, and cancer of the larynx have fallen, while there have been major increases in liver, thyroid, skin, prostate, oral and kidney cancers, as well as both Hodgkin's lymphoma and non-Hodgkin's lymphoma. Some of these clearly reflect lifestyle changes, whereas others are more difficult to pinpoint. (Data from Cancer Research UK.)

are based on the UK data, similar figures are seen throughout the developed world. Figures are different in less developed countries and can vary quite widely, as we can see looking at the WHO site (http://apps.who.int/gho/data/node.main.A864). The figures for the UK that are shown illustrate important shared changes in male and female cancer, and also some important differences between males and females. In particular, the increased frequency of lung cancer in women (up 13%) compared with the reduction in men (down 14%). It is interesting to speculate on why this may have occurred, for example is this due to an increase in smoking in young women? The data presented also indicate the low frequency of non-solid tumors, which do not count in the top 10. Changes in the incidence of various cancers in the new millennium reflect the impact of the environment and societal changes on disease, and not just on cancers. They are unlikely to reflect genetic changes in our population.

CONCLUSIONS

There are several different forms of cancer – some cancers are familial Mendelian cancers, others are sporadic cancers. Genetics plays a role in both types. However, unlike other diseases, there are two genomes to consider in cancer—the host genome and the tumor genome. Looking at both early and more recent studies of cancer indicates that host genes are important in all forms of cancer. Genetic markers in the form of tagged SNPs and copy number repeat sequences have been used in GWAS to detect possible cancer genes and their location across and throughout the genome. Some, though not all, of the genetic risk factors previously detected by candidate gene analysis have also been confirmed by GWAS. This research is important because not only does it enable us to identify potential cancer-causing genes and networks, it may also help us to use genetics to understand the response to therapy in cancer and to develop better (more personal), novel treatment options.

The last 10 years has seen considerable progress in our understanding of the genetic basis of susceptibility and resistance to cancer. The application of GWAS in cancer has resulted in the identification of some well-replicated risk factors for a number of cancers; however, slightly disappointingly, the effect sizes remain small. Some may question the use of this new knowledge. However, there is hope. These findings help to inform the debate on the genesis of cancer and it is possible that, in the future, developments in the areas of genome sequencing and epigenomics may enable additional risk factors to be identified that will help us to develop more advanced screening programs. The future lies in integrating knowledge of both the patient's biology and the tumor biology at the levels of DNA, RNA, protein, and other molecules, so that the likely response to a wide range of treatments can be modeled mathematically. Ultimately, this information may be used to inform clinical management of the cancer. In addition, the information gathered will help to make systems biology a reality.

Further advances will require larger more extensive studies and the application of alternative technologies, such as exome sequencing and whole-genome sequencing, using a variety of **next-generation technologies**. In 2013, studies in breast cancer and prostate cancer, reported by the International Cancer Group (**iCOGS**), published the largest cancer genetic association study to date (approximately 200,000 SNP probes in approximately 200,000 case/control samples for breast, prostate, and ovarian cancer). If these numbers seem large there are plans underway to repeat the iCOGS with the larger **OncoArray**, which has approximately 600,000 SNP/copy number variation probes using approximately 400,000 case control samples. Lung and colorectal cancer may also be included in the next study. The iCOGS study represents an example of how the design in association studies has developed over the years and how there will be increasing reliance on large international collaborative networks.

Probably the most promising use of genetics in cancer so far has been in its use to inform the development of personalized drug treatments for the disease. This has been generally successful and is being used increasingly, especially in the treatment of advanced cancers that may be refractory to other treatments. It is likely that personalized approaches will become more common at all stages of cancer in the future.

FURTHER READING

Books

Spector T (2012) Identically Different: Why You Can Change Your Genes. Weidenfeld & Nicholson.

Maitland-van der Zee A-H & Daly A (2012) Pharmacogenetics and Individualized Therapy. Wiley.

Strachan T, Goodship J & Chinnery P (2014) Genetics and Genomics in Medicine. Garland Science.

Weinberg RA (2007) The Biology of Cancer. Garland Science.

Articles

Byrne HM (2010) Dissecting cancer through mathematics: from the cell to the animal model. *Nat Rev Cancer* 10:221–230.

Chung CC & Chanock SJ (2011) Current status of genome-wide association studies in cancer. *Hum Genet* 130:59–78.

Dowsett M & Dunbier AK (2008) Emerging biomarkers and new understanding of traditional markers in personalized therapy for breast cancer. *Clin Cancer Res* 14:8019–8026.

Figueroa JD, Ye Y, Siddiq A et al. (2014) Genome-wide association study identifies multiple loci associated with bladder cancer risk. *Hum Mol Genet* 23:1387–1398.

Hanahan D & Weinberg RA (2000) The hallmarks of cancer. *Cell* 100:57–70.

Heyn H & Esteller M (2012) DNA methylation profiling in the clinic: applications and challenges. *Nat Rev Genet* 13:679–692.

Hindorff LA, Gillanders EM & Manolio TA (2011) Genetic architecture of cancer and other complex diseases: lessons learned and future directions. *Carcinogenesis* 32:945–954.

Hosking FJ, Dobbins SE & Houlston RS (2011) Genome-wide association studies for detecting cancer susceptibility. *Br Med Bull* 97:27–46.

Parkin DM, Boyd L & Walker LC (2011) The fraction of cancer attributable to lifestyle and environmental factors in the UK in 2010. *Br J Cancer* 105 (Suppl 2):S77–S81.

Rothman N, Garcia-Closas M, Chatterjee N et al. (2010) A multi-stage genome-wide association study of bladder cancer identifies multiple susceptibility loci. *Nat Genet* 42:978–984.

Varghese JS & Easton DF (2010) Genome-wide association studies in common cancer-what have we learnt? *Curr Opin Genet Dev* 20:201–209.

Wheeler HE, Maitland ML, Dolan ME et al. (2012) Cancer pharmacogenomics: strategies and challenges. *Nat Rev Gen* 14:23–34.

Online sources

http://apps.who.int/gho/data/node.main.A864
A good source of data on cancer and other diseases worldwide from the World Health Organisation (WHO).

http://www.cancerresearchuk.org/home/
This is an excellent source of data for cancer in the UK.

http://www.cogseu.org
This is the study database quoted in the text section above regarding a Collaborative Oncology Gene-environment Study consortium that has focused specifically on identifying genetic risk factors for breast, ovarian, and prostate cancers.

CHAPTER

10

Genetic Studies on Susceptibility to Diabetes

Diabetes has a clear and demonstrable genetic component. There are two major forms of diabetes with similar signs and symptoms: type 1 diabetes (T1D) and type 2 diabetes (T2D). There is a considerable degree of clinical overlap between these two diseases. There is also some clinical overlap with other diseases. In this chapter, we will look at these two major forms of diabetes specifically focusing on their genetics. We will also briefly consider the less common form of diabetes referred to as maturity onset diabetes of the young (MODY). By comparing and contrasting the different forms of diabetes, we will be able to demonstrate how genotype can be used to illustrate phenotype and to help us understand disease pathogenesis.

10.1 DIABETES MELLITUS

Diabetes is amongst the biggest killers in the developed world. To put this into context, according to the World Health Organization (WHO) (http://www.who.int/diabetes/facts/en/), a total of 57 million people died in 2008. Of these deaths, 63% were due to non-communicable diseases (NCDs). The major causes of these NCDs were cardiovascular disease (17 million), cancer (7.6 million), respiratory disease (4.2 million), and diabetes (1.3 million) (**Figure 10.1**). The WHO figures show that prevalence is highest in higher-income groups (10% versus 8% in the lower-income groups). The healthcare burden in terms of resources for diabetes is considered to be two to three times higher.

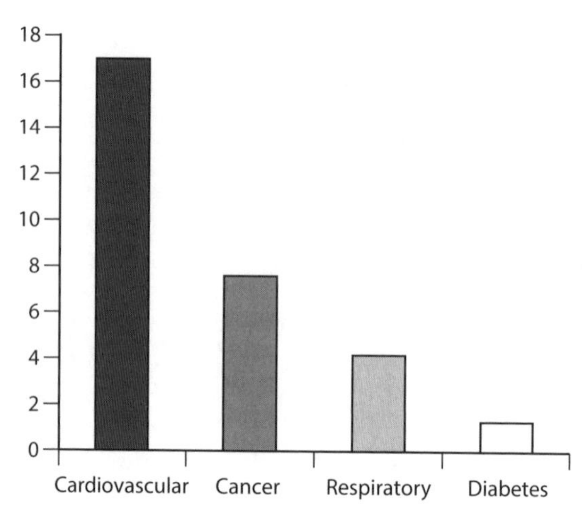

Figure 10.1: WHO data for deaths due to NCDs in 2008. The figure shows the impact of three major disease groups on mortality in 2008, reported in 2014. Though the diabetes column appears small compared with cardiovascular disease and cancer, these are very broad disease categories compared to diabetes, which is more specific. (Data from WHO.)

In terms of morbidity, diabetes is a major cause of non-traumatic lower limb amputations, which are 10 times more common in diabetes. Loss of sight and visual impairment are also more common in diabetics, especially in developing countries.

Diabetes mellitus is a medical term describing a group of metabolic diseases that can be characterized by higher than normal blood glucose levels due to either lack of **insulin** or failure to respond normally to insulin. There are two major forms of diabetes referred to as T1D and T2D; in addition, there is a group of mostly monogenic forms of diabetes referred to as MODY. T1D is much less common than T2D, affecting approximately 0.5% of the population compared with approximately 6% for T2D—a figure which is now being quoted in many developed countries. Overall T2D accounts for nearly 85% of diabetes cases, with T1D making up 10% and around 5% having MODY (**Figure 10.2**). Most cases of MODY tend to be Mendelian disorders,

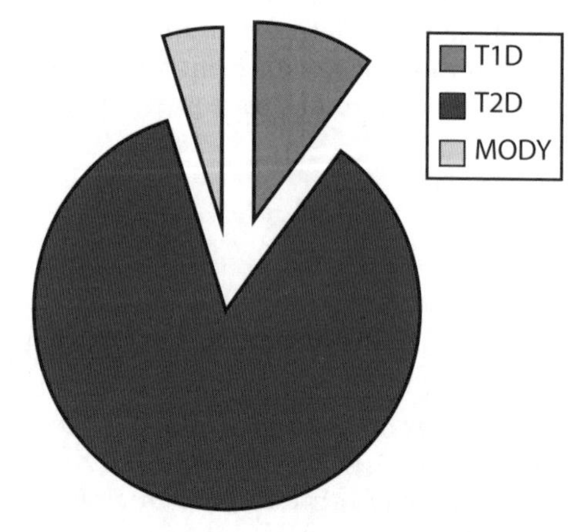

Figure 10.2: Worldwide incidence of T1D and T2D. Based on WHO data, there are approximately 347 million cases of diabetes worldwide with the majority being T2D (approximately 85%), as indicated in this pie chart, and the remainder being mostly T1D (10%) and the minority having MODY (5%).

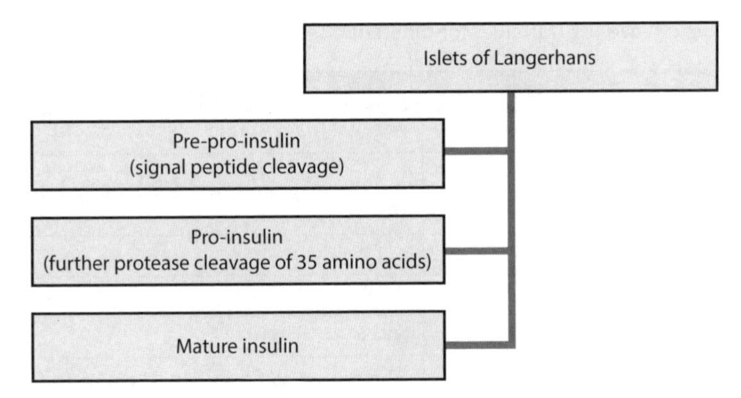

Figure 10.3: Three phases of insulin production. Insulin produced in the islet cells goes through three phases before it is functional.

while both T1D and T2D are genetically complex. To date, as with a number of other genetically complex diseases, including most autoimmune diseases and some cancers, only a proportion of the overall genetic risk for T1D and T2D has been accounted for. The amount of progress that has been made varies between the different types of diabetes, for example excellent progress has been made with respect to T1D, but progress has been slower for T2D.

10.2 GENETICS OF T1D

T1D occurs as a result of the immune destruction of the pancreatic β cells where insulin is produced (**Figure 10.3**). This autoimmune reaction begins early in life. Evidence for a genetic basis for T1D can be seen in studies of siblings of affected cases where the sibling relative risk (λ) may be as high as 15. This indicates a 15-fold increase in the risk of T1D for siblings of patients with this disease and suggests a considerable genetic component in T1D. Though this figure is lower than that for Crohn's disease, it is higher than for many other genetically complex diseases and on the same level as most other autoimmune disorders (**Table 10.1**). Studies of twins also show higher concordance rates in monozygotic twins than for dizygotic twins with values of 40 or more for monozygotic twins and less than 10 for dizygotic twins. However, caution needs to be applied in twin studies. Long periods of follow-up are important as the age of onset may vary between twins. If a study is performed for a short period only, the diagnosis of diabetes in the paired twin may fall outside the study period and this may be missed, giving false results.

Among complex diseases in general, the genetics of T1D is reasonably well understood. This is due to both the strength of the genetic component and the nature of the disease. T1D compares well with other autoimmune diseases and those where the immune response is known to play a major role in causality.

10.3 EARLY GENETIC STUDIES IN T1D

Early studies indicated a strong role for genetics in T1D. Studies in the 1970s first identified associations with a number of human leukocyte antigen (HLA) antigens; however,

Table 10.1 Sibling relative risk (λ) values for different diseases.

Disease	Estimated λ
T1D	15
T2D	3
Celiac disease	50
Ulcerative colitis	7–17
Crohn's disease	13–36
Multiple sclerosis	20–50
Primary biliary cirrhosis	10.5 (rising to 58 in daughters of affected mothers)

the first study on HLA found no association. This later proved to be incorrect, and associations with B7, B8, B15, B18, Cw3, DR2, DR3, and DR4 were soon identified. Most of these studies were classical case control association studies, though there were exceptions (**Figure 10.4**). In parallel with this work on the major histocompatibility complex (MHC), others were busy looking outside the MHC looking for genetic linkage using multi-case families.

HLA class II genotype is the strongest genetic risk factor for T1D

Early studies reported associations with HLA class I antigens, but in the 1980s the focus shifted to the HLA class II antigens and DR in particular. Studies identified associations with DR3 and DR4 as major risk factors, and DR2 (later DR15) as a protective factor. The development of better technology for HLA typing, including the introduction of both restriction fragment length polymorphism (RFLP) analysis and polymerase chain reaction (PCR) analysis (Chapter 6) led to the confirmation of the associations with DR2,

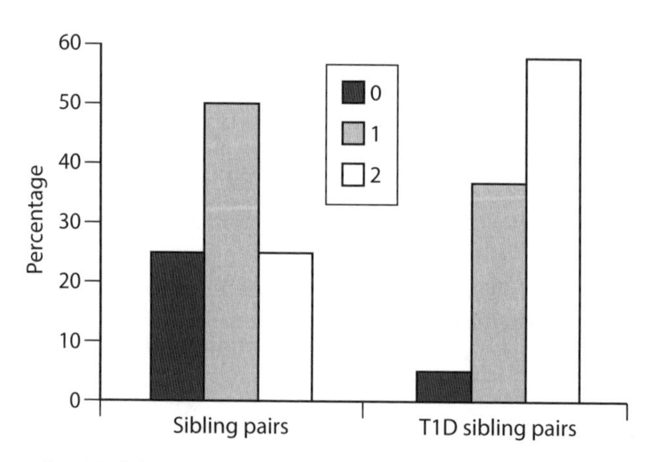

Figure 10.4: Sibling pair analysis in T1D looking at HLA haplotype sharing. The figure shows haplotype sharing in two groups: a group of non-diabetic siblings and a group of diabetic sibling pairs. This study illustrates the strong impact of HLA in T1D. This type of study is unusual, but it can be very useful as it does not identify specific alleles but concentrates on haplotype sharing. (Adapted from Parham P [2009] The Immune System, 3rd ed. Garland Science.)

DR3, and DR4 at a molecular level (now labeled *DRB1*03*, *DRB1*04*, and *DRB1*15*). In addition, the application of PCR enabled the associations with specific *DQB1* and *DQA1* alleles carried on the *DRB1*03*, *DRB1*04*, and *DRB1*15* haplotypes to be identified for the first time. Genotyping studies rapidly confirmed these associations in different cohorts of T1D patients. In 1987, Todd et al. published a groundbreaking study suggesting that HLA-encoded susceptibility to T1D is due to a specific amino acid at position 57 of the DQβ-chain. The amino acid aspartate is found on *DQB1* alleles that are associated with a reduced risk of T1D. In contrast, *DQB1* alleles that are associated with an increased risk encode alanine, valine, or serine at position 57 (see Chapter 6). This study was the first to consider HLA associations from a functional perspective. The study was published in the same year that the crystal structure of HLA-A2 was published in *Nature* and the authors of the T1D paper were able to show that this subtle structural difference determined whether a salt bridge would form over the antigen-binding groove of the expressed DQ molecule. Formation of the salt bridge involved the interaction between position 57 on the DQβ-chain and arginine at position 79 on the DQα-chain. This interaction may be critical in determining which antigens are preferentially presented to the T cell receptor (TCR) in the formation of the immune synapse and the orientation of the bound peptide antigen in that process.

This is certainly not the whole story for HLA and T1D. The associations established in the 1980s and 1990s have mostly been confirmed. In 1994, an early genome-wide study on affected sibling pairs using microsatellite markers confirmed a strong signal from the MHC. The 2007 Wellcome Trust Case Control Consortium (WTCCC1) study also found a very strong signal for the MHC [odds ratio (OR) = 5.49, $P = 2.42 \times 10^{-134}$]. The HLA class II haplotypes that are positively associated with T1D include *DRB1*03:01–DQA1*05:01–DQB1*02:01* (the **HLA 8.1 ancestral haplotype**) and *DRB1*04:01–DQA1*03:01–DQB1*03:02*. Approximately 95% of patients of European ancestry with T1D are positive for one or both of the two haplotypes. In contrast, only 40% of non-diabetics of European ancestry have one or both of these haplotypes. The HLA class II haplotype *DRB1*15:01–DQA1*01:02–DQB1*06:02* is significantly less common in affected individuals. The estimated contribution of HLA to genetic risk for T1D is between 30% and 50%. Interestingly, these three haplotypes, which are all common, are also associated with a wide variety of different diseases, especially immune-mediated and autoimmune diseases. The risk alleles do not always have the same effect in different diseases—the *DQB1*06:02* haplotype, which is protective in T1D, is associated with an increased risk of both multiple sclerosis and severe narcolepsy.

Not all of the risk for T1D above may be associated with the *DQB* allele or HLA class II

These extended haplotypes carry a number of other potential risk genes. For example, the HLA 8.1 ancestral haplotype carries *MICA*008*, which has been associated with an increased risk of the autoimmune liver disease primary sclerosing cholangitis and is involved in killer cell activation. The 8.1 haplotype also carries a null allele for *C4A*, thought to be the main factor in MHC-encoded susceptibility to sporadic cases of systemic lupus erythematosus. Therefore, it is possible that some of these haplotypes carry more than one risk allele. Such a multihit hypothesis would explain the strong effect of the MHC in T1D and other autoimmune diseases.

Finally, more recent evidence suggests an additional HLA class I-encoded contribution to susceptibility to T1D that is independent of HLA class II. This was reported in a study involving more than 5000 cases. Association signals from *HLA-A* and *HLA-B* were seen that were independent of HLA class II and not due to linkage disequilibrium. Based on this study, *HLA-B*39* appears to be the strongest independent class I risk allele, but additional *HLA-A* and *HLA-B* alleles also appear to contribute and may be important as determinants of age of onset of disease. Interestingly, an association with *HLA-B*39* was not listed in the compendium of early studies of HLA under the subtitle diabetes.

Other genetic risk factors for T1D include the genotype for the insulin gene

Evidence that the insulin gene was a predictor for susceptibility to T1D was reported in the early 1980s in a case control association study that used RFLP analysis. This association was subsequently confirmed using the transmission disequilibrium test (TDT) in families with at least one affected sibling pair. The 1994 genome-wide study that was based on the use of variable number tandem repeats (VNTRs or microsatellites as they are known) in affected sibling pairs also reported a signal in the insulin gene region, though it was not as significant as that seen for HLA. The signal was given the label IDDM2. The insulin gene codes for an insulin precursor known as **pre-pro-insulin**. This precursor is converted in the endoplasmic reticulum to **pro-insulin** and pro-insulin is converted to insulin by the enzymatic removal of a segment that connects the amino end of the α-chain to the carboxyl end of the β-chain. This produces the bipeptide chain of **mature insulin**. The segment that is removed is referred to as the connecting (C) peptide. Despite ongoing studies, the underlying mechanism for the association with the insulin gene remains unclear. However, it appears to involve variation in the 5′-non-coding sequence that results in decreased expression of the insulin precursor pre-pro-insulin. It is likely that this occurs in the **thymus**, but not in the pancreas, and it has been proposed that the decreased expression in the thymus leads to decreased immune tolerance to insulin resulting in an inappropriate immune response in the pancreas.

Candidate gene studies have identified a number of other non-MHC associations with T1D

Candidate gene studies have resulted in the detection of a number of associations for T1D with non-MHC immune-related genes. As with most complex diseases, not all of these associations have been confirmed. The best replicated associations include those with genes encoding protein tyrosine phosphatase, non-receptor type 22 (*PTPN22*), cytotoxic T lymphocyte associated protein 4 (*CTLA4*), and interleukin (IL)-2 receptor subunit α (*IL2RA*) (**Table 10.2**). These associations are all plausible because of the important roles of these gene products in T cell immunity. However, we need to be cautious about assigning causality on the basis of plausibility. Plausible should be read as good potential candidate and not as confirmed disease-causing mutation or allele.

PTPN22

PTPN22 encodes a lymphoid-specific intracellular phosphatase that is a negative regulator of TCR signaling. This occurs by direct de-phosphorylation of various cell signaling proteins including the Src family kinases LCK and FYN. Associations with this gene have been described for a number of different autoimmune diseases including rheumatoid arthritis.

Table 10.2 Pre-genome era regions identified as potential areas for candidate genes in T1D.

Gene	Location	Potential candidate (and estimated impact if available)
IDDM1	6p21.3	MHC ($\lambda = 3.1$: 30–40% of overall genetic risk)
IDDM2	11p15.5	*INS* ($\lambda = 1.3$: 10% of overall genetic risk)
IDDM3	15q26	β-2-Microgobulin
IDDM4	11q13	*LRP*
IDDM5/8/15	6q21–q27	No candidate
IDDM6	18q21	Kidd blood group
IDDM7/12/13	2q13–q33	*CTLA4*
IDDM10	10p11–q11	*IL2RA*
IDDM11	14q24.3	No candidate
IDDM17	10q25	No candidate identified; in large Bedouin family only

The table provides a summary of potential areas for investigation or known to be associated with T1D prior to the publication of the Human Genome Mapping Project. Among the candidates, IDDM1, IDDM2, and IDDM7 have all been confirmed.

CTLA4

CTLA4 encodes a co-stimulatory molecule expressed by activated T cells. It binds to CD80 and CD86 on antigen-presenting cells (APCs) (previously known as B7) and transmits an inhibitory signal, thereby deactivating the T cells and turning off the immune response. This process is competitive. In early T cell activation, T cells express CD28 which binds with CD80 and CD86 on the APCs, sending a positive signal that promotes the T cell immune response. As the immune response progresses, more CTLA4 protein is expressed on T cells and the immune response is downregulated (**Figure 10.5**). There are many polymorphisms in the *CTLA4* gene. Prior to the use of GWAS, Ueda et al. investigated 108 single nucleotide polymorphisms (SNPs) in and around the *CTLA4* gene in a large cohort of T1D cases and identified a specific SNP that they labeled CT60 as the strongest associated T1D *CTLA4* SNP. Recent studies indicate that *CTLA4* is associated with T1D, but have not pinpointed the specific risk allele. Associations with this gene have been described for several autoimmune diseases, most notably with the autoimmune thyroid disease Graves disease. Interestingly, Ueda et al. included Graves disease in their study and also found the strongest association with the so-called CT60 SNP. Finding associations with the same gene in several diseases is not surprising, especially when the gene in question has a broad (or non-specific) function, such as *CTLA4* that encodes a non-specific immunoregulatory protein.

IL2RA

IL2RA codes for a subunit of the IL-2 receptor. Like many receptors, the IL-2 receptor is a dimer composed of an α and a β subunit. **IL-2** is a major cytokine involved in T cell activation. *IL2RA* is constitutively expressed at high levels in $CD4^+$ T cells, which are also positive for the FoxP3 protein. These cells are believed to be important in immune tolerance to self-proteins. The association with T1D is protective and appears to involve a

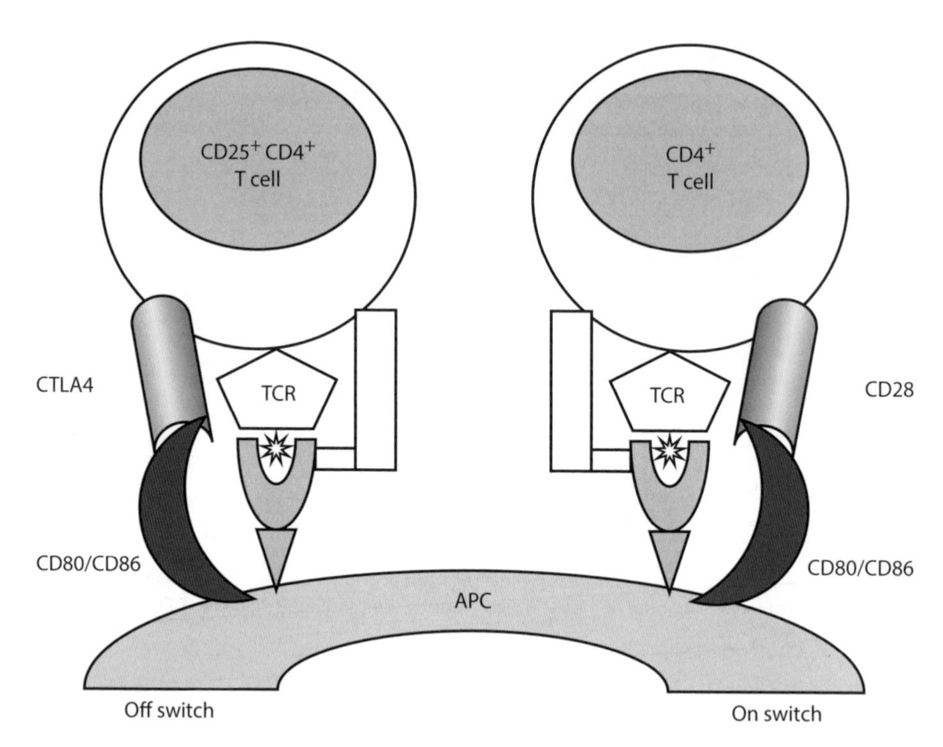

Figure 10.5: Suppression of auto-reactive T cells by CTLA4 signaling. The figure illustrates the process of CTLA4 signaling by the CD25$^+$ CD4$^+$ T cell. The process is the same as that for T cell activation; however, in this situation the T cells send out a negative response on activation by the APC. Antigenic peptides are represented here as stars. Antigenic peptides are presented to the TCR by MHC class II proteins (shown in gray as an inverted triangle supporting a curved structure), which is mounted on the APC (also shown in gray). There is co-recognition of this molecular complex by CD4 (shown as a lozenge with a short leg). The first part of the process is the same for both T cells. However, a second interaction occurs which is different. The T cell on the right expresses CD28, which interacts with the CD80/CD86 molecule (B7 in some texts; shown here as dark curves) and this causes T cells to induce immune activation. The T cell on the left expresses CTLA4, which interacts with CD80/CD86, sending a signal that downregulates the immune response. CD28 and CTLA4 are both illustrated as gray tubes on the surface of the T cells.

gain-of-function polymorphism. However, we should be cautious and not over-interpret such data. Biological systems are complex and often encompass some degree of redundancy. Increases and reductions in receptor expression can both influence the level of the immune response leading to greater or lesser levels of organ damage. Inappropriately high levels of receptor expression may lead to over-activation, whereas inappropriately low levels may prevent activation. However, high receptor expression does not necessarily lead to higher levels of activation because there may be low levels of the ligand protein, which in this case is IL-2. This is considered in more detail in **Figure 10.6**.

IFIH1

The interferon (IFN)-induced helicase C domain-containing protein 1 gene (*IFIH1*) codes for a protein that is believed to be able to interact with viral RNA and to induce the expression of IFNs. IFNs are key agents in both innate and adaptive immunity, and play an important role in immune regulation. Numerous genetic polymorphisms with IFN genes have been associated with genetically complex disease and it is not surprising to find genes from the interferon pathways associated with T1D.

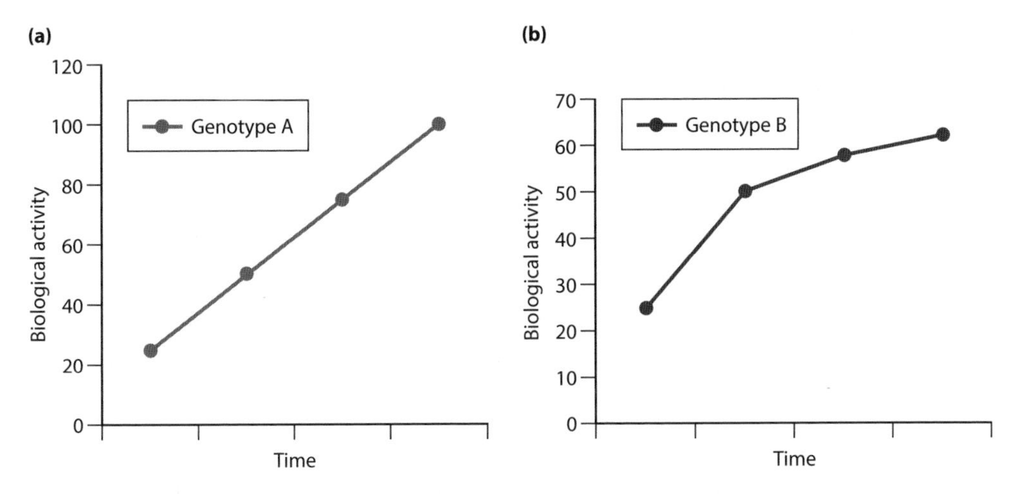

Figures 10.6: Illustration of the reason why both elements of a system need to be considered when considering the impact of a polymorphism on a trait. The two graphs illustrate the interaction between genetic polymorphisms in a biological pathway. If we consider a model where we have only two components, e.g. IL-2 and its receptor, biological activity can increase to infinity as long as there is adequate production of both the receptor and its ligand. If genotype A represents those with high ligand expression, activity will increase for as long as there is sufficient receptor expression. Similarly, if genotype A is associated with high receptor expression, activity will increase provided there is a continuous supply of ligand. Genotype B, however, indicates a more reasonable proposition. In the second example activity begins to reach an upper limit because one or other of the two components reaches saturation. In real biological systems high expression of a ligand or receptor like IL-2 does not necessarily result in activity. Thus, we need to be careful about interpreting data that indicates increased expression of a gene or protein associated with a specific polymorphism.

10.4 GWAS STUDIES IN T1D

There had already been some GWAS in T1D prior to the conclusion of the Human Genome Project (HGP), but these were few and far between. One using VNTRs performed in 1994 on T1D has been discussed already (Table 10.2). Most of the early studies were limited to a few hundred markers and though they represented a step forward in their time, they now appear primitive compared with today's technologies. Most importantly, with so few markers tested, the possibility of missing significant associations was very high in the pre-genome studies. It is therefore reasonable to say that the application of GWAS as a means of detecting risk alleles in complex disease in the post-genome era was a major step forward with great potential. The transition from a **hypothesis-constrained** approach, which applied to most case control studies, to a **hypothesis-generating** approach and the technology that enabled large numbers of cases and controls to be genotyped for upwards of 500,000 tagged SNPs was quite amazing. Even though new technologies involving higher throughput and higher resolution are currently being applied and developed, which will make current GWAS seem primitive in the future, the impact of the first GWAS is still fresh.

The 2007 WTCCC1 study was one of the first GWAS in T1D

The WTCCC1 2007 GWAS included 2000 T1D cases. Several regions were identified as major risk determinants including all of the genes or regions listed above: MHC, *CTLA4*, *PTPN22*, *IL2RA/CD25*, and *IFIH1/MDA5*. It is important to note that of these previously identified regions and genes only two had *P* values above the high statistical threshold $P < \times 10^{-7}$ (MHC and *PTPN22*) and two were in the lower statistical threshold region

Table 10.3 Six major risk genes and regions for T1D in the post-genome era.

Alleles or Gene(s)	Location	Function	*P* value and/or OR
*DQB1*02:01* (DQ2) *DQB1*03:02* (DQ8) *DQB1*01:02* (DQ1)	6p21.3	Antigen Presentation	WTCCC1 study $P < 2.42 \times 10^{-134}$ OR = 5.49
8.1 haplotype 7.2 haplotype *B*39* (independent)	6p21.3	Antigen Presentation	
INS	11p15.5	Insulin	
PTPN22	1q13	Protein tyrosine phosphatase non-receptor-type 22	WTCCC1 study $P < 1.82 \times 10^{-26}$ OR = 5.43
CTLA4	2q33	Cytotoxic T lymphocyte protein 4 Downregulates immune response	WTCCC1 study $P < 1.8 \times 10^{-5}$
IL2RA	10p15	IL-2 receptor subunit α. Cytokine receptor subunit	WTCCC1 study $P < 4.3 \times 10^{-5}$
IFIH1	2q24	IFN-induced helicase C domain-containing protein 1	WTCCC1 study $P < 7.6 \times 10^{-3}$

Four of the listed genes or regions were identified long before the publication of the Human Genome Mapping Project and use of GWAS: MHC, *INS*, *CTLA4*, and *IL2RA*. Apart from *INS*, all of these genes or regions (MHC) have associations with other diseases (i.e. they are not specific to T1D).

around $P < \times 10^{-5}$ (*IL2RA* and *CTLA4*) **(Table 10.3)**. In addition to the associations above a number of new regions were identified as showing strong risk ($P < 10^{-7}$); these include 12q13, 12q24, and 16p13 together with other regions with similar levels of significance, including 4q27, 12p13, 18p11, and 10p15 (CD25). Of these, the associations with 12q13, 12q24, 16p13, and 18p11 have all been confirmed in other studies. A number of potential functional signals can be identified in this group, e.g. 12q13 is close to the *ERBB3* gene that encodes the receptor tyrosine kinase erbB-3 precursor, and 12q24 is close to *SH2B3/LNK* (SH2-B adaptor protein 3), *TRAFD1* (TRAF-type zinc-phosphatase domain containing 1), and *PTPN11* (protein tyrosine phosphatase, non-receptor type 11).

PTPN11 is a particularly interesting candidate for T1D as it is a member of the same family of regulatory phosphatases as *PTPN22*. As discussed earlier, *PTPN22* is associated not only with T1D but also with Crohn's disease and rheumatoid arthritis, suggesting overlapping pathology. The 12q24 association reported above is associated with a combined signal (measured as probability value *P*) for T1D, Crohn's disease, and rheumatoid arthritis of 9.3×10^{-10}. Overlapping genetic associations are reported for a number of different complex diseases (**Figure 10.7**).

The association with the 10q15 region that contains the *CD25* gene is found in Graves disease, rheumatoid arthritis, and T1D. Graves disease was not included in the WTCCC1 study, but rheumatoid arthritis and T1D were both included. The study reported separate independent associations for both diseases. CD25 encodes a high-affinity receptor for IL-2, and this association may highlight the importance of the IL-2 pathway in T1D and other autoimmune diseases.

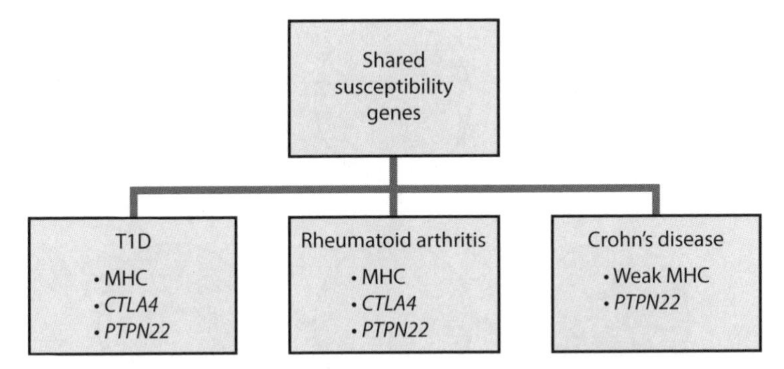

Figure 10.7: Genetic overlap in complex disease–finding the same associations in different diseases. The figure illustrates some of the shared susceptibility genes for three very common complex diseases: T1D, rheumatoid arthritis, and Crohn's disease. The shared susceptibility regions or loci are the MHC (a region not a locus), the *CTLA4* genes, and the *PTPN22* gene. The list is not exclusive as many other genes could be included; it is only intended to illustrate a point.

Following the introduction of GWAS in 2007, research has resulted in the identification of at least 40 further potential T1D alleles

Subsequent GWAS analysis has resulted in the detection of more novel genetic associations. These tend to be weaker than those discussed so far. The current count suggests there may be more than 40 risk alleles in all (a selection of the strongest of these is shown in **Table 10.4**). However, some of these are synonymous polymorphisms in specific genes and others are in non-coding sequences. Synonymous polymorphism in the exome sequence does not lead to sequence variation in the encoded polypeptide and therefore is not functionally significant. Synonymous polymorphism in the non-coding sequence is also unlikely to be functionally significant. However, the associations with SNP markers should not be written off; in both cases the associations may act as surrogates for functional associations elsewhere on the haplotype.

Table 10.4 Other risk alleles for T1D.

Potential candidate	Location
CD25 (high-affinity receptor for IL-2)	10p15
ERBB3 (receptor tyrosine protein kinase erbB-3 precursor)	12q13
SH2B3/LNK (SH2-B adaptor protein 3) *TRAFD1* (TRAF-type zinc finger domain protein 3) *PTPN11* (protein tyrosine phosphate, non-receptor type 11)	12q24 *PTPN11* is the most likely candidate with associations with both rheumatoid arthritis and Crohn's disease in addition to T1D
KIAA0350 (dexamethasone-induced transcript)	16p13 Function of these genes is unknown
Possibly *PTPN2,* a member of the same family as *PTPN11* and *PTPN22* protein tyrosine phosphatase, non-receptors	18p11

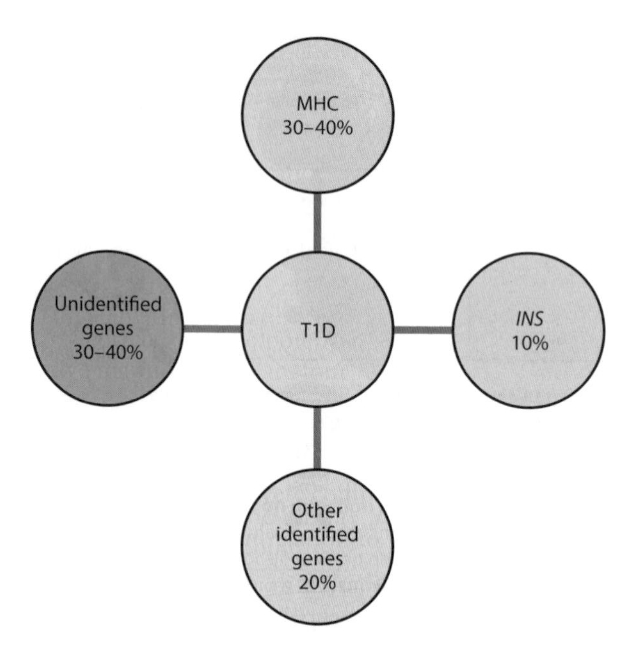

Figure 10.8: Genetic accounts book for T1D. The figure shows the current estimate of the total genetic impact of genes identified as risk markers for T1D. The MHC has the strongest impact, with the insulin (*INS*) gene second. The other group includes numerous genes involved in immune regulation, but there is still a substantial portion of the genetic risk to be identified if these figures are accurate.

The addition of replicated novel associations discovered in GWAS to those already well-established associations from a mix of family studies and candidate gene analysis has further increased the extent of the genetic account for T1D to approximately 70% (**Figure 10.8**). This is much higher than for the majority of complex diseases and is a very different situation to that seen with T2D (below).

10.5 EARLY GENETICS OF T2D

The vast majority of diabetes cases are of T2D, once called non-insulin-dependent diabetes mellitus (NDDM), and the WHO estimates that there are currently more than 340 million people suffering from this disease worldwide, with numbers expected to rise further over the next 20 years. The rise in T2D reflects differences in the environment and in our lifestyles. In addition, there is thought to be a substantial genetic contribution to T2D.

As the name T2D implies, this is not the same as T1D. It is a disease with a later average age of onset than T1D, the genetics of T2D is more complex than that for T1D, and this is a disorder that is less easy to investigate than T1D. Despite all of this, there have been intensive efforts to identify the genetic risk factors in T2D over the past 20–30 years. Initially, linkage analysis and candidate gene case control studies were used. More recently, GWAS has been employed. This has resulted in the identification of more than 60 genetic risk alleles for T2D.

Prior to the introduction of GWAS, it was simply not possible to identify risk alleles for many complex diseases, such as some sporadic cancers. However, in T2D, some early linkage and candidate gene case control association studies successfully identified risk alleles. In particular, three specific genes were identified as carrying potential risk alleles that were subsequently confirmed by GWAS. These included significant associations with non-synonymous SNPs in the *PPARG*, *KCNJ11*, and *TCF7L2* genes. A number of other associations originally described in the same era were not confirmed by GWAS. Confirmation in later studies is a mark of quality and reflects study design, and especially sample size. Unlike T1D, there are no associations with HLA or the MHC in T2D. This may indicate that the disease pathogenesis is very different from that for T1D and does not involve the same immunological (or other biological) processes.

There have been different interpretations of the associations with *PPARG*, *KCNJ11*, and *TCF7L2*

Studies have shown that carriage of the relatively rare alanine-encoding allele *PPARG* variant is associated with a reduced risk of disease. This association has been widely confirmed in a number of studies. The *PPARG* gene is a plausible candidate gene for T2D as it encodes the peroxisome proliferator-activated receptor-γ—a nuclear receptor that is a regulator of adipocyte differentiation and glucose homeostasis. The ligands for this receptor include prostaglandin metabolites and glitazone drugs. The PPARG protein has been used as a target for treatment by the thiazolidinedione drugs, such as rosiglitazone.

In contrast to *PPARG*, the situation with *KCNJ11* seems to be less clear. The *KCNJ11* gene encodes the inwardly-rectifying KIR6.2 component of the pancreatic β cell ATP-sensitive potassium channel (KATP), E23K. KIR6.2 is found in islet cells of the pancreas. These KATP channels couple cell metabolism to membrane excitability. In the pancreatic β cells the channel is an octameric complex of KCNJ11 subunits and SUR1 (encoded by the adjacent *ABCC8* gene) subunits. Four subunits form the channel pore. Each subunit is associated with a SUR1 subunit that regulates gating. Initial reports suggested an association with a non-synonymous variant of *KCNJ11*. Looking at the function of this gene's product, it appears to be a very plausible candidate for T2D. However, it has been reported that it is not possible to distinguish between the risk associated with this SNP and another non-synonymous SNP in the adjacent *ABCC8* gene that codes for the sulfonylurea receptor (SUR1). SUR1 has been used as a target for another group of widely used anti-diabetes drugs.

The third genetic risk factor for T2D identified prior to GWAS was *TCF7L2*. This gene was identified from a signal on chromosome 10q in linkage analysis of families affected by T2D. Fine mapping showed that the signal related to an intronic SNP in *TCF7L2*. *TCF7L2* codes for the high mobility group (HMG) box-containing transcription factor TCF4, which modulates the Wnt signaling pathway and possibly the secretion of insulin by the pancreas. In addition, this gene may have a role in glucose production in the liver. *TCF7L2* has a very strong association with T2D. The per allele OR of 1.4 remains the strongest risk factor for T2D identified so far. However, the underlying mechanism for the effect of this intronic SNP on *TCF7L2* gene expression remains unclear. It is possible that the location in an open chromatin site in pancreatic β cells is important and there is also some evidence that the risk allele is associated with increased transcription.

10.6 GWAS STUDIES IN TYPE T2D

The first GWAS in T2D were reported in 2007. These studies confirmed some previously reported associations and detected a range of additional genetic risk factors. The previously reported risk factors *PPARG*, *KCNJ11*, and *TCF7L2* were all confirmed, and a number of novel risk loci were identified (**Table 10.5**). These initial studies typically involved

Table 10.5 Selected well-replicated risk loci for T2D.

Name	Gene, location, OR, and *P* value
Peroxisome proliferator-activated receptor-γ	*PPARG* (3p25) WTCCC1 OR = 1.23, $P = 5.4 \times 10^{-3}$
Potassium channel, inwardly rectifying, subfamily J, member 11	*KCNJ11* (11p15) Close to *ABCC8* gene WTCCC1 OR = 1.15, $P = 5.2 \times 10^{-3}$
Transcription factor 7-like 2	*TCL7L2* (10q25) WTCCC1 OR = 1.36, $P = 5.2 \times 10^{-12}$
Solute carrier family 30 (zinc transporter), member A8	*SLC30A8* (4 GWAS) OR = 1.12–1.15
CDK5 regulatory subunit associated protein 1-like 1	*CDKAL1* (4 GWAS) OR = 1.12–1.25
Hematopoietically expressed homeobox	*HHEX* (3 GWAS) OR = 1.13–1.18
Hepatocyte nuclear factor-1α HFN1 homeobox A	*HNF1A* (2 GWAS) OR = 1.07–1.14
HFN1 homeobox B	*HNF1B* (3 GWAS) OR = 1.1–1.17
Insulin-like growth factor 2 mRNA-binding protein 2	*IGF2BP2* (3 GWAS) OR = 1.14
Potassium voltage-gated channel subfamily, member 1	*KCQN1* (3 GWAS) OR = 1.08–1.23
Cyclin-dependent kinase inhibitor 2A	*CDKN2A/2B* (3 GWAS) OR = 1.2

a number of centers, including the WTCCC1 study, and the number of cases varied from around 2000 to approximately 8000 cases. Most cases were of European origin. There was good agreement between the various studies in terms of risk factors detected and the findings were also replicated in additional cases in most studies.

Examples from the WTCCC1 study

SLC308A

The *SLC30A8* (solute carrier family 30, member A8) gene encodes the zinc transporter ZnT8—a protein with a role in the secretion of insulin by the pancreas. It contributes to insulin maturation and/or storage in the pancreatic β cells. This association was only weakly identified in the WTCCC1 study, but was also identified in a study of French T2D cases. Several GWAS have now confirmed this association.

FTO

The *FTO* (fat-mass and obesity-associated) gene on chromosome 16q12 was found to be very strongly associated with T2D in the WTCCC1 study (OR 1.34, $P < 5.24 \times 10^{-8}$), but has not been universally reported. Surprisingly, the *FTO* association is considerably stronger than that for *SLC30A8*. However, a number of studies have replicated the association with *SLC30A8*, which contrasts with the situation for the *FTO* gene. These differences between studies are important. They may reflect genuine evidence of false positives or they may reflect differences in study design, e.g. the use of different commercial chips (Illumina or Affymetrix) with different tagged SNPs. One solution to this would be to use specifically designed chips for studying specific diseases and disease groups. However, while many companies used to be willing to produce specific chips at a reasonably low price, this is no longer considered such a viable option and it may not be possible in the future. Observed differences in associations reported can arise from differences in the close proximity between tagged SNPs and actual disease alleles for which they are markers.

CDKAL1

The third strong association signal in the WTCCC1 GWAS was with the *CDKAL1* (CDK5 regulatory receptor subunit associated protein 1-like 1) gene. The product of this gene shares homology with the protein domain-level CDK5 regulatory subunit associated protein 1 (CDK5RAP1), which is known to inhibit the activation of CDK5. CDK5 is a cyclin-dependent kinase that has been implicated in normal β cell function. The association has been confirmed in several GWAS.

Other risk alleles for T2D from other studies

To date, the number of replicated genetic risk factors for T2D has increased to more than 60. This has been achieved by larger studies and particularly by meta-analysis. The list of risk loci includes *CDKN2A*, *HHEX*, *HNF1A*, *HFN1B*, *IGF2BP2*, and *KCNQ1*, among many others. Despite this work, our understanding of the underlying mechanism by which these genes increase or reduce the risk of T2D remains unclear. Each of the identified risk alleles is associated with relatively small though significant ORs. Such low ORs indicate that we have probably identified less than 10% of the genetic risk for T2D so far. These figures also serve as a reminder that the environment plays a major role in T2D (**Figure 10.9**). However, there are likely to be many more risk alleles yet to be identified in T2D—a situation similar to that for a range of different complex diseases.

The genes that have been identified in T2D are mostly involved in insulin secretion by the pancreas rather than insulin action. This suggests that these are genuine candidate genes for T2D. The absence of genes associated with autoimmunity in T2D helps to validate the case cohorts used in these studies. In addition, the absence of immune response genes in this list also helps to validate studies in T1D where such associations are found. This also proves the point that such comparisons between diseases with quite different etiologies can be useful.

Attempts have been made to score individual risk for the development of T2D based on the findings of these studies. Overall, knowledge of genotype is likely to be of limited value in diagnosis, adding little to the list of already well-established clinical criteria and tests. This is not surprising when we consider the low level of disease risk that is associated with each of these alleles. However, some of the genes that have been recently identified could encode potentially useful drug targets. One way forward could be stratification on the basis of genotype. By using selected genetic (and other) risk factors we could develop personalizing treatment to suit the needs of the individual patients. There is currently some, albeit limited, evidence that this may work in T2D.

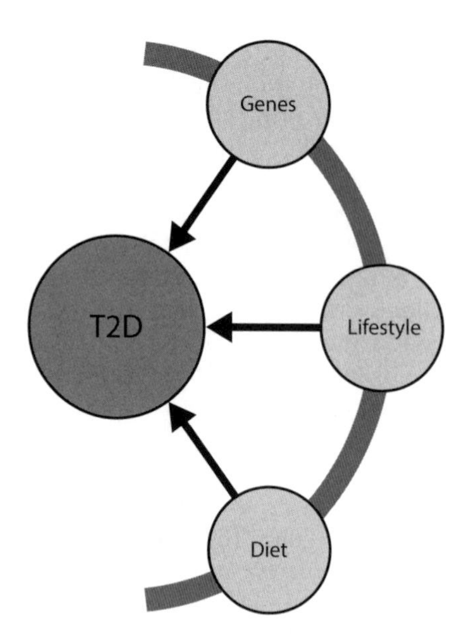

Figure 10.9: Factors influencing risk of T2D. The figure illustrates the main factors associated with an increased risk of T2D.

10.7 THE FUTURE OF GENETICS IN T2D

Unlike T1D, for which a very significant component of the genetic risk has now been identified, studies indicate that the pattern is likely to be much more complicated and challenging for T2D. As T2D represents a greater clinical case load than T1D, it is essential that these studies continue. Two developments are likely to have a large impact on future studies in diabetes: the introduction of **genome sequencing** as a routine research procedure and the expansion of the use of epigenetics.

Future prospects in T2D research involve genome sequencing

Rare genetic variants whose contribution cannot be assessed by GWAS are likely to be detected in genome-sequencing studies. One of the first studies on detection of rarer variants that contribute to disease used **exome sequencing** on 1000 Danish cases and 1000 controls. Polymorphisms identified as significant risk markers were replicated in a series of approximately 80,000 Europeans. Some novel associations were seen when comparing the genotypes identified with the metabolic traits in T2D, but the overall conclusions were slightly disappointing. The study found that coding region polymorphisms with frequencies in the range 1–5%, which would not have been covered by the previous GWAS, did not show large risk effects (OR values) for diabetes or related metabolic traits. Further studies may need to focus on very rare sequence variants, where effect sizes could be larger, but such studies will require very large numbers of cases.

Epigenetics may be important in diabetes

The missing 90% of disease risk is likely to involve a substantial contribution from non-genetic factors, epigenetic factors which are currently more difficult to detect than genetic factors, and epistasis involving gene–gene interactions. **Epigenome-wide association studies (EWAS)** where **DNA methylation** patterns across the genome are compared between cases and controls have recently been applied to T2D. Currently, such studies are technically more difficult to perform than GWAS for several reasons. These include the fact that patterns of DNA methylation may vary considerably between tissues, so using DNA from accessible tissues such as blood may not be representative of methylation patterns in disease tissues, e.g. the pancreas. A second issue is that accurate assessment of methylation requires DNA sequencing following treatment with bisulfite, though it is possible to narrow down the chromosomal regions differentially methylated prior to this sequencing. **Methylation chips** that cover methylation sites across the genome are also now available. Methylation patterns may also be influenced by a variety of environmental factors and so can change over time. In addition, studying DNA methylation represents only one measure of **epigenetic regulation**, which also involves processes such as **histone acetylation**. Despite these caveats, studies based on both methylation of candidate genes and EWAS for T2D have now been reported. One EWAS used DNA from diabetic pancreatic islets and detected differential methylation at promoter regions of 254 genes. These methylation changes were not present in DNA from blood samples. Another recent EWAS reported lower levels of methylation in the region of genes such as *TCF7L2* in DNA from diabetic blood samples, but did not study methylation patterns in other tissues. However, both approaches may be valid. Further discussion of **epigenomics** is beyond the scope of this chapter, but it seems likely that this area of research may provide novel insights into the risk of T2D and other diseases in the near future (see chapter 12, section 12.5).

10.8 GENETICS OF MONOGENIC DIABETES

There are a number of monogenic forms of diabetes that have some features in common with either T1D or T2D, but they represent distinct forms of the disease. In these cases the familial component is stronger, with at least one affected parent, and the age of onset is usually very young (typically ranging from infancy to early adulthood). The collective term for many of these disorders is MODY. This term is not applied to all early-onset forms of diabetes as there are some forms with more specific names, e.g. Donohue syndrome and Rabson–Mendenhall syndrome. Based on the location of the disease-causing mutation, there are at least 10 different forms of MODY currently listed in the OMIM (Online Mendelian Inheritance in Man) database, accounting for up to 5% of all cases of diabetes (**Table 10.6**). The phenotypes differ from those of classical T1D or T2D and will depend

Table 10.6 Chromosomal locations of genes causing MODY.

Gene	Location
MODY1	20q13
MODY2	7p13
MODY3	12q24.3
MODY4	13q12.2
MODY6	2q31.3
MODY7	2p25.1
MODY8	9q34.2
MODY9	7q32.1
MODY10	11p15.5
MODY11	8p23.1

on the specific gene defect. These are monogenic diseases. Two subtypes of monogenic diabetes are of particular interest in relation to our understanding of genetic susceptibility to T2D: neonatal diabetes, which usually has an onset during the first 6 months of life, and young onset diabetes, where the disease is due to mutations in transcription factor genes such as *HNF1A*, which codes for hepatocyte nuclear factor (HNF)-1α. This mutation has also been found to be associated with an increased risk of T2D.

A significant proportion of neonatal diabetes cases are due to gain-of-function mutations in either *KCNJ11* or *ABCC8*, which encode separate subunits of the pancreatic β cell potassium channel. These mutations typically increase the electrostatic current through the channel, preventing depolarization in response to glucose metabolism and impaired insulin secretion. Polymorphisms in these genes are also associated with increased susceptibility to T2D. The relatively recent finding that many cases of neonatal diabetes were due to mutations in potassium channel genes was an important advance in terms of treatment. These patients would have been previously treated with insulin on the grounds that their disease was insulin-dependent, but they are now normally treated with a high dose of an orally administered sulfonyl urea that targets potassium channels directly. Interestingly, another potassium channel gene, *KCNQ1*, has been found to be associated with an increased risk of T2D.

Mutations in *HNF1A* are a common cause of young-onset diabetes. *HNF1A* encodes the HNF-1α, a transcriptional activator that regulates tissue-specific expression of a range of genes, particularly in the liver and pancreatic islet cells. The mutations reported are dominant mutations which result in deterioration of pancreatic β cell function including the ability to secrete insulin. *HNF1A* is now well established as a genetic risk factor with a modest effect in T2D. Mutations in a gene encoding a separate transcription factor with homology to *HNF1A*, *HNF1B*, are also associated with young-onset diabetes and T2D.

There is considerable overlap between the genes associated with these monogenic forms of diabetes and those associated with increased risk of T2D. The individual mutations involved in the monogenic disease differ from the polymorphisms predicting susceptibility to T2D and the overall phenotypes also differ in a number of respects. However, this is different to the example of breast cancer considered in Chapter 9, where the genes contributing to the familial disease do not appear to make a significant contribution to susceptibility in sporadic disease.

CONCLUSIONS

In this chapter we have considered the results of genetic studies in one of the most common forms of disease in the developed world, i.e. diabetes mellitus (T1D, T2D, and MODY). There are profound differences between the genes identified in T1D versus T2D, corresponding to the different phenotypes for these two diseases. T1D is an autoimmune disease (see Section 10.1) with all of the features that we would expect with autoimmunity. There is a strong association with the MHC and especially, but not only, with HLA class II *DQB1*. There is also a strong association with a number of other genes, especially the insulin gene *INS* and a number of immune-regulatory genes, including *CTLA4* and *IL2RA*. Overall, the genetic studies support the hypothesis that this is an autoimmune disease. There is significant overlap with many other similar autoimmune and immune-mediated diseases. However, there are also some differences, e.g. the same *HLA-DQB1* allele that is associated with protection from T1D is associated with an increased risk of multiple sclerosis and narcolepsy. The picture presented is not a simple one and yet a high proportion of genetic heritability (up to 70%) may now be explained by a relatively small to moderate group of genetic risk factors that have a mix of very strong to moderate effect sizes.

Compare this to the situation in T2D where the genetic risk factors are very different from those identified in T1D. There are no associations with HLA in T2D and there are no major associations with other immune-regulatory genes. This almost certainly reflects the different pathogenesis of T1D compared with T2D. Furthermore, most of the identified and confirmed risk alleles have small effects. The risk genes are different from T1D, but there is some overlap with some of the early-onset monogenic forms of the disease, for example the potassium channel genes *KCNJ11* and *KCNQ1*.

Compared with T1D, where current estimates indicate a very significant portion of the total genetic risk has been identified, in T2D only 10% of the genetic risk has been identified. However, this is not because there has been a lack of effort to identify the risk genes in T2D. On the contrary, over 60 different candidate genes have been identified so far.

In the absence of strong associations it is going to be difficult to use the current data from T2D studies in disease diagnosis or in patient management. Studies will continue and it is likely that at some point, through a better understanding of the genetic components of this disease, it will be possible to use this knowledge in patient treatment and management, and to develop new treatments for this disease. When considered on a global scale, the ability to offer individualized risk assessment for T2D using genetic information remains an important target. Future studies will use a combination of new and old technologies, and there will be an increased use of epigenetics to understand diabetes, especially T2D.

Genotyping has led us this far in many diseases, but phenotyping will produce the next piece in the jigsaw. It is OK to be selective in terms of deciding which genes to look at as long as all of the potential candidates are considered at some time. Otherwise the genotyping will have been a waste of time and the phenotyping work will fall into the same trap that applied to early genetic association studies, i.e. rejection of positive associations, but not based on numbers or study design (as in the past), but this time based on absence of knowledge or absence of an obvious (simple) link.

FURTHER READING

Books

Armstrong L (2014) Epigenetics. Garland Science. This is a useful informative book for students wanting to understand the application of epigenetics to complex disease and epigenetics in general.

Holt RIG, Cockram C, Flyvbjerg A & Goldstein BJ (2010) Textbook of Diabetes, 4th edn. Wiley-Blackwell.

Parham P (2009) The Immune System, 3rd edn. Garland Science. This is an excellent basic immunology book for the student of human immunology including links to immunogenetics and some useful data on diabetes.

Wass JAH, Stewart PM, Amiel SA & Davies MC (2011) Oxford Textbook of Endocrinology and Diabetes, 2nd ed. Oxford University Press.

Articles

Ashcroft FM & Rorsman P (2012) Diabetes mellitus and the β cell: the last ten years. *Cell* 148:1160–1171.

Billings LK & Florez JC (2010) The genetics of type 2 diabetes: what have we learned from GWAS? *Ann NY Acad Sci* 1212:59–77.

Davies JL, Kawaguchi Y, Bennett ST et al. (1994) A genome-wide search for human type 1 diabetes susceptibility genes. *Nature* 371:130–136. This describes one of the first genome-wide genetic studies. This was a pre-genome map study and limited, but nevertheless inspirational.

de Miguel-Yanes JM, Shrader P, Pencina MJ et al. (2011) Genetic risk reclassification for type 2 diabetes by age below or above 50 years using 40 type 2 diabetes risk single nucleotide polymorphisms. *Diabetes Care* 34:121–125.

Drong AW, Lindgren CM & McCarthy MI (2012) The genetic and epigenetic basis of type 2 diabetes and obesity. *Clin Pharmacol Ther* 92:707–715.

McCarthy MI, Rorsman P & Gloyn AL (2013) *TCF7L2* and diabetes: a tale of two tissues, and of two species. *Cell Metab* 17:157–159.

Murphy R, Ellard S & Hattersley AT (2008) Clinical implications of a molecular genetic classification of monogenic beta-cell diabetes. *Nat Clin Pract Endocrinol Metab* 4:200–213.

Pal A & McCarthy MI (2013) The genetics of type 2 diabetes and its clinical relevance. *Clin Genet* 83:297–306.

Polychronakos C & Li Q (2011) Understanding type 1 diabetes through genetics: advances and prospects. *Nat Rev Genet* 12:781–792.

Steck AK & Rewers MJ (2011) Genetics of type 1 diabetes. *Clin Chem* 57:176–185.

Todd JA, Bell JI & McDevitt HO (1987) HLA-DQβ gene contributes to susceptibility and resistance to insulin dependent diabetes mellitus. *Nature* 329:599–604. This is a landmark paper in understanding T1D and the relationship between HLA and disease.

Toperoff G, Aran D, Kark JD et al. (2012) Genome-wide survey reveals predisposing diabetes type 2-related DNA methylation variations in human peripheral blood. *Hum Mol Genet* 21:371–383.

Ueda H, Howson JMM, Esposito L et al. (2003) Association of the T-cell regulatory gene *CTLA4* with susceptibility to autoimmune disease. *Nature* 423:506–511. This is a very important study on *CTLA4* highlighting the complexity of association studies even when targeted to a specific gene.

van de Bunt M & Gloyn AL (2010) From genetic association to molecular mechanism. *Curr Diab Rep* 10:452–466.

Volkmar M, Dedeurwaerder S, Cunha DA et al. (2012) DNA methylation profiling identifies epigenetic dysregulation in pancreatic islets from type 2 diabetic patients. *EMBO J* 31:1405–1426.

Online sources

http://www.who.int/diabetes/facts/en
This is an excellent site for finding information on disease prevalence, incidence, morbidity, and mortality, and especially projections for the future.

CHAPTER 11

Ethical, Social, and Personal Consequences

In the previous chapters we have considered how, why, and what we hope to achieve by investigating the genetics of complex disease, and used different example diseases to illustrate these points. Each chapter contains several examples of how common genetic variation increases the risk of common genetically complex (i.e. non-Mendelian) disease. The issue of common genetic variation being linked to common disease is critical in modern society. One of the key elements in this consideration is ethics. In this chapter, we will consider different ethical aspects of enquiry into complex disease with a brief overview of the general philosophy behind ethical enquiry. We will consider lessons from the past, looking at how genetics has been misused and why it is important to safeguard ourselves against the genetic inquisition. We will consider the use, and potential misuse, of complex disease genetics in a modern society and why ethical guidance is so important in genetics. We will also consider the use of genetic data, what is personal and confidential, and where the boundary of confidentiality fits into the picture? Finally, we will consider who owns the genome, and the relationship between commerce and academia. This chapter is designed to open the doors to ideas and concepts not yet dealt with in other areas of the book; therefore, unlike the other chapters in this book, this chapter is not crammed full of examples of genetic studies, except where they are appropriate. Instead, it is designed to stimulate debate and reading for those with an interest in ethical issues and in the history of genetics. The chapter includes some references to the history of genetics that may be sensitive issues for some readers. Nevertheless, it is thought important to include them. The chapter does not discuss the ethics of work on animals. This latter omission does not reflect the views of the authors, but simply the fact that the majority of the book reflects recent genetic research based on human subjects.

11.1 DEFINING ETHICS

The *Oxford English Dictionary* defines ethics as relating to morals or a set of principles of morals; the latter may be considered as a series of rules of conduct. Ethics is a philosophical concept and the word is derived from the Greek word *ethos*. Christopher Tollefsen, in *Biomedical Research and Beyond: Expanding the Ethics of Inquiry* (2008), refers to the ethics of enquiry as a term that covers the philosophy of ethics across a range of disciplines. Ethics applies to many aspects of human life. We are concerned with the ethics of enquiry as it applies to biomedical research from the bench to the bedside. Tollefsen states there is no unified ethics of enquiry; however, he goes on to say that by applying a single systematic approach we can consider all moral action is action for the sake of human fulfillment and well-being. At first this seems complicated, but what this statement allows us to do is to focus on one ethical principle: fulfillment and well-being. From this we are able to consider both the ethical aspects of the conduct of biomedical research, and the issues pertaining to the value of enquiry for individuals and different cultures.

There are philosophical arguments for and against ethical constraint in biomedical research

There are some who argue that to apply moral (ethical) judgment to science is wrong. However, others argue that as enquiry is a fundamental aspect of human behavior and we are born to question, it must be controlled through the application of ethical guidelines. Others believe in the autonomy of science (**Figure 11.1**). Scientists are not alone in this view; some other academics have also expressed a wish for pursuit of the truth at any price. Those against ethical/moral constraints argue that:

- Research could be more effective and perhaps cheaper if we put aside ethical considerations.
- Research should be judged on the results alone (ethical philosophers refer to this as **consequentialism**).
- Knowledge alone is sufficient to justify inquiry (this is a type of Aristotelian ethical philosophy).

Further possibilities involve discussions about the nature of the state's relationship with individuals and politics. Neither of these are strictly ethical issues, but they do impact on biomedical research. However, a detailed discussion of ethical philosophy is beyond the scope of this book. We are concerned with the practical implication of ethics in current research in genetics of complex disease only and readers wishing to delve into ethics in more detail are directed to Tollefsen (2008) in the first instance.

Figure 11.1: Ethical control versus autonomy in science. The figure is designed to represent the balance between those who argue that applying moral ethical judgment in science is wrong, i.e. scientific autonomy (the need for truth at any price - black bowl), versus those who say there must be control or regulation (white bowl). The reader must make his/her own mind up as to how the scales hang and whether to champion the dark or the white bowl as this is a matter of debate and personal moral judgment.

What are the practical ethical implications in the study of genetics of complex disease?

Science is public and, like medicine, law, and journalism, is a genuine profession. It is public in three respects: the methods applied, the respect obtained, and the use of the results in terms of products and power. Tollefsen tells us that for any practice to become a profession it must provide benefits not just to the practitioners, but to a wider society. There is no doubt that science provides abundant benefits in healthcare as well as in agriculture and engineering. Therefore, there is no questioning that under this definition science is a profession. The importance of this is that professions are regulated and have rules that govern them. In biomedical sciences we are concerned with the investigation and the application of the science of medicine, and very strict rules are applied. In practice, all research applications undergo scrutiny by ethical committees, all new practices are thoroughly considered by ethical boards and committees, and even ethical committees are reviewed to ensure appropriate procedures are followed. Scientific results are reviewed by groups of peers prior to publication and mutual criticism has become an essential brick in the wall for scientists. Furthermore, scientific information cannot be hidden, especially when advances rely on collaboration and public funding. This openness assures a level of ethical quality that could otherwise be missing.

Ethical approval and counseling: good practice

Many problems and issues arise in researching complex disease, but most of them can be dealt with through good practice in setting up studies and in provision of high-quality counseling services. When testing patients they must be made aware of the reasons for the test, the potential outcomes from the test, and the personal impact of the knowledge of the results on themselves and their families and on their social group. Most ethical committees will require this as standard practice before permitting the introduction of new practices or clinical trials. Sponsors, including the National Institutes of Health (NIH) in the USA, the Medical Research Council (MRC) and Wellcome Trust (both in the UK), and multinational drug companies, will all expect ethical approval before they accept and fund research and clinical trials. Seeking ethical approval can be quite a complex process. Practice and procedure vary between different centers and countries. As this is not the subject of this book, then for those needing help, the local ethical committee is the correct starting place.

As with ethical approval, counseling practice varies from country to country and even between different regions. However, it is becoming increasingly important that counseling is considered if genotype information is to be released to the tested individuals. In most clinical research centers genotype information has not been released to the patients and controls, and the samples are stored as anonymous samples. This means that back-tracking for identity is not an option. If, however, the intention is to provide the tested patient or individual with the test data, then counseling is essential both before and after the test. It may also be appropriate to collect evidence that the individuals understand the potential impact of this information in the form of signed papers registering their consent.

The age of computers poses a new problem regarding genetic counseling. Today we can consent to tests online. Can this be ethical? How can the tester know that the client truly understands the results of any test performed? The answer is that they cannot. Some may argue we can never tell whether a client understands the potential impact of a test and even if they say they do, the results can take them by surprise. However, the matter of whether the client is present with the counselor or the process occurs online may be irrelevant.

Others may say it is a step too far. Opinions will vary and it may be easier for us to ride this issue out than become entangled within it, but online testing and application is part of the future.

11.2 ETHICS IN GENETICS: WHAT WE CAN LEARN FROM THE PAST?

Historically, genetics has had something of a bad reputation. This branch of biological science has been misused. Genetics itself could be described as the result of twinning Gregor Mendel's observations with those of Charles Darwin. Mendel and Darwin were both very rigorous in the pursuit of their science, but others were far less cautious, willing to create or stretch hypotheses beyond reasonable limits. It was not long before the idea of **survival of the fittest** developed from Darwin's own observations (by others) was adopted as pure fact and from this sprang the concept that within populations some were genetically fit, while others were unfit. This idea led to the development of **eugenics** by Francis Galton (incidentally Darwin's cousin) and **social Darwinism** promoted by Herbert Spencer. The **Eugenics Movement** was mostly based on the idea that the fit members of the population were those who prospered (i.e. the upper classes), while the unfit were the lower classes. In Victorian and Edwardian England the differences between the upper classes and the lower classes were marked, and extreme poverty was widespread amongst the lower classes (**Figure 11.2**). Spencer's idea was based on the notion that poverty and wealth are inevitable as they reflect the biological rules that govern society—the biological rules in this case being the genes. Many of the upper-class intelligentsia were of the opinion that poverty was a natural byproduct of laziness and a lack of thrift among the poor. The Eugenics Movement, in particular, fed off this idea. In the warped view of the Eugenics Movement, being lower class and poor must be a genetic trait.

The consequence of the Eugenics Movement and the ideas it spread were extremely bad news for the developing science of genetics

Eugenics is not a science, it is a misuse of science. There are many examples of the use of eugenics to justify extremely inhumane activities. Steve Jones, in the introduction to his book *The Language of the Genes* (2000), writes a very informative section on eugenics with examples that include enforced (sometimes secret) sterilizations all the way to the

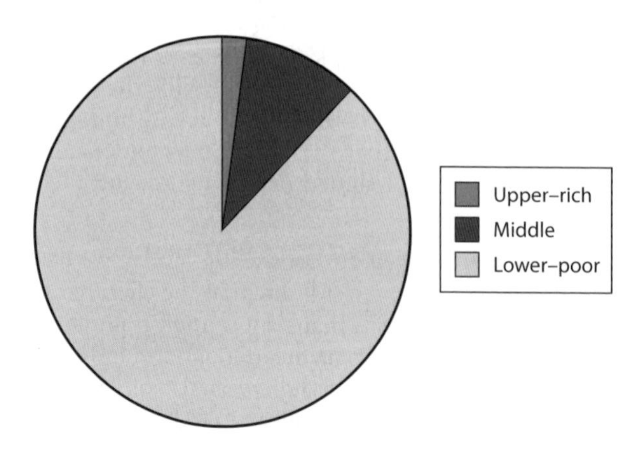

Figure 11.2: Class division in the UK around 1900. The figure illustrates the division between the rich (upper classes) and the poor (lower classes) in the UK around 1900. The members of the Eugenic Movement were convinced that poverty was an inherited trait.

death camps of the holocaust (see below). He talks of how Galton supported the idea of breeding from the best and sterilizing those individuals whose inheritance did not meet with his approval and how the Eugenics Movement joined the gentle concern for the unborn with a brutal rejection of the rights of the living. Eugenics was an idea shared by many members of society from the political left to the far right. The list includes George Bernard Shaw, Winston Churchill, and Charles Davenport (Professor of Evolutionary Biology at Harvard).

New Germania

At the lowest level, isolated communities such **New Germania** in Paraguay (which was set up by Bernhard Förster and his wife, Elisabeth, the sister of the philosopher Friedrich Nietzsche), were founded based on selected groups of individuals who set up home in isolation from their brethren to maintain and develop more pure communities and strengthen the stock. A glance at the peoples of this area today indicates that the experiment failed, and the descendents of the original settlers are a poor and sickly population. Unfortunately, that was not the limit of the application of this pseudo-science. In North America, 25,000 Americans were sterilized because they might pass feeble-mindedness or criminality to a future generation.

The Monist League

The German embryologist Ernst Haeckel founded the Monist League, which promoted the use of the biological rules of society for the survival of some races versus others. Of course Haeckel was himself amongst the select. However, it would be wrong at this stage to suggest that the idea that one race may be superior to another has only occurred once in human history. Human history is peppered with examples where one race or subpopulation has considered itself higher, better, or more worthy than others. This idea has been the most common cause for war between nations since historical records were kept. However, what the Monist League was promoting was the idea that eugenics provided a scientific basis for this and it was not long before this idea fell into the wrong hands, i.e. those of Hitler, who wrote that whoever is not bodily and spiritually healthy and worthy shall not have the right to pass his suffering in the body of his children. Mass forced (mostly secret) sterilizations followed. Worse was still to come. The German Society for Race Hygiene, many members of which carried university doctorates, was formed. This science was used to justify genocide on a hitherto unimaginable scale.

There is no doubt from the historical record that Hitler was aware of the Eugenics Movement and used the concept of inherited racial purity to justify the selection processes behind his most hideous activities. However, as Jones tells us, Hitler was not the only person to think in this way. One quote, in particular, is surprising: "The unnatural and increasingly rapid growth of the feeble-minded and insane classes, coupled as it is with steady restriction among the thrifty, energetic and superior stocks constitutes a national and race danger which is impossible to exaggerate.... I feel that the source from which the stream of madness is fed should be cut off and sealed before another year has passed." These words of Winston Churchill in 1910 would today cause a riot and his immediate dismissal from parliament, and rightly so, but this latter quote does put things into context. These ideas, which seem strange now and are obviously flawed when we look at them in 2015, did not seem so strange in 1910. However, this is in no way justifies the actions of the past (**Figure 11.3**).

One idea that eugenics helped spawn was the idea of a link between genetics and criminality. This began in the early years of the Eugenics Movement, but surprisingly lasted into the 1960s when studies of males in penal institutions indicated a higher frequency

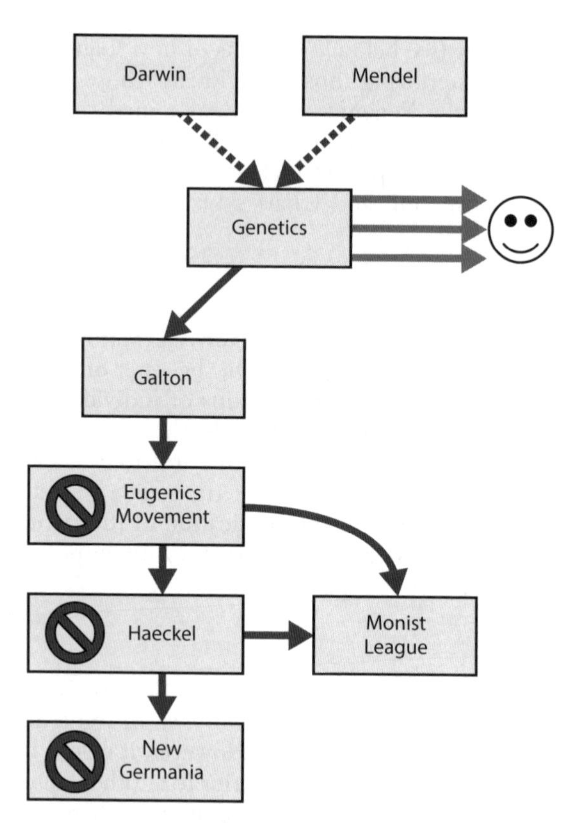

Figure 11.3: From Darwin and Mendel to eugenics: how science can be misused. The figure illustrates the period of eugenics, and the main characters and events involved in the misinterpretation and application of genetics.

of male inmates with an extra Y chromosome (XYY). In fact, this condition occurs in 1:1000 males and is not associated with hyper-aggressiveness as originally suggested, but is associated with a mildly reduced level of intelligence. The idea was seen once again in 1993 with a report on the *MAOA* gene (chromosome X) which encodes monoamine oxidase A. Monoamine oxidase A is an enzyme that plays a key role in the catabolism of a range of neurotransmittors. The study reported a link between this gene and criminal activity in a large Dutch family. Unlike the XYY story, this report has stood the test of time with the caveat that aggressive behavior only occurs in those males carrying the polymorphism who have also been maltreated during childhood. Such was public and scientific ethical concern about promoting the idea of a link between criminal behavior and genetics that a conference planned to discuss the idea was canceled to avoid raising controversy that may have had public and societal consequences.

Eugenics has left a scar across the heart of genetics as a science—one that has been hard to shake off. Given this, it is perhaps most surprising that some of these screwball ideas have lasted into the new millennium, but even today these ideas occasionally surface as Lone Frank reports in her book *My Beautiful Genome* (2011). The message from the past is clear. We must not allow this ideology to raise its ugly head in the future.

We need to consider several other ethical issues as more and more samples undergo genome-wide scanning for clinical or personal enquiries:

- How are we going to use the data?
- Who does the data belong to?
- Who should be able to access the data?

11.3 LOOKING INTO THE FUTURE USE OF GENETIC DATA

There are multiple uses for data from studies of complex disease and these can be divided into two main subgroups: those that impact on the disease itself (i.e. in clinical practice) and those that are of a more personal nature.

Genetic studies of complex disease will have a major impact on clinical medicine

This subject has been discussed at length elsewhere in this book and therefore requires only a brief discussion here. The Human Genome Project (HGP) stated at the outset that one of the goals of the study was to enable disease genes to be identified so that the information could be used in disease diagnosis and patient treatment, and to help us understand disease pathology (**Figure 11.4**). Though the results so far are mixed, there are some examples illustrating at least a degree of success in each of these areas. However, not all of the genetic associations we have dealt with were reported after the HGP report. Some, like that of *HLA-B*27* in ankylosing spondylitis, were reported long before the project began. Despite this, the association is a great illustration of what is possible. Though the *HLA-B*27* alleles are quite common, the genetic association is so strong that testing can be helpful in cases where a differential diagnosis is required. The use of genetics in informing treatment options is also starting to be applied in clinical practice, e.g. the use of abacavir in the treatment of human immune deficiency virus (HIV-1) induced acquired immune deficiency syndrome (AIDS). In addition studies in diseases such as Crohn's disease have

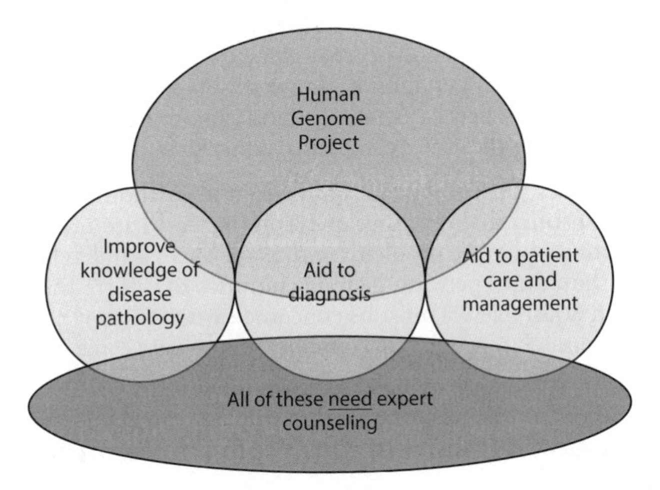

Figure 11.4: Promises of the HGP. The figure shows three of the major promises of the HGP with respect to complex disease. In each case, there is a need for expert counseling as the impact of risk alleles and haplotypes is very different from that seen in Mendelian disease.

identified a large number of risk alleles and this information is helping to advance our understanding of the pathology of this disease.

The ethical issues concerning these promises

This is all very well, but what are the ethical issues arising from this. One of the major problems with using genetic polymorphisms in disease diagnosis in complex disease is that the polymorphisms are often common in the healthy population and have weak associations with the disease. Therefore, the value of the polymorphisms as diagnostic tools is limited. Where they have larger effects, the presence of an allele may be a useful adjunct in diagnosis and particularly in making a differential diagnosis between two diseases, e.g. *HLA-B*27* in ankylosing spondylitis (see above). However, we are faced with a dilemma. If the allele is common we have to be careful not to over-promote the concept that the allele is important in disease pathogenesis, because in most complex diseases possession of the risk allele is neither necessary nor sufficient for the disease to occur. In some cases there are strong genetic associations with very common polymorphisms, e.g. some human leukocyte antigens (HLA), where up to 20% or more of the healthy population can carry the risk allele or one member of the allele family. For example, the *DRB1*04* family is the most common group of the *HLA-DRB1* alleles in the UK. This group is associated with susceptibility and resistance to a range of autoimmune diseases, including rheumatoid arthritis and autoimmune hepatitis (increased risk), and primary sclerosing cholangitis (reduced risk). We cannot announce to the population at large that they carry a four- to five-fold greater risk of a disease without a proper explanation of what this means. The outcome of simply broadcasting risk values could create global (and unnecessary) panic. The statistics need to be put into context.

When it comes to therapy there are some clear links between outcome measures and genetic polymorphism. It is very helpful in some cases to perform genetic testing before a patient is treated with a specific drug. In the science of pharmacogenetics, where the associations are strong and the differences (i.e. in response to treatment) are clear, genetic testing will be increasingly used prior to treatment. Here, the reason for using genetic information and testing is obvious, e.g. consider the non-response to codeine (for pain relief) or to β-blockers (for cardiovascular problems), or the adverse reaction to abacavir used to treat HIV-1 infection. However, where the strength of the association is weaker, but the outcome potentially lethal (e.g. in the case of some adverse drug reactions), we find ourselves in an ethical dilemma. To genotype or not to genotype, that is the question. If the incidence of a severe reaction is 1:100,000 new cases per annum, do we genotype everyone before we prescribe the drug or do we monitor all new cases for signs and symptoms of adverse reactions? It is a societal question and one for the ethics committee rather than one that we can answer.

Considering ethical issues is easier when it comes to understanding the disease pathogenesis. It is our human instinct that makes us question the world around us and our humanity that makes us want to solve the problems of disease. There is unlikely to be anyone who would feel that we should not perform genetic enquiries to inform the debate on disease pathogenesis. The question that we must be concerned with is how we handle the data and use the outcomes from such research. Early publication of data can cause fear amongst the "at risk" public.

The potential personal impact of data from studies in complex disease is considerable

Excluding all of the above, which all have personal implications; there are many other ways in which these data may impact on all of us (**Figure 11.5**). Lone Frank (2011) describes

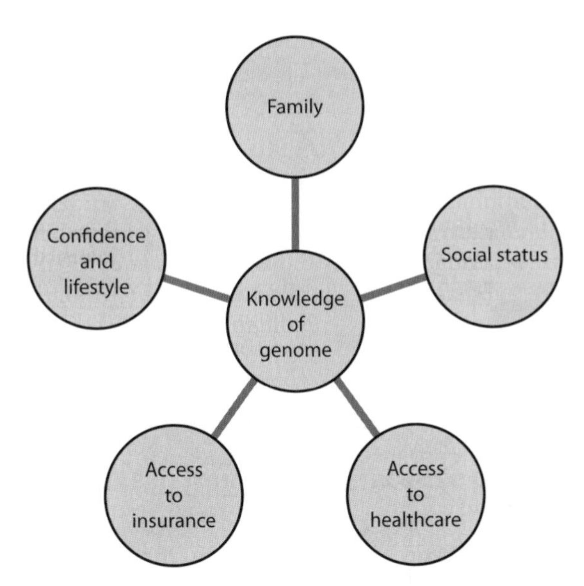

Figure 11.5: The personal impact of knowing your own genotype. The figure illustrates five of the factors that may be impacted by genetic testing. Individual responses to genetic testing can have negative effects on self-confidence and on family members as well as those tested. In addition, there is potential for testing to have a negative impact on insurance and healthcare costs depending on local rules and regulations. Social status can also be impacted. However, there can also be positive impacts in all of these areas.

waiting patiently (or not so patiently) for the postman to deliver the results of genetic tests for common polymorphisms identified in complex disease. Her book gives a personal view of how we ourselves may feel waiting for such results. She describes joy when low-risk alleles are reported, but deep concern when higher-risk alleles are reported. Even with counseling, the author reports being nervous on some occasions. Of course it depends on what is being tested; consider, for example, *BRCA* versus the interleukin (IL)-2 receptor gene *IL2R*. It depends on the size of effect and on the disease or diseases that are associated with the polymorphism (or occasionally the mutation). It depends on your knowledge of your family medical history, and your understanding of medicine and science. For both the academic elite and the general public, the level of concern depends on how much we understand the subject. Do we have an adequate grasp of the concepts before us? Do we understand genetic risk? Even if we do understand the science there is no promise that we will be better equipped to cope with the news. Therefore, it is also important to note that when doctors become patients they deserve the same level of consideration and should receive the same level of counseling as the less informed member of the public receives.

Patients being tested for diagnostic reasons and those making independent enquiries are one group. But, what do we do about volunteers? Here the ethical dilemma is simple—when testing samples, should we release data to the volunteers? The question may be simple, but the solution is far from it. We get around this by making all samples anonymous, so that back-tracking to the original ID is not possible (**Figure 11.6**). However, this has limitations. What if we find a polymorphism in our studies that identifies an allele that is very strongly associated with a severe disease, but a disease that can be easily treated? If we find that 2% of the healthy controls are carriers for the risk allele, then 2% of our healthy control volunteers are at increased risk of the disease. As the DNA samples will be handled as anonymous samples there is no possibility of alerting these volunteers to the risk and giving them prophylactic treatment. The same individuals may suffer the disease in later life.

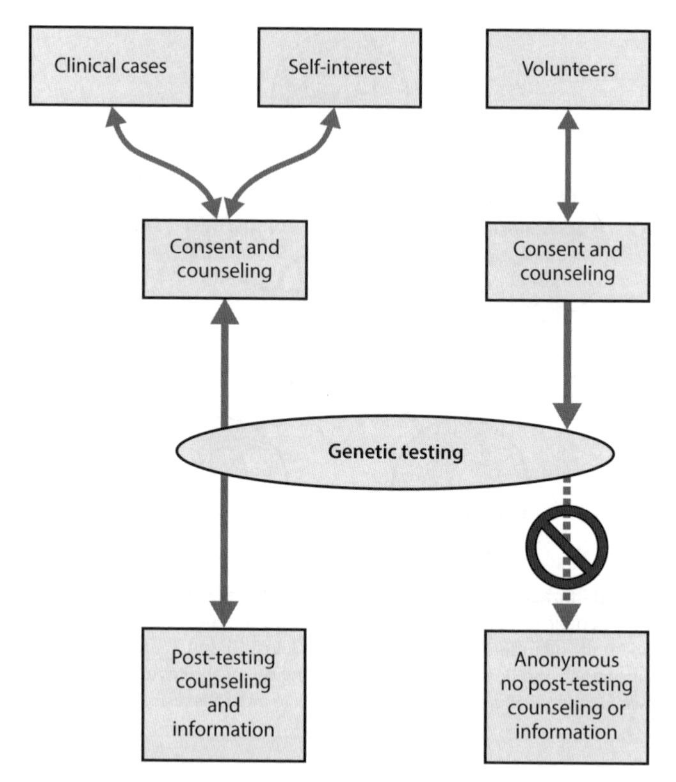

Figure 11.6: Common (good) practice counseling in different groups. The figure illustrates the importance of consent and counseling in all groups undergoing genetic testing. Those being tested for clinical purposes need to give consent for testing and receive appropriate counseling prior to consent being given. Those requesting tests out of self-interest should be treated in the same way. Those who are being tested as part of a research program (i.e. volunteers) must also give consent and should also be counseled. However, normal practice in research is to make all samples anonymous and therefore volunteers should not expect to receive information in return or counseling after testing. In contrast, both clinical cases and those being tested out of self-interest should be counseled following testing as well as before.

In complex disease it is always difficult to know how to handle the data, because the size of effect is small and some of the genetic relationships are very complicated. Nowhere is this better illustrated than in the major histocompatibility complex (MHC). The majority of the genetic associations with this area involve odds ratios (ORs) of less than 10, though values of between 0.02 and 150 have been quoted for different diseases. The problem is that most HLA polymorphisms and haplotypes associated with disease are common in the healthy population (often as high as 10–20% for the major allele families, e.g. *DRB1*04*). In contrast, the diseases are not as common. This means that the majority of those with the risk allele, family of risk alleles or haplotype do not develop the disease. Therefore, being positive for an allele or haplotype is not useful in a diagnostic setting. However, it is easy for the public to get the impression that these associations are diagnostic.

The example of *HLA-B*27*

We can consider the example from the 1980s of a middle-aged healthy volunteer who agreed to be involved in a study of HLA and disease. The control was found to have the *HLA-B*27* phenotype by standard serological assay. The close-knit community from which the volunteers were recruited meant that after HLA typing they would frequently ask to know their HLA type. On this occasion time passed without any issue until one day the volunteer rounded on the immunogenetics team declaring their anger that they had not been told they carried an allele that promotes ankylosing spondylitis. The team's defense, that though the association between ankylosing spondylitis and *HLA-B*27* is strong it is not found in all cases, and more importantly only one in approximately 20 *HLA-B*27*-positives develop the diseases, was no comfort. The volunteer had recently discovered that a younger member of their family, also an *HLA-B*27* carrier, had developed ankylosing spondylitis. The team were not even able to defend their position with the observation that the volunteer had not indicated a history of ankylosing spondylitis and was over the normal age at which they it would develop. Classical geneticists would use two words: anonymity and counseling. Anonymity may have helped here, but probably not. Rigorous refusal to disclose the original data would not have been easy in these circumstances. Counseling would have helped, but then that means either all those undergoing genotyping would need to be informed of every HLA association that had been reported and the risk with diseases before testing, or that all those with any high risk alleles and haplotypes would require counseling after being typed, or both. If we restricted counseling to the second group, when we consider HLA genotypes this would mean almost everyone as nearly all the common HLA allele families have some positive risk associations with one or more diseases. Considering *HLA-B*27*-positives alone, out of 20 cases given counseling, only one would really need it. The effect in terms of stress for the remaining 19 would be enormous. This would be a huge burden and represents an impossible task.

Interestingly, Francis Collins reports in his book *The Language of Life: DNA and the Revolution in Personalized Medicine* (2010) of having had himself tested by three companies for a number of well-known risk alleles. He said "there was one test result that I thought about just not looking at … the one for Alzheimer's disease risk." (We should note Collins does not specify which Alzheimer's gene.) Given that we are talking about one of the best-informed members of the scientific community with regard to genetics and disease, this is perhaps something of a surprise and it sets the question for the community at large. However, in considering his statement we need to remember that as yet there is no cure for Alzheimer's disease and that may be driving the concern expressed here. Certainly the issue of whether there is a cure for this disease, or any other disease, is one that is likely to affect an individual's choice of whether to be tested at all and if so which genes to be tested for.

Do we really want to know?

The answer to this question is yes and no. Some people do want to know and they are finding out through direct consumer testing (DCT) as Collins puts it. There is quite a debate about this in the USA at present concerning whether DCT should be allowed; some say yes, others (e.g. the American College of Medical Genetics) say no. It is interesting that Collins himself is among the yes group. The American Society for Human Genetics considers that DCT is OK provided adequate information is available about the limitations of the tests. As Collins says, personal genomics is here, but *caveat emptor*.

Confidence and lifestyle

All of the above can influence our confidence. In a stressful world more stress is something no one wishes for. Knowledge that we have inherited a polymorphism that increases the risk of disease, no matter how small, can have a big effect on our personality. Some individuals are more susceptible than others. This trait could also be genetic. Confidence is hard to measure, but it impacts in all our lives, everyday, at work and at home.

The immediate impact of reduced confidence may be stress (as above), but it may also lead to changes in lifestyle, not all of which may be beneficial. Some individuals when diagnosed with chronic disease become determined to defeat the illness. They become champions in the crusade against the disease, getting involved in fundraising events and even starting new charities. Many will become more self-aware, and diet and exercise may become higher priorities than previously. However, not everyone reacts positively nor is increased activity and diet the right path for everyone. Though being told you have an increased risk of a common disease is not the same as being diagnosed with the illness, some members of society will react to this type of information as though it were. In extreme cases, some become withdrawn—almost developing a Munchhausen's-like state. This is why and where we need to consider the ethical side of genetics of complex disease. We need to reduce and avoid this problem, and avoid sending out a negative message. Counseling is essential.

Family issues

When genetics is mentioned most people think of familial disease, but this is not always the case in complex disease. Very few complex diseases have large numbers of affected families and even though there is increased risk for family members, it is not on the scale of risk associated with Mendelian disease. However, knowledge of genetic susceptibility can have multiple effects in a family. Even though selected individuals may be aware of the impact of the risk allele, and its biological and personal significance, there is certainly no possibility that all family members will be aware. Suddenly we may find family members selling up and cruising round the world, living every day as if it were their last, only to run out of money and find themselves impoverished with another 30–40 years of life to go. Others may take more radical action. However, not everyone will react in this way and radical responses can be prevented with careful counseling.

One other family issue for more immediate consideration may be in the choice of a mate or the decision to start a family. It is one thing to be aware of your own risk of a complex disease; it is another thing to know you are likely to pass that risk on to your offspring. If you are made aware that you carry an increased risk of a particular disease, you may decide not have a family. Some people may decide not to marry into families with a high risk of certain common diseases and others may decide to adopt rather than have children themselves. With a growing world population, some might say this is a good thing. However, when we say this we are ignoring the fact that the reason for the decision may be completely wrong. The issue is risk and size of risk. In complex disease there is incomplete penetrance of the allele in the disease and having the risk allele is neither necessary nor sufficient for the disease to develop. The size of the risk varies between diseases, and between risk alleles and groups of risk alleles. For example, the Ueda et al. (2003) paper on *CTLA4* gene polymorphism suggested the maximum OR (risk) in type 1 diabetes was 1.2 and in Graves disease was closer to 1.6. Neither of these is particularly high. Would you consider not marrying your bride-to-be on the basis of having the *CTLA4* risk allele? How high does the risk need to be? As our knowledge grows, we will be better equipped to offer counseling and help individuals make informed decisions. These decisions should not be made without careful counseling. It is also important to be aware that the impact of this information may be different in different societies and ethnic groups.

There are of course broader issues when it comes to genetics and families. Though we are concerned with complex disease, we cannot ignore the other uses of genetic testing. In the third millennium tracking our kindred through multiple generations has become a worldwide business, not just because of links with disease, but also because of links with our history. However, understanding our ancestry can be used in both positive and negative ways. Paternity testing can be used to test for fidelity. There are strong financial and legal reasons why testing is often requested by the mother or father of a child. For example, fathers who prove to be surrogates as a consequence of such testing may withdraw their support for the mother's child or children. Ultimately it is the child who suffers.

On a more positive note, historical searches can reveal interesting details about our ancestors. In the USA, more than 50 companies are engaged in genetic genealogy studies. Companies such as African Ancestry look specifically for ethnic and geographic matches between mostly American clients and those of racial and tribal groups from various parts of Africa. However, not all reports are confirmed and in some cases initial findings have turned out to be incorrect. At least one high-profile case came to light recently where the wrong link was proposed. Cases such as this illustrate the need for counseling here too. Overall, we cannot get away from the need to inform the patient, client, or individual fully about the consequences of testing as well as the strengths, weaknesses, and limitations of any testing.

One of the most difficult issues in genetic testing is that of prenatal testing. Different countries have different laws regarding prenatal testing and the consequences of testing. In addition, opinions vary between different social, ethnic, and religious groups within societies. In complex disease, genetic testing is not currently used to predict the health of the unborn and indeed this would be an error. Due to low penetrance, the risk of disease is most often very low and there is no justification for using such testing. However, we need to make sure that this type of idea does not sneak in through the back door. We have the example of prenatal testing for male status as an example of bad practice highlighted in Steve Jones's book *The Language of the Genes*. Termination of pregnancy based on testing for the sex chromosomes is now less likely than it was 10 years ago, but undoubtedly it still occurs in some countries and there is much written about this on the web and in papers cited there.

Social status

The societal impact of genetics in complex disease cannot be underestimated. Knowledge of our genomes and how they may determine our lifestyle, health, even our wealth and happiness can have a major affect both on our own view of our social standing and on how others view us. One recent report questioned the value of knowing our genome, with a gloomy prediction that the cost to our collective mental health is incalculable. Others disagree saying most people would not be burdened. Opinions are divided.

Just as the Victorian and Edwardian Eugenics Movement was set against the survival of the unfit, so too can modern society suddenly turn upon those who are seen to be less able. Anyone with a disability will tell us that for them life is made more difficult through the actions and attitudes of some members of society. In the context of this book we are not considering the illness per se. We are considering just the potential impact of carrying the knowledge of an increased risk of a common trait.

Some people when given bad news will develop a coping strategy based on ignoring the information (or denial) or on accepting the information and using it to their benefit. Some court sympathy. Others are activated and become associated with those in similar circumstances. Other people when given bad news collapse. They go into a mental meltdown of variable proportions and find it hard to cope with the information. Some become hypochondriacs. In society as a whole, it is easier to cope with those who have a positive response to a situation. No one wants to spend time with negative individuals. However, it is not easy to maintain a positive response.

Of course with high risk genes, e.g. *BRCA1* and *BRCA2* that are linked to breast cancer and also to ovarian cancer, many of the patients tested positive opt for preventative surgery rather than risk the illness itself. At its most extreme this can mean having both breasts and ovaries removed. The personal consequences of this are considerable, with prolonged surgical and medical treatment, post-surgical medication, and potentially early menopause. The personal and physical impact of all of this together can be quite considerable and the individual's societal status can change. There is a question about whether preventative surgery is appropriate. This question is one that only the individual is equipped to answer. It can be very positive or it can be very negative. On one hand, an individual may take the view that they can deal with the problem in a positive way through surgery; on the other hand, a different individual may take the view that surgery is the only option (this may be seen by some as a negative response).

How we handle our genetic portfolio is going to be one of the most interesting challenges of the post-genome era. We need to understand the implications of risk and accept our situation. We cannot change our genome. Everyone will have some risk alleles that are protective from and others that are predictive of common traits. The genome information itself is not the same as having the trait. Society can relax and we can relax within it. There is no great societal threat and no threat to our social status unless we allow it to happen. We need to avoid creating a **genetic underclass** as in Huxley's *Brave New World*, whereby the human race is divided into subgroups, with the upper class being the alphas and the lower classes being the betas, etc., all the way down to the epsilons at the bottom of the (genetic) caste system. To those who scoff at this idea, look at the past. A mathematician will tell us that by definition, "if anything is possible" then "anything" includes this possibility.

Access to healthcare and health insurance

One of the downstream consequences of knowing our genomes is that we could face a problem with healthcare planning and cost. In the UK, the majority of individuals currently rely on the National Health Service (NHS), which provides healthcare free at the point of entry for all members of the population. The NHS is paid for through Tax and National Insurance contributions. Though there is some debate about this provision and who has rights to automatic NHS treatment, at present no one is suggesting that access be based on genetic testing. In addition, no one is suggesting that genetic tests be used to determine how much individuals each pay in Tax or National Insurance. However, that does not mean this will always be the case, but such change is unlikely. Other countries have different systems of healthcare, many based on a personal subscription and opt-in insurance schemes. It is easy to see why a privately funded system is more likely to be interested in an individual's genetic portfolio.

Of more concern is the potential application of genetic testing for health and life insurance. Currently in the UK, genetic information does not have to be given to insurance providers and they are prohibited from asking customers to provide such information. The situation in the USA is different as the health system is not based on the same plan as the NHS in the UK. Instead, the vast majority of health provision is through private health companies. Practice also varies elsewhere and as most insurance companies are global or at least international, access to genetic data needs to be carefully monitored. In the future it may be seen as reasonable (by some) for genetic data to be used to set insurance premiums. Insurance companies understand risk and work with risk models. They are well equipped to assess the likelihood of adverse circumstances on a personal basis given the appropriate information. Information on drinking, smoking, and other health issues is routinely gathered when setting up policies. Extending the list to include genes would be a simple step.

Overall, this is an ethical issue. Currently there are constraints in the UK and in some other countries, and genotype information does not need to be provided on request. In the USA, for instance, legislation is in place to outlaw the discriminatory use of predictive genetic information in health insurance and the workplace. However, that does not mean that the elevated risk of disability or illness cannot be used against an individual when setting insurance costs. Privacy is a major issue and defending our individual freedom is important, i.e. we need to defend our rights. These include the right to equality and freedom in the matter of our genome. Allowing this very minor misuse of genetics could be seen as the first tiny step on the descent into madness.

Other issues

Imagine there was a gene for addiction, fidelity, sexual orientation, intelligence, or spirituality. In each of these cases there may well be genes that predispose to these traits. Each gene may carry a bad allele and a good allele. The question is do we want to know the answer about which alleles we carry. How would we use the data? The answer lies in each example. These are difficult issues. With regard to sexual orientation, there are those who wish to be able to say their sexual orientation is all (or partly) in their genes, and others who want to be able to say it is a matter of choice and genes have nothing to do with it. Underlying this are often different political agendas and there are implications whichever explanation one applies. Spirituality is another difficult issue; again, there are some for and against a genetic explanation for this. Personal agendas apply. None of the above are illnesses but all of them can influence behavior and for that reason alone all are worthy of consideration in this chapter. These are both individual and societal issues.

11.4 WHO DOES THE DATA BELONG TO? INTERACTING WITH COMMERCE

One of the biggest questions we are faced with is who does the genome belong to and who owns our individual genomes? One of the big developments in the past 25 years has been the increasing interaction between academia and industry. In some cases this has taken on a gladiatorial perspective with the giants of industry and commerce apparently hammering out their differences in a public arena. This was best illustrated by the competitive nature of the publication of the Human Genome Mapping Project (HGMP) in 2001: on one side, academia with the NIH (USA), the Wellcome Trust and Sanger Centre (UK), and others; on the other side, commerce with J Craig Venter and Celera. The project started with a proposal in 1990 and the NIH, Sanger Centre and others began working on mapping the genome. J Craig Venter then offered a more rapid solution. In the end it worked out fine with the parallel publications in *Science* and *Nature*, but there were times of controversy.

The struggle is also well illustrated in the story of deCODEme. Kari Stefannson, backed by a number of venture capitalists, set up the company to genotype the entire population of Iceland. There was bitter local opposition to the idea that the genome could be owned by private business. Interestingly, the idea not only offended the Icelanders, but anger spread far and wide. In the end the company had great success identifying a large number of risk alleles for complex disease. For a short period at the end of the 2010s, deCODEme worked as a diagnostics company offering direct to consumer genetic testing focusing on that part of healthcare associated with assessing and preventing individual disease risk.

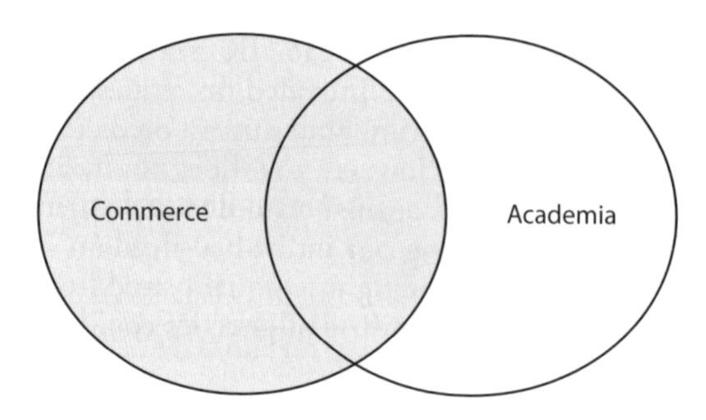

Figure 11.7: Interaction between commerce and academia. The figure illustrates the increasing collaboration between commerce and academia. Almost all genetic testing is now being partly outsourced and commerce has long since played a very significant role in genetic research.

However, more recently the parent company deCODE (the major research base for the company) has been purchased by Amgen. The service offered by deCODEme may not be offered in future. The problem for commercial companies like deCODE and deCODEme is that they need to generate payback from their investments. From a pure academic point of view this can be a negative relationship, though individuals have different opinions (**Figure 11.7**).

Many academics would rather see genomes mapped by university research laboratories than in commercial companies. Unfortunately, there is no longer any option to exclude commerce. Commerce has provided the technology for the great advances we have made. Commerce has been involved frequently at the center of all we have done. We do not exist in ivory towers in isolation any more. Our materials, apart from the DNA samples we have collected, are all purchased. The equipment we use has been invented and developed by commerce, and today it is cheaper to contract out genotyping for all but a few genes than it is to genotype in-house. As a consequence, we are inevitably tied to commerce as commerce is tied to us. Outsourcing or contracting out is going to be a major part of the future.

Once more there are ethical issues, but there is no problem provided the anonymity of the material sent out is maintained. This needs to be very strictly governed and assessed to ensure compliance with ethical standards.

Do I own my genome?

Accepting that commerce is part of the current package, we need to know if our genomes are our own possession or belong to another. If we allow others to genotype or sequence our genomes at no cost it is possible that the sequence/genotype produced will belong to the group who did the work. This could be a commercial company or it could be a research group. However, just as with some tumor cell lines that have been grown in laboratories for many years, if the research group or commercial company failed to get permission or explain the situation clearly and obtain written consent, then this could be contestable. Different countries have different regulations on this and it could be costly to contest such an issue.

It is certainly true that a number of the genetics companies would like to develop genotyping techniques for the detection of common risk genes with large effects and thereby be able to offer a commercial genetic testing service. Some companies concentrate on just a few genes, e.g. the company Myriad Genetics (Salt Lake City, USA) have a patent to test the *BRCA* genes, others such as deCODEme (Iceland) and 23andME (USA) will assess 500,000 to 1 million single nucleotide polymorphisms (SNPs) and provide us with a **personal profile**, but they have not yet obtained a patent for the genes. Myriad

Genetics not only tests the genes, but they recently (February 2013) won a court case enabling them to patent a number of cancer genes in Australia. Appeals against this ruling in have been dismissed. Protesters in Australia say that this should not be allowed and it could have a major affect on future research in cancer. The depth of debate about patents is reflected in Luigi Palombi's book *Gene Cartels: Biotech Patents in the Age of Free Trade* (2009). The author states that "no matter how important it is to identify a gene linked to a disease, it's still not something that Myriad or anyone else has invented;" he goes on, "Politicians must not change the law to prevent patenting of genetic materials" (http://www.bloomberg.com/news/2013-02-15/).

The American Association for Molecular Pathology and the American College of Medical Genetics have been quoted as saying they are worried that the company is trying to get legal ownership of part of the human body. However, the situation in the USA is quite different. A Supreme Court ruling in June 2013, authored by Justice Clarence Thomas, states that naturally occurring DNA segments are not patentable. This will apply to a vast range of testable genes. However, the ruling permits edited forms of genes not found in nature to be patented. In the USA there are two major implications from this ruling. Firstly, genetic testing will become cheaper as companies will compete in the testing market. Second, the cost of whole genome sequencing is likely to fall. This is because the restriction that is imposed by patenting sequences, in the genome, is no longer an issue and therefore whole genome sequence is not likely to infringe patents.

One of Myriad's fact sheets states that under USA patent law:

> "No-one can patent anyone's genes. Genes consist of DNA that is naturally occurring in a person's body and as products of nature [they] are not patentable. In order to unravel the mysteries of what genes do, researchers have had to separate them from the rest of the DNA by producing man-made copies of only that portion of the gene that provides instructions for making proteins (only about 2% of the total DNA in your body). These man-made copies, called "isolated DNA," are unique chemical compositions not found in nature or the human body. The U.S. Patent and Trademark Office has been granting patents on isolated DNA to universities, hospitals, patient advocacy groups and companies for over 30 years."

Currently this is correct. However, there have been efforts to appeal some of these laws in the USA but so far they have failed.

Some scientists argue that their work has been affected because of the costs of paying royalties and in some cases they have even received letters demanding they stop using patented inventions. In response, the biotech companies may reply arguing that if they cannot protect their inventions, then they cannot compete with others in the market place.

Counseling the customer

In terms of ethics, it may be ethical for commercial companies to hold patents and provide services, but there is a problem with how the data are presented to the individuals they test. Should individuals simply be provided with the data and be left to get on with it without counseling? There is a very strong argument against this. In the UK, the Human Genetics Commission is quite concerned that the lay public will simply not understand the information and the effect, without counseling, could be similar to that described above. The situation is more complex in some states in the USA, which have prohibited testing unless prescribed by a medical practitioner.

11.5 WHO SHOULD BE ABLE TO ACCESS THE DATA?

There is a clear overlap between sections in this chapter and some of the ideas in this section have also been considered in previous sections. There are those who would like to be able to use genetic tests to determine our general fitness. Where this information is to be used for productive reasons, there is no problem (e.g. for health advice etc). However, if the data were then to be used to determine our financial contribution to the NHS or equivalent scheme and our health and life insurance, then there would be problems. Currently, in the UK and most other European countries, insurers cannot ask for this data, but in the future they may be able to. The UK Data Protection Act is supposed to act as a foundation for any company to follow when requesting access to personal information, protecting the individual, and conserving their personal rights. Most importantly, it requires companies to obtain individual consent before obtaining personal information.

Once again, by playing with our genes we are potentially opening Pandora's Box. There is an echo from the past here. In the 1970s, there was a great deal of concern about the use of recombinant DNA to develop sequences that are not produced in nature for use in the agrochemical industry. In the past we have seen examples ranging from the genetic engineering of strawberries, antifreeze genes to the development of pigs with genes for Alzheimer's disease. The concerns of the 1970s and 1980s have not gone away. At the present time strict guidelines exist for both current and future research and practice. The use of these guidelines goes far beyond humans.

CONCLUSIONS

Looking at the genetics of complex diseases from a social point of view allows us to consider the ethical issues associated with it. Ethics is best defined as the fulfillment of well-being. Some would suggest this is not part of science and argue for the pursuit of truth at any price. Others would argue that ethics is part of science, and that we should apply more rigid rules and laws. When it comes to the genetics of complex disease, ethics refers to the issues around testing, counseling, data handling and storage, publication, and access to data.

Genetics has been misused in the past and for that reason there is a higher level of suspicion regarding genetics than for many other branches of science. Through the dark pseudo-science of eugenics, genetics became attached to some of the darkest chapters in human history. Eugenics illustrates the misuse of science extremely well. It also reminds us not to allow our science to be misused in the future.

Given the promises of the HGP, it is important to understand how ethical and societal issues are themselves composite parts of the puzzle. Using genetics in diagnosis is a key example. Should we, or should we not, test. Using genetics in disease treatment and patient care also raises the same issues. Testing can be very advantageous for a patient for whom a new treatment is seen as having a high likelihood of a good response, but it can be to the disadvantage to a patient who, after testing, finds out that they are genetically less likely to respond to a newly developed potent drug. No one would argue that using genetic data to understand disease pathogenesis is not worthwhile. However, if as part of the testing risk alleles are identified in healthy individuals we are once again faced with an ethical and societal dilemma. We need to be aware of personal and societal issues relating to confidence, confidentiality, social status, access to healthcare, and health insurance in relation to all of these promises.

What is the relationship between the individual, commerce, and public bodies? Who owns the genome, who owns an individual's DNA sequence, and who should be able to access an individual's personal genetic data? Science and commerce are strongly interlinked in genetics—more so now than ever before. Separating them is not possible, but protecting one's data is. Laws and regulations vary in different countries; however, most countries currently do have some form of regulation to restrict access, especially the USA and UK. All in all, the ethical issues are relatively straightforward in complex disease research at present. There are rules and guidelines in the UK; rules on consent and the storage of human tissue (e.g. the Human Tissue Act) are designed to prevent the misuse of materials and enable the best practice to be followed. In the USA, it took 12 years to get a bill passed that prohibited DNA being used to discriminate against individuals. The bill, a genetic information non-discrimination act, was signed in the Oval Office on 4 May 2008.

Ethics is a philosophical activity, and therefore views and ideas change, and the debate about the ethics of enquiry will continue. New challenges will arise with new procedures and new opportunities. We need to maintain a balanced opinion. We have seen from the past what happens when that balance is lost and the extremists have free reign. We cannot allow that to happen again. Nor should we stifle scientific advancement. Simple measures, such as informed consent, not only for invasive research but also for data collection, are important. In many ways the ethics of scientific investigation is far more advanced than other areas of enquiry and altogether this is a very positive sign for the future.

FURTHER READING

Books

Collins F (2010) The Language of Life: DNA and the Revolution in Personalised Medicine. Harper Collins. This is excellent book to delve into to look at the whys and wherefores of gene hunting in complex disease.

Frank L (2011) My Beautiful Genome: Exposing Our Genetic Future, One Quirk at a Time. Oneworld. This is an excellent book for the lay public and scientist alike. It is interesting, controversial, and humorous, and perfect for the student of complex disease genetics.

Jones S (2000) The Language of the Genes, 3rd ed. Harper Collins. This book has a very good introduction for those with an interest in the history of genetics.

Maynard-Moody S (1995) The Dilemma of the Fetus: Fetal Research, Medical Progress and Moral Politics. St Martins.

Palombi L (2009) Gene Cartels: Biotech Patents in the Age of Free Trade. Edward Elgar. This book by patent law academic Luigi Palombi is a timely contribution to heated global debates about the ownership of human genes.

Parham P (2009) The Immune System, 3rd ed. Garland Science. Though the general subject of this book is immunology, it is a good source for those interested in immunogenetics and the impact of MHC alleles/haplotypes in complex disease.

Spector T (2012) Identically Different: Why You Can Change Your Genes. Weidenfeld & Nicholson. This book champions the subject of epigenetics in complex traits and provides a different perspective for the scientist. The study of identical twins represents a gold standard for geneticists, but it also presents a challenge because they are rare and because they are genetically identical at birth. Therefore, testing inherited polymorphisms is of little value, but identifying concordance levels between identical versus non-identical twins is valuable. Novel sequences (new mutations) are informative and epigenetic changes are also informative. In the era of sequencing, more use may be made of identical twin sets. Spector also tackles a series of occasional controversial topics in this book, some of which are covered briefly in this chapter.

Strachan T, Goodship J & Chinnery P (2014) Genetics and Genomics in Medicine. Garland Science.

Tollefsen CO (2008) Biomedical Research and Beyond: Expanding the Ethics of Inquiry. Routledge. This book provides insight into the ethical basis of scientific enquiry from a philosophical perspective.

Articles

Ueda H, Howson JM, Esposito L et al. (2003) Association of the T-cell regulatory gene *CTLA4* with susceptibility to autoimmune disease. *Nature* 423:506–511. This is a very important study on *CTLA4* highlighting the complexity of association studies even when targeted to a specific gene.

Online sources

http://23andme.com

This site offers genotype information about health and ancestry for a relatively low cost with a rapid response. The site enables the reader to see what is offered commercially by such companies.

http://www.decode.com

The correct current site for deCODEme is that shown above. deCODE is part of the Amgen company.

http://www.myriad.com

Myriad Genetics is a company from Salt Lake City, USA with interesting patents on a number of cancer genes. They offer testing for a variety of common cancers and are working on genetic diagnosis for a number of other conditions. Their patent rights are a matter of some controversy at the time of writing, and serve as an example of the division between academia and commerce.

http://www.gtglabs.com

The Genetic Technologies website offers a variety of services for genotyping common diseases in humans and animals. Another informative site that illustrates the potential for commercial exploitation of genetic research, whatever our own individual views are.

http://www.hta.gov.uk

This site contains details about the Human Tissue Act (or links to) and is updated regularly.

http://www.patient.co.uk/doctor/Medical-Ethics.htm

This site and the one above are two sites for the UK Human Genetics Commission: one deals with general issues, and the other deals specifically with ethics and good medical practice. These are excellent sites for medical students and trainees in medical research and medical sciences.

http://www.familytreeDNA.com
This website offers to test the Y chromosome, mitochondrial genomes, and other chromosomal markers, providing genealogy data using a databank from close to 500,000 individuals.

http://www.oxfordancestors.com
A UK-based company that offers a range of services from paternity and maternity testing to genealogy testing.

http://www.africanancestry.com
A US-based company that traces African ancestry aimed specifically at the African-American market.

http://www.patientslikeme.com
This is a website for patients that allows them to share their experience and knowledge with others who suffer from the same and similar diseases. Some may question the use of this and the openness of the site. There are certainly ethical issues about this type of site concerning the presence or absence of counseling, and the accuracy and peer review of information added to it.

http://www.snpedia.com/index.php/Promethease
A web-based program that uses a tool to build a report on a system called SNPedia that can be applied to data from customers of 23andMe.com, FamilyTreeDNA.com, Ancestry.com, and others. It allows the customer to link SNPs to diseases and traits.

CHAPTER 12

Sequencing Technology and the Future of Complex Disease Genetics

In this chapter, we will discuss the present and future of complex disease genetics, looking at what needs to be done to bring us closer to meeting the promises of the Human Genome Project (HGP). Among the many promises of the HGP were the future use of genetics in disease diagnosis, disease treatment, patient management, the development of novel therapies (including new personalized therapies), and a better understanding of the pathogenesis of the genetics of common complex non-Mendelian disease. Throughout this book we have consistently focused on these points and the extent to which these outcomes have or have not been achieved.

Many of the techniques and technologies discussed in this chapter are also discussed in previous chapters of this book, for example genome-wide association studies (GWAS). The purpose of this chapter is to look into the future, and consider how techniques and technologies such as **next-generation sequencing (NGS)**, genotyping the transcriptome, the expanding field of epigenetics, the use of imputation analysis, the future for GWAS, and the value of the HapMap, 1000 Genomes Project, and ENCODE may be applied to further advance our understanding of complex disease. In addition, we will also consider **metagenomics** as applied to the bacteria in our guts.

12.1 DNA SEQUENCING: THE PAST, PRESENT, AND FUTURE

The key to understanding genetics in complex disease lies with our ability to sequence DNA, either searching for allelic variation in individual genes or across the whole genome. However, prior to 1977, sequencing was very difficult and considered as an impossible task by some scientists. Two things changed this: the development of more robust sequencing methods by Frederick Sanger and colleagues in 1977, and the advent of the use of the polymerase chain reaction (PCR) analysis (whereby a polymerase enzyme is used to amplify short DNA sequences) by Mullis et al. in 1986. PCR amplification was rapidly adopted as the method of choice for genotyping in both clinical and research laboratories, and automated sequencing transformed DNA sequencing from a dream to a reality.

Since then the technological developments applied to genetic research in complex disease have been exceptional. Many of the studies discussed elsewhere in this book concern genotyping single nucleotide polymorphisms (SNPs) as markers of disease risk, either at individual or multiple candidate loci, for chromosomal regions (including whole chromosomes), or across the whole genome. However, while this work has been going on, other researchers have been increasingly focusing their efforts on the possibilities of sequencing whole genomes. This approach is completely different to that used in more conventional studies of SNPs where only selected SNPs are genotyped. In whole-genome sequencing, the entire genome is sequenced and nothing is left out. This technology has now become a reality and the pace of development within this technology has increased year on year. For example, it took almost 10 years and an uncountable number of man-hours to map the first human genome. In 2008, complete human genomes belonging to the three main population groups were sequenced: one European, one Asian, and one African. Two years later, in 2010, the first genome of an ancient human was published in *Nature* by Rasmussen et al.. In November 2012, the 1000 Genomes Project announced the publication of 1092 human genome sequences (http://www.1000genomes.org) and it is now possible using NGS technology to produce millions of DNA sequence reads, providing a full genome sequence in a much shorter period of time (weeks and months, not years). At present, the study of genetics is evolving so rapidly that today's technologies may be out of date by tomorrow.

Due to the massive progress in genetics in the early part of the twenty-first century it is difficult to accurately predict what technological trends there will be in the future. High-throughput technologies are rapidly changing the landscape of genetics, providing the ability to answer questions with exceptional speed. New DNA-sequencing technologies are expected to be widely employed in research into the genetic basis of common disease. It is quite clear reading from the past that progress has been made, but there is still much to do. Will we reach our final goals? Or will we end up with more questions than answers?

The development of DNA sequencing using the Sanger sequencing technique opened the way to sequencing the genome

The most direct method for determining the information contained in a gene is DNA sequencing. Sequencing allows the genetic information contained in a DNA molecule to be read, and can provide an enormous amount of information about protein sequence, gene structure, and function.

In 1977, Sanger et al. described a method to chemically decode the sequence of DNA based on DNA replication relying on the incorporation of dideoxynucleotide triphosphates

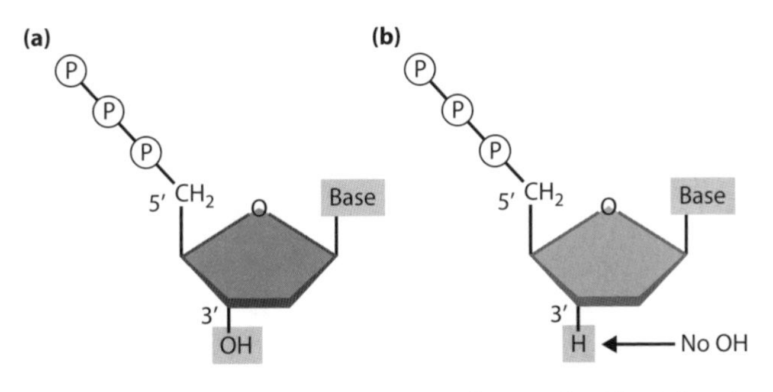

Figure 12.1: Decoding the DNA sequence during DNA replication using ddNTPs. The figure shows the structure of (a) the normal substrate dNTP, which carries a hydroxyl group, and (b) the abnormal substrate ddNTP, which does not carry the hydroxyl group. The interaction with the dNTP allows the DNA strand to grow, whereas the interaction with the ddNTP substrate terminates the growth.

(ddNTP) during the synthesis of a DNA strand. The structures of dNTP and ddNTP are illustrated in **Figure 12.1**. Dideoxynucleotide triphosphate molecules are identical to deoxyribonucleoside triphosphates (dNTPs), except that they lack a hydroxyl group (-OH) at the 3′ end. Sanger reasoned that if ddNTP is added to a growing DNA strand, the strand can no longer grow because the dideoxyribonucleotide is missing the 3′ hydroxyl group. The absence of a hydroxyl group prevents the formation of a phosphodiester bond between the DNA and the new nucleotide. Consequently, after ddNTP has been incorporated into the DNA strand, no more nucleotides can be added and thus the incorporation of ddNTPs terminates DNA synthesis.

In 1986, automated DNA sequencing methods based on the Sanger method were developed. These machines used fluorescent dyes to label each ddNTP, allowing the sequencing reactions to be quickly monitored. The mechanics of Sanger sequencing are simple and they are illustrated in **Figure 12.2**. First, the target DNA molecule is denatured into two single strands by heating. A primer or a short fragment of DNA is then used to trigger the sequencing reaction. The primer anneals adjacent to the sequence of interest and is extended by a DNA polymerase. During the extension reaction the sequence chain is terminated by the random incorporation of fluorescently labeled dideoxynucleotides, which have complementary identity to the base on the opposite strand. Each ddNTP is labeled with a different color fluorescent dye corresponding to one of the four bases (A, T, C, and G). Once the extension reaction has been terminated, the resulting mixture containing fluorescently labeled DNA strands of varying length is separated by capillary electrophoresis. The final results are read as fluorescent peaks and are used to determine the DNA sequence.

Sanger sequencing was used to map the first human genome; it has a base accuracy of approximately 99.9% and can sequence DNA fragments up to 1 kb in length. Sanger sequencing is used clinically to identify mutations in selected Mendelian disease genes. However, the level of sensitivity of testing using the Sanger technique (generally estimated at 10–20%) may be insufficient for the detection of somatic gene mutations in solid tumors and in acute leukemia, and for the characterization of complex microbiological specimens. In addition, it is not able to achieve efficient high throughput for analyzing complex diploid genomes at low cost. Since the late 1990s, researchers in both academia and industry have revisited DNA sequencing methods, creating a new generation of DNA sequencing methodologies—the so-called NGS technologies.

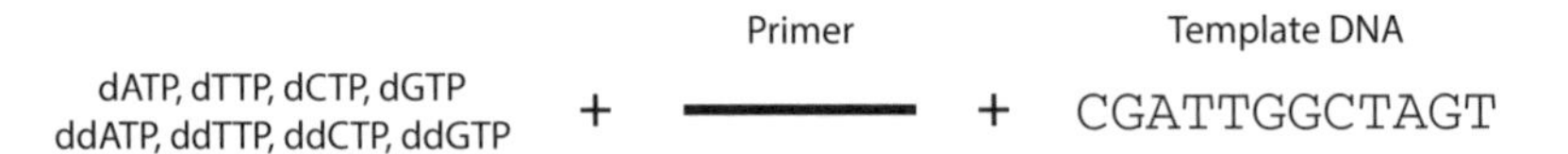

(a)

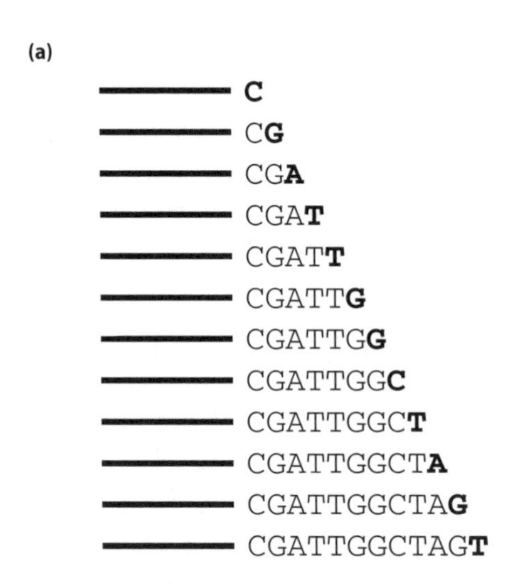

(b)

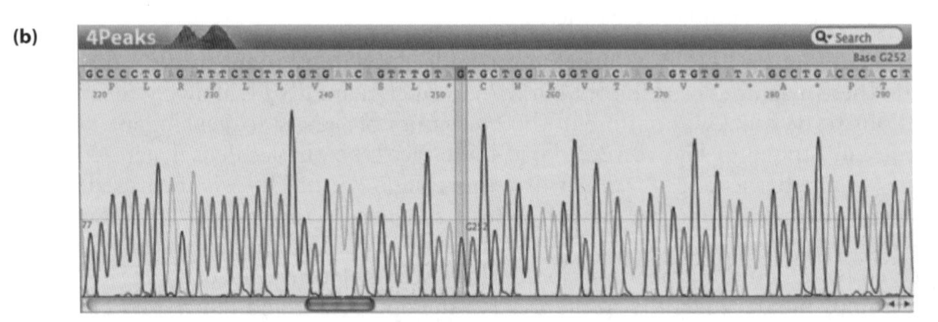

Figure 12.2: Sanger sequencing. The figure illustrates the principles of Sanger sequencing. (a) DNA is denatured into two strands by heating and is treated with a primer to trigger sequencing. The primer anneals with the DNA sequence of interest and is extended by a polymerase enzyme. The sequence chain may be terminated at any time during this extension phase by the addition of random fluorescently labeled ddNTPs. Each ddNTP is labeled with a different color corresponding to one of the four nucleotide bases. (b) Once the extension reaction has been terminated, the DNA strands of varying length are separated by capillary electrophoresis and the fluorescent peaks can be used to interpret the sequence. (From Thomas A [2015] Introducing Genetics, 2nd edn. Garland Science.)

The new era: next-generation DNA sequencing

The sequencing revolution began in 2005 with the introduction of the **sequencing-by-synthesis** technology developed by 454 Life Sciences and the multiplex polony sequencing protocol. Following on from this, a number of different commercial next-generation DNA sequencing systems started to appear in the market place.

These new sequencing machines all have the capacity to deliver significantly cheaper and faster sequencing, and are having a significant impact in bioscience. Most NGS technologies do sequencing in parallel, which means that hundreds of thousands or even millions of DNA fragments are simultaneously sequenced, resulting in very high throughput levels. It is likely that in the future NGS will become more widespread and better established. A comprehensive technical overview of each platform is beyond the scope of this book, but we will look briefly at the fundamentals of a number of them. However, as this field is rapidly advancing, readers are advised to refer to the different manufacturer's websites and other sources referenced in this chapter for the most up-to-date information on these technologies.

In each of the cases described below, the output of genome sequencing is not the whole DNA sequence as a single item. Instead a series of sequences of short DNA fragments called reads is created. These need to be assembled using bioinformatics software in order to determine the complete sequence of the whole DNA sample. The bioinformatics step is challenging and it needs to be performed with powerful computers able to handle the very large amounts of data generated by the sequencing platforms. A comparison of some of the different platforms can be found in **Table 12.1**.

Table 12.1 A comparison of some commercially available second-generation DNA sequencing platforms.

	Roche/454	Life Technologies SOLiD	Illumina HiSeq 2000
Library amplification method	Emulsion PCR on bead surface	Emulsion PCR on bead surface	Enzymatic amplification on glass surface
Sequencing method	Polymerase-mediated incorporation of unlabeled nucleotides	Ligase-mediated addition of two-base-encoded fluorescent oligonucleotides	Polymerase-mediated incorporation of end-blocked fluorescent nucleotides
Detection method	Light emitted from secondary reactions initiated by release of pyrophosphate	Fluorescent emission from ligated dye-labeled oligonucleotides	Fluorescent emission from incorporated dye-labeled nucleotides
Post-incorporation method	Not applicable (unlabeled nucleotides are added in a base-specific fashion, followed by detection)	Chemical cleavage removes fluorescent dye and 3′ end of oligonucleotide	Chemical cleavage of fluorescent dye and 3′ blocking group
Error model	Substitution errors rare, insertion/deletion errors at homopolymers	End of read substitution errors	End of read substitution errors
Read length (fragment/ paired end)	400 bp/variable length mate pairs	75 bp/50 + 25 bp	150 bp/100 + 100 bp

From Mardis ER (2011) *Nature* 470:198–203. With permission from Macmillan Publishers Ltd.

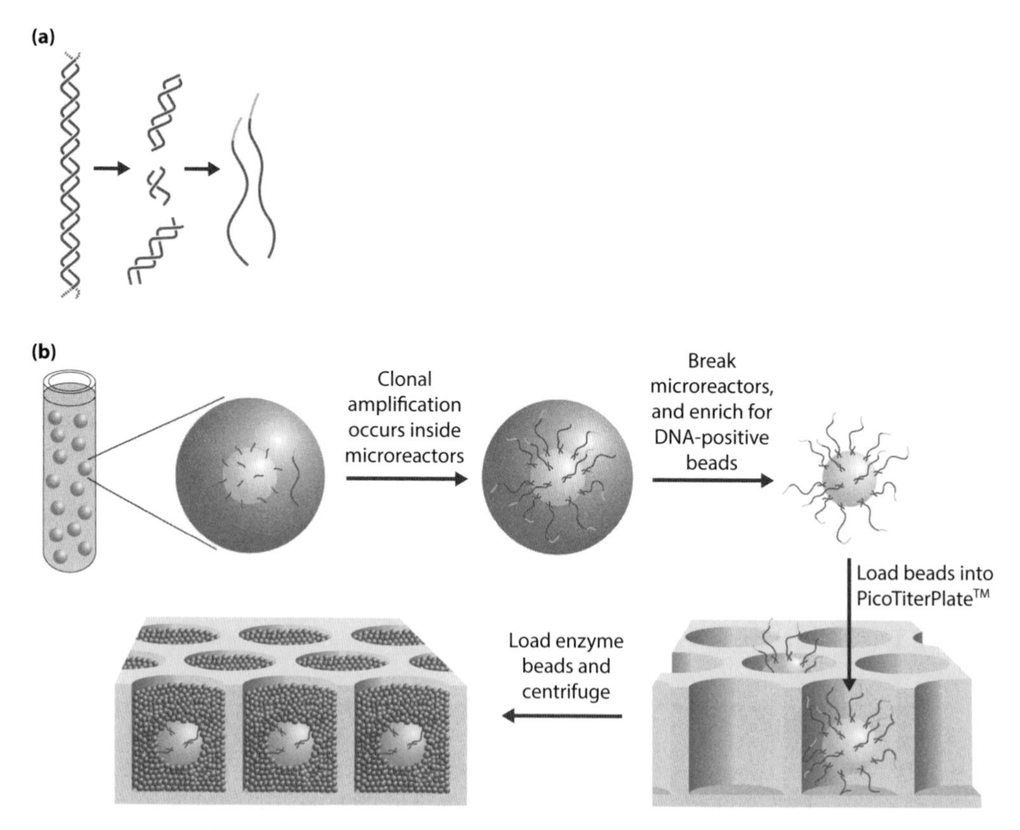

Figure 12.3: An overview of the Roche/454 sequencing technique: 454 GS FLX sequencer workflow. (a) Isolated genomic DNA is fragmented, ligated to adapters, and denatured into single strands. (b) Each fragment is bound to beads in a proportion of 1:1 and each fragment-bead complex is isolated in droplets of a water/oil mixture. PCR amplification is then performed within each droplet, resulting in beads carrying millions of copies of a unique DNA template. The emulsion is broken down and the DNA strands are denatured; beads carrying single-stranded DNA templates are enriched and are randomly deposited into wells of a fiber-optic slide. Clonally amplified beads generated by emulsion PCR serve as sequencing features where the pyrosequencing is carried out. (Adapted from Margulies M, Egholm M, Altman WE et al. [2005] *Nature* 437:376–380. With permission from Macmillan Publishers Ltd.)

The 454 technology (http://www.454.com) was launched in 2005 by 454 Life Sciences when Margulies et al. published the entire genome of the bacterium *Mycoplasma genitalia* with 96% coverage and 99.96% accuracy in a single GS 20 run. In 2007, Roche Applied Science acquired 454 Life Sciences and introduced the second version of the 454 sequencer, the GS FLX sequencer, which has the ability to sequence a longer sequence per run with longer read length. The 454 GS FLX sequencer is illustrated in **Figures 12.3 and 12.4**.

An alternative platform is the Applied Biosystems **SOLiD (Supported Oligonucleotide Ligation and Detection)** platform (http://www.appliedbiosystems.com). This uses short-read sequencing technology based on ligation. This approach was applied in 2005 to re-sequence an evolved strain of *Escherichia coli* and is similar to the 454 approach (above). One of the advantages of the SOLiD technique is that it uses an **offset sequencing primer strategy**. Therefore, each nucleotide in the sequence is interrogated twice. A given nucleotide in the template sequence will generate two different fluorescent

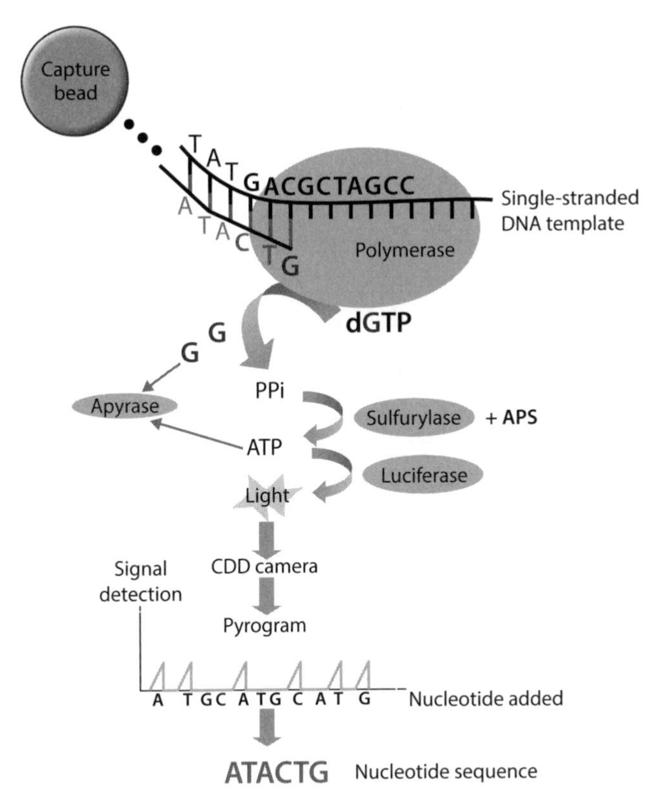

Figure 12.4: The principles of pyrosequencing technology. The figure illustrates the basic principles of pyrosequencing technology. DNA templates are linked to a capture bead and then exposed to multiple sequencing rounds. Only one nucleotide is added in each round. As each nucleotide (dGTP in this example) is incorporated into the DNA sequence, inorganic pyrophosphate (PPi) is released which reacts with adenosine 5′-phosphosulfate (APS) and sulfurylase to generate ATP. ATP is then used as a substrate for luciferase to generate light, which can be detected and quantified. (From Armougom F & Raoult D [2009] *J Comp Sci Syst Biol* 2:74–92.)

signals based on the identity of the neighboring base. As a result, the false-positive rate for mutation detection is reduced, as a SNP will generate two color changes when compared with the reference sequence. At the end of a 6-day run, the SOLiD instrument is capable of generating 4 Gb of sequencing data. The SOLiD method is illustrated in **Figure 12.5**.

The **Illumina Genome Analyzer** produced by the Illumina Corporation (http://www.illumina.com) is a short-read sequencing platform. This sequencing-by-synthesis process on an Illumina GA IIx appears to be the most widely used platform (at the time of writing). The method is said to generate read lengths of 35 bp with greater than 99% raw base accuracy and an overall throughput of approximately 5 Gb over a 3-day run (**Figure 12.6**).

The upcoming era: third-generation sequencing

To date, most of these advances have been applied within the research setting; however, these technologies will increasingly be used in the future to provide accurate, rapid, and low-cost information in clinical practice.

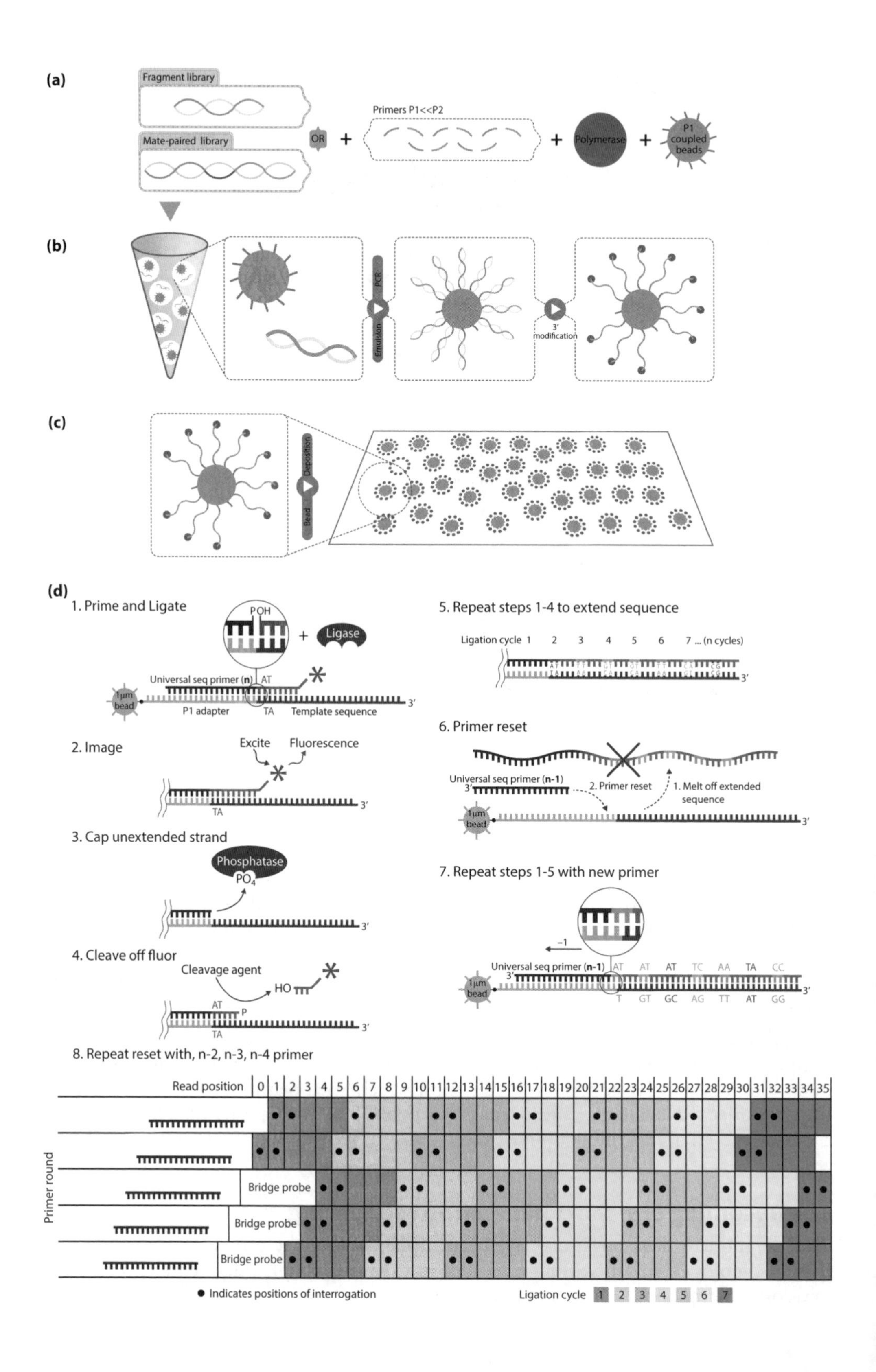

(a)
Fragment library
Mate-paired library
OR
Primers P1<<P2
Polymerase
P1 coupled beads
(b)
PCR
Emulsion
3′ modification
(c)
Bead Deposition
(d)
1. Prime and Ligate
P OH
Ligase
Universal seq primer (n)
AT
1μm bead
P1 adapter
TA
Template sequence
3′
2. Image
Excite
Fluorescence
TA
3. Cap unextended strand
Phosphatase
PO4
4. Cleave off fluor
Cleavage agent
HO
AT
P
TA
5. Repeat steps 1-4 to extend sequence
Ligation cycle 1 2 3 4 5 6 7 ... (n cycles)
6. Primer reset
Universal seq primer (n-1)
2. Primer reset
1. Melt off extended sequence
7. Repeat steps 1-5 with new primer
−1
Universal seq primer (n-1)
AT AT AT TC AA TA CC
T GT GC AG TT AT GG
8. Repeat reset with, n-2, n-3, n-4 primer
Read position
0 1 2 3 4 5 6 7 8 9 10 11 12 13 14 15 16 17 18 19 20 21 22 23 24 25 26 27 28 29 30 31 32 33 34 35
Primer round
Bridge probe
Bridge probe
Bridge probe
Indicates positions of interrogation
Ligation cycle 1 2 3 4 5 6 7

The third generation in the history of sequencing is now emerging and is expected to bring new insights on DNA sequences. Third-generation sequencing has two main advantages compared with second-generation technology (above). The first main advantage is that the PCR step is no longer needed before sequencing. Though PCR amplification increases the number of copies of DNA fragment, it can introduce errors into the sequence. The ideal DNA sequencing platform, which is what the third-generation platform aims to be, would combine the advantages of high-throughput and rapid-sequence technology with the capability to sequence long stretches of DNA directly from a single DNA molecule, without any pre-sequencing PCR step. This would prevent errors introduced by PCR amplification being introduced into the sequence. The second advantage is that the signal is captured in real-time.

Though this technology is still developing, several prototypes exist such as nanopore DNA sequencing, single-molecule real-time (RNA polymerase) sequencing, and the HeliScope sequencer, which has pioneered an innovative methodology able to perform high-throughput DNA sequencing, starting from a single-molecule. The **HeliScope sequencer** relies on the principle of **true single-molecule sequencing** that does not require a preclonal amplification step. The absence of an amplification step during sample preparation circumvents the problems of sequencing errors attributable to PCR artifacts, as stated above. This system is capable of sequencing up to 28 Gb in a single sequencing run in about 8 days, but it can also generate short reads with a maximal length of 55 bases. The Helicos platform was first used in 2008 to sequence the 6407-base genome of the bacteriophage M13 and a year later to sequence an individual human genome (**Figure 12.7**). Currently there is major competition in the sequencing arena and in 2015 not all of the original companies are still trading. Some have been subsumed into other companies and some have closed down completely. Consequently the reader needs to be mindful that the availability of information from websites changes over time. Nevertheless websites when open offer a good source of information.

There are a number of companies with an interest in nanopore (a tiny hole) sequencing and nanopore technologies, amongst which Oxford Nanopore Technologies (https://www.nanoporetech.com) have declared an interest in DNA sequencing and other activities. The US government recently released a number of grants for the further development and use of this technology with five new grants in September 2013 alone. Though there is great interest, there have been a few problems. The speed at which DNA is tracked through the nanopore makes the process of sequencing especially tricky. However, it appears that altering the wavelength of light that the system employs can effectively slow down the flow and this may solve the problem. The potential for this technology in DNA sequencing is great.

Figure 12.5: The Applied Biosystems SOLiD sequencing-by-ligation method. Fragments of DNA are ligated with oligonucleotide adaptors at each end and hybridized to complementary oligonucleotides attached to magnetic beads (a). Beads are then placed in a water/oil emulsion where DNA amplification is performed (b). Once amplification is finished, the beads are placed on a glass surface and entered into the sequencer (c). In the sequencer a universal sequencing primer, complementary to the adaptor sequence, is used to trigger the sequencing reaction where ligation cycles with fluorescently labeled degenerate probes is performed. Once the probe anneals to the DNA template, a DNA ligase covalently binds the probe to the sequencing primer and the fluorescence is recorded. The probe is then cleaved and another cycle starts. After seven rounds of sequencing, the extended universal primer is removed and a new universal primer is added that is offset by one base (d). The sequence of the read is inferred by interpreting the ligation results for the 16 possible dinucleotide interrogation probes. (Courtesy of Applied Biosystems.) (See also color plate.)

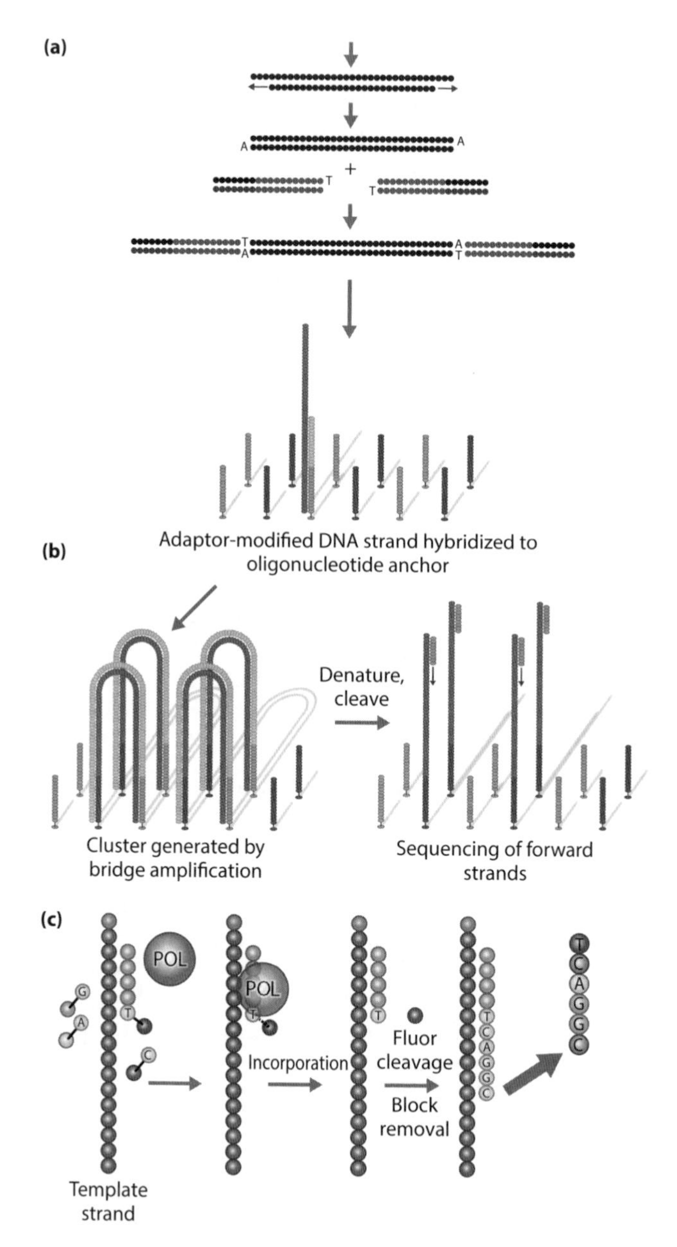

Figure 12.6: The principles of the Illumina sequencing technique. DNA fragments are first ligated onto oligonucleotide adaptors to form double strands (a). Adapter-modified, single-stranded DNA is then added to a flow cell and immobilized by hybridization. Bridge amplification generates clonally amplified clusters (b). The cluster fragments are denatured, annealed with a sequencing primer, and subjected to sequencing-by-synthesis employing DNA polymerase and four reversible dye terminators. Once incorporated, the terminator stops the sequencing reaction, which restarts immediately by cleavage of the incorporated dye terminator. Post-incorporation fluorescence is recorded (c). (See also color plate.)

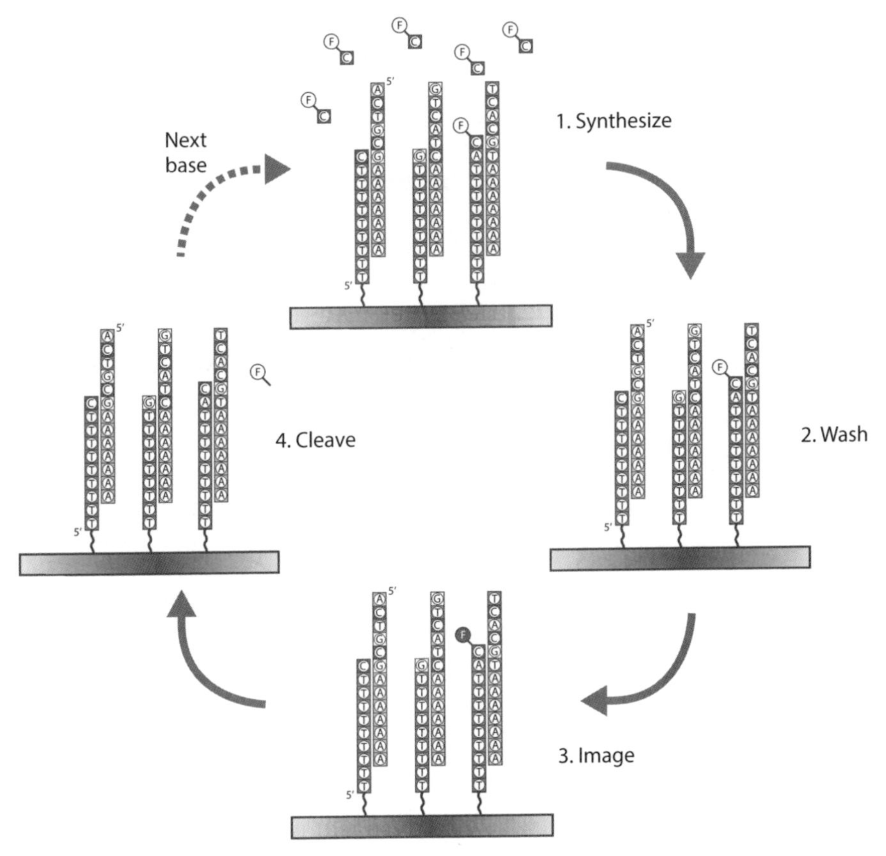

Figure 12.7: The principles of the Helicos sequencing technique. DNA fragments are captured by poly-T oligomers linked to an array. During each sequencing cycle, a DNA polymerase and a single fluorescent labeled nucleotide is added to the array. The array is imaged and the fluorescent label is then removed, and the cycle repeated.

12.2 THE FUTURE OF NGS IN CLINICAL PRACTICE AND RESEARCH

In the recent past, the size and complexity of the human genome (3 billion base pairs) was seen as a burden for every research group that aimed to sequence the genome. Consequently, any proposal to use genome sequencing in clinical practice was considered impossible; however, the idea of genomic sequencing has become a reality since the publication of the first draft of the Human Genome Map (HGM) in 2001. To generate the first draft of the HGM took more than 10 years and cost several millions of dollars. Since then, a number of developments, including the advent and feasibility of NGS, have not only dramatically reduced the cost and the time required to sequence genomes, but massively increased the scale of data output (**Figure 12.8**). We may therefore be approaching the time, in the near future, when sequencing whole genomes of patients will be practical in the clinical setting. In this section, we will briefly discuss some of the possible applications of NGS technology in both research and clinical practice.

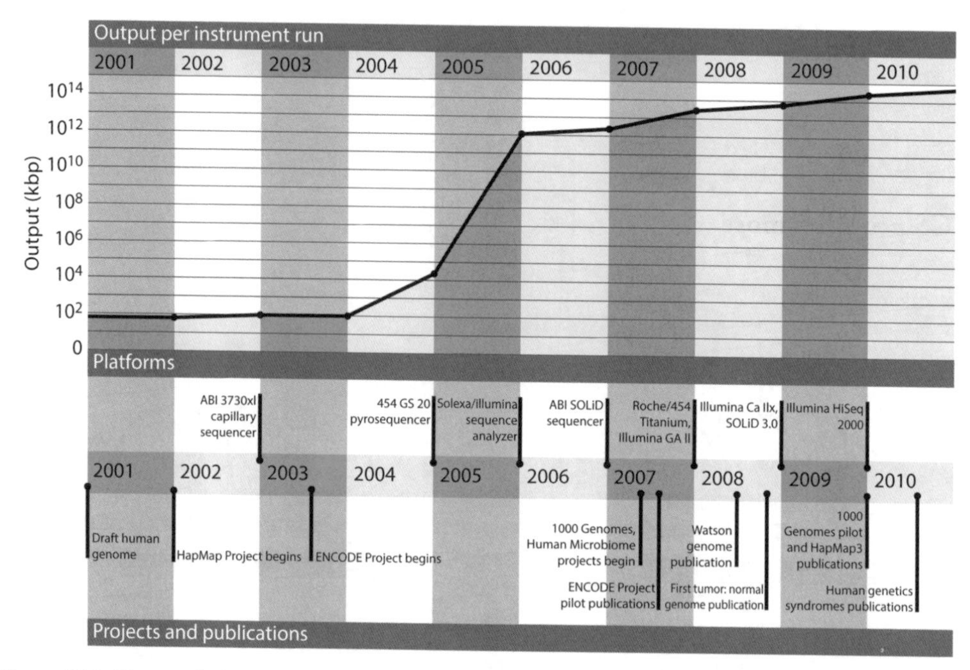

Figure 12.8: Changes in instrument capacity over the past decade and the timing of major sequencing projects. The figure provides a perspective on a decade of advances in technologies applied to complex disease and a time-line of the major research projects during the same period. The figure illustrates how technology and science are linked. (From Mardis ER [2011] *Nature* 470:198–203. With permission from Macmillan Publishers Ltd.)

Using NGS will enable high-resolution genotyping for SNPs in complex disease

Genotyping and association studies have long been employed to catalog single nucleotide variation across the genome and to associate genetic variation with phenotype. In recent studies, high-density oligonucleotide arrays have been the predominant methodology for SNP genotyping in complex disease; however, the identification of specific risk alleles remains challenging and the results of both **genome-wide linkage analysis (GWLA)** and GWAS have been mixed. Some GWLA and GWAS studies have been very successful, but others have not been so productive.

One problem with GWLA and GWAS is that the early studies were conducted mainly on genotyping arrays that, though high throughput, test for a limited number of selected SNPs. In addition, the SNPs to be tested were selected to identify polymorphisms where the least frequent variant was present in 5% or more of the population. This was done to maintain a reasonable level of statistical power. As a consequence, in many common diseases, such as cancer, and some clinically important quantitative traits, which may have very complex genetic and epigenetic (in the broadest sense) architectures, this standard genotyping format may not have been sufficient to identify the risk alleles or causative variants. Relevant marker SNPs may have been absent from the original (standard) genotyping array. More recently, with the larger-scale studies, it has become possible to investigate less common SNPs, i.e. those were the least common variant is 1% or less, with a reasonable degree of statistical confidence.

Beware the fly in the ointment

Though the selection of which SNPs to investigate can depend on the size of the study population, selection is currently more likely to be determined by the commercial testing platform used. Most commercial platforms offer a pre-set selection and selection is not a free-for-all process based on the most interesting SNPs identified by the researchers.

NGS promises more

The next-generation high-throughput sequencing approach promises more. NGS technologies provide higher resolution genotyping designed to detect and characterize SNPs, rare variants, and also **mosaic mutations**. The power of high-throughput sequencing to identify unknown causative mutations in human disease has recently been demonstrated in a family with a recessive form of Charcot–Marie–Tooth disease and in a family with both Miller syndrome and primary ciliary dyskinesia. Whether these examples, which are based on Mendelian (i.e. non-complex) disease, prove the potential for this type of analysis in some complex diseases, where the risk alleles may be present in a small percentage of the disease group, is a matter of debate. If we consider a genetic variation that is not associated with reproductive fitness, we cannot expect this to be present at high levels in a population. Therefore, any such genetic variant under the old GWAS model design, which generally selected tagged SNPs where the least common allele is found at 5% or more in the population, would be unlikely to be successful. Francis Collins suggests this may be particularly important in relation to diseases involving mental health, such as autism, bipolar disorder, and schizophrenia, where reproductive fitness may be low. Genome sequencing is more likely to be useful in these cases. As genome sequencing is not restricted by the frequency of a specific marker and it also includes variations that are not SNPs, such as copy number variations (CNVs), which may also be important in determining disease and disease traits (e.g. response to commonly used pharmaceutical agents such as codeine), it will be capable of identifying a greater range of risk alleles and sequences.

The key application of NGS technologies therefore is studying genetic variation between individuals using whole-genome or targeted re-sequencing in order to discover new associations with disease and to improve the resolution for the identification of previously detected associations so that the precise risk allele can be identified. Such comprehensive SNP identification will undoubtedly improve the predictive power of GWAS and GWLA, and impact on our understanding of complex trait loci in complex disease, but it will not necessarily solve all of our problems—only increasing sample size can do that.

The greater the number of markers tested or sections of the genome sequenced, the finer the map will become. Fine mapping refers to the level of resolution being achieved in a study. Fine mapping is not a single technique—it is an idea that can be achieved by applying one or more techniques. Fine mapping has not been the first step in most studies, as stated in Chapter 4. Instead, it is usually performed to confirm and extend a study. Often it involves focusing on a particular area or series of chromosomes rather than the whole genome. However, with direct sequencing and the possibility to sequence the whole genome, fine mapping may increasingly be the first step in a study, not a second or later step. There is also a link between fine mapping and imputation. Imputation can be used to find missing data based on known patterns of linkage disequilibrium within haplotypes. Imputation analysis is discussed in further detail below.

NGS can also be applied for the discovery of **novel somatic mutations** in cancer, leading to a better understanding of the molecular pathogenesis of cancer. This will undoubtedly result in novel diagnostic and therapeutic approaches.

Using NGS will enable better identification of CNVs

CNVs include deletions, duplications, inversions, and rearrangements of pieces of the genome. These range from small insertions and deletions to very large cytogenetic changes that involve entire chromosomes. CNV has been associated with a number of genetically complex diseases, for example Alzheimer's disease and schizophrenia. Many of these findings are based on the use of array technologies and comparative genomic hybridization, which though powerful for the detection of large CNVs (with a resolution of approximately 1 kb) is unable to detect small balanced structural variations such as inversions.

High-throughput sequencing technology is able to detect any kind of CNV including balanced and unbalanced variation by employing a strategy called **paired-end mapping**. Using this approach, a single DNA fragment is sequenced from each end and thus the same fragment produces two sequencing reads that could overlap at their extreme ends provided they are long enough. Paired-end sequencing is emerging as a key technique for assessing genome rearrangements and structural variation on a genome-wide scale. This technique is particularly useful for detecting copy-neutral rearrangements, such as inversions and translocations, which are common in cancer where novel fusion genes can also be produced. Paired-end mapping was demonstrated by Korbel et al. in 2007 as a means of detecting deletions, inversions, and insertions with an average resolution of 644 bp, and the clinical potential for this technique has been demonstrated in both lung and breast cancer. However, sequencing-based approaches to detect CNVs are at the moment still too expensive and difficult for routine clinical diagnostics, but lower costs and the allure of more rapid results in future may enable sequencing techniques to replace array-based genomic hybridization in the clinical laboratory.

Sequencing the RNA transcript and the whole transcriptome is an alternative way forward

High-throughput sequencing technology has also been adapted to enable the study of DNA transcripts. This RNA sequencing technique is called **whole-transcriptome shotgun sequencing (WTSS)**, which aims at mapping and quantifying DNA transcripts in biological samples. This approach was first applied to report the human transcriptome from a human embryonic kidney and a B cell line, and later to detect gene fusions in cancer from chronic myelogenous leukemia cells. The goal of transcriptome sequencing is to sequence all transcribed genes.

Transcriptome sequencing is similar to exome sequencing (see Section 12.3) as it does not detect mutations in non-coding regions of the genome, but it has the main advantage of obtaining quantitative information about gene expression levels as well as detecting changes in gene expression (e.g. alternative splicing) and fusion transcripts produced by chromosomal rearrangements. One limitation of transcriptome sequencing is that it is biased toward abundantly expressed transcripts and therefore the detection of sequences or genes expressed at lower levels can be low or absent.

12.3 WHOLE-GENOME VERSUS EXOME SEQUENCING

As noted above, whole-genome sequencing may soon be available in all clinical laboratories where it may be used as a standard genetic test. However, in order to answer a specific clinical question, targeted analysis of specific genomic regions may be preferable

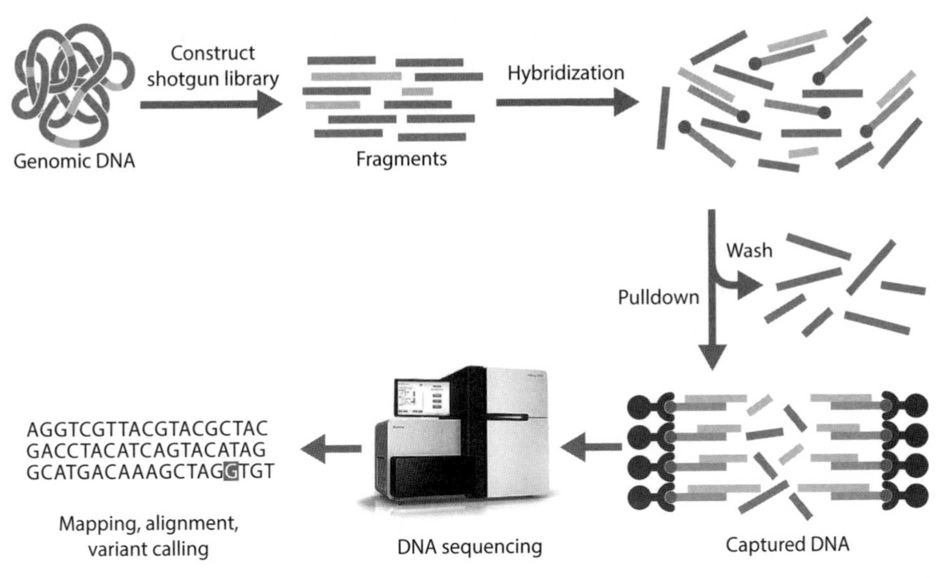

Figure 12.9: Diagram showing workflow for whole-exome sequencing. The figure shows the sequence of events in whole-genome sequencing. The DNA is broken up into fragments that are hybridized, washed, and captured on beads before undergoing sequencing. (From Bamshad MJ, Ng SB, Bigham AW et al. [2011] *Nat Rev Genet* 12:745–755. With permission from Macmillan Publishers Ltd.)

to sequencing the whole genome. When applied to the whole genome, restricting analysis to the protein-coding regions, that is exome sequencing, is more cost-effective compared with sequencing the entire genome. Whole-exome sequencing has been applied in the identification of germ-line mutations underlying some Mendelian disorders, some somatic mutations in a variety of different cancers, and *de novo* mutations in some neuro-developmental disorders. In the future the greatest application may be in cancer therapy where it may be useful to screen genes encoding protein targets before applying chemotherapy. These so-called **enrichment strategies** that target specific genomic regions will save time and money, and because smaller data bundles are formed than in whole genome analysis, which includes non-coding regions, whole-exome analysis can be performed more rapidly. Use of exome as opposed to genome sequencing may speed up the transfer and application of sequencing to clinical practice.

Restricting analysis to the exome does introduce some limitations. This analysis does not look at mutations and polymorphisms within non-coding regions, and it is not able to detect most structural variants, such as chromosomal translocations happening within intronic break points. In contrast, whole-genome sequencing will identify all of the variation in the genome not just that found in protein coding regions (**Figure 12.9**).

12.4 THE NEXT GENERATIONS OF GENOME/EXOME-WIDE ASSOCIATION STUDIES

The same basic principles apply to all investigations whether a study design is based on an investigation of the whole genome or the whole exome. The limitations and the ultimate goals have not changed, but the methods used have.

Missing and non-genotyped SNP data can be imputed using large databases and known patterns of linkage disequilibrium

Genotype imputation is a technique increasingly used in GWAS that allows geneticists to evaluate the evidence for associations with genetic markers that are not directly genotyped. There are two uses for this method: to fill in missing data and to assign genotypes for markers not typed in the study in order to increase the size and range of the study in terms of the number of potential risk alleles investigated. For whatever reason imputation is applied, the method is based on linkage disequilibrium. In statistical analysis, imputation is a method used for handling missing data (**Figure 12.10**). Missing data are simply

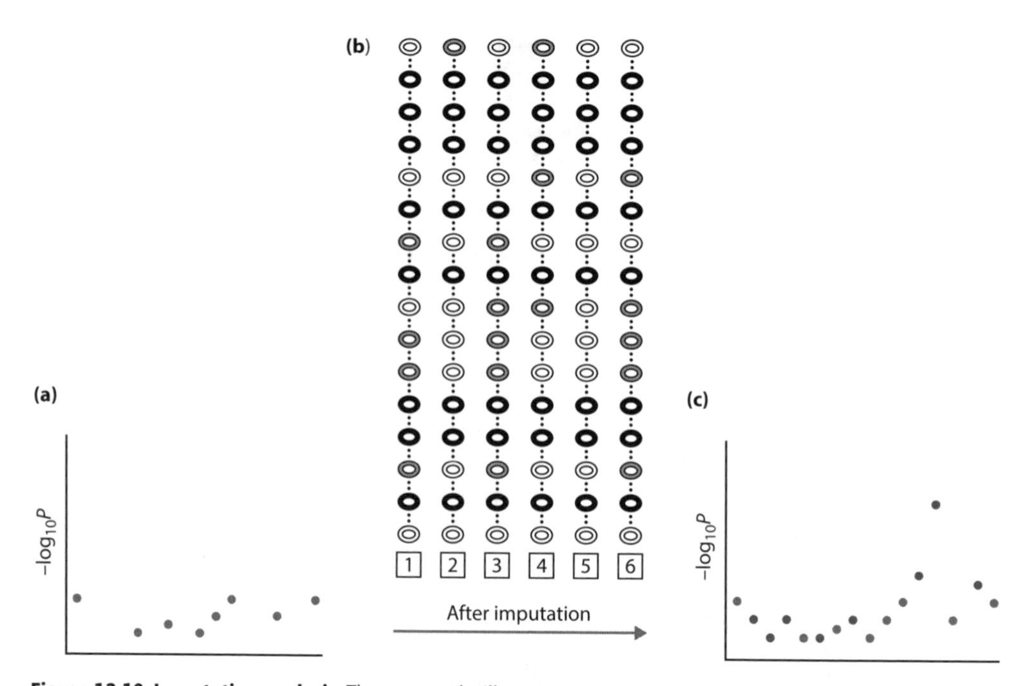

Figure 12.10: Imputation analysis. The two graphs illustrate a genotype sequence before (on the left) and after (on the right) impute has been applied. The central section of the figure illustrates in detail six different haplotypes numbered 1 to 6 for a combination of sixteen bi-allelic genes illustrated as double rings. Each gene is polymorphic, eight of the genes have been successfully genotyped. There are two alleles for all of genes. The alleles are illustrated by gray shading (A allele) or no shading (white–B allele) in those genes that have been successfully genotyped. The eight genes illustrated in black have not been genotyped. Imputation analysis can be applied to enable the genotypes of the black genes on this haplotype to be estimated based on the known pattern of linkage disequilibrium for the surrounding genes which have been successfully genotyped. To do this, data are extracted from large genotype databases to fill in the "missing" genotypes. For example if we know that persons with haplotype 1 which runs top to bottom; white, unknown, unknown, unknown, white, unknown, gray, unknown, white, gray, gray, unknown, unknown gray, unknown, white (or B-?-?-?-B-?-A-?-B-A-A-?-?-A-?-B) carries the white (B) alleles at positions 1, 5, 9, and 16 and the gray (A) alleles at position 7, 10, 11, and 14 in the sequence, then the full sequence for this haplotype can be assigned. Using the known genotypes from positions 1, 5, 7, 9, 10, 11, 14, and 16 we can identify the unknown genotypes for the other eight genes on the remaining five haplotypes (haplotypes 2 to 6). This imputed data can be included in the analysis. Because linkage disequilibrium is so strong in certain areas of the human genome, imputation analysis from initial GWAS data can massively extend the studies. Thus an initial GWAS study of based 300,000 to 400,000 SNPs can generate data for up to one million or more markers. (From Marchini J & Howie B [2010] *Nat Rev Genet* 11:499–511. With permission from Macmillan Publishers Ltd.) (See also color plate.)

substituted with predicted values based on the known levels of linkage disequilibrium. When all missing values have been imputed, the resulting data set can then be analyzed using standard techniques. The sort of databases used for this exercise include those produced by the 1000 Genomes Project and HapMap.

GWAS identifies both synthetic (false) associations and direct (real) associations

To date, GWAS have identified hundreds of alleles at gene loci that may be involved in determining susceptibility and resistance to common diseases. Most early GWAS restricted analysis to common variants where the minor allele frequency (MAF) was greater than 5%. This is discussed above.

However, very few GWAS have identified causal variants and this has led many geneticists to assume that the causal variants lie elsewhere in the genome, but in close proximity to the identified associated alleles and their markers. In general, many feel that with a few exceptions the genetic polymorphisms identified so far are unlikely to explain the majority of phenotypic variation in disease susceptibility. If susceptibility is determined by multiple rare or low-frequency variants, then these may not be identified in a GWAS even with lower statistical thresholds. Consequently, some, perhaps even many, of the current GWAS signals could reflect linkage with multiple rare variants. In this case, associations will be wrongly credited to a common allele, while in reality the effect that is seen relates to inheritance of other alleles with which they are in linkage disequilibrium. These associations with common markers resulting from multiple (unobserved) low-frequency causal variants are referred to as **synthetic associations**—a form of **indirect association**. Thus, in a typical GWAS with large sample sizes, rare variants may create multiple significant associations, but some may be synthetic. The impact of synthetic associations can be dramatically different from that which occurs when the causal variant is identified. For example, the causal variants can be megabases away from the SNPs found to be significant in GWAS and the real risk can be several-fold greater than that for a synthetic association. More recent studies have suggested that the contribution of rare variants is likely to be small and thus the importance of synthetic associations generated by multiple variants has perhaps been overstated. It is not yet known how many GWAS signals are due to synthetic associations; perhaps we will only know the answer when all of the associations have been tested in functional studies and a link between genotype and phenotype established. In the meantime, we should question some of the inferences from GWAS and consider the use of refined approaches for follow-up studies. The possibility of false positives (synthetic associations) provides strong support for the growing need to follow-up GWAS and GWLA studies with sequencing. In cases where the sequencing is restricted to a specific chromosomal region, the study design should consider expanding the region of interest beyond the target block. These follow-up studies will employ faster and better NGS technologies.

The importance of linking genotype to phenotype

When associated marker SNPs are located a long distance away from the nearest gene, it is often assumed that a regulatory variant with a small effect is the cause of the association. If we consider the case of SNPs on chromosome 8q24 that have been found to be associated with colorectal cancer, these SNPs map within range of the *MYC* oncogene and it is thought that the associated SNPs may play a role in influencing *MYC* function. However, these SNPs are located hundreds of kilobases away from the *MYC* oncogene, and no connection between the associated SNPs and the expression of this gene has yet been demonstrated.

Genotyping on new arrays provides a focus and higher level of resolution for GWAS

As stated above, the basic criterion used in the first phase of GWAS was to catalog common variants (5% or greater) and genotype them either directly or indirectly (through linkage disequilibrium) in cases and controls, and identify associations with particular traits and diseases. However, if the frequency of a SNP contributing to a phenotype is less common in the normal population, for example in the range of 1–5%, then we need to broaden our SNP selection criteria and test for less common variants, otherwise significant associations may be missed. This is one of many reasons for the 1000 Genomes Project, which is extending the catalog of known human variants down to a frequency near to 1%.

With respect to the first phase of GWAS that scanned primarily for common variants in complex disease, it seems reasonable in 2015 to suggest that even if we have not yet reached the end of the first phase, we have at least reached the end of the beginning. A new wave of higher-resolution GWAS is now being applied that will investigate less common variants. At present, this new wave of GWAS is based on one of two approaches: GWAS on new genotyping arrays and GWAS on NGS technologies.

Companies such as Illumina and Affymetrix have already designed chips containing both common and low-frequency variants discovered in the 1000 Genomes Project. The new analysis will identify millions of genotypes on these new chips and it is likely that they will identify new associations. However, there is some concern that the causative variants may be very rare variants. There are questions over how such data can be used. Though not useful in disease diagnosis, identifying rare associations with complex disease does help us to understand disease pathology, and also has the potential to help in patient management and therapy. Thus, this is not wasted time, even though the diagnostic use of this data may be limited. However, as stated above to accurately identify association signals for rare variants we will need even larger, sample pools compared with previous studies in order to achieve statistical significance at an appropriate significance threshold.

Sample size is dictated by the expected size of the effect, i.e. the expected genetic impact of the allele set as an expected risk ratio [either odds ratio (OR) or relative risk (RR)]. For rare alleles with large risks, sample sizes can be smaller. An example of this is seen with the 32-bp deletion in the C-chemokine receptor 5 gene (*CCR5*-Δ32), where approximately 1% of Europeans are homozygous for the mutation and as many as 10% may be heterozygous. The homozygous individuals are virtually immune to early HIV-1 infection. As this mutation has a large effect in HIV-1/AIDS, identifying this association with a reasonable level of statistical confidence does not require large numbers.

Different forms of NGS technology will impact on how GWAS is used

NGS will be used to conduct whole-genome and whole-exome sequencing of individuals rather than genotyping and cataloging specific variants. Such studies will be performed in a manner similar to current GWAS sequencing, with both cases and healthy controls being studied, and similar statistical analysis for associations applied.

Whole-exome sequencing

Whole-exome sequencing has already been used to identify causal alleles for several Mendelian disorders, such as those shown in **Table 12.2**. These studies looked for rare alleles or novel alleles that are more common in affected individuals compared with unaffected individuals (controls). Using public databases, this method could be used to filter out potential causative variants by assuming that any allele found in both cases and

Table 12.2 Some Mendelian and a few complex diseases identified up to 2011 via exome sequencing.

Disorder	PubMed ID	Mode of inheritance	N[a]	Strategy	Gene(s)
Comparison of unrelated cases					
Kabuki	20711175	AD	10	10 cases/10 kindred	*MLL2*
Schinzel–Giedion	20436468	AD	4	4 cases/4 kindred	*SETBP1*
Fowler	20518025	AR	2	2 cases/2 kindred	*FLVCR2*
Sensenbrenner	20817137	AR	2	2 cases/2 kindred	*WDR35*
Comparison of related cases					
Miller	19915526	AR	4	4 cases/3 kindred	*DHODH*
Retinitis pigmentosa	21295283	AR	3	3 cases/1 kindred	*DHDDS*
Spinocerebellar ataxia	21106500	AD	4	Linkage + 4 cases/1 kindred	*TGM6*
Primary failure tooth eruption	21404329	AD	4	Linkage + 4 cases/1 kindred	*PTH1R*
TARP (talipes equinovarus, atrial septal defect, Robin sequence, and persistent left superior vena cava)	20451169	XLR	2	Linkage + 2 cases/2 kindred	*RBM10*
X-linked leukoencephalopathy	21415082	XLR	2	Linkage + 1 case/1 kindred	*MCT8*
Homozygosity mapping					
Autoimmune lymphoproliferative syndrome	21109225	AR	1	1 case/1 kindred	*FADD*
Complex I deficiency	21057504	AR	1	1 case/1 kindred	*ACAD9*
Non-syndromic mental retardation	21212097	AR	2	2 obligate carrier parents	*TECR*
Identification of *de novo* mutations					
Sporadic mental retardation	21076407	Complex	30	10 parent–child trios	Multiple
Autism		Complex	60	20 parent–child trios	Multiple

The table lists a number of diseases, the traits associated with them, and the genes thought to carry the responsible mutations.

[a]Number of exome-sequenced individuals.

AD, autosomal dominant; AR, autosomal recessive; XLR, X-linked recessive.

MLL2, myeloid/lymphoid or mixed-lineage leukemia 2; *SETBP1*, SET binding protein 1; *FLVCR2*, feline leukemia virus subgroup C cellular receptor family, member 2; *WDR35*, WD repeat domain 35; *DHODH*, dihydroorotate dehydrogenase; *DHDDS*, dehydrodolichyl diphosphate synthase; *TGM6*, transglutaminase 6; *PTH1R*, parathyroid hormone 1 receptor; *RBM10*, RNA binding motif protein 10; *MCT8*, monocarboxylate transporter 8; *FADD*, FAS-associated death domain (also called *MORT1*); *ACAD9*, acyl-CoA dehydrogenase family, member 9; *TECR*, *trans*-2,3-enoyl-CoA reductase.

The numbers in the studies are small and therefore the statistical inference is not shown. These are conditions from rare kindred and mostly they are Mendelian autosomal diseases; however, there are two complex diseases included in the list.

From Bamshad MJ, Ng SB, Bigham AW et al. (2011) *Nat Rev Genet* 12:745–755. With permission from Macmillan Publishers Ltd.

healthy controls cannot be causative. Though this strategy can be exceptionally powerful for rare Mendelian disorders, it is less useful in complex disease because the assumption that a causative allele must be absent from the healthy population does not work in complex disease, where many potentially causative alleles have been found to be expressed in both diseased and healthy individuals. In addition, not all cases are positive for the causative allele in complex diseases.

Exome sequencing reduces the number of candidate genes tested. However, as with GWAS, these studies still require large sample pools to provide strong statistical evidence for association and even though costs are currently falling, they are still too high. Therefore, two different strategies have been proposed to reduce these costs: family-based sequencing and basic extreme-trait design.

Family-based sequencing

Family-based studies are impractical for most complex diseases. Even when familial cases occur we cannot assume that these rare cases represent the whole disease population. Therefore, making assumptions based on testing families in complex disease is a rare option. However, if families are tested, the strategy is to sequence the index case (i.e. the first identified case) and any co-affected family member(s) to identify overlapping variants. The closer the family cases are to the index case, then the greater the number of candidate genes that will need to be tested as closer family members will share more genes. In contrast, the more distant related cases are compared with the index case, then the lower the level of genetic similarity is and therefore a smaller number of candidate alleles will need to be genotyped.

Basic extreme-trait design

Basic extreme-trait design offers more potential. In this situation cases are carefully selected from one or both ends of a phenotype distribution and then sequenced. This strategy is based on the studies of Cohen et al., who in 2004 demonstrated the effectiveness of sequencing candidate genes at the extreme ends of a phenotype distribution to find rare alleles involved in the risk for a complex trait. An obvious example of this extreme phenotype study is seen in individuals who are highly exposed to HIV-1, but remain uninfected. As there is survival advantage in those who carry genetic variants that contribute to the exposed/uninfected trait, these alleles will be enriched in the population. This means that studies of small sample size can be performed with a higher degree of confidence than if the alleles were less common. Identified candidate alleles can then be genotyped for confirmation in much larger samples. In HIV-1, where genetic variants contribute to protection against HIV-1 infection and HIV-1/AIDS, the high impact of the protective variants/alleles means that whole-exome (or whole-genome) sequences of a few individuals may be powerful enough to identify candidate variants/alleles associated with the disease trait. Once identified, the genotype of the identified candidate can be re-investigated in a much larger group to confirm or refute the findings of the sequencing report.

The use of NGS technology applied to the new wave of GWAS has the potential to discover the entire spectrum of sequence variations in a sample of well-phenotyped individuals. However, the use of NGS platforms in GWAS studies has raised some concerns. The error rate of the NGS platforms is higher than the Sanger sequencing methods and, because not all errors are random, they may obscure true associations or generate false-positive associations if they are frequent enough. In addition, NGS technologies can also produce data pools with high missing rates. In practice, however, high missing rates are less of a problem because imputation methods may be used to recover (or back-fill) the missing data. Authors using imputation should make sure that any data generated by this means are clearly acknowledged in their publications.

12.5 EPIGENETICS: A COMPLIMENTARY STRATEGY IN COMPLEX DISEASE STUDIES

Epigenetics is based on the word "epi-" meaning "above", "outside", or "in addition to" genetics and is usually used to refer to heritable effects not produced by changes in the DNA sequence. The term epigenetics is applied in a number of different research strands all around genetics. It can have several meanings, as discussed earlier in this book. However, the main one which concerns us is that where it refers to reversible biochemical modification of DNA that does not involve any change in the nucleotide sequence. These modifications may persist through cell divisions and represent memories or **molecular signatures** over multiple cellular generations. Examples of such modifications are DNA methylation and **histone modification**, both of which serve to regulate gene expression without altering the underlying DNA sequence. Two alleles with the same sequence may have different states of methylation and this may result in different phenotypes even though the sequence is the same. Though these changes do not alter the DNA sequence, they may have major effects on the expression of the gene. Some of these changes are heritable, though they do not affect the DNA structure. Methylation establishes epigenetic inheritance as long as methylase acts to restore the methylated state after each cycle of replication. Thus, a methylated state can be perpetuated through an indefinite series of somatic meiosis.

Determining how proteins interact with DNA to regulate gene expression is essential for the full understanding of many biological processes and disease states. A prominent genomics approach for detecting biochemical modifications to DNA is the **chromatin immunoprecipitation assay (ChIP-chip)**. This technique has been used primarily to detect locations where a protein of interest is bound to DNA *in vivo* (**Figure 12.11**).

ChIP sequencing has been used to characterize histone and transcription factor binding sites in human $CD4^+$ T cells, in a cervical carcinoma cell line, and also in pluripotent murine stem cells undergoing differentiation. While high-throughput sequencing has improved our ability to detect and characterize DNA–protein interactions, further work will be required to determine how these biological changes result in a clinical disease phenotype.

12.6 METAGENOMICS AND THE BACTERIAL GENOME

Laboratory culture-based approaches have been traditionally used for characterizing microbe genomes. However, microbes are difficult to culture in the laboratory because they normally live in mixed communities, and different species interact both with each other and with their habitats, including the host organism. Metagenomics describes genomics on a huge scale and it is particularly important in relation to the study of microbial communities because it can be used on mixed samples and does not require pure cultures for genotyping.

Metagenomics is based on the genomic analysis of microbial DNA directly extracted from microbial communities present in samples such as soil, water, or feces. This technology has had a tremendous impact on the study of microbial diversity in environmental and clinical samples, and may lead to major advances in medicine, agriculture, and also energy

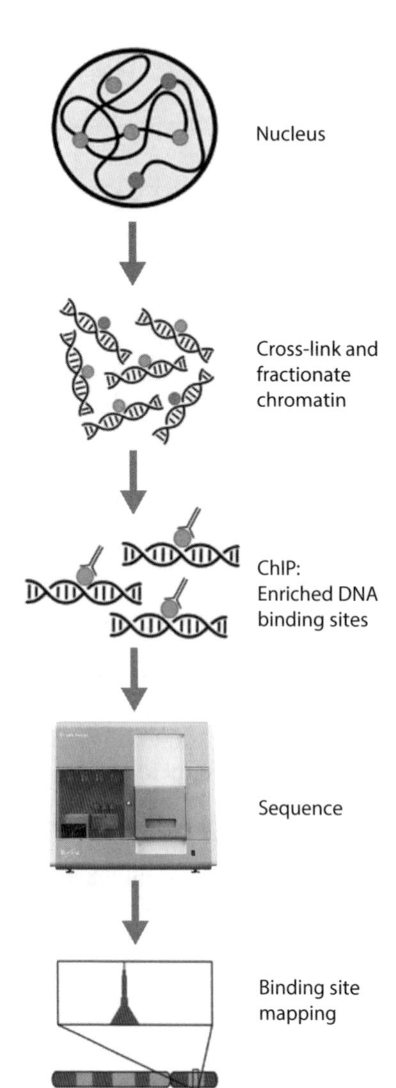

Figure 12.11: Illumina ChIP sequence experiment workflow. DNA fragments are first cross-linked *in situ*, fractionated, and then immunoprecipitated. DNA is then sequenced using an Illumina platform to identify genome-wide sites associated with proteins of interest. (Courtesy of Illumina.) (See also color plate.)

production. In a typical metagenomic study, the genomic content of all the microbes in a sample is sequenced using NGS technology. This results in millions of sequences of DNA fragments. Computational methods are then employed to predict the taxonomic affiliation of these DNA sequence fragments, which are then compared against a reference database such as those of the National Center for Biotechnology Information (http://www.ncbi.nlm.nih.gov) where the sequences of microorganisms are collected. Sequences are then affiliated to the corresponding microorganism and compared with those of others to identify the level of biodiversity present in the sample. This approach has been used to identify novel viral pathogens, detect viral drug resistance mutations, and diagnose bacterial infections. However, given the relatively high cost of high-throughput sequencing, these techniques are unlikely to replace traditional microbiological techniques for routine pathogen identification in the immediate future, but it may not be long before metagenomic screening is standard. Metagenomic technology is at present the main approach employed by the Human Microbiome Project (http://commonfund.nih.gov/hmp/)—an initiative launched in 2008 that aims to characterize the microbial communities associated with both health and disease in different parts of the human body, such as nasal passages, oral cavities, skin, and gastrointestinal and urogenital tracts.

To put metagenomics into context, we need to consider the impact it may have

Tim Spector, in his book *Identically Different: Why You Can Change Your Genes* (2012), discusses the impact of bacterial genes on human health. He states "... within us all there lies a parallel universe of warring personalised bacterial colonies, each of them with genes and proteins with a possible role in our health of which we remain largely ignorant." He continues "We are only just scratching the surface of this hidden universe ..." and suggests "... the type of bacteria and extra genes we have in our guts could explain around 50 per cent of obesity, compared with a paltry 1–2 per cent explained by the common human genes we have discovered so far." The evidence for this hypothesis is, as the author suggests, currently limited, but it is a subject that is worthy of serious investigation. Bacteria react closely with the gut wall, and modifications in the permeability of the gut wall and host immune tolerance to certain bacteria have been shown to be particularly important in common diseases of the bowel (Crohn's disease and ulcerative colitis, in particular), but also in some cancers (gastric and colon cancer). This interaction between host and bacteria illustrates how we cannot consider the human genome in isolation. Not only is it essential to consider host systems and how they interact (including epigenetics), but we must also consider the environment and how environmental factors interact (**Figure 12.12**). This idea is not new. However, there has been a tendency to consider invading bacteria as passive, almost inactive, in the context of predisposition to and protection from disease. Of course pathogens can cause disease, but not all bacteria are malignant—some may be beneficial. It is clear from the findings reported above that bacteria are not simply the passive targets for the host immune response nor are they simply acting as malicious

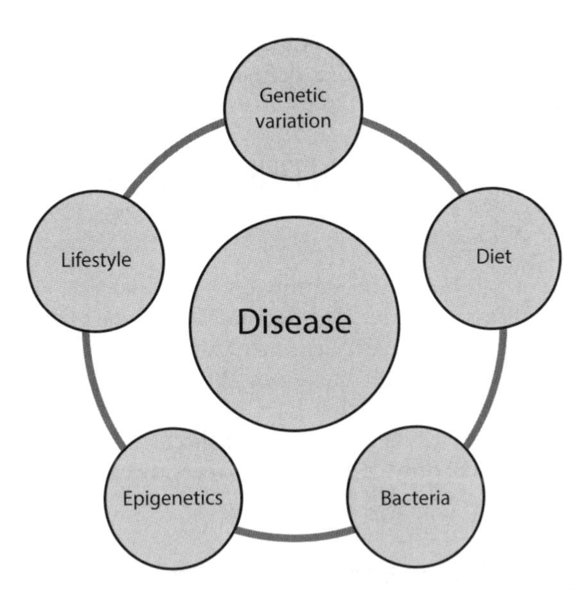

Figure 12.12: Genetic and non-genetic factors impacting on disease risk. The figure illustrates the idea that multiple factors influence susceptibility to complex disease and not all are genetic. The factors interact in complex ways. In order to understand the genetics of complex disease we need to apply a holistic approach that takes account of these other factors.

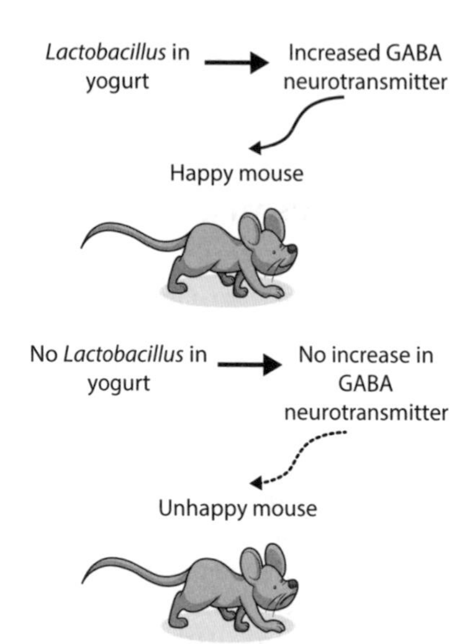

Figure 12.13: *Lactobacillus*-rich yogurt and mice. The figure shows the effect of nutritional factors on biosystems and reminds us that our genes are influenced by our environment. In this example, mice fed a diet with *Lactobacillus* appear to be happier and have higher measurable levels of GABA neurotransmitters than those not on the same diet.

pathogens. The close interaction suggests there may be a level of symbiosis. Once again, this is not a new concept. For example: millions of people consume food supplemented with *Lactobacillus* which is thought to be beneficial in maintaining homeostasis between the host and the bacterial flora in the gut. Looking at this from a genetic perspective provides a different focus. Simply altering the diet of mice by giving *Lactobacillus*-enriched yogurt can modify the levels of expression of the γ-aminobutyric acid (GABA) neurotransmitter receptor. The mice with the yogurt supplement appear to become happier mice (**Figure 12.13**). The *Lactobacillus* must gain something from this relationship, which appears to be a true symbiotic relationship. The balance of such relationships will determine whether a bacterium is beneficial to the host or not and may lead to some traits being promoted, while others may be restrained. Once again we are talking about susceptibility and resistance, i.e. risk—the one word that defines complex disease.

12.7 MAJOR ONGOING INTERNATIONAL GENOME PROJECTS

As discussed throughout this book, the worldwide web provides a wealth of resources for the student of complex disease genetics. The databases developed from major international research projects, such as the Human Genome Mapping Project, the Haplotype Map (HapMap), the SNP Map, the 1000 Genomes Project, and ENCODE, are especially useful. These are not the only sources that have been used, but all of them have important functions for the future. These have all been discussed elsewhere in the book and therefore in this section we will only consider three recently complete or ongoing projects: HapMap, 1000 Genomes Project, and ENCODE.

HapMap is a project with major significance in current research, especially GWAS

HapMap was launched in 2002 by a consortium of academic centers and private companies of six nations (Canada, China, Japan, Nigeria, the UK, and the USA) aiming to pinpoint the genes behind common diseases. Phase 1 typed 270 individuals from four geographically diverse populations (**Table 12.3**), identified approximately 1.3 million SNPs, and was published in 2005 (http://www.hapmap.org). Phase 2 of the project aimed to increase the resolution of the haplotype map identified in phase I by adding a further 2.1 million SNPs to the test pool.

The third phase of the HapMap project (HapMap3) published findings in 2010 based on 1.6 million common SNP genotypes in 1184 individuals from 11 different populations (Table 12.3). In addition, 10 regions of 100 kb were sequenced in 692 genomes. HapMap3 includes both SNPs and copy number polymorphisms, and pinpoints population-specific differences for low-frequency variants. The idea behind phase 3 was to produce high-resolution haplotype maps (including both the most common and some of the less common variations from different populations) aimed at helping us to examine patterns of linkage disequilibrium within different populations. Haplotypes are an essential piece in the jigsaw for imputation analysis and the design of association studies.

The map produced by the HapMap project describes common patterns of human genetic variation and thus it has been extremely valuable in the design of association studies, accelerating the search for genetic factors in human disease. The value of the HapMap should not be underestimated and it is likely to guide research for years to come. With approximately 38 million SNPs existing in the human population (in a biallelic locus there are in principle $2n$ haplotypes, where n is the number of SNPs), there is a tremendous potential for diversity; however, fewer haplotypes than expected are observed in practice due to the extensive linkage disequilibrium across the genome. This fact has an important consequence in association studies whereby if a causal variant is not directly tested it could still be identified indirectly. The haplotype map also confirms the idea that the human genome contains recombination hot spots. This means that linkage disequilibrium is not continuous (or equal) across the genome, but there are regions where the recombination rate is

Table 12.3 Populations genotyped in HapMap phases 1–3.

Population	Code	Phase		
		1	2	3
Utah residents with ancestry from Northern and Western Europe	CEU	☑	☑	☑
Yoruba in Ibadan, Nigeria	YRI	☑	☑	☑
Japanese in Tokyo, Japan	JPT	☑	☑	☑
Han Chinese in Beijing, China	CHB	☑	☑	☑
African ancestry in South-Western USA	ASW			☑
Chinese in metropolitan Denver, CO, USA	CHD			☑
Gujarati Indians in Houston, TX, USA	GIH			☑
Luhya in Webuye, Kenya	LWK			☑
Maasai in Kinyawa, Kenya	MKK			☑
Mexican ancestry in Los Angeles, CA, USA	MXL			☑
Tuscans in Italy	TSI			☑

higher than expected and others where it is lower than expected (cold spots). The human major histocompatibility complex (MHC) is a noted area for cold spots with higher than expected levels of conservation of haplotypes.

All of the data produced by the HapMap project are freely available for unrestricted public use at the HapMap website (http://www.hapmap.org). This site offers bulk downloads of the data set, as well as interactive data browsing and analysis tools for data mining and visualization. In 2015, the HapMap is a key resource for researchers designing any genetic association studies aimed at identifying genetic variants associated with common disease or investigating responses to common therapeutic drugs as well as other environmental factors.

The 1000 Genomes Project has major potential in studies of complex disease

The 1000 Genomes Project is another challenging project aimed at producing an extremely detailed catalog of human DNA variation that can be used in future studies. Specifically, the goal of the project is to employ high-throughput sequencing technologies to characterize over 95% of human genomic variants with allele frequency of 1% or higher in each of five major population groups (populations in, or with ancestry from, Europe, East Asia, South Asia, West Africa, and the Americas) using a sample of at least 1000 individuals (**Figure 12.14**). The project was launched in January 2008 by a collaborative agreement of three world-leading genome Institutes: the Beijing Genomics Institute in Beijing (China), the Wellcome Trust Sanger Institute (Cambridge, UK), and the National Human Genome Research Institute (Bethesda, MD, USA). Early phase 1 and phase 2 studies have now been published, and the study is now in phase 3, aimed at extending the pool to 1500 and eventually 2500.

In November 2012, whole-genome sequences for 1092 individuals from 14 populations were published in *Nature*. The results provided a validated map of 38 million SNPs, 1.4 million insertion/deletions (indels), and 14,000 larger deletions. In addition, the results showed that there is considerable variation in genetic structure between populations and this can best be seen by looking at the low-frequency variants. Individuals from different populations carry different profiles of both rare and common variants. The number of the rare alleles within each individual genome varies substantially and many of these are rare non-coding variants localized at conserved sites, such as motif-disrupting changes in transcription factor binding sites.

The 1000 Genomes Project is not yet complete, but the impact of the project on medical genetics is expected to be massive. Data from the studies so far are available at the 1000 Genomes public database (http://www.1000genomes.org), and can be freely used for comparisons to screen variants in exome data from individuals with genetic disorders and/or to screen cancer genomes for mutations that might suggest new therapeutic approaches, and/or to speed diagnoses of diseases. Another major use of the 1000 Genomes Project is imputing genotypes in existing GWAS; the large data set provided by the 1000 Genomes Project will allow more accurate imputation of variants in GWAS and thus better localization of disease-associated polymorphisms. The information contained within the 1000 Genomes database will be employed by worldwide researchers to analyze and interpret DNA variations between people in order to understand which SNPs are implicated with a disease. In addition, the 1000 Genomes Project has stored cell samples from all of the people sequenced and this will allow future functional studies to determine the effects that DNA variations may have on a phenotype. The 1000 Genomes Project represents a step toward a complete description of all human DNA polymorphism and though the project

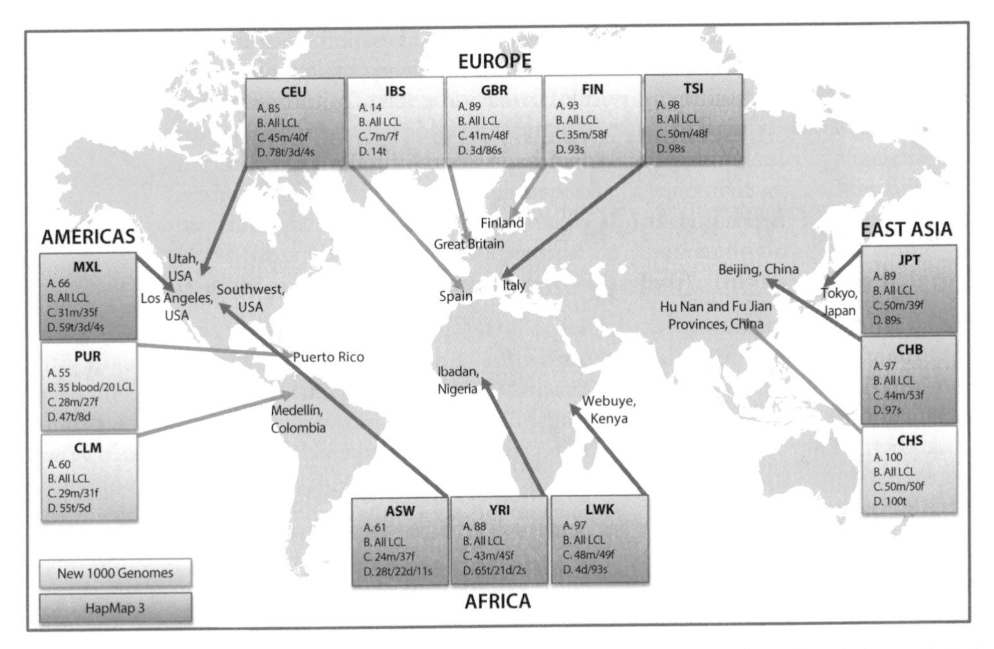

Figure 12.14: Populations collected within the HapMap and 1000 Genomes Projects. Populations collected as part of the HapMap Project are shown in blue and for the 1000 Genomes Project in green. The populations involved include European: IBS (Iberian Populations in Spain), GBR (British from England and Scotland), CEU (Utah residents with ancestry from Northern and Western Europe), FIN (Finnish in Finland), TSI (Tuscans in Italy); East Asian: JPT (Japanese from Tokyo), CHB (Han Chinese in Beijing, China), CHS (Han Chinese South); African: ASW (African ancestry in Southwestern USA), YRI (Yoruba in Ibadan, Nigeria), LWK (Luhya in Webuye, Kenya); and Americans: MXL (Mexican ancestry in Los Angeles, CA, USA), PUR (Puerto Ricans in Puerto Rico), CLM (Colombians in Medellín, Colombia). (A) Total number of samples sequenced. (B) Source of DNA [blood or lymphoblastoid cell lines (LCL)]. (C) Gender composition (male/female). (D) Number that are part of mother-father-child trios (t), parent-child duos (d), or singletons (s); for trios and duos, only parent samples were sequenced. (From The 1000 Genomes Project Consortium [2012] *Nature* 491:56-65. With permission from Macmillan Publishers Ltd.)

has cataloged a large number of human genomes, it is not yet complete. The extended project will only be complete when the scientists have sequenced 2500 individuals.

ENCODE will help to link genotype to phenotype in complex disease

The second phase of the ENCODE (Encyclopedia of DNA Elements) project was published in *Nature* in 2012. The ENCODE project is designed to identify and catalog functional elements across the genome. Only 3% of the total human genome encodes functional proteins. The majority of the genome has been said to be made of "junk." Of course, this is a gross over-simplification. However, with so little of the genome encoding functional sequences it is essential that we map and identify all of the functional elements across the genome. The ENCODE project was designed to do this. We need to understand how our genes are organized, and how they interact and regulate each other.

The first phase of ENCODE looked at 1% of the genome; the second stage (completed in 2012) looked at 5% of the genome. The project has gone well beyond sequencing, and is assessing the chromatin state and structure, level of DNA methylation, and protein expression, among other things, in multiple cell types from frequently used cell lines, a human embryonic cell line, and some primary cells. The ultimate aim of this project is to bring together genotype and phenotype across the human genome. The project is likely to continue for several years. The data generated so far can be viewed at https://www.encodeproject.org/publications. Some scientists are critical of the ENCODE project and not all believe that it will achieve all of its aims. However, time will tell.

12.8 SYSTEMS BIOLOGY

Considering systems biology allows us to look into the future

Biological systems are complex. Proteins interact. Protein agonists interact with specific receptors and ligands, but many systems also include regulatory proteins in the forms of receptor antagonists and decoy receptors. Systems also include a degree of redundancy allowing our biology to make up for functional deficiencies in a specific product. Altogether, we are complex beings. On top of this our genetics is complex, as we have seen, and we produce variable levels of proteins depending (to a greater or lesser extent) on our inherited genotype. We could say we each inherit different genotypes therefore we each inherit different phenotypes.

We can illustrate the potential of genetic variation to impact on a biological system with the following simplified example. Consider the interleukin (IL)-1 gene family that maps to chromosome 2q. IL-1 is an important pro-inflammatory cytokine and genetic variation in the *IL1* genes has been associated with a variety of diseases, most notably with gastric cancer. IL-1β is important in gastric acid production and gastric acid is important in determining resistance to infection with the common bacteria *Helicobacter pylori*. There is quite a large family of *IL1* genes (**Figure 12.15**). To simplify the scenario, we will concentrate on one of the two major agonist (functional) forms of IL-1, IL-1β (encoded by the *IL1B* gene), the active receptor gene *IL1R1*, a decoy receptor gene *IL1R2*, and a receptor antagonist *IL1RN*. We can exclude the *IL1A* gene because the protein product, IL-1 α, is mostly (if not exclusively) active within the cell.

All of the genes above are known to be polymorphic with multiple possible polymorphisms. Once again, for the sake of simplicity, let us consider each gene to have only two alleles, which we can label allele *1* and *2*. Let us also assume in this scenario that the two alleles are associated with different levels of protein expression, with in every case high levels of expression being encoded by allele *1* (high-producer allele) and low levels with allele *2* (low-producer allele). In this scenario there are three genotypes for each gene: *1/1*, *1/2*, and *2/2*. Consider first *IL1B*: those with the *1/1* genotype are high producers, those with the *1/2* genotype are moderate producers (a combination of the effect of high- and low-producer alleles), and those with the *2/2* genotype are low producers. The impact of different levels of IL-1β production is, however, dependent on the other components of the system. Thus, an individual with the *IL1B 1/1* genotype and *IL1R1 1/1* genotype can have very high levels of IL-1β activity compared with someone with *IL1B 2/2* and *IL1R1 2/2*. This is because these individuals have high levels of the agonist form of IL-1 (IL-1β) and high levels of the active receptor IL-1R1. In those with the IL1B 1/1 genotype (high producers) and the *IL1R1 1/2* and *IL1R1 2/2* genotypes, which are associated with lower levels of *IL1R1* expression, overall IL-1β activity may be restricted. This is because even though there is surplus IL-1β produced, the active receptor is less abundant and may become saturated more rapidly. In this situation the effect of producing high levels of the agonist IL-1β is reduced.

Add to this the further complications of a decoy receptor encoded by the *IL1R2* gene and a receptor antagonist encoded by the *IL1RN* gene. If the *IL1R2* and *IL1RN* genotypes are

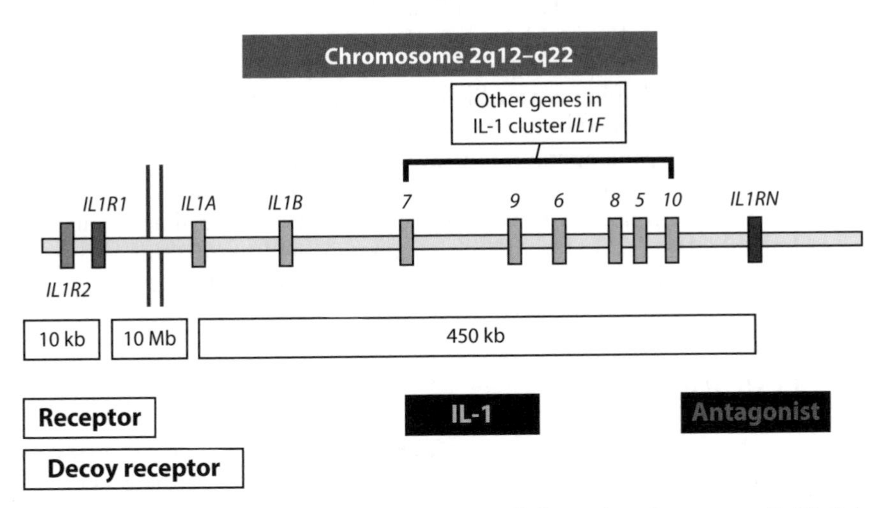

Figure 12.15: A map of some of the members of the IL-1 family located on chromosome 2q22. Only a selected few of these genes are discussed in the example of a complex genetic system. The figure illustrates that there are many more family members to consider, though not all are likely to be functional.

both low-level expression genotypes (*2/2* in each case), then the potential activity of IL-1β will remain high. However, where the genotypes for *IL1B* and *IL1R1* are both high expression genotypes (*1/1* and *1/1*). IL-1β activity may be lower even when the genotype is that associated with high IL1-B production (*IL1B 1/1*). In this model, high levels of the decoy receptor will effectively reduce IL-1β binding to the active receptor and higher levels of the receptor antagonist will interfere with the activity of the functional receptor. Overall, IL-1β activity will be determined by the complex interaction of these four elements. This interactive genetic minefield is illustrated in **Figure 12.16**.

Of course this is an oversimplified model for illustration only (though the genes do exist). There are further levels of complexity in this system, e.g. IL-1β is produced in the cell in the form of pre-IL-1β and needs to be converted by the enzyme caspase 1 [also known as IL-1-converting enzyme (ICE)] before it can be exported from the cell. ICE may also be polymorphic and so conversion of the pre-IL-1β is likely to be affected by this.

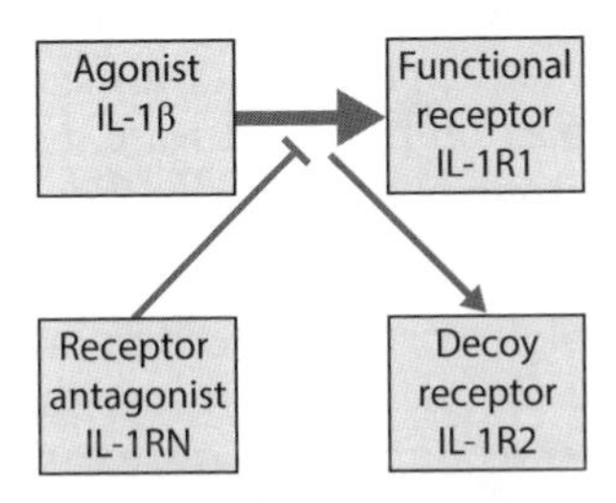

Figure 12.16: Simplified illustration of interaction in the IL-1 system. The figure illustrates the activity of the products of four IL-1 genes: two genes encoding the agonist IL-1β and IL-1R, the receptor antagonist IL-1RN, and the decoy receptor IL-1R2. IL-1β interacts with the receptor agonist IL-1R1 but this can be blocked by either IL-1RN (the receptor antagonist) or by the decoy receptor (IL-1R2). This simple figure indicates how systems may interact and coupled with the knowledge that the genes that encode these proteins may be polymorphic we can produce a simple mind model to consider the potential complexity of this system. The gene family as illustrated in Figure 12.15 is more complex.

Even in this grossly simplified model it is possible to understand how systems biology is important in helping us to understand the genetics of complex disease. It helps us to understand why not all polymorphisms have functional significance and why in many cases the risk associated with specific polymorphisms is small. Multiple interacting polymorphisms and extended haplotypes are likely to be important. Systems often also include a level of redundancy whereby the effects of a dysfunctional gene or genotype can be compensated for by another gene operating in the same, or in some cases another, pathway. When either of these situations exist it is possible that effects of the susceptibility alleles are only seen when the system is stressed by other factors, such as diet, illness, and infection. Thus, most individuals carrying the risk alleles can cope with normal life and only when stressed does the system collapse, leading to an increased risk of illness.

CONCLUSIONS

The past is part of the future, and the present is also part of the past and the future. Science goes through phases that illustrate this concept well. Understanding the genetics of complex disease began with old-fashioned technologies testing single genes and haplotypes for case control association studies in the 1970s, 1980s, and early 1990s. Then came RFLP and PCR analysis, which were faster and more efficient, but equally challenging. Just as we were settling down, the new millennium was upon us with the HGM, HapMap, and SNP Map. From HapMap and SNP Map, rapid genome wide case control and linkage studies evolved (GWAS and GWLS). We are now entering a new phase. This is the beginning of the third age in complex disease genetics. One in which techniques like whole genome sequencing will be commonplace. These new technologies will produce data that will increasingly be applied in clinical practice. However, as always with developments, there are some pros and some cons associated with new technologies. The choice in the future will be whether to use whole-genome or whole-exome sequencing. Choices over low- and high-resolution genotyping will be applied in initial studies, but are unlikely to be on the menu for large-scale studies. Missing data will be handled by systems like impute, which will fill in the blanks.

There is clearly more to do in the study of the genetics of complex disease. The past 10 years have revolutionized the way we investigate the genetics of complex disease, but not the reasons why we do it. We are beginning to understand just how complex the genetics of common non-Mendelian disease may be. It is clear in 2015 that scientific isolationism cannot be tolerated in genetics. If we are to understand how our genome contributes to disease and use this information to the patient's advantage, we must also understand how factors other than our genes impact on human disease. Future strategies will also need to include epigenetics and the environment, especially diet (discussed earlier in this book), bacteria, and metagenomics.

We need to think of systems in biology and we need to interact more with our fellow bioscientists in order to identify the complex relationships that make the human body work. A systems approach to genetics is required so that we can understand how genetic variations can predispose to, or protect from, complex traits. It is not going to be simple, but we have made a start, and with hard work and perseverance great advances in biomedicine are possible.

Studies will be able to take advantage of ongoing projects such as the 1000 Genomes Project and ENCODE (and modENCODE), which were designed to inform studies of human genetics, including those involving disease. Of course we cannot ignore the cornucopia of information provided by the HGP and the disease studies performed so far. Better, faster, and more precise technologies will also contribute.

Ultimately, progress will be measured in terms of the extent to which our research fulfils the three major promises of the Human Genome Mapping Project, i.e. that genetics will be used in disease diagnosis, that genetics will be used in patient management (especially in the development of individualized and novel therapies), and that genetics will be used to inform the debate on disease pathology.

These three promises are linked. Understanding disease pathogenesis goes cap in hand with better patient management and development of individualized and novel therapies. The use of complex disease genetics in diagnosis is perhaps questionable in all but a few cases, but there is certainly evidence that studies are helping us to understand disease pathology and there are some examples where allelic variation may determine the selection of specific therapeutic options. Some would call this translational medicine, but it is not a new idea; scientists in medicine have always sought to relate basic findings to medical practice. "Bench-to-bedside" is a community goal and not just the dream of a few.

Overall, great advances have been made in understanding complex disease. This is especially true in some areas, such as pharmacogenetics and in immune inflammatory diseases such as Crohn's disease. However, we should not be misled by wishful science, and we must remember the words of satirist and philosopher Hugo Mencken who stated in 1917 "there is always an easy solution for every human problem (that is) neat, plausible and wrong". The future lies with unpicking and reassembling these complex human problems and in doing so reaching our final goal, i.e. to advance medical practice through patient diagnosis and treatment, with a hope that some diseases (at least) may one day be eradicated.

FURTHER READING

Books

Armstrong L (2014) Epigenetics. Garland Sciences. This is an excellent book on epigenetics for the graduate and final-year undergraduate scientist with an interest or need to study this subject.

Collins F (2010) The Language of Life: DNA and the Revolution in Personalised Medicine. Harper Collins. This is an excellent book to delve into to look at the whys and wherefores of gene hunting in complex disease.

Frank L (2011) My Beautiful Genome: Exposing Our Genetic Future, One Quirk at a Time. Oneworld. This is an excellent book for the lay public and scientist alike. It is interesting, controversial, and humorous, and perfect for the student of complex disease genetics.

Maitland-van der Zee A-H & Daly AK (2012) Pharmacogenetics and Individualized Therapy. Wiley.

Spector T (2012) Identically Different: Why You Can Change Your Genes. Weidenfeld & Nicholson.

Strachan T & Read AP (2011) Human Molecular Genetics, 4th edn. Garland Science.

Strachan T, Goodship J & Chinnery P (2014) Genetics and Genomics in Medicine. Garland Science.

Articles

Anderson CA, Soranzo N, Zeggini E & Barrett JC (2011) Synthetic associations are unlikely to account for many common disease genome-wide association signals. *PLoS Biol* 9 (1):e1000580. This paper suggests that the importance of synthetic associations generated by multiple variants has perhaps been overstated; see also Wray et al. (2011).

Barski A, Cuddapah S, Cui K et al. (2007) High-resolution profiling of histone methylations in the human genome. *Cell* 129:823–837. This paper describes the use of the ChIP sequencing method to characterize histone and transcription factor binding sites on CD4 T cells; see also Mikkelsenet et al. (2007) and Robertson et al. (2007).

Bentley DR, Balasubramanian S, Swerdlow HP et al. (2008) Accurate whole human genome sequencing using reversible terminator chemistry. *Nature* 456:53–59. This paper describes the first complete genome sequence from an African DNA sample; see also Rasmussen et al. (2010), Wang et al. (2008), and Wheeler et al. (2008).

Blencowe BJ, Ahmad S & Lee LJ (2009) Current-generation high-throughput sequencing: deepening insights into mammalian transcriptomes. *Genes Dev* 23:1379–1386. This paper reports on sequencing the transcriptome; see also Wang et al. (2009).

Branton D, Deamer DW, Marziali A et al. (2008) The potential and challenges of nanopore sequencing. *Nat Biotechnol* 26:1146–1153.

Braslavsky I, Hebert B, Kartalov E & Quake SR (2003) Sequence information can be obtained from single DNA molecules. *Proc Natl Acad Sci USA* 100:3960–3964.

Campbell PJ, Stephens PJ, Pleasance ED et al. (2008) Identification of somatically acquired rearrangements in cancer using genome-wide massively parallel paired-end sequencing. *Nat Genet* 40:722–729. This is one of three papers on pair-ended mapping that demonstrate the use and application of this technique; see also Korbel et al. (2007) and Stephens et al. (2009).

Cohen JC, Kiss RS. Pertsemlidis A et al. (2004) Multiple rare alleles contribute to low plasma levels of HDL cholesterol. *Science* 305:869–872. This study demonstrates the effectiveness of sequencing candidate genes at the extreme ends of the phenotype distribution to find rare alleles.

Dean M, Carrington M, Winkler C et al. (1996) Genetic restriction of HIV-1 infection and progression to AIDS by a deletion allele of the CKR5 structural gene. Hemophilia Growth and Development Study, Multicenter AIDS Cohort Study, Multicenter Hemophilia Cohort Study, San Francisco City Cohort, ALIVE Study. *Science* 273:1856–1862. This report is of great importance in understanding the interplay between host genetics and the pathogen. Here the host is protected from infection and development of the disease by the presence of the 32-bp deletion *CCR5*-Δ32; see also Liu et al. (1996) and Samson et al. (1996).

de Souza N (2012) The ENCODE Project. *Nat Methods* 9:1046.

Dickson SP, Wang K, Krantz I et al. (2010) Rare variants create synthetic genome-wide associations. *PLoS Biol* 8 (1):e1000294. This paper, along with Maher (2008) and Schork et al. (2009), poses the question of whether studies so far have identified the true susceptibility alleles for complex disease or are they synthetic associations?

Eid J, Fehr A, Gray J et al. (2009) Real-time DNA sequencing from single polymerase molecules. *Science* 323:133–138.

Freeman JL, Perry GH, Feuk L et al. (2006) Copy number variation: new insights in genome diversity. *Genome Res* 16:949–961. This paper illustrates how second-generation sequencing will be used to identify CNV in the genome; see also Redon et al. (2006) and Zhang et al. (2009).

Harris TD, Buzby PR, Babcock H et al. (2008) Single-molecule DNA sequencing of a viral genome. *Science* 320:106–109. This paper describes the use of the Helicos platform (third-generation sequencing).

Hayden E (2008). International genome project launched. *Nature* 451:378–379. This paper announces the launch of the 1000 Genomes Project.

Howie BN, Donnelly P & Marchini J (2009) A flexible and accurate genotype imputation method for the next generation of genome-wide association studies. *PLoS Genetics* 5:e1000529.

International Human Genome Sequencing Consortium (2001) Initial sequencing and analysis of the human genome. *Nature* 409:860–921. This paper, together with that of Venter et al. (2001), describes the preliminary sequence of the first whole human genome.

Johnson PL & Slatkin M (2008). Accounting for bias from sequencing error in population genetic estimates. *Mol Biol Evol* 25:199–206. This report discusses some of the concerns that have been raised about the use of NGS in GWAS.

Korbel JO, Urban AE, Affourtit JP et al. (2007) Paired-end mapping reveals extensive structural variation in the human genome. *Science* 318:420–426. This is one of three papers on pair-ended mapping that demonstrate the use and application of this technique; see also Campbell et al. (2008) and Stephens et al. (2009).

Le T, Chiarella J, Simen BB et al. (2009) Low-abundance HIV drug-resistant viral variants in treatment-experienced persons correlate with historical antiretroviral use. *PLoS One* 4:e6079.

Li Y, Willer C, Sanna S & Abecasis G (2009) Genotype imputation. *Ann Rev Genomics Hum Genet* 10:387–406. This paper discusses the use of genotype imputation, which is increasingly used in studies of complex disease either to fill in the gaps or to extend the study; see also Marchini et al. (2007).

Liu R, Paxton WA, Choe S et al. (1996) Homozygous defect in HIV-1 co-receptor accounts for resistance of some multiply-exposed individuals to HIV-1 infection. *Cell* 86:367–377. This report is of great importance in understanding the interplay between host genetics and the pathogen. Here the host is protected from infection and development of the disease by the presence of the 32-bp deletion *CCR5*-Δ32; see also Dean et al. (1996) and Samson et al. (1996).

Lupski JR, Reid JG, Gonzaga-Jauregui C et al. (2010) Whole-genome sequencing in a patient with Charcot–Marie–Tooth neuropathy. *N Engl J Med* 362:1181–1191.

Maher B (2008). The case of the missing heritability. *Nature* 456:18–21. This news feature along with papers by Dickson et al. (2010) and Schork et al. (2009), poses the question of whether studies so far have identified the true susceptibility alleles for complex disease or are they synthetic associations?

Maher CA, Kumar-Sinha C, Cao X et al. (2009) Transcriptome sequencing to detect gene fusions in cancer. *Nature* 458:97–101. This study demonstrates the potential use of transcriptome sequencing in clinical practice.

Marchini J, Howie B, Myers S et al. (2007) A new multipoint method for genome-wide association studies by imputation of genotypes. *Nat Genet* 39:906–913. This paper discusses the use of genotype imputation, which is increasingly used in studies of complex disease either to fill in the gaps or to extend the study; see also Li et al. (2009).

Margeridon-Thermet S, Shulman NS, Ahmed A et al. (2009) Ultra-deep pyrosequencing of hepatitis B virus quasispecies from nucleoside and nucleotide reverse-transcriptase inhibitor (NRTI)-treated patients and NRTI-naive patients. *J Infect Dis* 199:1275–1285. This paper describes a study of a viral drug resistance mutation using a metagenomics approach; see also Nakamura et al. (2008) and Palacios et al. (2008).

Margulies M, Egholm M, Altman WE et al. (2005) Genome sequencing in microfabricated high-density picolitre reactors. *Nature* 437:376–380. The paper describes the use of a novel system for rapid "shotgun" gene sequencing of the genome of the bacterium *Mycoplasma genitalium*.

Mikkelsen TS, Ku M, Jaffe DB et al. (2007) Genome-wide maps of chromatin state in pluripotent and lineage-committed cells. *Nature* 448:553–560. This paper describes the use of the ChIP sequencing method to characterize histone and transcription factor binding sites in a cervical carcinoma cell line; see also Barski et al. (2007) and Robertson et al. (2007).

Mullis K, Faloona E, Scharf S et al. (1986) Specific enzymatic amplification of DNA *in vitro*: the polymerase chain reaction. *Cold Spring Harbor Symp Quant Biol* 51:263–273. This represents a major development in genetics.

Nakamura S, Maeda N, Miron IM et al. (2008) Metagenomic diagnosis of bacterial infections. *Emerg Infect Dis* 14:1784–1786. This paper describes a study of bacterial pathogens using a metagenomics approach; see also Margeridon-Thermet et al. (2009) and Palacios et al. (2008).

Palacios G, Druce J, Du L et al. (2008) A new Arenavirus in a cluster of fatal transplant-associated diseases. *N Engl J Med* 358:991–998. This paper describes a study of viral pathogens using a metagenomics approach; see also Margeridon-Thermet et al. (2009) and Nakamura et al. (2008).

Pool JE, Hellmann I, Jensen JD & Nielsen R (2010) Population genetic inference from genomic sequence variation. *Genome Res* 20:291–300. This paper discusses population variation in relation to whole-genome sequencing and the use of inference (imputation) to assign missing data.

Prokunina-Olsson L & Hall JL (2009) No effect of cancer-associated SNP rs6983267 in the 8q24 region on co-expression of *MYC* and *TCF7L2* in normal colon tissue. *Mole Cancer* 8:96. This report indicates that tagged SNPs can be located a long way from putative causative mutations in complex disease, but establishing a relationship with function can be difficult; see also Tomlinson et al. (2007).

Pushkarev D, Neff NF & Quake SR (2009) Single-molecule sequencing of an individual human genome. *Nat Biotechnol* 27:847–852. This paper describes the use of the Helicos platform (third-generation sequencing); see also Harris et al. (2008).

Rasmussen M, Li Y, Lindgreen S et al. (2010) Ancient human genome sequence of an extinct Palaeo-Eskimo. *Nature* 463:757–762. This is a particularly interesting report of a human genome sequence from an ancient sample. This study proves the concept of rapid genome sequencing and the ability to sequence not only fresh samples, but also ancient samples. This paper describes the first complete genome sequence from a European DNA sample; see also Bentley et al. (2008), Wang et al. (2008), and Wheeler (2008).

Redon R, Ishikawa S, Fitch KR et al. (2006) Global variation in copy number in the human genome. *Nature* 444:444–454. This paper illustrates how second-generation sequencing will be used to identify CNV in the genome; see also Freeman et al. (2006) and Zhang et al. (2009).

Ren B, Robert F, Wyrick JJ et al. (2000) Genome-wide location and function of DNA binding proteins. *Science* 290:2306–2309. This paper describes the chromatin immunoprecipitation assay (ChIP-chip) used to detect locations where the target protein is bound to DNA.

Roach JC, Glusman G, Smit AF et al. (2010) Analysis of genetic inheritance in a family quartet by whole-genome sequencing. *Science* 328:636–639.

Robertson G, Hirst M, Bainbridge M et al. (2007) Genome-wide profiles of STAT1 DNA association using chromatin immunoprecipitation and massively parallel sequencing. *Nat Methods* 4:651–657. This paper describes the use of the ChIP sequencing method to characterize histone and transcription factor binding sites in pluripotent murine stem cells undergoing differentiation; see also Barski et al. (2007) and Mikkelsenet et al. (2007).

Samson M, Libert F, Doranz BJ et al. (1996) Resistance to HIV-1 infection in Caucasian individuals bearing mutant alleles of the CCR-5 chemokine receptor gene. *Nature* 382:722–725. This report is of great importance in understanding the interplay between host genetics and the pathogen. Here the host is protected from infection and development of the disease by the presence of the 32-bp deletion *CCR5*-Δ32; see also Dean et al. (1996) and Liu et al. (2006).

Sanger F, Nicklen S & Coulson AR (1977) DNA sequencing with chain-terminating inhibitors. *Proc Natl Acad Sci USA* 74:5463–5467. This represents a major development in genetics and, in combination with the development of PCR by Mullis et al. (1986), this enabled automated sequencing of DNA to become a reality.

Schork NJ, Murray SS, Frazer KA & Topol EJ (2009) Common versus rare allele hypotheses for complex diseases. *Curr Opin Genet Dev* 19:212–219. This paper, along with Maher (2008) and Dickson et al. (2010), poses the question of whether studies so far have identified the true susceptibility alleles for complex disease or are they synthetic associations?

Stephens PJ, McBride DJ, Lin ML et al. (2009) Complex landscapes of somatic rearrangement in human breast cancer genomes. *Nature* 462:1005–1010. This is one of three papers on pair-ended mapping that demonstrate the use and application of this technique; see also Korbel et al. (2007) and Campbell et al. (2008).

Sultan M, Schulz MH, Richard H et al. (2008) A global view of gene activity and alternative splicing by deep sequencing of the human transcriptome. *Science* 321:956–960. This paper reports on the application of transcriptome sequencing in human kidney and B cell lines; see also Maher et al. (2009).

The 1000 Genomes Project Consortium (2010) A map of human genome variation from population-scale sequencing. *Nature* 467:1061–1073.

The 1000 Genomes Project Consortium (2012) An integrated map of genetic variation from 1,092 human genomes. *Nature* 491:56–65. This is the most recent report from the 1000 Genomes Project at the time of writing. More up-to-date information will be available on the website cited below.

The ENCODE Project Consortium (2012) An integrated encyclopedia of DNA elements in the human genome. *Nature* 489:57–74. This is the original report from the ENCODE Project reporting from phase II, which started in 2007.

The International HapMap Consortium (2005) A haplotype map of the human genome. *Nature* 437:1299–1320.

The International HapMap Consortium (2007) A second generation human haplotype map of over 3.1 million SNPs. *Nature* 449:851–861.

The International HapMap Consortium (2010) Integrating common and rare genetic variation in diverse human populations. *Nature* 467:52–58. The HapMap data can be accessed through the weblink in the Online sources and provides a very useful database for all studies in complex disease. The impact of these maps and studies cannot be overestimated.

Tomlinson I, Webb E, Carvajal-Carmona L et al. (2007) A genome-wide association scan of tag SNPs identifies a susceptibility variant for colorectal cancer at 8q24.21. *Nat Genet* 39:984–988. This report indicates that tagged-SNPs can be located a long way from putative causative mutations in complex disease, but establishing a relationship with function can be difficult; see also Prokunina-Olsson L & Hall JL (2009).

Venter JC, Adams MD, Myers EW et al. (2001) The sequence of the human genome. *Science* 291:1304–1351. This paper, together with that of the International Genome Consortium (2001), describes the preliminary sequence of the first whole human genome.

Wang J, Wang W, Li R et al. (2008) The diploid genome sequence of an Asian individual. *Nature* 456:60–65. This paper describes the first complete genome sequence from an Asian DNA sample; see also Bentley et al. (2008), Rasmussen et al. (2010), and Wheeler et al. (2008).

Wang Z, Gerstein M & Snyder M. (2009) RNA-Seq: a revolutionary tool for transcriptomics. *Nat Rev Genet* 10:57–63. This paper reports on sequencing the transcriptome; see also Blencowe et al.

Wheeler DA, Srinivasan M, Egholm M et al. (2008) The complete genome of an individual by massively parallel DNA sequencing. *Nature* 452:872–876. This paper describes the first complete genome sequence from a European DNA sample; see also Bentley et al. (2008), Rasmussen et al. (2010), and Wang et al. (2008).

Wray NR, Purcell SM & Visscher PM (2011) Synthetic associations created by rare variants do not explain most GWAS results. *PLoS Biol* 9:e1000579. This paper suggests that the importance of synthetic associations generated by multiple variants has perhaps been overstated; see also Anderson et al. (2011).

Zhang F, Gu W, Hurles ME & Lupski JR (2009) Copy number variation in human health, disease and evolution. *Annu Rev Genomics Hum Genet* 10:451–481. This paper illustrates how second generation sequencing will be used to identify copy number variation in the genome; see also Redon et al. (2006) and Freeman et al. (2006).

Online sources

http://www.hapmap.org
This is a most useful source for studies of the genetics of complex disease and is essential in study design.

http://www.1000genomes.org
Data from this public database can be used freely for research purposes. The data can help with imputation studies and identifying common haplotypes in different populations.

http://www.ncbi.nlm.nih.gov
This website provides a reference database for metagenomic studies and bacterial genomes.

http://commonfund.nih.gov/hmp
This website refers to the Human Microbiome Project.

http://www.encodeproject.org/publications/
This database shows the data produced by the encode project so far.

http://www.454.com
This website provides information on the Life Sciences NGS technology owned by Roche.

http://www.appliedbiosystems.com
This website provides information on the Applied Biosystems SOLiD NGS system owned by Applied Biosystems.

http://www.illumina.com
This website provides information on the Illumina Genome Analyzer—an NGS sequencer produced by the Illumina corporation.

Glossary

Acentric – term describing a chromosome that does not possess a centromere.

Adaptive Darwinian selection – selection of specific genotypes within a population as a result of selection pressure.

Adaptive immunity – modified immune response following initial stimulation with an antigen.

ADP receptor – adenosine diphosphate receptor. This is an important receptor for many drugs.

Affymetrix gene chip – involves light-directed *in situ* synthesis of oligonucleotide probes on a microarray that can produce approximately 6 million oligonucleotides on a chip array of less than 2 cm^2.

Allele(s) – alternative forms of the same gene.

Allele frequency – frequency of a gene variant at a specific locus.

Allelic heterogeneity/variation – existence of several different mutations or polymorphisms, all in the same gene, in unrelated individuals with the same phenotype.

Allo-reactive antibodies – antibodies that recognize antigens of the same species.

Allotropic expression – most often referred to regarding genes that are normally expressed from the mitochondrial genome.

Alternative hypothesis (H_1) – the opposite idea to the **null hypothesis**, which suggests there is no significant difference in a study. The alternative hypothesis is incompatible with the null hypothesis. The alternative hypothesis suggests there is a difference between the two groups.

Amyloid plaques – caused by deposits of β-amyloid protein in the brain.

Ancestral haplotype (e.g. HLA 7.2 and 8.1) – a group of alleles located on a series of different genes that are inherited as a single unit and from a common ancestor. Haplotypes like this are preserved and their component alleles are found together more often than expected by chance (i.e. they are in **linkage disequilibrium**).

Aneuploidy – an extra copy or missing copy of one or more chromosomes. Aneuploidies include trisomy 21 (Down syndrome) and monosomy X (Turner's syndrome).

Antibody-dependent cytotoxicity – the use of antibodies to kill cells as in an HLA serotyping assay.

Antigen-binding groove (HLA-binding groove) – the location on an HLA class I or II molecule where antigenic peptides are bound and presented to the T cell receptor.

Antigen-presenting cells (APCs) – specific cells that present antigenic peptides to T lymphocytes and activate them. These include dendritic cells, macrophages, and B cells, and are often referred to as professional antigen-presenting cells. All of these three cell types express HLA class II molecules and a number of other co-stimulatory molecules on their surface required for activation of naive T cells.

Antigenic peptides – refers to short antigen protein sequences that are presented to T cells for the immune response. They can be derived from self-antigens or from non-self-antigens (e.g. pathogens).

Appreciable frequency – can be any value suggested to be significant for a study. Generally refers to numbers in studies, size of difference reported, and the level of expected probability value. Also a term used in the definition of **polymorphic** genes as opposed to mutations.

Ascertainment bias – distorted recruitment (usually of individuals for a study, either as controls, disease cases, or both).

Association(s) – tendency of a marker and a trait to occur together more often than expected by chance. Association is technically a statistical observation and is not a genetic phenomenon; however, association is used in genetic studies and can also indicate **linkage disequilibrium**.

Association analysis (studies) – attempts to find genetic associations between genetic markers i.e. alleles, SNPs, mutations, or other genetic variation and diseases by comparing the frequencies of markers in two different groups.

Autophagy – a form of programmed cell death in which the cell components are broken down as a means of coping with adverse circumstances or during infection.

Autosomal dominant disorder or disease – a disorder or disease (discord is also used) showing a Mendelian pattern of inheritance whereby only one parental genome needs to pass on the disease-causing mutation to their offspring for the trait to occur.

Autosomal recessive disorder or disease – a disorder or disease (discord is also used) showing a Mendelian pattern of inheritance whereby both parental chromosomes need to pass on the disease-causing mutation to their offspring for the trait to occur.

Autosome– any chromosome other than the sex chromosomes.

Balanced structural abnormalities – occur when there is no loss or gain of function associated with an abnormality.

Balancing selection – occurs where mutant alleles confer increased fitness on the population, but only in heterozygotes and not in homozygotes.

Basal cell carcinoma – common form of skin cancer; rarely fatal.

Bayesian – a form of mathematical calculation based on Bayes' theorem.

Bonferroni's correction – a mathematical correction applied by statisticians to correct for multiple testing in different studies. It is mostly used in **case control studies**. It is much criticized, but it has been widely used in the past.

Bottleneck effect – this occurs when a founder population is reduced in size as a result of a catastrophic event and it leads to reduced genetic diversity in the population.

Candidate gene-based association study – a genetic study based on investigation of one or more identified "candidates." This is a hypothesis-constrained study.

Carcinogenesis – literally the creation of cancer. The process by which normal cells become cancer cells.

Cardiotoxicity – dysfunction in the electrophysiological processes in the heart or heart muscle.

Case control (association studies/population) studies – an investigation based on the comparison of two or more groups where one (or more) group consists of patients (cases) and the other consists of controls.

Cataplectic narcolepsy – severe form of a chronic neurological disorder (narcolepsy) caused by failure to regulate the sleep/wake cycle. Cataplexy is caused by a sudden loss of voluntary muscle tone during which time an individual is unable to move.

CD4⁺ T cells – specific T lymphocytes that express CD4 on their cell surface interacting with HLA class II.

CD8⁺ T cells – specific T lymphocytes that express CD8 on their cell surface interacting with HLA class I.

Centimorgan (cM) – measure of genetic distance on a chromosome based on the recombination fraction (named after geneticist TH Morgan); 0.01 or 1% recombination fraction is equal to 1 cM.

Centromere – point at which the two strands of the chromosome (p-short and q-long) merge. It is important in the construction of the chromosome and in providing an anchor point at which spindle fibers attach to pull chromatids apart during cell division.

Chemokines – a group of small proteins that guide white blood cells to the sites where they are needed.

Chimeric – a functional mix of genetic or other biological material from one or other members of the same or different species.

Chromatin – packaged DNA in a cell. Double coils of approximately 30 nm and histones.

Chromatin immunoprecipitation assay (ChIP-chip) – used to detect biochemical modifications to DNA and locations where proteins of interest are bound to DNA, such as histone and transcription-binding sites on human CD4⁺ T cells.

Chromosomes – linear structures comprised of long DNA molecules coupled with a variety of structural and regulatory proteins.

Chromosomal abnormalities – changes in chromosomal structure or number, e.g. the Philadelphia chromosome or trisomy 21.

Chronic myelogenous leukemia – one form of cancer of the white blood cells.

***Cis* expression** – the expression of two different alleles at different loci on the same paternal chromosome. Resulting in the formation of a heterodimer in *cis* as opposed to a heterodimer in *trans*. See ***Trans* expression**.

Clinical heterogeneity – variation in the clinical conditions in a test population.

Clones – copies of a cell or molecule (e.g. a short DNA or longer chromosomal sequence) produced under laboratory conditions.

Cochran–Armitage test – a statistical method used to test for significance when comparing data for a variable with two possibilities (e.g. a gene with two alleles) against a variable with multiple variables. The test is very similar to the **χ^2 test**.

Codon – a nucleotide tri-repeat sequence that encodes an amino acid.

Common disease common variant hypothesis (CDCVH) – hypothesis that predicts that common disease-causing alleles will be found in all human populations that express specific traits and diseases.

Common disease rare variant hypothesis (CDRVH) – hypothesis that predicts that if a disease is common in the population the genetic risk may not be common in the population but may be due to the effect of multiple interacting risk alleles.

Compare and contrast model – a simple model where the details of two sides are compared to seek similarity and differences between them.

Complement (complement activation) – a group of plasma proteins that interact to neutralize pathogens and activate pathogen destruction and eradication.

Complement-mediated cell lysis – one of the major functions of the complement proteins; the destruction of bacteria and also cells infected by viruses.

Complex disease/trait – a disease/trait or character where one or more alleles acting alone or in concert increase the risk.

Concordance – the degree of agreement (harmony) between two observations, ideas, or arguments.

Confidence interval (CI) – a statistical measure of the range of probability within which any quoted value falls.

Consequentialism – an ethical theory that states the consequences of an individual's conduct are the ultimate basis for any judgment of the right of that conduct.

Constitutional chromosomal abnormalities – chromosomal abnormalities present in every cell in the body.

Copy number variation (CNV) – inherited variation in the number of copies of a specific gene between individuals. Not all CNVs have functional consequences.

Co-segregation – the transmission of two or more genes on the same chromosome to the daughter cell.

Crossover – occurs during meiosis whereby fragments of parental chromosomes interact and combine to form new combinations (see **Recombination**).

Cyto-adherence – a process that can be used to block adherence of virus to the cell. The process involves antibodies.

Cytochrome P450 – a family of phase I drug-metabolizing enzymes (see ***CYP2D6***, ***CYP2C9***, and ***CYP2C19***).

Cytokine – proteins made by cells that influence the behavior of other cells. If made by lymphocytes, they are called lymphokines or interleukins.

Degrees of Freedom (d.f.) – a statistical measure of the number of tests performed in an analysis. Probability values for some commonly used tests are routinely adjusted taking into account the number of degrees of freedom and values of probability can be found in published sources for tests such as the **χ^2 test**, etc..

Dicentric – a chromosome with two centromeres.

Diploid – having two copies of the whole chromosome set.

Direct sequencing – early use of the term "direct sequencing" referred to DNA sequencing based on use of PCR to amplify the gene or region of interest. More recently, it has been used to describe sequencing without the need for cloning of the DNA fragment under investigation.

Directional selection – a situation in natural selection where a particular phenotype is favored and the allele frequency moves in one direction.

Disease cassette – the concept that a number of risk alleles carried on the same haplotype may each contribute to disease susceptibility.

Disease-causing mutation – a mutation that shows a direct causal link with disease (see **Mutation**).

Disease phenotype – the pattern of a disease, including signs, symptoms, progression, and response to treatment.

Disease profile – idea that a disease may arise when we inherit series of different alleles with different patients having different personal profiles of risk alleles in complex disease (see also **personal portfolio**).

Dispermy – two sperm fertilizing a single egg. A common cause of triploidy.

Dizygotic twins – twins who arise as a result of independent fertilization of two different eggs by two different sperm at the same time. Dizygotic twins are as genetically identical as other siblings in the same family; they do not carry an identical genome. Also see **Monozygotic twins**.

DNA – abbreviation for deoxyribonucleic acid.

DNA methylation – conversion of cytosine in DNA to 5-methylcytosine. This is an important process in the regulation of gene expression.

Duplications – increase in the number of copies in a gene or a DNA sequence. Where there are duplications in specific genes, individuals may carry more than one copy of a gene. This does not always have any influence on the phenotype such as in circumstances where the duplicated copy is not expressed. However, duplications can have interesting effects on phenotype.

Electrostatic charge – the charge on an amino acid: positive, negative, or neutral. Charge is important in considering protein structures, formation, and function.

ENCODE – Encyclopedia of DNA Elements. An ongoing project targeting functional elements in selected model organisms (new name modENCODE).

Endocytosis – uptake of extracellular materials into cells. Endosomes are formed from segments of the cell membrane and capture extracellular materials. These are then transported through the cell. This process is particularly important in immunity.

Endoplasmic reticulum (ER) – an intercellular organelle that forms an interconnecting network with the cell membrane. There are two forms of the ER: smooth (SER) and rough (RER). The RER is involved with protein synthesis, while the SER is involved with lipids and detoxification.

Enrichment strategies – targeting of specific selected genomic regions in the treatment of disease by using data from studies of somatic and *de novo* mutations identified in **whole-exome sequencing**.

Epidemiological – in the context of this book this is the study of health and disease in populations – considering causes, response, and outcomes.

Epigenetic(s) – means above, outside or in addition to genetics and is usually used to refer to heritable effects not produced by changes in the DNA sequence. DNA methylation is the most often quoted form of epigenetics. The wider context for this growing sub-branch of genetics is to be found in all things outside of "genetics."

Epigenetic regulation – DNA methylation, in particular.

Epigenome-wide association studies (EWAS) – epigenetic studies performed on a genome-wide basis have started and more will follow in the future.

Epigenomics – the growing interest in the wider consideration of genetics, especially epigenetics.

Epistasis – means standing above. In this case, we are concerned with the interaction of genes in pathways. Where there is epistasis, there is a sequence of activation of genes. Failure in the early stages due to genetic variation would be expected to cause failure further down the pathway. This is an important concept in systems biology.

Eugenics (Eugenics Movement) – an idea conceived by Francis Galton (the cousin of Charles Darwin) based on the principle that social status is an inherited trait and therefore the upper classes should do everything possible to conserve themselves, whereas the lower classes are simply the unfortunate consequence of their genes. This misconception had major historical consequences.

Exome – the region of the genome that encodes the genes (protein sequences).

Exome sequencing – restriction of genome sequencing to these coding regions only; this reduces cost and time.

Exon(s) – DNA sequences in genes that encode the protein/peptide sequences. Also see **intron(s)**.

Extended major histocompatibility complex (xMHC) – the **major histocompatibility complex (MHC)** after 2004, at which point the MHC was extended to include the hemochromatosis gene *HFE* (telomeric of *HLA-A*). The extension added approximately 3.5 Mb to the existing MHC.

Factorial (!) – mathematical value whereby the value is the sum of that value multiplied in sequence with lower values, e.g. $5! = 5 \times 4 \times 3 \times 2 \times 1 = 120$. Used in Fisher's exact probability test. The symbol "!" is the mathematical sign for factorial.

False discovery rate – the number of false positives discovered in a study compared with the total number of tests performed and the total number of true positives.

False negative – acceptance of the null hypothesis for an allele that later turns out to be incorrect. In GWAS, this means dismissing a genetic association that later turns out to be true.

False positive – the rejection of the null hypothesis for an allele that later turns out to be true. In GWAS, this means accepting the presence of a genetic association that is later proven to be incorrect.

Familial (family studies/familial disease) – studies based on the collection of informative families.

Family-based association test (FBAT) – a form of **transmission disequilibrium testing (TDT)**.

Fisher's exact probability test – a statistical method for testing for associations usually applied when any individual value is less than 5 though not all studies follow this rule. Fisher worked with Yates to set up the original χ^2 **test**.

Fixation – the situation that occurs when through natural selection in a population there is a change in the gene pool, such that where there were once two alleles, there is now only one allele. In this situation the gene is said to be fixed, i.e. invariant.

Fluorescence *in situ* hybridization (FISH) – *in situ* hybridization using fluorescently labeled DNA or RNA.

Founder effect – a high frequency of a particular allele due to small numbers in the original population (**founder population**).

Frameshift – a shift in the remaining DNA when part of a DNA sequence is deleted. This can have consequences for the gene product as, for example, with the ***CCR5*-Δ32** mutation.

Gametocytes – precursors to the male and female gametes.

Gene – a basic unit of inheritance. A functional DNA unit that determines the phenotype of an individual and segregates in pedigrees according to Mendel's laws.

Gene duplications – see **Duplication**.

Gene flow – transfer of alleles or genes between populations through migration.

Gene pool – all the genes (or a selection) in a specified population.

Genetic anticipation – the tendency of the severity of a condition to increase over generations, often due to the increase in copy number from generation to generation.

Genetic divergence – accumulation of genetic variation from two or more ancestral populations over time giving rise to even more genetic variation.

Genetic drift – random changes in gene frequencies over generations.

Genetic imprinting – when the expression of a gene is determined by the parental origin of that gene. Thus, expression can be different if the same allele is inherited from the paternal or maternal route.

Genetic portfolio – the sum of an individual's genotype, including the high-, low-, and medium-risk alleles for all of the genes in the genome or in research reports for all of the alleles tested.

Genetic protection – alleles that reduce the risk of a disease.

Genetic selection – the concept that survival is determined in part by genetic variation in the population. This is a central pillar of Darwinism.

Genetic underclass – the distinction of a subgroup as inferior based on their genetic makeup. Also see **Eugenics Movement** and Chapter 11.

Genome – the complete set of human DNA stored in our chromosomes (22 pairs plus X and Y) and our mitochondria.

Genome sequencing – sequencing the entire genome as opposed to sequencing only selected sections.

Genome project – an international project to sequence the genome of humans and other species, and to catalogue variation within the human genome.

Genome-wide association analysis/studies (GWAA/GWAS) – analysis of genetic markers across the whole genome to identify significant differences in allele distribution between groups (usually healthy controls and disease cases, though not always). GWAS has been highly successful in recent years.

Genome-wide linkage analysis/studies (GWLA/GWLS) – same as above, except uses linkage analysis based on families or trios.

Genomic imprinting – see **Genetic imprinting**.

Genotoxic stress – damage to the genome of an organism attributable to a toxin.

Genotype – a combination of two or more alleles in an individual at a single locus or multiple loci (e.g. a haplotype). Individuals are either heterozygous (carrying both alleles) or homozygous (carrying two copies of one allele) for an individual gene.

Genotype portfolio – the sum of genetic data held on or by an individual (may also be considered under **Risk portfolio**).

Genotyping – identification of genes or alleles individually or in groups using a variety of methods.

Glial cells – small cells in the brain involved in the repair of neurons.

Golgi stacks – the Golgi apparatus is a membranous organelle where proteins and lipids are sorted for transport within the cell and where newly synthesized HLA molecules are held prior to their expression at the cell surface.

Gram-positive bacteria – bacteria that stain dark blue, purple, or violet when stained with Gram stain.

Haploid – a cell with one single copy of each chromosome (usually a **gamete**).

Haplogroups – formation of a combination of different alleles into haplotypes and haplotype families; this is a useful tool in analysis. Haplogroups and haplotypes are not always exactly the same – haplogroups is a more relaxed broader term.

Haplotype(s) – a series of alleles at linked loci that are inherited on the same chromosome, usually in strong linkage disequilibrium.

Haploview – commonly used bioinformatics software designed to identify patterns of linkage disequilibrium, estimate haplotype frequencies, and help select tagged **SNPs** for analysis.

HapMap – map of common human haplotypes created as part of the aftermath of the **Human Genome Project**.

Hardy–Weinberg equilibrium (HWE) – the equation most often used to determine whether observed values depart from the expected norm.

Hardy–Weinberg principle (HWP) – the simple principle based on the relationship between allele frequencies and genotype frequencies that is found in many populations, and is broadly used to verify studies in complex disease.

HeliScope sequencer – pioneered as an innovative rapid sequencing technology, it relies on **true single-molecule sequencing** without a preclonal amplification step.

Heritability (heritable component) – the proportion of a trait that is inherited it is not always genetically determined, but many heritable traits have a considerable genetic component.

Heritable trait – any inherited characteristic determined by genes or other heritable factors.

Heterogeneous – a mixed genotype where both alleles are present–one from each parent. If there is a single variation (i.e. two alleles at a locus), one may be the original (**wild-type**) and the other may be a mutant. Where there are multiple alleles, this type of naming is less easy to apply.

Heteroplasmy – mosaicism, usually in a single cell.

Heterozygote advantage – a situation that can occur when an individual or individuals in a population who are heterozygous for a specific genotype have a reproductive or survival advantage over those who are homozygous.

Heterozygotes or heterozygous – individuals who carry different alleles for the same gene on each of the parental chromosomes (see also **Genotype**).

Histone(s) – a family of small basic molecules that complex with DNA to form nucleosomes.

Histone acetylation – histones are acetylated on the core of the nucleosome as part of the normal process of gene regulation.

Histone modification – histones can be modified by a number of biochemical reactions usually involving methylation or acetylation (see **Histone acetylation**).

HIV escape epitopes – immunodominant epitope in *HLA-B*27* that is associated with changes in HIV viral loading.

HLA – abbreviation for **human leukocyte antigen**; named "leukocyte antigens" because they were first described expressed on leukocytes, but not found on red blood cells.

HLA 7.2 (ancestral haplotype) – see **Haplotypes(s),** also **Linkage Disequilibrium (LD)**.

HLA 8.1 (ancestral haplotype) – see **Haplotypes(s),** also **Linkage Disequilibrium (LD)**.

HLA association – genetic associations with HLA. See **Association(s)**.

HLA antigen-binding groove – the site at which HLA molecules bind antigenic peptides for presentation to the T cell receptor in the formation of the immune synapse.

HLA class I (loci) – a distinct group of HLA genes encoding a series of molecules with a similar structure and function. These molecules present short antigenic peptides to the T cell receptor of $CD8^+$ T cells in the formation of the T cell synapse. The use of the term loci indicates that there are several HLA class I genes.

HLA class II (loci) – a distinct group of HLA genes encoding a series of molecules with a similar structure and function. These molecules present short antigenic peptides to the T cell receptor of $CD4^+$ T cells in the formation of the T cell synapse. The use of the term loci indicates that there are several HLA class II genes.

Homoplasmic – a cell where all the copies of the mitochondrial DNA are identical.

Homozygotes or homozygous – individuals where the gene or genes of interest carry the same allele on both chromosomes (see also **Genotype**).

Hot spot positions – recombination hot spots are areas where the frequency of recombination is greater than expected. Recombination occurs naturally during prophase in the first phase of meiosis when homologous chromosomes bind and exchange DNA segments.

Human Genome Map (HGM) – the first map of the positions of the majority of the human genome was published in 2001 in *Nature* and *Science*.

Human Genome Project (HGP) – the term describing the entire genome project, including mapping the genome, SNP identification (SNP Map), HapMap, and, more recently, the development of **next-generation sequencing (NGS)** and the 1000 Genomes Project.

Human leukocyte antigen (HLA, HLA typing) – these are the protein products of the HLA class I and class II genes in the MHC. HLA typing refers to the use of serum in a microcytotoxicity assay to identify the antigens on the surface of lymphocytes prepared from whole blood. This was normal during the 1970s to the late 1990s; in the late 1980s and through the 1990s, genotyping with molecular-based methods started to take over. Most laboratories now use molecular genotyping.

Humanized antibodies – modification of antibodies produced by non-human species to increase their similarity to natural human variants (see **Chimeric**).

Hypothesis constrained (hypothesis driven) – research that is based on testing a specific idea.

Hypothesis generating – research that is likely to generate ideas for testing (see also **Hypothesis non-constrained**).

Hypothesis non-constrained (hypothesis free) – research that is open to all possibilities and is not set up to test a predetermined idea.

iCOGS – a custom-designed Illumina array designed to test genetic variation in three major cancers: breast, ovarian, and prostate cancer.

Identical by descent (IBD) – alleles in an individual are identical because they are inherited from a common ancestor. In this case, the exact same allele and haplotype is inherited.

Identical by state (IBS) – alleles in an individual appear to be identical to those of another individual but there is no common ancestor. In this case, the allele is likely to be passed through the family from different parents.

Illumina genome analyzer – produced by the Illumina Corporation. A short-read sequencing platform like **SOLiD**. It is currently very widely used.

Immune regulatory pathway (genes) – biological pathways involved in turning the immune response on and off (regulation). A pathway will involve the interaction of a group of different proteins with different functions encoded by a range of different genes.

Immune stress – situation when the immune system is placed under stress such that systems begin to be overwhelmed by demand and breakdown can occur with potentially severe consequences.

Immune synapse – region of contact between receptor and ligand – in this book most often used in relation to HLA peptide–T cell receptor interaction.

Immune tolerance – a situation in which the immune system does not respond to an antigen (i.e. tolerate the antigen). This can apply to self-antigens and non-self-antigens.

Immunogenetics – branch of immunology that is concerned with the study of genes that encode proteins involved in immunity and immune regulation.

Immunoglobulin domain – part of a protein structure composed of approximately 100 amino acids folded into β-pleated sheets and α-helices. IgG-like domains are found in HLA class I and class II molecules.

Immunohistochemistry – the use of chemical staining by antibodies (often monoclonal antibodies) to identify antigens in tissues (see also **Fluorescence *in situ* hybridization**).

Imprinting – the determination of the expression of a gene by its parental origin.

Imprinting control elements – a short sequence within an imprinted gene cluster where methylation controls the imprinting status of the genes.

Impute (imputed, imputation analysis) – literally meaning to attribute or ascribe. In the context of the genetics of complex disease, it means to assign genotypes based on a known pattern of linkage disequilibrium. This method will back-fill missing data for which genotyping was not performed using the expected genotype at each locus as suggested by the data for the genes that were fully genotyped, and the known and expected genotype combinations in other databases.

Inbreeding depression – when inbreeding occurs there is a higher frequency of identity by descent (IBD) compared with identity by state (IBS).

Incidence – the number of new cases of a disease in a study population over a set period of time (see also **Prevalence rates**).

Incomplete penetrance – the co-occurrence of an allele and a trait. When an individual possesses the allele, but has not developed the trait, then there is "incomplete penetrance."

Incorrect segregation – a process that can lead to aneuploidy, altering the number of chromosomes in the cell (see **Aneuploidy**).

Independently assorted – according to Mendel's laws, traits are passed independently from parents to their offspring, thereby being independently assorted.

Index case – first reported case within a family.

Indirect associations – see **Synthetic associations**.

Individual missingness – missing data in analysis. Usually due to poor SNP selection in GWAS where a specific SNP has failed to give an adequate signal. This leaves a hole in the data set – such holes can sometimes be back-filled using imputation. It can also refer to failure of a sample from one of the test groups to work in a GWAS; the "missing" data in this case cannot be back-filled (also relates to sample call rate).

Insulin – product of pre-pro-insulin and pro-insulin produced in the islets of Langerhans in the pancreas; regulates glucose metabolism.

Intercellular neurofibrillary tangles (NFT) – aggregates of the hyperphosphorylated tau proteins that occur in Alzheimer's disease. Tau is a microtubule-associated protein and there other diseases where this may occur.

Interferons – a family of cytokines with a variety of functions acting on a variety of different cell types depending on function. Interferons are key agents in both innate and adaptive immunity, and also play an important role in immune regulation.

Intermediate metabolizers – a group who express the intermediate phenotype based on their individual genotype for a gene that metabolizes a specific drug.

Intron(s) – the space between the protein-coding regions of the DNA sequences (known as **exons**). Intronic sequences have a range of roles, not all of which are fully understood.

Invariant chain – polypeptide that associates with HLA class II molecules in the endoplasmic reticulum where it acts as a molecular chaperone to prevent inappropriate peptide binding until the HLA molecule can be transported to the endosomes where it is degraded and peptides can compete for binding.

Inversions – a re-arrangement where a segment of a chromosome is reversed from end to end.

Killer cell immunoglobulin-like receptor (KIR) – family of receptors on natural killer cells that bind HLA class I on target cells and either activate or downregulate the natural killer cell.

Kinship coefficients – a measure of consanguinity between individuals in complex disease studies; this can be used to estimate bias.

Least-squares criterion – statistical method of looking for errors in a study and corresponds with the maximum-likelihood ratios if the experimental error has a normal distribution.

Leber hereditary optic neuropathy – disorder inherited from the mitochondrial DNA causing degeneration of retinal ganglia cells and their axons.

Leukocyte (antigens, typing, antibodies) – white cells. Antigens refers to proteins on white cells; typing refers to phenotyping using serological testing; antibodies refers to the agents required to perform phenotyping.

Leukotriene – a family of inflammatory mediators produced by leukocytes and possibly some other cells. They are involved in immune inflammation and vascular adhesion.

Linear regression model – mathematical model used in statistical analysis to define a relationship between a dependent variable (on the *y*-axis) and one or more other explanatory variables (on the *x*-axis).

Linkage – a physical link between two or markers on a chromosome (see **Linkage analysis**). It can also refer to a link between a genetic marker and disease or disease trait.

Linkage analysis – one of two major strategies for identifying "disease alleles" based on the likelihood of the marker alleles, mutations, copy number variants of a gene or genes being inherited with the trait. Linkage analysis is mostly based on informative families, though occasionally sibling pairs are used, and it works well for diseases that present in early life, have high penetrance, and a near-Mendelian pattern of inheritance. It does not work in diseases that are transmitted horizontally through the population, such as infectious diseases.

Linkage disequilibrium (LD) – the co-occurrence of alleles from two or more gene loci more often than expected by chance [see also **Ancestral haplotype (HLA 7.2 and 8.1)**]. For example, two alleles each present at 10% in a population should be found together at a frequency of 1% if there is equilibrium. However, when there is "dis"-equilibrium the alleles are likely to be found together more often (or less often) than expected by chance. In this latter case there is non-random segregation of the alleles.

Locus (plural: loci) – the location (or site) of a gene in the genome.

Locus heterogeneity – variation at a locus that may exceed the normal biallelic variation.

LOD score – log of odds (a statistical value usually to base 10). This is a measure of the degree of linkage between a marker gene and a trait. It is used in **linkage analysis**.

Logistic regression (model) – statistical method used for determining the outcome of categorically dependent variables.

Logit – a mathematical test. There are various forms of this test.

Lymphatic system – a circulatory system comprised of a network of vessels through which lymph (a fluid derived from blood plasma is transported). The lymphatic system is not a closed system, and there is free movement of plasma into the system from the blood as and when required.

Macrophage(s) – large mononuclear phagocytic cells important in innate and adaptive immunity.

Major histocompatibility complex (MHC) – region of approximately 7.2 Mb of DNA on chromosome 6p21.3 encoding over 252 expressed genes, including the HLA genes. HLA matching can have a major effect on the survival of transplanted tissues, though other factors are also important. The term MHC should be used only to identify the region and sub-regions within it.

Manhattan plots – a plot of probability values for SNPs tested across the genome in a **GWAS** or **GWLA** study. In these plots, data for each SNP is entered chromosome by chromosome 1 to 22, and then X and Y. The individual dots for the low *P* value SNPs merge to give colored squares at the foot of each chromosome column, but the SNPs above the significance threshold are shown in a bright rising column. It is not unusual to see *P* values of less than 10^{-30}.

Marker locus – a locus that is used as a trait marker in a linkage or association study. The locus may or may not encode a gene of potential interest or may be in close proximity to another more likely candidate. However, some candidates are more difficult to genotype than others and so surrogate markers are selected that are known to have tight linkage with alleles on the gene of interest.

Mature insulin – the product of pro-insulin produced in the Golgi apparatus of the β cells of the islets of Langerhans.

Maximum-likelihood estimation – a statistical method used to determine the parameters of a model from a data set.

Meiosis – a special form of cell division that gives rise to gametes.

Mendelian autosomal dominant – a single copy of the disease-causing mutation causes the disease.

Mendelian autosomal recessive – two copies of the disease-causing mutation are required to cause the disease (one from each parent).

Mendelian diseases/disorders – see **Mendelian**.

Mendelian monogenic – Mendelian diseases by definition involve a single gene, though different individuals with the same disease may have different disease-causing genes or gene variants (alleles or mutations) (disease-causing mutations). There are several patterns of Mendelian disease, referred to above.

Meta-analysis – the analysis of combined studies. This is performed to increase the statistical power of genetic studies and identify variables not identified in initial testing.

Metagenomics – genomic studies on a large scale; used particularly in the study of the bacterial genome and in the **Microbiome project**.

Metastasis – the ability of cancer or a tumor to migrate from one region of the body to another.

Methylation chips – bead-chip technology currently being used to identify all of the methylation sites in the human genome.

MHC class I chain (MIC)-related (MICA, MICB, MICC, MICD, and MICE) – a group of stress-induced molecules that are expressed on the surface of epithelial cells in response to infection or damage – expression is recognized by the NKG2D receptor on **natural killer** cells. MICA and MICB are functional MICC, MICD and MICE are essentially pseudogenes.

Microbiome Project – launched in 2008 to characterize the genetics of bacterial communities associated with health and disease in the human body.

Microcytotoxicity or microlymphocytotoxicity test – assay to identify different HLA phenotypes using lymphocytes from whole blood and an array of different polyclonal antibodies from human serum. The assay used complement to induce lysis where the lymphocytes had interacted with the antibodies. Toxicity was assessed by the use of a fluorescent dye. The early assays were refined to allow micro-scale testing in 1964 and again in 1978. Following the 1978 modification, the suffix **NIH** was added to the name. The assays were performed on microtiter trays of 60, 72, or 96 wells each approximately 10 μl deep.

Microsatellites – short number tandem repeat sequences. These **variable number tandem repeats (VNTRs)** are short DNA sequences that are repeated in the DNA sequence. Individuals inherit different numbers of repeat sequences and these can be used to genotype regions of the genome for genetic associations.

Minor allele frequency (MAF) – alleles with low frequencies. A minor frequency can be set at any limit provided the numbers in the study are large enough. Early GWAS based on studies of 2000 cases and controls tended to use a 5% cutoff, but with studies using case and control series sometimes 25 times larger (i.e. as many as 50,000 cases for example) than the 2000 previously used, limits are now being set at 1% or even less. It is all a matter of setting appropriate statistical thresholds.

Missing – genotypes that are not determined in analysis or are not identified in the original test. The reason for these failures is often unknown, but when there are large numbers it can impact on the study. However, if there are not too many they can be assigned through imputation in some cases.

Mitochondrial diseases – diseases and traits that have been mapped to mutations and polymorphism in the **mitochondrial genome**.

Mitochondrial genome (mtDNA) – the DNA encapsulated in the mitochondria (see **Nuclear genome**).

Mitosis – process during embryonic development in which there is constant production of new cells. These new cells are made by rounds of cell division called mitosis and the daughter cells are genetically identical to the parental cell.

Mixed lymphocyte culture (MLC) – early method for HLA class II typing.

Molecular clock – this term refers to a specific hypothesis used in studies of evolution. Essentially if evolution is consistent and there is a constant rate of divergence between two sequences then this hypothesis can be applied to identify the likely time at which species diverge from a common ancestor.

Molecular signatures – this term has a number of uses. It refers to G-coupled proteins and is also used in discussions of natural selection. Both are relevant here.

Monocytes – blood cells with a single nucleus.

Monoclonal antibodies – antibodies that are mono-specific; made from cloned cells that are identical.

Monogenic – involving a single gene.

Monosomy – having one copy.

Monozygotic twins – identical twins created from the same gametes. Also see **Dizygotic twins**.

Morbidity – something affecting the quality of life, usually associated with some tissue damage or loss of function.

Mosaic mutations – a mosaic mutation is one that does not occur in every cell (see **Mosaicism**).

Mosaicism – situation when only some of the cells in the body possess the mutant gene. It can also apply to a situation when only a proportion of the mitochondria in a cell possess the mutant gene. This is especially relevant in cancers.

Multicellular eukaryotic organisms – an organism that has both nucleated cells (i.e. each cell has a nucleus – eukaryote) and is composed of more than one cell (multicellular).

Multiple linear regression model – attempts to model the relationship between two or more explanatory variables and a response variable by fitting a linear equation to observed data.

Multihit hypothesis – the idea that a haplotype may carry more than one risk allele–thus common haplotypes may have several "hits" perhaps accounting for the strong association with certain haplotypes (e.g. HLA 8.1).

Multipoint mapping – this may involve actual genotyping on a large scale or imputation (see **Impute**) on a large scale (i.e. **multipoint imputation**).

Multiple sclerosis – an inflammatory autoimmune disease in which the myelin sheath that insulates the nerves is degraded, leading to a variety of symptoms.

Mutation – technically, the majority of genetic variation arises as a result of mutation irrespective of frequency. However, in clinical genetics some use the term mutation when genetic variation gives rise to a disease (i.e. it is a **disease-causing mutation**). Others prefer to use the term for genetic variations that are found at a low frequency in the population (usually less than 1%). Genetic variations at a frequency of greater than 1% tend to be called polymorphisms. Both terms are used interchangeably in places in this book, but polymorphism is preferred in complex disease studies.

Myelin sheath – a material that insulates the electrically charged neurons and is broken down in **multiple sclerosis**.

Myopathy – general term used to describe muscle pain or adverse reactions involving the muscle.

Myositis – general term for the inflammation of the muscles which can be common in some autoimmune diseases.

Natural killer (NK) cell – large granular cytotoxic lymphocytes, particularly important in innate immunity.

Natural selection – Darwin's major work; the idea (or hypothesis) of evolution driven by natural selection.

New Germania – a region in Paraguay where a group of German settlers set up a community, with a view to maintaining race purity and thereby improving the breeding stock. Originally suggested and supported by Wagner and Elisabeth Nietzsche, the sister of the famous philosopher.

Next-generation sequencing (NGS) technologies – technologies used to sequence DNA in ever-shorter periods of time. This does not refer to one technique, but to all of the recent and some of the future developments to come.

Non-Mendelian complex diseases – diseases that do not have a known pattern of inheritance, but clearly have a heritable (genetic) component.

Non-parametric linkage analysis – linkage analysis in which the pattern of inheritance is not predetermined or preset. This can be used for linkage analysis in complex disease.

Non-random segregation – the distribution of allelic sequences between daughter cells at meiosis. Segregation should be random, but it is not due to the presence of hot spots and cold spots in the genome.

Non-synonymous SNPs – changes in the DNA sequence that result in amino acid differences. Unlike synonymous (conservative mutations) these SNPs do have a biological effect. See **Single nucleotide polymorphism**.

Novel somatic mutations – new mutations such as those that occur in cancer. It is predicted that **next-generation sequencing (NGS)** techniques will have a greater ability to detect these.

Nuclear genome – the DNA encapsulated in the nucleus (i.e. the 22 pairs of chromosomes plus X and Y); not the mitochondrial DNA.

Nucleosome – see the **Nuclear genome**.

Null hypothesis (H_0) – the null hypothesis states that there is no difference (null) between the tested groups or populations (see **Alternative hypothesis**).

Neutrophils – granular white cells. They often have segmented nuclei and are the most abundant of all white cells in the body. They are important in innate immunity, in particular.

Odds ratio (OR) – a crude measure of relative risk, though more commonly used than relative risk (see **Relative risk**).

Offset sequencing primer strategy (OSPS) – part of the **SOLiD** short-read sequencing method. This is illustrated in Figure 12.5.

Oligogenic – many forms (oligo) of a gene (genic). In complex disease this term refers to a disease or subgroup of patients with a disease where there is more than one susceptibility allele, but the number of susceptibility alleles is limited. There is no numerical value that can be applied to the term "oligo" (see also **Polygenic**).

OMIM (Online Mendelian Inheritance in Man) – a comprehensive and free-to-access database of human genes and phenotypes.

OncoArray – a new chip array that is designed to carry 500,000 tagged SNPs for genotyping patients with cancers, including breast, ovarian, prostate, colorectal, and lung cancers (see also **iCOGS** Illumina chip).

Over-dominance – heterozygous advantage, where the heterozygote is at greater advantage than the homozygote.

Pair-wise kinship estimates – a measure of the level of genetic similarity with a kinship group and indicator of the level of inbreeding in ancestral populations.

Pair-wise probability (of identity by descent) – this test would be performed using **PLINK**.

Paired-end mapping – pair-ended sequencing or mapping is increasingly used to access genome rearrangements and structural variations on a global scale. Using this technique, a single DNA fragment is sequenced from each end, producing two reads that can overlap.

Paneth cells – major cell type in the epithelium of the small intestine.

Paracentric inversions – a stable outcome after the incorrect repair of two breaks in a single chromosome. Here, the fragments rejoin in reverse order and there is no loss of genes, only a change in the sequence – so this is an inversion. Paracentric inversion occurs when the repair does not involve the centromere.

Parametric linkage (analysis) – linkage analysis where the parameters are preset (or known), i.e. the pattern of inheritance has been proposed.

Pathology – the science of seeking, observing, and recording normal and damaged tissues in medicine. In practice, it is the detectable changes in and damage to the tissue caused by a disease.

Pearson's χ^2 test – the most frequently used form of the χ^2 test first applied in 1900, and later developed by Yates and Fisher. The test is based on the normalization of the sum of the squared deviations between observed and expected results.

Pedigree file – a collection of family data.

Penetrance – a measure of the occurrence of the trait in individuals with the risk allele or vice versa the risk allele in those with the trait.

Pericentric inversions – occur when two breaks occur at different ends of the same chromosome and the two fragments created rejoin the terminal fragments. The gene sequence along the chromosome is altered, but the overall content remains the same.

Permutation procedure – permutation procedures provide a computationally intensive approach to generating significance levels empirically.

Personal profile – idea that we each inherit a personal profile of risk alleles for complex disease (see also **Disease portfolio**).

Phagocytosis – internalization of matter from outside the cell by specialized cells called phagocytes.

Pharmacogenetics – the study of genetic variation and response to different pharmacological agents, including **adverse drug reactions**. This term was first used in 1959 by Freidrich Vogel.

Pharmacogenomics – the study of the whole genome from a pharmacological perspective (see **Pharmacogenetics**).

Phase (phasing) – a method that can be used to determine which SNPs are inherited together.

Phenotype – the result of genetic variation; this can be seen in the expression of a trait, character, or disease (**disease phenotype**).

Phenotyping – the observation and recording of the expression of a trait.

Philadelphia chromosome – this results from an unusual translocation between chromosome 9 and chromosome 22. It is found in 95% of patients with chronic myelogenous leukemia and 25% of patients with acute lymphoblastic leukemia.

Phylogenic tree – a diagram showing variation between species and populations.

pi-hat – a statistical test that can be performed in **PLINK**.

PLINK software – open-source GWAS toolkit designed to perform basic large-scale analysis: http://pngu.mgh.harvard.edu/purcell/plink/.

Ploidy – number of copies of the autosome (i.e. the chromosome set).

Polar body – the byproduct of uneven division of egg cells. After cell division, the polar body will apoptose or die.

Polygenic – involving several genes. (See also **oligogenic**.)

Poly(ADP) ribose polymerase inhibitors (PARPs) – a family of proteins associated with DNA repair and programmed cell death.

Polymerase chain reaction (PCR) (analysis) – a commonly used method for genetic studies involving the amplification of short DNA sequences.

Polymorphic – multiple (poly) forms (morph).

Polymorphism(s) – existence of many (poly) forms (morphisms). Strictly speaking, this should only be used if the least common variant/allele is found in more than 1% of the population under study. However, as with mutations, not all accept this definition even though it is widely used in complex disease. A further complication is that some genes have so many variants/alleles that this strict application of the term polymorphism becomes meaningless in relation to some genes.

Polyploidy – many copies of the autosome (see **Ploidy**).

Poor metabolizer – genetic variation that encodes a specific phenotype for drug metabolism.

Population – populations can be defined in geographical terms or in terms of study groups. In the context of this book, a population is most often a group of individuals with shared characteristics.

Population genetics – the study of genetics in populations.

Population risk – the risk within a population. Sometimes termed **population attributable risk (PAR)**, which epidemiologists use to assess the contribution of a factor to the overall incidence of a trait.

Population stratification – the subdivision of a population.

Population vigor – the health of a group or population.

Positional cloning – identifying a gene using knowledge of its chromosomal location. This mechanism has been widely used to identify genes responsible for the vast majority of Mendelian disorders.

Positive (Darwinian/natural) selection – selection that favors reproductive success, survival, or increase in advantageous traits within a population.

Power calculation (or power) – the power referred to here is "statistical power". By ensuring there is adequate statistical power we can reduce the likelihood of false-negative associations in a case control association study or in a GWAS.

Pre-pro-insulin – primary translational product of the ***INS*** gene (insulin gene). It is produced in a biologically inactive form and converted to **pro-insulin**.

Prevalence rates – the total number of cases expressing a trait at a particular time point (see also **Incidence**).

Primary immune deficiency – immune deficiency occurring as a result of inherited traits manifest from birth.

Principal components analysis (PCA) – a mathematical procedure used to convert a set of possible correlated observations into a set of linearly uncorrelated observations.

Probability values (*P*) – the statistical measure of the likelihood of an event occurring.

Pro-drug – a drug that is produced in a pre-active form and converts to the active form after administration (e.g. codeine).

Product rule – a statistical rule based on calculus that is used to identify the derivative of the products of two or more functions.

Pro-insulin – precursor to insulin made in the β cells of the islets of Langerhans. Pro-insulin is converted to mature insulin in the Golgi apparatus. Also see **Pre-pro-insulin**.

Proteasome – a large multi-subunit protease found in the cell. The proteasome degrades proteins in the cytoplasm, generating smaller peptide units that are suitable for binding and presentation by HLA class I molecules.

Protein – an organic compound containing carbon, hydrogen, oxygen, nitrogen, and other elements.

Pulmonary fibrosis – fibrosis and scarring of the connective tissue of the lungs, often secondary to other diseases.

Punnett Square – a simple table of data used in the **χ^2 test**.

Purifying (natural) selection – selection against unfavorable alleles or genotypes.

Quantile–quantile (Q–Q) plot – a probability plot used to compare two probability distributions using their quantile values.

Quantitative trait – a universal character rather than a character that some individuals have and others do not.

Quantitative trait locus – locus that contributes to the phenotype. Such a locus encodes one or more genetic variants to allow for this phenotypic variation.

Random genetic drift – random changes in gene frequencies over generations.

Reactive arthritis – autoimmune response that follows from an infection episode; swelling in the joints is the classical symptom.

Recessive trait – an inherited trait whereby both parental haplotypes carry the same genotype (mutation or polymorphism). In other words, two copies are required for the trait (disease or characteristic) to be expressed – one of Mendel's patterns of genetic inheritance.

Recombination – natural process that occurs during prophase in the first phase of meiosis when homologous chromosomes bind and exchange DNA segments.

Recombination cold and hot spots – recombination occurs naturally during prophase in the first phase of meiosis when homologous chromosomes bind and exchange DNA segments. A single recombination event produces two recombinant and two non-recombinant chromatids. Double crossovers can occur, but the overall average is 50%. There are said to be cold and hot spots at which the frequency of recombination is either higher than expected (hot spots) or lower (cold spots).

Recombination fraction (values) (θ) – the measure of recombination, which never exceeds 50% (or 0.5).

Recruitment bias – the potential introduction of bias into cohorts under study.

Reductionist – taking apart an idea–dividing a question into its constituent pieces–often involves over-simplifying an idea or task.

Redundancy – a mechanism by which biological systems protect their host from disaster. This occurs through the use of multiple pathways for the same function, gene duplication, and a host of other mechanisms.

Regression (analysis) – a statistical process for estimating relationships between variables. In complex disease this means genes and traits or populations.

Regulatory cytokines – cytokines that regulate the immune response. This term usually refers to those cytokines that switch off the immune response rather than those which promote inflammation, such as tumor necrosis factor-α.

Relative risk (RR) – a measure of the size of effect of an allele in terms of increased or reduced susceptibility to a trait. This is often presented as an **odds ratio (OR)**, but these are not exactly the same.

Restriction enzyme – an enzyme that cuts at a specific DNA sequence. These enzymes have been used extensively in genotyping studies.

Restriction fragment length polymorphism (RFLP) (analysis) – one of most commonly used methods to identify genetic variation and genetic associations. Alleles can be assigned based on the pattern of fragments identified by electrophoresis after digestion with an enzyme which cuts at a specific sequence (i.e. is restricted). Most (but not all) methods require amplification of the DNA sequence of interest by **polymerase chain reaction (PCR)** prior to digestion with the enzyme–this increases the level of target DNA in the sample and gives a clearer signal.

Ribosomal RNA – there are two mitochondrial rRNA molecules (12S and 16S) and four cytoplasmic molecules. rRNA was not included in the **Human Genome Mapping Project (HGMP)** and therefore we are currently unsure of the total number of rRNA genes in the genome.

Ring chromosomes – potential outcome following incorrect repair of a chromosome. The ring chromosome is a stable chromosome and can propagate to other daughter cells.

Risk (λ) – see **Sibling relative risk**.

Risk allele(s) – a variation in a gene that increases; is associated or linked with an increased likelihood of a trait occurring.

Risk portfolio – an individual's personal genotype data for susceptibility and resistance alleles.

"rs" number – location of a SNP; true name: **refSNP cluster ID number**.

Salt bridge – a biochemical structure or a molecular bridge between two locations with different electrostatic properties.

Sample call rate – the number of samples working (i.e. giving results) in an assay. Used in **PLINK**.

Sclerotic plaques – these occur in a number of diseases and involve the formation of fibrous plaques on the walls of different tissues that disrupt normal function.

Segregate – the distribution of allelic sequences between daughter cells at meiosis. Segregation analysis is used to determine the likelihood that a child will inherit a trait from a parent.

Selective pressure – any factor that reduces reproductive success within a population. Environmental factors include infectious agents, e.g. the malaria parasite.

Sequencing-by-synthesis – developed by 454 Life Sciences using what is called a polony sequencing protocol. This was an early form of **next-generation sequencing (NGS)** moving towards rapid methods for DNA sequencing. See also **SOLiD**.

Semi-dominant – same as co-dominant; heterozygotes have an intermediate phenotype compared with homozygotes.

Serotyping (serological typing) – HLA typing using antibodies in human serum to identify specific HLA antigens. Multiple sets of microtiter trays with 60, 72, or 96 wells are used to determine each phenotype at varying degrees of resolution.

Sex-linked – Mendelian disease involving the two sex chromosomes X or Y.

Shared epitope – said to occur when two or more alleles share a sequence that encodes amino acids in a region of functional importance within a molecule. It is most frequently applied to studies of HLA to explain why there are often several different HLA alleles associated with susceptibility to the same disease.

Sibling relative risk (λ) – a crude calculation of the increased risk of disease that a sibling of an affected case is likely to have. The calculation is based on a comparison of the incidence or prevalence in cases versus the healthy population (controls). The quality of the result depends on the size of the population surveyed and the frequency of the trait.

Significance threshold – the preset measure at which a result will be considered as significant. For example, in the **WTCCC1** study any numerical value lower than the significance threshold is significant (i.e. less than 5×10^{-7}) and any value higher is not significant.

Single nucleotide polymorphism (SNP) – self-defined units, variations in single nucleotides commonly polymorphisms used extensively in genetic studies.

Slow acetylators – a group of individuals who inherit genetic variation that causes them to express this phenotype.

Small call rate – the number of samples that are successfully genotyped in a study.

SNP – see **Single nucleotide polymorphism**.

SNP Map – a map of the positions of all SNPs in human genome.

Social Darwinism – Social Darwinists generally argue that the strong should see their wealth and power increase, while the weak should see their wealth and power decrease. The precise definition of who belongs in either group is a matter of debate amongst those who hold these outdated beliefs.

SOLiD (Supported Oligonucleotide Ligation and Detection) – a **next-generation sequencing (NGS)** platform developed by Life Technologies. Performs short-read sequencing based on "sequencing-by-ligation" (see also **Sequencing-by-synthesis**). It also uses the **offset sequencing primer strategy**.

Somatic chromosomal abnormalities – abnormalities in any chromosome found in a somatic cell.

Statistical confidence – the level of acceptance of a probability value. Similar to the statistical threshold, but can be expressed as a range of **confidence intervals**.

Statistical error (type I and type II) – errors in statistical calculation caused by low sample size causing false positives (type 1) and false negatives (type 2). These are not errors in the application of the statistical test, but errors introduced in the planning of the study that are picked up through statistical analysis (see **Statistical power**).

Statistical power – in the context of this book, the likelihood of finding true-positive associations and reporting true-negative associations.

Stratification score analysis – measure of the degree of stratification within a population.

Structural abnormalities – a missing, extra, or irregular fragment of DNA.

Structured association – relates to population stratification.

Sum rule – a complex mathematical calculation used in genetic studies; best left to the statistical geneticist.

Super-coiled – the multiple rounds of coiling of DNA strands.

Survival of the fittest – concept based on Darwin's theory of evolution. The term was first coined by Herbert Spencer.

Synthetic association – low-frequency causal variants. A form of indirect association.

Tagged SNPs – selected **single nucleotide polymorphisms (SNPs)** used in **GWAS** in particular. Each SNP is selected based on the likelihood of amplification during testing and proximity to a marker gene.

TAP – abbreviation for transporter associated with antigen processing.

T cells – white cells or leukocytes that mature and undergo selection in the thymus.

T cell receptor (TCR) – the receptor for peptide antigens on mature T cells. A critical element in the formation of the T cell synapse.

Telomerase – an enzyme that adds DNA repeat sequences at the 3′ end of the DNA molecule in the telomere region.

Telomeric/telomere – the end of the chromosome.

Tetraploidy – possession of four copies of one or more chromosomes. All cases are lethal.

Thalassemias – a group of inherited autosomal recessive blood disorders.

Thrifty gene hypothesis – hypothesis proposed by JV Neel in 1962 to explain the growing incidence of diabetes in the Western World.

Thymus – area of the body where T cells develop, mature, and go through selection.

Toll-like receptors (TLRs) – receptors important in innate immunity.

Trait – a term used to describe a character or phenotype, which can be a disease, response to treatment, subgroup within a disease, or a behavioral character in a study population.

Transcriptome – that part of the genome that encodes transcribed genes.

***Trans* expression** – the expression of two different alleles from different loci on the same chromosome, but from different parents. Resulting in the formation of a heterodimer in *trans* as opposed to a heterodimer in *cis*. This can create a different molecular structure than expected if proteins were only able to interact in *cis*. See ***Cis* expression**. Particularly relevant in creating diversity of variation in HLA DQ and DP molecules.

Transfer RNA – serves as a physical link between DNA and RNA and the amino acid sequence by loading amino acids onto the messenger RNA in peptide generation.

Translocations – transfer of chromosomal regions between non-homologous chromosomes.

Transmission disequilibrium testing (TDT) – a statistical test that can be used to identify genetic associations with candidate genes.

Triploidy – possession of three copies of one or more chromosomes.

True single-molecule sequencing – the basis of the **HeliScope sequencer**. It does not require a preclonal amplification step.

Tumor genome (tumor DNA) – the genetic makeup of the tumor cells that may develop mutations during cycles of division.

Tumor necrosis factor (TNF) – pro-inflammatory cytokine produced by macrophages that increases the permeability of the vascular epithelium.

Tumor suppressor gene – a gene that protects a cell from developing into a cancer cell.

Twin studies – method frequently used to determine the level of heritability for a trait or disease.

Ultrarapid metabolizers – phenotype in some individuals treated with specific drugs.

Unbalanced chromosomal abnormalities – involve increased numbers of chromosomes (numerical abnormalities) or when there are structural abnormalities (involving deletions and insertion or ring chromosomes, etc.).

Unicellular eukaryotic parasites – a parasite consisting of a single cell with a nucleus.

Variable number tandem repeats (VNTRs; microsatellites or minisatellites) – short repeat sequences of DNA of variable length. The terms mini or micro are used depending on the sequence length. They have been used since the 1990s as markers for genetic analysis.

Wahlund's principal – reduction in heterozygosity causes changes in population substructure due to genetic drift or migration.

Whole-exome sequencing (WES) (or exome sequencing) – sequencing the expressed genes in the genome and not the whole genome. The majority of the human genome considered to be "genetic junk" perhaps acquired through evolution. Exome sequencing ignores this "junk".

Whole-genome sequencing (WGS) – sequencing of the entire genome, not just the coding regions.

Whole-transcriptome shotgun sequencing (WTSS) – an RNA sequencing technique that aims to map DNA transcripts in whole-blood samples.

Wild-type – the product of the standard, "normal" allele at a locus, in contrast to the non-standard, "mutant" allele. However, as some genes have a large number of alleles assigning the wild-type can be difficult. Equally, the wild-type is not always the most common allele.

X-linked dominant or recessive – Mendelian traits that are linked with genes on the X chromosome.

Y-linked – Mendelian traits that are linked with genes on the Y chromosome.

***Z* score** – statistical measure used in linkage analysis.

Zygote – the fertilized egg cell from which all other cells in the body are derived.

Index

Note: Page numbers followed by F refer to Figures, those followed by T refer to Tables and those followed by B refer to Boxes.

D

E

I

K

L

M

Q

R

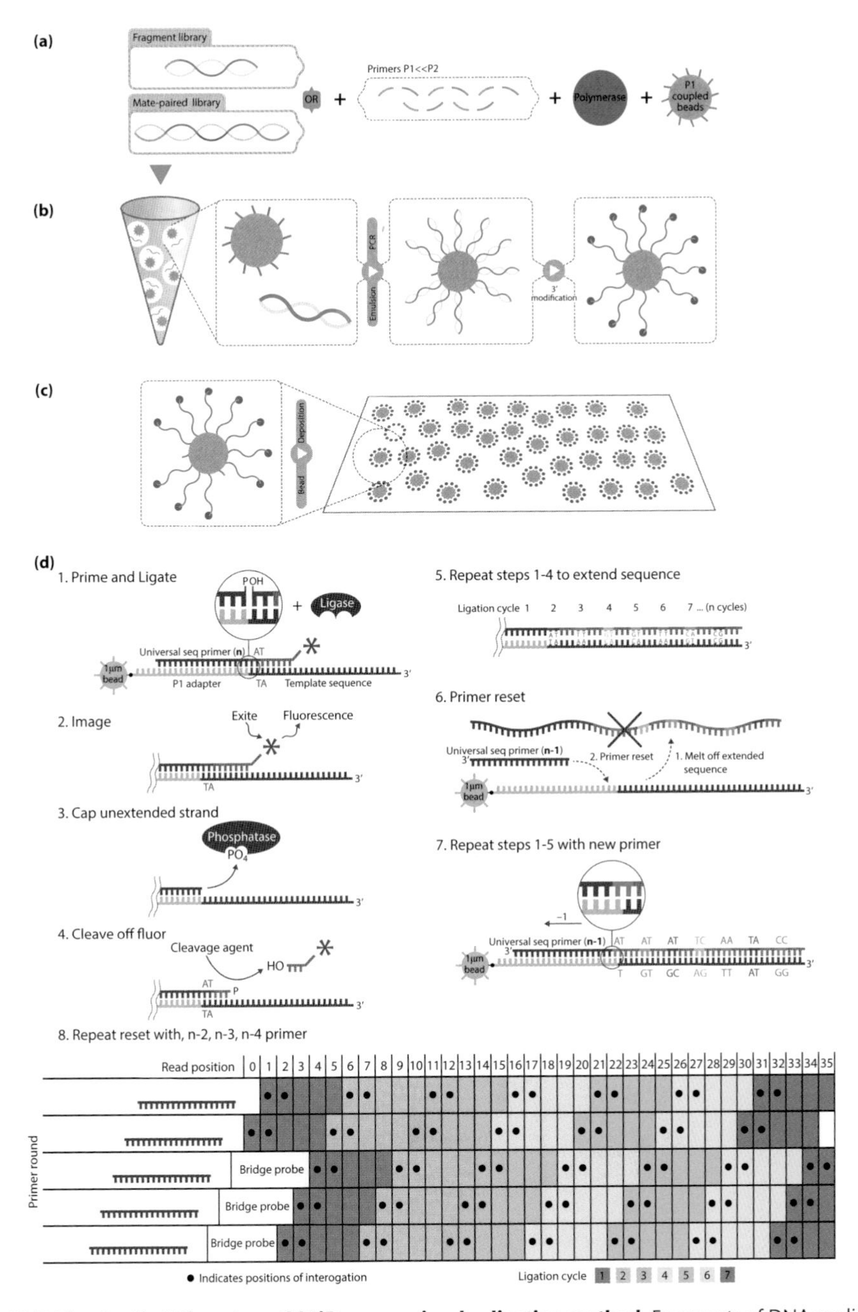

Figure 12.5: The Applied Biosystems SOLiD sequencing-by-ligation method. Fragments of DNA are ligated with oligonucleotide adaptors at each end and hybridized to complementary oligonucleotides attached to magnetic beads (a). Beads are then placed in a water/oil emulsion where DNA amplification is performed (b). Once amplification is finished, the beads are placed on a glass surface and entered into the sequencer (c). In the sequencer a universal sequencing primer, complementary to the adaptor sequence, is used to trigger the sequencing reaction where ligation cycles with fluorescently labeled degenerate probes is performed. Once the probe anneals to the DNA template, a DNA ligase covalently binds the probe to the sequencing primer and the fluorescence is recorded. The probe is then cleaved and another cycle starts. After seven rounds of sequencing, the extended universal primer is removed and a new universal primer is added that is offset by one base (d). The sequence of the read is inferred by interpreting the ligation results for the 16 possible dinucleotide interrogation probes. (Courtesy of Applied Biosystems.)

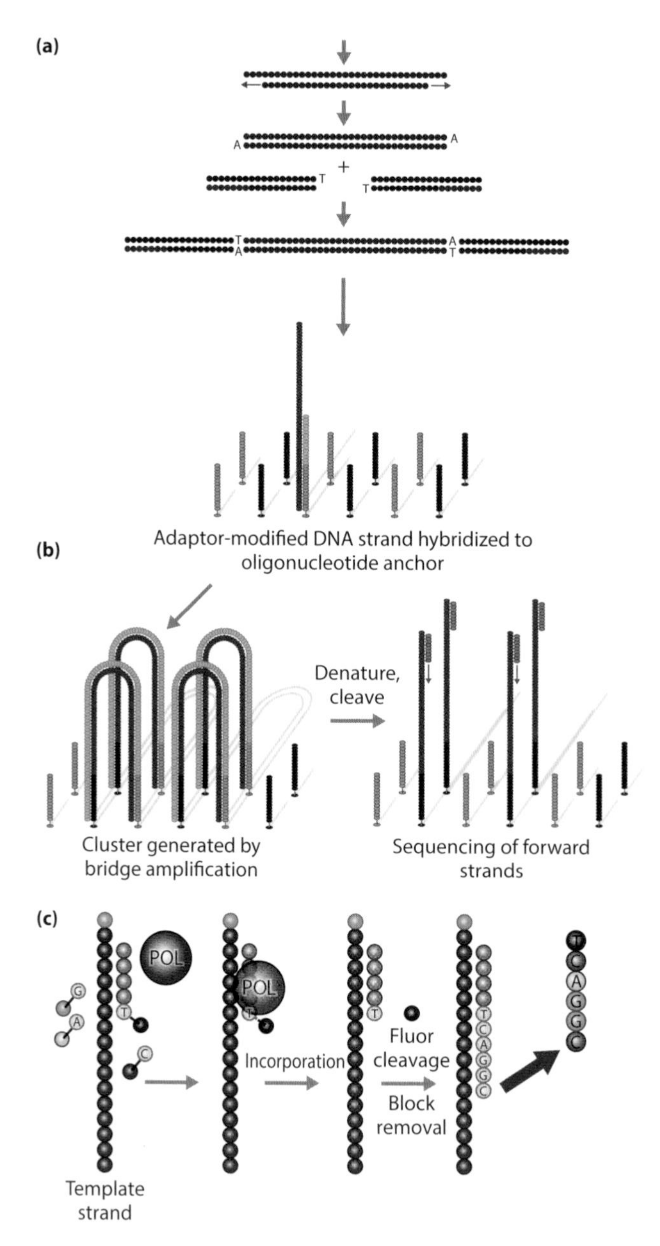

Figure 12.6: The principles of the Illumina sequencing technique. DNA fragments are first ligated onto oligonucleotide adaptors to form double strands (a). Adapter-modified, single-stranded DNA is then added to a flow cell and immobilized by hybridization. Bridge amplification generates clonally amplified clusters (b). The cluster fragments are denatured, annealed with a sequencing primer, and subjected to sequencing-by-synthesis employing DNA polymerase and four reversible dye terminators. Once incorporated, the terminator stops the sequencing reaction, which restarts immediately by cleavage of the incorporated dye terminator. Post-incorporation fluorescence is recorded (c).

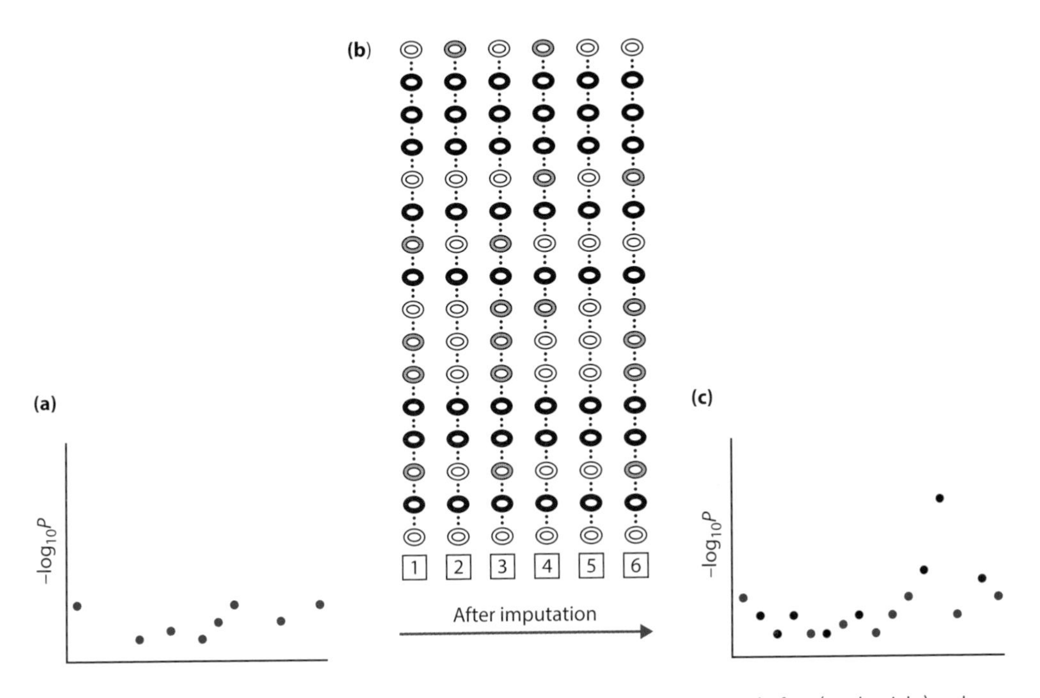

Figure 12.10: Imputation analysis. The two graphs illustrate a genotype sequence before (on the right) and after (on the left) impute has been applied. The central section of the figure illustrates in detail six different haplotypes numbered 1 to 6 for a combination of thirteen bi-allelic genes illustrated as double rings. Each gene is polymorphic six of the genes have been successfully genotyped. There are two alleles for all of genes (both genotyped and unknown - not genotyped). The alleles are illustrated by gray shading (A allele) or no shading (white - B allele) in those genes that have been successfully genotyped. The seven genes illustrated in black have not been genotyped. Imputation analysis can be applied to enable the genotypes of the black genes on this haplotype to be estimated based on the known pattern of linkage disequilibrium for the surrounding genes which have been successfully genotyped. To do this, data are extracted from large genotype databases to fill in the "missing" genotypes. For example if we know that persons with haplotype 1 which runs top to bottom; white, unknown, unknown, unknown, white, unknown, gray, unknown, white, gray, gray, unknown, unknown (or B-?-?-?-B-?-A-?-A-B-A-?-?) carries the white (B) alleles at positions 2, 3, 4, and 6 and the gray (A) alleles at position 8, 12, and 13 in the sequence, then the full sequence for this haplotype can be assigned. If we use the same alphabetical system (A and B) to name the alleles of the seven unknown genes the sequence for the thirteen genes of haplotype 1 would be: B-B-B-B-B-B-A-A-B-A-A-A-A. Using the known genotypes from positions 1, 5, 7, 9, 10, and 11 we can identify the unknown genotypes for the other seven genes on the remaining five haplotypes (haplotypes 2 to 6). This imputed data can be included in the analysis. Because linkage disequilibrium is so strong in certain areas of the human genome, imputation analysis from initial GWAS data can massively extend the studies. Thus an initial GWAS study of based 300,000 to 400,000 SNPs can generate data for up to one million or more markers. (From Marchini J & Howie B [2010] *Nat Rev Genet* 11:499–511. With permission from Macmillan Publishers Ltd.)

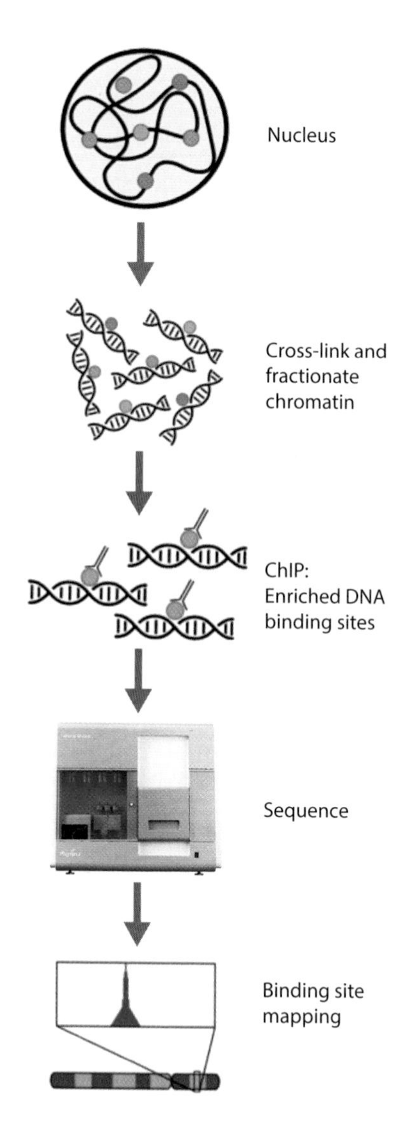

Figure 12.11: Illumina ChIP sequence experiment workflow. DNA fragments are first cross-linked *in situ*, fractionated, and then immunoprecipitated. DNA is then sequenced using an Illumina platform to identify genome-wide sites associated with proteins of interest. (Courtesy of Illumina.)